# 1st International Conference on Engineering and Industrial Technology (ICEIT 2020)



IOP Conference Series: Materials Science and Engineering
Volume 965

Pattaya, Thailand
11 - 13 September 2020

ISBN: 978-1-7138-3202-7
ISSN: 1757-8981

**Printed from e-media with permission by:**

Curran Associates, Inc.
57 Morehouse Lane
Red Hook, NY 12571

**Some format issues inherent in the e-media version may also appear in this print version.**



Institute of Physics
Dirac House, Temple Back
Bristol BS1 6BE UK

Phone: 44 1 17 929 7481
Fax: 44 1 17 920 0979

techtracking@iop.org

**Additional copies of this publication are available from:**

Curran Associates, Inc.
57 Morehouse Lane
Red Hook, NY 12571 USA
Phone: 845-758-0400
Fax: 845-758-2633
Email: curran@proceedings.com
Web: www.proceedings.com

# TABLE OF CONTENTS

## CHEMICAL, MECHANICAL, AND ENERGY ENGINEERING



## MATERIAL ENGINEERING

**ELECTRICAL ENGINEERING, ELECTRONICS COMPUTERS, TELECOMMUNICATIONS AND INFORMATION TECHNOLOGY**

IOP Conf. Series: Materials Science and Engineering **965** (2020) 011001 doi:10.1088/1757-899X/965/1/011001

PAPER • OPEN ACCESS

# Preface and Welcome Speech (ICEIT 2020)

To cite this article: 2020 *IOP Conf. Ser.: Mater. Sci. Eng.* **965** 011001

Chaicharn Thavaravej [1], Arunsri Leejeerajumnean [2], Tomohiko Tagawa [3], Qingyuan Wang [4], Hiromori Tsutsumi [5], Suttichai Assabumrungrat [6,7], Armando T. Quitain [8], Maizatul Shima Shaharun [9], Sathit Krongsut [10], Worapon Kiatkittipong [2]

[1] *Office of the President, Silpakorn University, 22 Borommarachachonnani Rd., Talingchan District, Bangkok 10170, THAILAND*
[2] *Faculty of Engineering and Industrial Technology, Silpakorn University, Nakhon Pathom 73000, THAILAND*
[3] *President, National Institute of Technology (Kosen) Toyota College, Eisei-cho, Toyota, Aichi 471-8525, JAPAN*
[4] *Chengdu University, Chengdu 610106, CHINA*
[5] *Graduate School of Sciences and Technology for Innovation, Faculty of Engineering, Yamaguchi University, JAPAN*
[6] *Center of Excellence on Catalysis and Catalytic Reaction Engineering, Chulalongkorn University Biorefinery Cluster, Department of Chemical Engineering, Chulalongkorn University, Bangkok, 10330, THAILAND*
[7] *Bio-Circular-Green-economy Technology & Engineering Center, BCGeTEC, Department of Chemical Engineering, Faculty of Engineering, Chulalongkorn University, Bangkok, 10330, THAILAND*
[8] *Division for Multidisciplinary Global Education, Center for International Education, Kumamoto University 2-40-1 Kurokami, Chuo-ku, Kumamoto shi 860-8555, JAPAN*
[9] *Department of Fundamental and Applied Sciences, Universiti Teknologi PETRONAS, 32610, Seri Iskandar, Perak Darul Ridzuan, MALAYSIA*
[10] *Research and Innovation Division, Electricity Generating Authority of Thailand (EGAT), Nonthaburi, 11130, THAILAND*

IOP Conf. Series: Materials Science and Engineering **965** (2020) 011001 doi:10.1088/1757-899X/965/1/011001

# PREFACE

(The 1st International Conference on Engineering and Industrial Technology, ICEIT 2020)

The 1st International Conference on Engineering and Industrial Technology 2020 (ICEIT 2020) is organized by Faculty of Engineering and Industrial Technology, Silpakorn University, Thailand together with 7 co-hosts: 1. Chengdu University, China, 2. National Institute of Technology, Toyota college, Japan, 3. Yamaguchi University, Japan, 4. Kumamoto University, Japan, 5. Universiti Teknologi PETRONAS, Malaysia, 6. Bio-Circular-Green-economy Technology & Engineering Center (BCGeTEC) and Center of Excellence on Catalysis and Catalytic Reaction Engineering (CECC) of Chulalongkorn University, Thailand and 7. Electricity Generating Authority of Thailand.

Since the affected by COVID-19 outbreak, the conference organizers offer the hybrid meeting which is both onsite at the conference venue, the ZIGN Hotel, Pattaya, Thailand, and virtual online to those authors who are unable to attend the conference due to any traffic difficulties in September 11-13, 2020.

The options presented here were intended to assist those in need of special arrangements. For information of plenary and keynote sessions, discussions and Q&A sessions, it was already shown in the website ICEIT2020 (http://iceitsu.org/Home/Home.aspx) as follow,

1. For Plenary and Keynote speakers, we were actively working so we can set up a video-teleconference connection with Plenary and Keynote speakers during the session. We ensured there was a laptop running Zoom (preferred). Before the conference, Plenary and Keynote speakers were invited to rehearse for the conference. The talk was recorded as a spare in case we had a problem on connection on the conference date.
2. For normal presenter, the authors can present a video-teleconference connection during the session or submit a video and PowerPoint of the authors' presentation via Admin message in your account in advance before the conference date. The video presentation must be in English with displaying presenter view, and should not be longer than 12 min. If there are any question in case of video presentation, the chair of session can send the questions to the authors and the authors can respond via Admin message.

IOP Conf. Series: Materials Science and Engineering **965** (2020) 011001 doi:10.1088/1757-899X/965/1/011001

The 1st International Conference on Engineering and Industrial Technology 2020 (ICEIT 2020)
11-13 September 2020, Thailand

# Welcome Speech: President, Silpakorn University, Thailand

**Asst. Prof. Chaicharn THAVARAVEJ**

First of all, on the behalf of Silpakorn University, I would like to express my warm welcome and sincere appreciation for your participation in the 1st International Conference on Engineering and Industrial Technology 2020 (ICEIT 2020).

For the year 2020, I would like to sincerely thank to Faculty of Engineering and Industrial Technology, Silpakorn University and co-host universities from China, Japan and Malaysia for organizing the ICEIT 2020. We strongly believe that we will have a fruitful cooperation in the areas of engineering and technology and our friendly relationship will be strong concerted development. I am sure that the conference will be a successful venue for knowledge exchange among researchers, and also strengthen the network of national and international researches in the future.

Finally, I sincerely wish The 1st International Conference on Engineering and Industrial Technology 2020 (ICEIT 2020) has a great success, and I hope that our exchange and cooperation with ICEIT conference as well as with other activities will be continued and reach a new level.

IOP Conf. Series: Materials Science and Engineering **965** (2020) 011001 doi:10.1088/1757-899X/965/1/011001

The 1st International Conference on Engineering and Industrial Technology 2020 (ICEIT 2020)
11-13 September 2020, Thailand

# Welcome Speech: Dean, Faculty of Engineering and Industrial Technology, Silpakorn University, Thailand

**Asst. Prof. Dr. Arunsri LEEJEERAJUMNEAN**

Welcome all of you to the The 1st International Conference on Engineering and Industrial Technology 2020 (ICEIT 2020). The conference was held on September 11–13, 2020 at the ZIGN Hotel, Pattaya, Thailand. ICEIT 2020 was a major hosted by Silpakorn University, Thailand together with 7 co-hosts: 1. Chengdu University, China, 2. National Institute of Technology, Toyota college, Japan, 3. Yamaguchi University, Japan, 4. Kumamoto University, Japan, 5. Universiti Teknologi PETRONAS, Malaysia, 6. Bio-Circular-Green-economy Technology & Engineering Center (BCGeTEC) and Center of Excellence on Catalysis and Catalytic Reaction Engineering (CECC) of Chulalongkorn University, Thailand and 7. Electricity Generating Authority of Thailand.

ICEIT 2020 aims to bring together researchers, engineers and practitioners who work in the area of engineering and industrial technology. The scientific theme of ICEIT 2020 has a wide spectrum of six tracks, Track 1 Food and Biotechnology, Track 2 Chemical, Mechanical and Energy Engineering, Track 3 Material Engineering, Track 4 Electrical Engineering, Electronics, Computers Telecommunications and Information Technology, Track 5 Manufacturing Systems, Engineering management and Logistics, and Special Track – which is established from the esteem research grants i.e. e-Asia, IRN, NSTDA Chair Professor.

On behalf of the conference organizers, I would like to give a special thanks to the special issue of *Processes* (Q2 WOS, Impact factor 2.753), *Engineering Journal* (Q2 Scopus, SJR 0.23), *Science, Engineering and Health Studies* (Scopus, SJR 0.11), a proceeding *IOP Conference Series: Material Science and Engineering* (CPCI of WOS and Scopus SJR 0.2) for acceptance those papers for publications.

Finally, on behalf of the Faculty of Engineering and Industrial Technology, Silpakorn University, I would like to say deeply thank you to co-hosts and acknowledge the conference organization team for hard work during the past year. I am sure that the conference will be a successful venue for knowledge exchange among researchers, and also strengthen the network of national and international researches in the future.

IOP Conf. Series: Materials Science and Engineering **965** (2020) 011001 doi:10.1088/1757-899X/965/1/011001

IOP Conf. Series: Materials Science and Engineering **965** (2020) 011001 doi:10.1088/1757-899X/965/1/011001

The 1st International Conference on Engineering and Industrial Technology 2020 (ICEIT 2020)
11-13 September 2020, Thailand

# Welcome Speech: President, National Institute of Technology (NIT, Kosen) Toyota College, Japan

**Prof. Dr. Tomohiko TAGAWA**

On behalf of National Institute of Technology (NIT, Kosen) Toyota College, I would like to congratulate the opening of International Conference on Engineering and Industrial Technology (ICEIT) 2020. I would also like to thank all the participants and committee members with great respect. Without the enthusiasm of all, it is not possible for ICEIT 2020 to be conducted under COVID-19 pandemic situation.

Now, whole the world recognized the importance of sharing scientific information properly and that it is possible not only getting together at one place but also in front of your computer system when the border is still closed. I hope ICEIT 2020 would bring together researchers, engineers, and practitioners over the world who work in the areas of engineering and industrial technology with a wide spectrum of topics.

For us, NIT(Kosen) Toyota College, it is great honor to be one of the co-host institutes of this conference. I appreciate our MOU with Silpakorn University helped us to contribute ICEIT. Lastly, with great appreciation to committee and supporting staff members, I wish all the participants would enjoy this conference very much.

IOP Conf. Series: Materials Science and Engineering **965** (2020) 011001 doi:10.1088/1757-899X/965/1/011001

The 1st International Conference on Engineering and Industrial Technology 2020 (ICEIT 2020)
11-13 September 2020, Thailand

# Welcome Speech: President, Chengdu University, China

**Prof. Dr. Qingyuan WANG**

Dear President Chaichan Thavaravej, dear chairperson, partners and supporters of ICEIT, dear researchers and engineers all over the world, ladies and gentlemen, Thanks to the host, Silpakorn University, a warm, strong and enduring friend of Chengdu University since 2015, I am honored to be invited to give the speech and also have teachers from our School of Mechanical Engineering take part in this comprehensive forum to learn about the most advanced trends and achievements in engineering and industrial technology. Firstly, I would like to introduce Chengdu University. Established in 1978, it grows fast and has got very good rankings in 40 years: this year, Chengdu University ranks the 266th in Best Chinese Universities and 903rd globally in the International Natural Index Ranking. With over 115 overseas partners from 32 countries and regions, it attaches great importance to internationalization. Particularly for our cooperation with Thailand, we have invited a great many excellent Thai experts, including Princess Sirindhorn and Mr. Chaphiporn who won the 2018 Chinese Government Friendship Award for lecturing; CDU Sichuan Provincial Research Institute of Thai Studies was selected in the National and Regional Research Center List of the Ministry of Education. Moreover, we have boosted the establishment of International Sister City Relationship between Chengdu and Chiang Mai. As you may know, CDU will be the Athlete's Village of the 31st FISU Summer World University Games Chengdu 2021. Meanwhile, an International Presidents Forum will be held in Chengdu University. Hence, we sincerely invite presidents and delegations from your universities to visit, enjoy and participate next year. Last but not the least, please allow me to extend my sincere appreciation to all the attendees. With your unremitted efforts and generous sharing of novel and innovative thoughts, this conference will definitely direct us on how we should collaborate in the post-pandemic era. May this conference a big success!

IOP Conf. Series: Materials Science and Engineering **965** (2020) 011001 doi:10.1088/1757-899X/965/1/011001

The 1st International Conference on Engineering and Industrial Technology 2020 (ICEIT 2020)
11-13 September 2020, Thailand

# Welcome Speech: Dean of Graduate School of Sciences and Technology for Innovation, Faculty of Engineering, Yamaguchi University, Japan

YAMAGUCHI UNIVERSITY
山口大学

**Prof. Dr.Hiromori TSUTSUMI**

It is my great pleasure and honor to give you all a congratulatory address to ICEIT 2020, Thailand. I hope that the conference will be a good opportunity to build good relationships with the scientists, the engineers, and the students who attend from Thailand, Japan, and other countries.

I would like to introduce our Faculty, Faculty of Engineering, Yamaguchi University to you all. The original school of our Faculty was established about 80 years ago, 1939. We have acquired an excellent reputation for the high-level research and for providing a high level of quality education. We have the seven departments dedicated to fundamental engineering and the Graduate School of Sciences and Technology for Innovation which has offered excellent post graduate programs through its Masters and Doctorate courses since 2016.

Our Faculty is also focusing on international exchanges of the researchers and students based on the cooperation agreements with universities in many countries. We also have partnership agreements with the universities in Thailand. We signed an MOU in 2019 with Silpakorn University. Asst. Prof. Dr. Worapon Kiatkittipong, one of the members of the organization committee of ICEIT 2020, is a key person of the MOU collaboration between Faculty of Engineering and Industrial Technology, Silpakorn University, and our Faculty. Dr. Worapon Kiatkittipong visited our Faculty last year and discussed with Prof. Dr. Izumi Kumakiri, our Faculty member, about bio-ethanol separation and purification technology. This opportunity led to conclusion of the MOU.

I hope that the good relationships established through the conference will be handed over to the next generation researchers and students of the countries who attend the conference.

Finally, I would like to also express my sincere thankfulness to the contributors of the conference who made all this possible and I wish you all have fruitful time with the conference.

UTP EGAT

IOP Conf. Series: Materials Science and Engineering **965** (2020) 011001 doi:10.1088/1757-899X/965/1/011001

The 1st International Conference on Engineering and Industrial Technology 2020 (ICEIT 2020)
11-13 September 2020, Thailand

# Welcome Speech: Director, Bio-Circular-Green-economy Technology & Engineering Center (BCGeTEC), Thailand

**Prof. Dr. Suttichai ASSABUMRUNGRAT**

On behalf of Bio-Circular-Green-economy Technology & Engineering Center, Department of Chemical Engineering, Faculty of Engineering, Chulalongkorn University, it is our great honor to be one of the co-host institutes of the International Conference on Engineering and Industrial Technology (ICEIT) 2020. Our strong collaboration between Chemical Engineering Departments at Chulalongkorn University and Silpakorn University has lasted for more than 15 years with very fruitful research outputs. We appreciate this long-term collaboration.

As a co-host institute, it is a great pleasure to welcome you to the ICEIT 2020 which brings together experts, researchers, professors, scientists, engineers, postdoctoral fellows, and students, providing them with the opportunity to share and discuss their findings and achievements, and to address challenges in Engineering and Industrial Technology field. We would like to thank all speakers, delegates, participants and guests for their participation, and also thank the committee members and staffs for making the ICEIT 2020 possible on both conventional and online platforms in response to the unusual COVID-19 pandemic.

We wish that all the participants of the ICEIT 2020 enjoy the scientific program and take this great opportunity to expand their research networks.

IOP Conf. Series: Materials Science and Engineering **965** (2020) 011001 doi:10.1088/1757-899X/965/1/011001

The 1st International Conference on Engineering and Industrial Technology 2020 (ICEIT 2020)
11-13 September 2020, Thailand

# Welcome Speech: Deputy Director, Division for Multidisciplinary Global Education, Center for International Education, Kumamoto University, Japan

**Prof. Dr. Armando T. QUITAIN**

Firstly, on behalf of the director of our newly established Center for International Education and our faculty members, I congratulate the organizers for the successful preparation and execution of this first ever "International Conference on Engineering and Industrial Technology" despite the many challenges brought about by the global COVID-19 pandemic!

Silpakorn University, the host of this international conference, has been an active partner of Kumamoto University since the signing of the Memorandum of Understanding for academic and student exchanges in 2018. For over 2 years, we have been actively exchanging students and faculties for academic and research collaboration. This active collaboration will be further supported by our newly established Center for International Education.

Here at the center, we train future global leaders that will take lead in facing the challenges of our globalized modern society. We offer multilingual, multicultural, and multidisciplinary studies courses aimed at equipping graduates with the skills needed to work on a global scale. These subjects help our students develop deep comprehension of social and natural sciences, learn modern languages, and understand the values of various foreign countries and cultures. Thus, providing them the tools needed to think both comprehensively and globally. The multidisciplinary spectrum of topics in this conference suits the kind of mindset we would like to develop in our students.

Let this event be a venue for us and among participating universities to further strengthen research collaboration and student exchanges with the aim of developing future STI leaders for regional sustainability and improvement of lives in our society.

We wish you success in this first ever "International Conference on Engineering and Industrial Technology"!

IOP Conf. Series: Materials Science and Engineering **965** (2020) 011001 doi:10.1088/1757-899X/965/1/011001

The 1st International Conference on Engineering and Industrial Technology 2020 (ICEIT 2020)
11-13 September 2020, Thailand

# Welcome Speech: Chair of Fundamental & Applied Sciences Department, Universiti Teknologi PETRONAS, Malaysia

UNIVERSITI
TEKNOLOGI
PETRONAS

**Assoc. Prof. Dr.Maizatul SHIMA SHAHARUN**

I would like to congratulate Silpakorn University for organising the International Conference on Engineering and Industrial Technology (ICEIT 2020), scheduled on September 11–13, 2020 in Pattaya, Thailand.

ICEIT 2020 is timely in this unprecedented time. The conference makes the perfect platform in sharing insights into the recent research and cutting-edge technologies by renowned speakers and scientists across the globe. It is hoped that through the sharing of knowledge, ideas, and innovations, we can address the latest trends and challenges facing the world and question how we can prepare present and future generations to thrive in this transforming world.

On behalf of the Department of Fundamental and Applied Sciences (FASD) of Universiti Teknologi PETRONAS (UTP), Malaysia, we are honoured to be a part of ICEIT 2020. Established on 1st April 2009, FASD is one of the departments under the Faculty of Science and Information Technology. Currently FASD is offering the following programmes:

- **Undergraduate Studies Programmes**
  - BSc (Hons) in Applied Chemistry
  - BSc (Hons) in Applied Physics
- **Postgraduate Studies Programmes**
  - MSc of Science (by Research)
  - Doctor of Philosophy of Science

This year, FASD will be organising the Southeast Asia Global Innovation Challenge (SEA-GIC 2020) with the theme "Battling Climate Change Through Chemistry" on the 3rd December 2020. This competition is open to all undergraduate students in the ASEAN region. The main objective of this event is to allow participants to pitch their innovative ideas and proof of concept to solve real-world problems involving climate change. We welcome your participation and would like you to visit our website for more information https://www.utp.edu.my.

I convey my warm greetings and felicitations to the organising committee and the participants and extend my best wishes for the success of the conference.

UTP EGAT

IOP Conf. Series: Materials Science and Engineering **965** (2020) 011001 doi:10.1088/1757-899X/965/1/011001

The 1st International Conference on Engineering and Industrial Technology 2020 (ICEIT 2020)
11-13 September 2020, Thailand

# Welcome Speech: Director of Research and Innovation Division, Electricity Generating Authority of Thailand

EGAT
Power for Thai Happiness

**Sathit KRONGSUT**

Dear Distinguished Delegates, Sponsors and Exhibitors,

Welcome,

I would like to congratulate to Silpakorn University for organising the International Conference on Engineering and Industrial Technology (ICEIT 2020), scheduled on September 11–13 , 2020 in Pattaya, Thailand.

The influence of disruptive technology and the approach to the age of digital transformation led to this conference (ICEIT 2020), which can be shared analytical data, including the results of experiments from the researches in wide variety of disciplines.

According to the rapid world technology today that we have challenged to manage them for our better life in present and the future. By sharing the innovations knowledge and technology from the famous scientists, engineers and well-known speakers in this ICEIT2020, it can be guided us to cope well with our future trends.

On behalf of the Division of Research and Innovation of Electricity Generating Authority of Thailand (EGAT), Thailand, we are honoured to be a part of ICEIT 2020. EGAT has continuously supported the research and development projects to promote electrical innovation for prosperity in the country.

To improve Thai people's living quality, EGAT employees and outside institutions worked on the research with emphasis on six main purposes as following:

- To replace import of electrical power technologies. materials, and equipment.
- To contribute for peak cuts or improved energy efficiency.
- To develop new energy alternatives for electricity generation.
- To improve efficiency and/or lower production cost.
- To prevent and mitigate environmental and social impacts caused by power operations and associated activities.
- To support social research for the quality development of community life around EGAT's power plants.

I convey my warm greeting to the organizing committee and participants and best wishes for happiness to the success of this international conference.

IOP Conf. Series: Materials Science and Engineering **965** (2020) 011001 doi:10.1088/1757-899X/965/1/011001

The 1st International Conference on Engineering and Industrial Technology 2020 (ICEIT 2020)
11-13 September 2020, Thailand

# Committee

**Organizing Committee**

Asst. Prof. Dr. Arunsri Leejeerajumnian (Chairperson of Conference)
Asst. Prof. Dr. Worapon Kiatkittipong (Secretary of Conference)
Dr. Woraruthai Choothian (Treasurer of Conference)
Asst. Prof. Dr. Thosapon Katejanekarn
Asst. Prof. Dr. Choowong Chaisuk
Asst. Prof. Dr. Rujikarn Nasanit
Asst. Prof. Dr. Sujin Wootthichaiwat
Dr. Phamorn Silapan
Asst. Prof. Dr. Porametr Arromdee
Asst. Prof. Dr. Nattawut Tharawadee
Asst. Prof. Dr. Niti Yongvanich
Asst. Prof. Dr. Suched Samuhasaneetoo
Dr. Chonlatep Usaku
Asst. Prof. Patipat Hongsuwan
Asst. Prof. Dr. Teerasak Hudakorn
Assoc. Prof. Dr. Prakorn Ramakul
Asst. Prof. Dr. Rapeepan Keaw-on

**International Committee**

***Canada***
Prof. Derick Rousseau, Ryerson University
***Japan***
Prof. Shuji Adachi, Kyoto University of Advanced Science
Prof. Tetsuya Kida, Kumamoto University
Prof. Armando T.Quitain, Kumamoto University
Dr. Hiroshi Yamada, Nagoya University
Prof. Tomohiko Tagawa, National Institute of Technology, Toyota College
Prof. Izumi Kumakiri, Yamaguchi University
***Korean***
Prof. Daeyong Jeong, Inha University
***Malaysia***
Dr. Lam Man Kee, Universiti Teknologi PETRONAS
Dr. Lim Jun Wei, Universiti Teknologi PETRONAS
Dr. Leong Siew Yoong, Universiti Tunku Abdul Rahman
***Turkey***
Prof. Bekir Karasu, Anadolu University
***United Kingdom***
Dr. Farid Aiouache, Lancaster University
Dr. Vesna Najdanovic, Aston University
***United States of America***
Prof. Manbir S. Sodhi, University of Rhode Island
Prof. Sushil Adhikari, Auburn University

IOP Conf. Series: Materials Science and Engineering **965** (2020) 011001 doi:10.1088/1757-899X/965/1/011001

The 1st International Conference on Engineering and Industrial Technology 2020 (ICEIT 2020)
11-13 September 2020, Thailand

**Scientific Committee**

***Track 1 Food and Biotechnology***

Assoc. Prof. Dr. Sopark Sonwai (Track Chair)
Assoc. Prof. Dr. Pramote Khuwijitjaru
Asst. Prof. Dr. Prasong Siriwongwilaichat
Dr. Chonlatep Usaku
Dr. Nardrapee Sanchez

***Track 2 Chemical, Mechanical, and Energy Engineering***

Asst. Prof. Dr. Thosapon Katejanekarn (Track Chair)
Assoc. Prof. Dr. Kasama Sirisomboon
Dr. Mutsee Termtanan (Track Chair for Chemical Engineering)
Dr. Nutchapon Chotigkrai

***Track 3 Material Engineering***

Asst. Prof. Dr. Pat Sooksaen (Track Chair)
Asst. Prof. Dr. Niti Yongvanich
Asst. Prof. Dr. Chanchai Thongpin
Asst. Prof. Dr. Bussarin Ksapabutr
Dr. Narit Triamnak

***Track 4 Electronic, Electrical, and Computer Engineering***

Asst. Prof. Dr. Yutana Jewajinda
Dr. Natapong Wongprommoon (Track Chair)
Dr. Kantida Pancharoen
Dr. Sopon Phumeechanya
Dr. Jhirat Mearnchu

***Track 5 Manufacturing Systems, Engineering Management and Logistics***

Asst. Prof. Dr. Choosak Pornsing (Track Chair)
Asst. Prof. Dr. Kanate Pansawat
Asst. Prof. Patipat Hongsuwan

PAPER • OPEN ACCESS

# Introduction and Abstract of Plenary and Keynote Speakers (ICEIT 2020)

To cite this article: 2020 *IOP Conf. Ser.: Mater. Sci. Eng.* **965** 011002

ICEIT 2020
THE ZIGN HOTEL
11-13 SEPTEMBER 2020
PATTAYA,THAILAND
UTP
EGAT
The International Conference on Engineering and Industrial Technology
ICEIT
PLENARY SPEAKERS

UTP EGAT

IOP Conf. Series: Materials Science and Engineering **965** (2020) 011002 doi:10.1088/1757-899X/965/1/011002

The 1st International Conference on Engineering and Industrial Technology 2020 (ICEIT 2020)
11-13 September 2020, Thailand

# Introduction of plenary speaker

**Prof. Dr. Tomohiko TAGAWA**

President National Institute of Technology (NIT Kosen), Toyota College
Professor Emeritus, Nagoya University

1982; Dr. Eng., Nagoya University, Graduate School of Engineering
1982; Research Associate, National Research Council of CANADA
1984; Assistant Professor, Tokyo Institute of Technology
1988; Associate Professor, Nagoya University
2004; Professor, Nagoya University
2015; Vice Dean, School of Engineering, Nagoya University
2017; Professor Emeritus, Nagoya University
2017; President, National Institute of Technology, Toyota College

Area of Research and Education
Chemical reaction engineering, Chemical engineering,
Applied chemistry, Green and sustainable process,
Engineering education

IOP Conf. Series: Materials Science and Engineering **965** (2020) 011002 doi:10.1088/1757-899X/965/1/011002

The 1st International Conference on Engineering and Industrial Technology 2020 (ICEIT 2020)
11-13 September 2020, Thailand

***Plenary Speaker***

# Early Engineering Education in National Institute of Technology (Kosen) Toyota College

Tomohiko Tagawa[1*]

[1] *President, National Institute of Technology (Kosen) Toyota College, Eisei-cho, Toyota, Aichi 471-8525, Japan*
* Corresponding address (tagawa@toyota-ct.ac.jp)

*HIGHLIGHTS*

- Early engineering education in NIT(Kosen) college, Japan
- Collaborative education program in Toyota college

**ABSTRACT**

National Institute of Technology Colleges (Kosen) is a unique Japanese higher educational institute for engineering education founded in 1961. There are 51 NIT colleges across Japan and recently several Kosen colleges were established in Asian countries including Thailand. The colleges accept students from junior high school at age of 16 and give a five years higher engineering education. One department consists of 40 students and one college usually operates 4-5 departments. Therefore only one percent of the same age population have a chance to get early engineering education in a small-size class of Kosen college. These colleges share main core curriculum as well as their original education contents. Inter-college contests such as famous Robot contest, Programming contest, Design competition and so on are highlights of Kosen education.

NIT(Kosen) Toyota College locates in the city of Toyota which is a home of Toyota Motor Corporation. With five engineering departments such as Mechanical, Electrical and Electronic, Information and Computer, Civil, and Architecture, Toyota College provides unique education in collaboration with local corporations such as long term internship. In this lecture, the characteristic early engineering education system of NIT(Kosen) college is summarized and several practical examples among education programs in Toyota College are introduced.

**Keywords**: Engineering education; NIT(Kosen) College, Japan; Collaborative education program

IOP Conf. Series: Materials Science and Engineering **965** (2020) 011002 doi:10.1088/1757-899X/965/1/011002

The 1st International Conference on Engineering and Industrial Technology 2020 (ICEIT 2020)
11-13 September 2020, Thailand

# Introduction of plenary speaker

**Prof. Dr. Qingyuan WANG**

President of Chengdu University, Sichuan, China. Elsevier highly-cited scholar in China (2014-2019).VHCF6 Chairman of the Sixth International Super High Cycle Fatigue Conference. Ph.D. in the major of Materials and Mechanics, Ecole Centrale Paris, Paris, France. Postdoctoral fellow at Purdue University and JSPS researcher at Kagoshima University. Prof. Qingyuan Wang has served and is still serving in many professional and academic organizations. Prior to Chengdu University, he was Dean of the School of Architecture and Environment at Sichuan University. He is also a significant member of the academic Evaluation Committee of the State Council Academic Degrees Committee in China. As an excellent researcher, he has put his focuses on investigating mechanical problems of new materials and structures, ultra-long life fatigue and reliability, new methods and apparatus for experimental mechanic, etc. Including key and major projects, he has accomplished 8 projects of National Natural Science Foundation, published more than 200 SCI papers, and been cited more than 2000 times from 2014 to 2019. He was the recipient of various awards, such as First Prize of the Ministry of Education Natural Science Award (2006, China), and First Prize of Sichuan Science and Technology Progress Award (natural science category) (2014, China), and Second Prize of National Natural Science (2018, China). He was also one of the winners of the Chinese National Outstanding Youth Fund and Chang Jiang Scholar and Leader of Innovation Team from the Ministry of Education, P.R.C.

The 1st International Conference on Engineering and Industrial Technology 2020 (ICEIT 2020)
11-13 September 2020, Thailand

***Plenary Speaker***

# Fundamentals of Very High Fatigue and Its Engineering Application

**Qingyuan Wang**

**ABSTRACT**

Mechanical structures of key projects in high-speed rail, aerospace, and automobile industries put forward higher requirements for reliability and long life. The problem of very high cycle fatigue (VHCF) has recently attracted significant attention and special research from scientists and engineers. This lecture summarized our research progress on the development of ultrasonic-accelerated vibrational fatigue testing methods in the loading modes of tension-tension, bending and torsion, which have been successfully coupled with complex environments such as corrosion, vacuum, high pressure and high temperature. Furthermore, the VHCF properties including S-N curves, crack initiation and propagation mechanism at multi-scales and life prediction models based on different physical mechanisms were presented for high strength metallic materials. The research results have been successfully applied to the life assessment of major equipment such as aero engines, gas turbines, bridges, etc.

**Keywords:** Very high cycle fatigue; ultrasonic fatigue test; fatigue crack initiation; life prediction.

IOP Conf. Series: Materials Science and Engineering **965** (2020) 011002 doi:10.1088/1757-899X/965/1/011002

The 1st International Conference on Engineering and Industrial Technology 2020 (ICEIT 2020)
11-13 September 2020, Thailand

# Introduction of plenary speaker

**Prof. Dr. Mamoru YAMADA**

Thermotolerant fermenting microbes bear the potential to efficiently produce useful materials under stress conditions, including heat stress. We have thus been exploring such microbes in tropical countries that are considered to adapt to high temperature environments, for over 20 years. The number of publications relating to thermotolerant microbes in the world is increasing to more than 200 articles per year, and their potential for industrial applications is further enhanced by combination with abiotec advanced technologies. In my talk, I will introduce high-temperature fermentation with thermotolerant microbes, bioethanol production, the improvement of their stress tolerance and several types of experiments for industrial applications.

Mamoru Yamada graduated from Biochemistry Department, School of Medicine, Yamaguchi University in 1984. He became the Assistant Professor of the same University in 1984, worked in University of California at San Diego from1985 to1987, and was promoted to Associate Professor in 1990 and Professor in 2001. He performed Director of Science Research Center from 2009 to 2012, Dean of Faculty of Agriculture from 2012 to 2016 and Director of Research Center for Thermotolerant Microbial Resources from 2019 in Yamaguchi University, and is also Senior Presidential Advisor (International Cooperation for Bioscience) from 2017 in the same University. He organized the Asian Core Program (2008-2012) among Japan, Thai, Lao and Vietnam, which was supported by JSPS and NRCT, the Asian Science & Technology Strategic Cooperation Promotion Program (MEXT-ARDA project) (2010-2012), which was supported by MEXT and ARDA, to develop high-temperature fermentation technologies, and the Core to Core Program (2014-2018) among 7 countries including Japan, Thai, Lao, Vietnam, Germany, Indonesia and United Kingdom, which was supported by JSPS, NRCT, MOST in Vietnam and Core Universities involved in its program.

IOP Conf. Series: Materials Science and Engineering **965** (2020) 011002 doi:10.1088/1757-899X/965/1/011002

The 1st International Conference on Engineering and Industrial Technology 2020 (ICEIT 2020)
11-13 September 2020, Thailand

***Plenary Speaker***

# High-temperature Fermentation with Thermotolerant Microbes for Industrial Applications

Mamoru Yamada*

*Graduate School of Science and Technology for Innovation, Department of Biological Chemistry, Faculty of Agriculture, and Research Center for Thermotolerant Microbial Resources, Yamaguchi University, Yamaguchi 753-8515, Japan*
Email: m-yamada@yamaguchi-u.ac.jp, +81-83-933-5869

*HIGHLIGHTS*

- High-temperature fermentation (HTF) and its several benefits
- Thermotolerant microbes for HTF and thermal adaptation
- Combination of HTF with abiotec advanced technologies

Bioconversion of biomass to useful materials is strongly recommended under global warming but it is still costly and inefficient. Utilization of biomass thus requires innovative ideas and technologies. As one of the technologies, high-temperature fermentation (HTF) is attracting attention, which is expected to bring several benefits including cooling cost reduction, water saving, hydrolytic enzyme cost reduction, and consequently overall running cost reduction. Moreover, in combination with abiotec advanced technologies, we can establish HTF-based new technologies. However, the development of thermotolerant and efficiently fermenting microbes is essential for HTF.

The global demand for bioethanol is growing rapidly, and a variety of biomass applications require energy-saving and efficient production processes. We have thus screened thermotolerant microbes from tropical environments and developed HTF technology for ethanol production and fermentation processes including temperature-non-controlled fermentation, combination of HTF with distillation under a low pressure (1-4) and combination of HTF with membrane separation. On the other hand, to improve thermotolerance and stress-tolerance for more stable fermentation, the thermal adaptation of isolated thermotolerant microbes was performed and genes responsible for the adaptation were analyzed (5-10).

1. N. Lertwattanasakul *et al*, *Biotechnol. Biofuels*, 18;8:47. doi: 10.1186/s13068-015-0227-x (2015)
2. M. Murata *et al.*, *J. Jpn. Inst. Energy*, 94:1154-1162 (2015)
3. M. Nurcholis *et al.*, *Appl. Microbiol. Biotechnol.* (Review), doi.org/10.1007/s00253-019-10224-3 (2019)
4. T. Kosaka *et al.*, *In* Fuel Ethanol Production from Sugarcane. ISBN: 978-1-78984-937-0 (print) 978-1-78984-937-7 (online) IntechOpen, pp121-154 (2019)
5. M. Murata *et al.*, *PLoS ONE*, 6(6): e20063 (2011)
6. W. Soemphol *et al.*, *Biosci. Biotechnol. Biochem.*, 75:1921-1928 (2011)
7. K. Matsushita *et al.*, *Biosci. Biotechnol. Biochem.* (Review), 13:1-14 (2015)
8. K. Charoensuk *et al.*, *Biotechnol.Biofuels*, 10:204. doi: 10.1186/s13068-017-0891-0 (2017)
9. M. Murata *et al.*, *PLoS ONE*, doi.org/10.1371/journal.pone.0189487 (2018)
10. T. Kosaka *et al.*, *PLoS ONE*, doi: 10.1371/journal.pone.0215614 (2019)

**Keywords:** thermotolerant yeast; high-temperature fermentation; thermotolerance

**Acknowledgment**

The study was supported by e-Asia joint research program (e-Asia JRP) and by Advanced Low Carbon Technology Research and Development Program (ALCA).

The 1st International Conference on Engineering and Industrial Technology 2020 (ICEIT 2020)
11-13 September 2020, Thailand

# Introduction of plenary speaker

**Prof. Dr. Suttichai ASSABUMRUNGRAT**

Dr. Suttichai is a professor at Department of Chemical Engineering, Faculty of Engineering, Chulalongkorn University. He received his B.Eng. degree from Chulalongkorn University, and M.Sc. and Ph.D. degrees from Imperial College London. His research interest is on process intensification with particular focus on multifunctional reactors. He has published more than 290 peer-reviewed journal-proceedings articles and book chapters. He is now working on several projects, for examples, hydrogen production technologies, biodiesel production, biorefinery and $CO_2$ capture and utilization. He has received a number of awards such as Outstanding Paper Award from Journal of Chemical Engineering of Japan, and Young Scientist and Outstanding Scientist Awards from the Foundation for the Promotion of Science and Technology under the Patronage of His Majesty the King. He is a working committee member of Thai Academy of Science and Technology Foundation.

IOP Conf. Series: Materials Science and Engineering **965** (2020) 011002 doi:10.1088/1757-899X/965/1/011002

The 1st International Conference on Engineering and Industrial Technology 2020 (ICEIT 2020)
11-13 September 2020, Thailand

*Plenary Speaker*

# Bio-Circular-Green-economy Technology & Engineering (BCGeTEC) Center for Supporting Development of Bio - Circular - Green (BCG) Economy

Suttichai Assabumrungrat [1,2]

[1] *Center of Excellence in Catalysis and Catalytic Reaction Engineering, Department of Chemical Engineering, Faculty of Engineering Chulalongkorn University, Bangkok 10330, Thailand*
[2] *Bio-Circular-Green-economy Technology & Engineering Center, BCGeTEC, Department of Chemical Engineering, Faculty of Engineering, Chulalongkorn University, Bangkok, Thailand 10330*
Email: Suttichai.a@chula.ac.th

**ABSTRACT**

Thailand is attempting to drive the country under the Bio – Circular – Green (BCG) economy model which offers a sustainable growth based on its immense wealth of natural and bio-resources and less dependence on fossil-based resources. Transforming the existing fossil-based economy to the BCG economy requires great effort, interaction and collaboration between different players on various levels. Although it provides a new opportunity for industry, research, policy and financing stakeholders to work together to design new values for the country, the successful implementation of this concept nowadays is still rare as it requires many key elements for achievement such as advanced and multidisciplinary technologies, talented human resource, and strong public and private partnership. BCGeTEC was recently established aiming at supporting this transformation to BCG economy. Based on our strong academic and research background on chemical engineering, BCGeTEC focuses on collaborative research and innovation between different disciplines and institutions nationally and internationally. It also aims at extending from fundamental research in laboratory to industrial scale to achieve high impact to the economy. In addition, BCGeTEC supports the human resource development through training and education. Our Objectives and Key Results (OKR) as well as activities will be introduced.

IOP Conf. Series: Materials Science and Engineering **965** (2020) 011002 doi:10.1088/1757-899X/965/1/011002

The 1st International Conference on Engineering and Industrial Technology 2020 (ICEIT 2020)
11-13 September 2020, Thailand

# Introduction of plenary speaker

**Prof. Dr. Armando T. QUITAIN**

Dr. Armando T. Quitain is currently a Professor at the Center for International Education, Kumamoto University (Kumamoto, Japan). His research interests are centered around development of green technologies based on microwave carbocatalysis and supercritical fluids as applied to biomass conversion. He is currently the Lead Principal Investigator of the e-ASIA Joint Research Project on Alternative Energy (Bioenergy): "Development of Algal Bioenergy Systems for Green and Sustainable ASEAN Region", which aims to convert algal biomass into biofuels and biochemicals. He is also collaborating with Kyoto University for Japan-ASEAN Science and Technology Innovation Platform (JASTIP) program to strengthen research collaboration among Japan and ASEAN universities. As a Japan Society for the Promotion of Science (JSPS) Scientist for Joint International Research, he is also collaborating with top researchers from Germany and Spain on $CO_2$ utilization and conversion technologies. All these research activities are in accord with the UN Sustainable Development Goals, which forms the "blueprint for a better and more sustainable future for all."

Student mobility is also in his portfolio, developing exchange programs (outbound and inbound) to cultivate "global mindset" to future global STI leaders that will promote sustainability in the region and beyond.

He has been exploring many challenges that Japan could offer for over 24 years to date.

UTP EGAT

The 1st International Conference on Engineering and Industrial Technology 2020 (ICEIT 2020)
11-13 September 2020, Thailand

***Plenary Speaker***

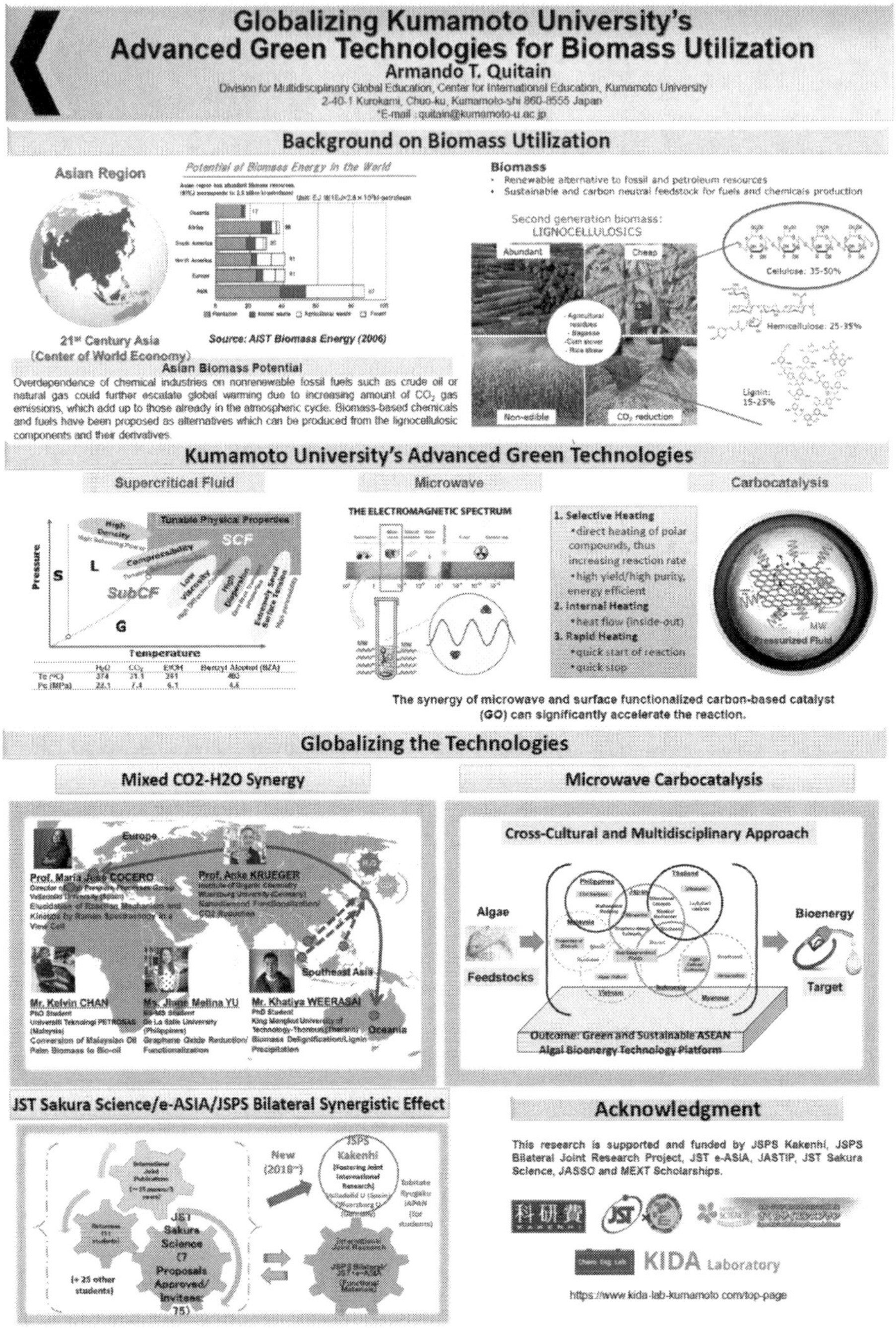


UTP EGAT

IOP Conf. Series: Materials Science and Engineering **965** (2020) 011002 doi:10.1088/1757-899X/965/1/011002

The 1st International Conference on Engineering and Industrial Technology 2020 (ICEIT 2020)
11-13 September 2020, Thailand

# Introduction of plenary speaker

**ChM. Dr. LIM JUN WEI**

ChM. Dr. LIM JUN WEI (AMIChemE, MRSC) was conferred with the Bachelor of Science (Hons) degree in Chemistry from Universiti Sains Malaysia in year 2009. He later received his Ph.D. qualification in Environmental Chemistry from the same university in year 2013. Currently, he is affiliated with the Department of Fundamental and Applied Sciences, Universiti Teknologi PETRONAS, serving as the Senior Lecturer and Cluster Leader of Applied Chemistry program. His major research interests are insect-based biological compounds, bioremediation of solid wastes and wastewaters and microbial biofuels. Accordingly, he has published more than 100 research papers inclusive of book chapters of late. In terms of professional associations, he is the Associate Member of Institution of Chemical Engineers (AMIChemE) and Member of The Royal Society of Chemistry (MRSC) at international level and Professional Chemist registered with Institut Kimia Malaysia at national level. He is also one of the Graduate Technologists under the Malaysia Board of Technologist. Besides, he has joined the Editorial Board Members of Chemical Science and Biomolecular Engineering under Boffin Access and Archives of Biochemical Engineering under Somato Publications.

The 1st International Conference on Engineering and Industrial Technology 2020 (ICEIT 2020)
11-13 September 2020, Thailand

***Plenary Speaker***

# BIOconversion to BIOremediate

Dr. Jun Wei Lim

*Department of Fundamental and Applied Sciences, HICoE-Centre for Biofuel and Biochemical Research, Institute of Self-Sustainable Building, Universiti Teknologi PETRONAS, Seri Iskandar 32610, Perak Darul Ridzuan, Malaysia*

**ABSTRACT**

The rapid progress of civilization has inevitably resulted in pervasive environmental pollutions. Off late, various anthropogenic pollutants have been directly or indirectly discharged into the soil, water and air mediums with the quantities depending on the extent of socioeconomic developments. To reciprocate in cushioning the untoward impacts on the environment, also myriad treatment approaches have been employed to remove, if not to degrade the pollutants until reaching the innocuous level. Unconsciously, some of the adopted physicochemical approaches may generate secondary pollutants or intermediates which are more debilitating to the environment. Viewing from the sustainability perspective, the imprudent options may as well entail intensive energy requirements which are not cost-effective in the long run. Ideally, the Green Chemistry mode of remediations via bioconversion has been extensively explored in proffering insights and novel solutions that will help addressing the pressing concerns associated to the environmental pollutions. However, the common trade-off incurred by the bioconversion such as long bioremediation time need has limited its feasibilities for real applications; albeit having a cost advantage. Therefore, it is of great interest to delve into spurring the metabolisms of targeted macroorganisms/microorganisms employed to execute the bioconversion processes. The pollutants in the form of wastes are anticipated to be bioconverted into the innocuous by-products with portions being assimilated to produce new biomass for growing mediators' cells. In closing the loop, the generated biomass should be further exploited for the production of valuable products, advocating the wastes to valuable resources model. While the research focus in recent years is all about green and environment, the future trend for bioremediation paves smoother path by the sustainable bioconversion.

IOP Conf. Series: Materials Science and Engineering **965** (2020) 011002 doi:10.1088/1757-899X/965/1/011002

The 1st International Conference on Engineering and Industrial Technology 2020 (ICEIT 2020)
11-13 September 2020, Thailand

# Introduction of plenary speaker

**Prof. Dr. Piyasan PRASERTHDAM**

Professor Piyasan Praserthdam is a professor at the Department of Chemical Engineering, Faculty of Engineering, Chulalongkorn University, Thailand. In 2019 Professor Piyasan served as the President of the Catalysis and Reaction Engineering Association, Thailand. He also served as the Dean of the Faculty of Engineering and Industrial technology during 2001-2002. In addition, he is the first president of the Thai Institute of Chemical Engineering and Applied Chemistry (TIChE).

About his research, he focuses on heterogeneous catalysis applied in industries and also catalyst deactivation. He published more than 400 international peer-reviewed articles and 3 patents and have been awarded the Thailand outstanding scientist award in 2009. In industry aspect, he has been a consultant in various chemical engineering related companies and has been working togerther with the Thai company in the development of new catalysts research with more than 15 million US of funding from industries.

Internationally, Prof. Piyasan is the Chairman of the 8th Asia Pacific Congress in Catalysis with more than 600 researchers attended.

IOP Conf. Series: Materials Science and Engineering **965** (2020) 011002 doi:10.1088/1757-899X/965/1/011002

The 1st International Conference on Engineering and Industrial Technology 2020 (ICEIT 2020)
11-13 September 2020, Thailand

***Plenary Speaker***

# ENTECH Moving Forward: Outside-in and Inside-out Approach

Piyasan PRASERTHDAM

*Center of Excellence in Catalysis and Catalytic Reaction Engineering, Department of Chemical Engineering, Faculty of Engineering, Chulalongkorn University, Bangkok, 10330, Thailand*

ABSTRACT

Here we present three principles for effective materials and energy utilization based on the modified second law of thermodynamics: (1) If the potential of the system change, the potential of the surrounding will change in opposite direction, (2) The process that has largest change in potential will be the best one., and (3) The product from the smallest change in potential will be the highest value. These new criteria are realized through the parameter called the potential index ($\Theta^*$) assigned to explain natural processes in the world and identify preferable processes used instead of the typical efficiency parameter. In addition, the term integrates the knowledge of mechanical, electrical and chemical engineering by describing satisfactory transformation processes that are well-known in mechanical and electrical engineering, and also the increasing potential processes familiar in chemical engineering. Interestingly, such principles can also be applied for the management of ENTECH, Silpakorn University

IOP Conf. Series: Materials Science and Engineering **965** (2020) 011002 doi:10.1088/1757-899X/965/1/011002

The 1st International Conference on Engineering and Industrial Technology 2020 (ICEIT 2020)
11-13 September 2020, Thailand

# Introduction of plenary speaker

**Assoc. Prof. Dr. Anongnat SOMWANGTHANAROJ**

Anongnat is an Associate Professor at the Department of Chemical Engineering, Faculty of Engineering, Chulalongkorn University, Thailand. She obtained a doctorate degree in Macromolecular Science and Engineering Program, University of Michigan, Ann Arbor, USA in 2003. Her research interests are in the field of polymer engineering, plastic packaging, bioplastic, adhesive for electronics, battery and plastic recycle. She has served as as Associate Dean in Research Affairs since July 2016. From her research, she has co-found Garden Fresh spin-off startup in November 2018. Garden Fresh offers the biodegradable plastic bags that can extend shelf life of fresh produce.

She has been active in research work since 2004 and until now she has been granted 33 projects from both domestic and international funding agencies and industries. She published 44 articles in international journals. Two US patents have been grants and 7 Thai patents are being in the process. She also serves as a reviewer for both international journals & conferences as well as domestic and international funding. She has served as an Associate Member, Thai Academy of Science and Technology Foundation since 2010 and as a reviewer board, Polymers, MDPI since early 2020.

She has received several awards including

- Gold medal, 47th International Exhibition of Invention of Geneva, Switzerland, April 2019
- The runner-up in the MIT Enterprise Forum, IDE Competition 2018, The Innovation-Driven Entrepreneurship (IDE) Center
- The runner-up in the Thailand Research Fund's Leaders in Innovation Fellowships "Final Business Model Pitch," Leaders in Innovation Fellowship (LIF) programme by TRF and Newton Fund
- Outstanding Research Utilization of social and/or industry Award, Ratchadaphiseksomphot Endowment Fund of Chulalongkorn University, 2014
- Invention award, National Research Council of Thailand, "Bioplastic packaging bags for extending shelf life of fresh produces and dried fruits," 2015
- Fellow in Materials Science, L'Oréal for Woman in Science 2009, L'Oréal Thailand and United Nations Education, Scientific and Cultural Organization (UNESCO) in the topic "Development of PLA/clay nanocomposite films to use in film packaging application," 2009
- The first runner up, PTT Chem Green Innovation Award, "Using polymer/clay nanocomposite as packaging film," December 12, 2008
- Royal Thai Scholarship, Ministry of Education, Royal Thai Government, 1997 –2003
- Xerox Corporation Award in recognition of academic excellence, 2002

UTP EGAT

IOP Conf. Series: Materials Science and Engineering **965** (2020) 011002 doi:10.1088/1757-899X/965/1/011002

The 1st International Conference on Engineering and Industrial Technology 2020 (ICEIT 2020)
11-13 September 2020, Thailand

*Plenary Speaker*

# Coverall and N95 Mask Innovation & Production's Journey in Thailand during COVID-19 Pandemic

Anongnat Somwangthanaroj

*Department of Chemical Engineering, Faculty of Engineering, Chulalongkorn University*
*Anongnat.s@chula.ac.th*

ABSTRACT

During COVID-19 pandemic in Thailand, we have found out that we did not have both factory and testing lab for the PPE. Fortunately, the governmental science service sector and the university have both practical professionals and sufficient equipment so we can adapt the available machines to perform several important tests to make sure that both N95 mask and coverall PPE meets the standards to protect our medics who are frontline in this situation. In addition, for the isolation gown and coverall PPE, we have all component in the supply chains starting from plastic production, film and laminated film factory and garment factories and all private sectors work closely for a greater good. However, there is still one more challenge, N95 mask production. From upstream (plastic production) midstream (meltblown nonwoven fabric production) and downstream (N95 mask fabrication line), at present, we have only one N95 mask production factory in Thailand which just began commissioning in late July 2020. We still have work to do to make sure that our medical staff will have high quality protective equipment.

The 1st International Conference on Engineering and Industrial Technology 2020 (ICEIT 2020)
11-13 September 2020, Thailand

# Introduction of plenary speaker

**Dr. Prapapong VANGTOOK**

**Education:** Doctor of Engineering (Energy Technology): Asian Institute of Technology (AIT), Pathumthani, Thailand, 2005.
Master of Mechanical Engineering: King Mongkut's Institute of Technology North Bangkok (KMITNB), Bangkok, Thailand, 1997.
Bachelor of Mechanical Engineering: Kasetsart University, Bangkok, Thailand, 1994.

**Employment/Experience:**
Present Engineer Level 10, Research and Innovation Division, Electricity Generating Authority of Thailand (EGAT), Nonthaburi, 11130 Thailand
E-mail prapapong.v@egat.co.th
Phone +66-2436-4911

**Speaker**
- Sharing session on good practice of inter-stakeholder collaboration – government sector's perspective, the 27th annual meeting of renewable energy sub-sector network, Asean Centre for Energy (ACE), 2020
- Thailand wind energy industry: The outlook in Future Energy Show Thailand at IMPACT Convention Centre, 2019

**Training**
- Join The 11th International Warsaw Invention Show (IWIS), Poland 2017
- Join The 8th International Exhibition of Inventions – Kunshan (IEIK), China 2014
- Technical site visited, Electric Supply Industry Technology, Smart grid Technology, High Voltage Power Laboratory and Research and Technology, France, Switzerland, Germany and, Netherland 2011
- Topic is Renewable Energy and Energy Efficiency by the ITEC (Indian Technical and Economic Cooperation programme), TERI (The Energy and Resources Institute), New Delhi, India 2010
- INSA (Institut National des Sciences Applique'es de Lyon), France 2002

**International Paper Published**
- Thermal comfort assessment and application of radiant cooling: A case study, Building and Environment, Vol.43, No.7, pp. 1185 – 1196 R.A.Memon, S.Chirarattananon and P.Vangtook, 2008
- Application of radiant cooling as passive cooling option in hot humid climate. Building and Environment, Vol.42, No.2, pp. 543-556 P.Vangtook and S.Chirarattananon, 2007
- An experimental investigation of application of radiant cooling in hot humid climate. Energy and Buildings, Vol.38, Issue 4, pp. 273 - 285 P.Vangtook and S.Chirarattananon, 2006
- Thermal comfort assessment in two tropical regions and radiant cooling as a passive cooling option, network for comfort and energy use in buildings, windsor conference, S.Chirarattananon, N.N.Htan, R.A.Memon, and P.Vangtook, 2006

**Association Committee**
- Committee: Electricity Supply Industry Association of Thailand (TESIA), 2009 – 2014
- Committee: Electricity Supply Industry Association of Thailand (TESIA), 2020 – 2022

The 1st International Conference on Engineering and Industrial Technology 2020 (ICEIT 2020)
11-13 September 2020, Thailand

***Plenary Speaker***

# Sharing Session on Good Practice on Research and Innovation of EGAT - Stakeholder Collaboration

Dr. Prapapong Vangtook

*Engineer Level 10, Research and Innovation Division, Electricity Generating Authority of Thailand (EGAT), Nonthaburi, 11130 Thailand*

ABSTRACT

The research and development department of Electricity Generating Authority of Thailand (EGAT) were established in 1987. Between 1987 - 2006, only EGAT staffs were the participants for the research to solve the work problem. From 2006 to the present, EGAT fund, for EGAT employees and other institutions, has set at least 3% of the net profit annually which also has continuously supported the research and development projects to promote the electrical innovation for prosperity in Thailand. The scopes of work are the generation system, transmission system, power plant development and renewable energy, power business, social and environment, and fuel. To improve Thai people's living quality, the research will be emphasized on six main purposes as following:

- To replace import of electrical power technologies. materials, and equipment.
- To contribute for peak cuts or improved energy efficiency.
- To develop new energy alternatives for electricity generation.
- To improve efficiency and/or lower production cost.
- To prevent and mitigate environmental and social impacts caused by power operations and associated activities.
- To support social research for the quality development of community life around EGAT's power plants

The Benefit of the innovation from the research that EGAT provide sponsorship is one of the most important strategy of the strategic plan to drive EGAT to be the national leader. At the present, EGAT has given financial support over 450 projects to support the innovation and development of technology in electric generating and environment conservation, sustainable energy, and social responsibility.

IOP Conf. Series: Materials Science and Engineering **965** (2020) 011002 doi:10.1088/1757-899X/965/1/011002

IOP Conf. Series: Materials Science and Engineering **965** (2020) 011002 doi:10.1088/1757-899X/965/1/011002

The 1st International Conference on Engineering and Industrial Technology 2020 (ICEIT 2020)
11-13 September 2020, Thailand

# Introduction to Keynote Speaker

**Prof. Dr. Tetsuya KIDA**

Department : Applied Chemistry and Biochemistry
Institution : Kumamoto University, Japan
Phone Office : +81-96-342-3664
e-mail : tetsuya@kumamoto-u.ac.jp

**Educational Backgrounds**

1994, B. Eng., Kyushu University, Japan
1996, Ms. Eng., Kyushu University
2001, Ph.D., Kyushu University (supervisor: Prof. Noboru Yamazoe)

**Field of Research**

Synthesis of inorganic nanomaterials such as quantum dots, oxide nanostructures, and graphene oxide, and their applications for energy devices, gas sensing, gas separation, etc.

**Biography**

2001-2003: Postdoctoral Fellow, Postdoctoral Fellow, National Institute of Advanced Industrial Science and Technology, Japan
2003-2006: Assistant Professor, Department of Chemistry and Applied Chemistry, Saga University, Japan
2006-2013: Associate Professor, Department of Energy and Materials Sciences, Kyushu University, Japan
2010-2011(13 months): Visiting Scholar, Materials Sciences Division, Lawrence Berkeley National Laboratory, USA
2013-present: Professor, Division of Materials Science, Faculty of Advanced Science and Technology, Kumamoto University, Japan
2016-2016 (1 month): Visiting Professor, Department of Chemical Engineering, Chulalongkorn University, Bangkok, Thailand
2017-2017 (6 months): Visiting Professor, Institute for Chemical Investigation of Catalonia (ICIQ – Institut Català d'Investigació Química), Tarragona, Spain.

**Latest publications**

1. N.L. Hamidah, M. Shintani, A.S.A. Fauzi, G.K. Putri, S. Kitamura, K. Hatakeyama, A.T. Quitain, M. Sasaki, **T. Kida**, Graphene Oxide Membranes with Cerium-Enhanced Proton Conductivity for Water Vapor Electrolysis, ***ACS Applied Nano Materials***, 3, 5, 4292–4304 (2020).
2. A.D. Pramata, K. Suematsu, A.T. Quitain, M. Sasaki, **T. Kida**, Synthesis of Highly Luminescent $SnO_2$ Nanocrystals: Analysis of their Defect-Related Photoluminescence Using Polyoxometalates as Quenchers, ***Adv. Funct. Mater.***, 28, 1704620 (2018).
3. **T. Kida**, Y. Kuwaki, A. Miyamoto, N.L. Hamidah, K. Hatakeyama, A.T. Quitain, M. Sasaki, A. Urakawa, Water Vapor Electrolysis with Proton-Conducting Graphene Oxide Nanosheets, ***ACS Sustainable Chem. Eng.***, 6, 11753-11758 (2018).

The 1st International Conference on Engineering and Industrial Technology 2020 (ICEIT 2020)
11-13 September 2020, Thailand

***Keynote Speaker***

# Energy Applications of Graphene Oxide Nanosheet Membranes

Tetsuya Kida*

*Division of Materials Science, Faculty of Advanced Science and Technology, Kumamoto University.*
*Corresponding address (Email: tetsuya@kumamoto-u.ac.jp, +81-96-342-3664)

*HIGHLIGHTS*

- A fast proton conductor based on graphene oxide (GO) is developed.
- Stacked GO nanosheet membranes sandwiched with electrocatalysts efficiently electrolyze water vapor into hydrogen with a 100% Faradaic efficiency.
- Mixed conducting GO membranes selectively separate hydrogen from mixed gases without electrical power.

**ABSTRACT**

A two-dimensional nanosheet of graphene oxide (GO) has attracted a great deal of attention due to its potential versatile applications. GO contains only earth abundant, non-toxic element of carbon. It can be synthesized from graphite by simple chemical processes including oxidation and exfoliation. These advantages make GO promising alternative to rare and expensive materials that are used for some applications. One of the novel properties of GO is its high proton conductivity, allowing GO to be used as proton exchange membranes for fuel cells. The proton conductivities arise from proton hopping at oxygen functional groups originally anchored to the GO surface. Epoxy groups are regarded as major proton hoping sites on graphene oxide. We also reported that the proton conductivities can be tuned by changing the oxygen content in GO. The partial reduction makes GO a mixed conducting material that shows protonic and electronic conductivities. Using GO's high proton conductivity that is comparable to Nafion, we are attempting to develop new carbon-based electrochemical devices such as water electrolysis devices, hydrogen sensors, and hydrogen separation devices, as schematically shown in Fig.1.

The membrane showed a good proton conductivity at room temperature, as revealed by concentration-cell measurements and complex impedance spectroscopy. The proton conduction in the GO membrane was confirmed by proton pumping experiments. The membrane was applied for water vapor electrolysis at room temperate. The GO device showed good hydrogen and oxygen evolution rates that are close to the theoretical values. The membrane with Pt/C also showed good sensitivity to hydrogen. The results successfully demonstrate the promising capability of GO as alternative to Nafion in producing hydrogen from water vapor by electrolysis and detecting hydrogen in air.

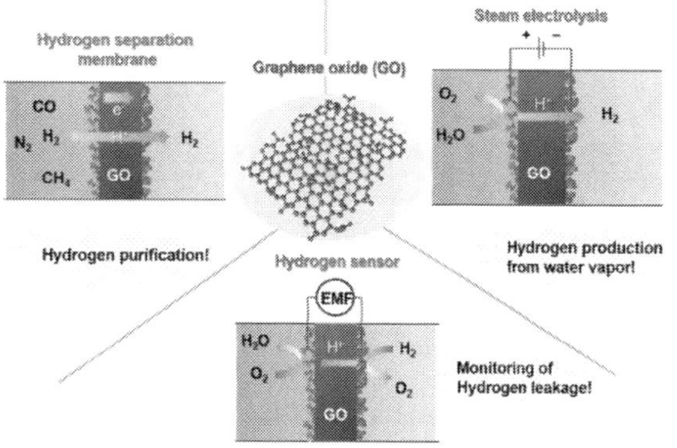


Fig.1 Developed applications of graphene oxide.

**Keywords**: Graphene oxide; proton conductor; water vapor electrolysis; separation membrane

IOP Conf. Series: Materials Science and Engineering **965** (2020) 011002 doi:10.1088/1757-899X/965/1/011002

The 1st International Conference on Engineering and Industrial Technology 2020 (ICEIT 2020)
11-13 September 2020, Thailand

# Introduction to Keynote Speaker

**Prof. Li JIAN**

Li Jian is a professor in the school of mechanical engineering, Chengdu University, China and he previously served as the dean of the school of mechanical engineering.

Professor Li Jian's main research focuses on the modern design of mechanical transmission mechanism, mechanical CAD/CAM, and integrated manufacturing technology, etc. In recent years, he studied different types of spatial cam mechanisms and established a unified mathematical model of spatial cam mechanisms cam, NC machining cutter path, machining error analysis, and surface contact stress analysis in 3D space. The research goal is to form a CAD/CAM system capable of designing and manufacturing digital cam mechanisms for engineering application. He has presided over or participated in numerous scientific research projects including a few funded with the National Natural Science Foundation of China, and his research have been applied in the industrial fields widely.

Recently, Professor Li Jian organized and completed the engineering education professional accreditation of mechanical design, manufacturing and automation specialty in Chengdu University. In June 2020, the Ministry of Education announced that it had passed the professional accreditation of China's higher education. He has presided over and completed multiple high education reform projects including the curriculum reform project of the Ministry of Education of China.

Professor Li Jian serves as an executive director of Southwest Automation Society of Chinese University. He is a member of China Society of Mechanical Engineering, Sichuan Mechanical Engineering Society of China, and Academic and Degree Evaluation Committee of Chengdu University. He is also an online evaluation expert of National Natural Science Foundation of China, and an expert of science and technology project evaluation in Sichuan, Fujian and Hebei provinces.

He received the B.S. and M.S. from Sichuan University in 1983 and 1998 respectively. He has been awarded the second prize of cultivation achievement by Sichuan provincial government and the famous mentor award of Chengdu University.

UTP EGAT

IOP Conf. Series: Materials Science and Engineering **965** (2020) 011002 doi:10.1088/1757-899X/965/1/011002

The 1st International Conference on Engineering and Industrial Technology 2020 (ICEIT 2020)
11-13 September 2020, Thailand

*Keynote Speaker*

# The Integrated International Education of Mechanical Engineering based on OBE Methodology

## ---A case study of the exchange student project between Chengdu University and Silpakorn University

Li, Jian

*Chengdu University, Chengdu, Sichuan Province, P.R. China. 610106*
*E-mail: lijian@cdu.edu.cn*

**ABSTRACT**

The standard of higher education is an important indication of a country's development level potential. As engineering disciplines are the driving force of the development of science and technology, it has always played an important role in international education.

In 2017, the Ministry of Education of China put forward the concept of "new engineering", which requires the National Engineering Colleges and majors to actively optimize the disciplinary layout, promote the integration of multi-disciplines in engineering and expand emerging disciplines. At the same time, it is necessary to learn from international advanced concepts and standards, and clearly point out the development direction of new engineering education. In addition, "new engineering" clearly requires strengthening international exchange and cooperation. To cultivate international talent of engineering education, the mode for undergraduates should be in line with international engineering education, and graduates should have the ability and qualification to participate in international engineering projects. Under the guidance of this mode, the school of mechanical engineering of Chengdu University has been exploring and seeking breakthroughs in the professional certification of mechanical engineering education and successfully received the accreditation in June 2020. Furthermore, under the theoretical guidance of OBE (Outcome Based Education), we have actively carried out a series of multi-angle and multi-level international cooperation with the engineering schools of Claustar University of technology in Germany, Gyeonggi University of Korea, and Silpakorn University of Thailand which constitutes an advancement on the professional accreditation of international engineering education.

From 17th November, 2019 to 18th December, 2019, the school of mechanical engineering of Chengdu University and Silpakorn University successfully held the first exchange project. In this project, the synthesis capabilities were developed, and both software and hardware skills of mechanical manufacturing were trained accordingly. After this project, the students mastered several software and mechanical manufacturing skills such as the operations of numerical control (NC), 3D prototype, and laser engraving machine. To enhance the multi-field cross capabilities of the students, several skills such as team work and cross-cultural communication skills were deliberately fostered in the sessions. Finally, a woodcut art of the badge for both universities was produced by the students and a complete report including the detailed description of the manufacturing procedures throughout the whole process was provided in Chinese and English.

IOP Conf. Series: Materials Science and Engineering **965** (2020) 011002 doi:10.1088/1757-899X/965/1/011002

The 1st International Conference on Engineering and Industrial Technology 2020 (ICEIT 2020)
11-13 September 2020, Thailand

# Introduction to Keynote Speaker

**Prof. Dr. Izumi KUMAKIRI**

Izumi Kumakiri received her Ph.D. degree from the University of Tokyo, Japan. Then, she worked at two research institutes in Europe, CNRS-France and SINTEF-Norway, for more than 10 years in total. Then she moved to a national university, Yamaguchi University, in Japan in 2011. In Europe, she worked on several international and domestic projects in the fields of carbon capture and storage (CCS), water cleaning and bio-energy by applying nano-functional materials and membranes. At Yamaguchi University, she collaborates with Prof. Yamada, who gives a plenary lecture in this conference, and has been working on the downstream of bio-fuel production.

The worldwide energy demand is growing, but we cannot continue the fossil-fuel based economy no longer. Combustion of fossil-fuel has a huge impact on the climate change and a quick action is required. Therefore, we need to shift from the fossil-fuel based economy to a more sustainable one. Bio-fuel is one of the potential energy sources and Southeast Asian countries have large biomass sources. Fermentation, which Prof. Yamada and others are working on, is a promising rather simple conversion technology of biomass to bio-fuels. However, the concentration of e.g. alcohol after fermentation is very low, e.g. a few percent. For fuel applications, the dilute alcohol solution needs to be concentrated over 99%. The conventional process, distillation, is an energy-demanding process for the concentration. By replacing distillation with more energy-efficient separation process, the advantage of using bio-alcohol will be greater.

In this talk, a new separation and concentration technology, membrane separation, will be introduced. The recent development on micro-porous inorganic membranes, their unique properties and simple operation will be discussed.

The 1st International Conference on Engineering and Industrial Technology 2020 (ICEIT 2020)
11-13 September 2020, Thailand

*Keynote Speaker*

# Application of Membrane Separation in Bio-ethanol Production

Izumi Kumakiri[1*], Naoyuki Aso[1], Ryutaro Kishibe[1], Yusuke Maruo[1], Marino Hara[2,3], Tomoyuki Kosaka[2,3], Mamoru Yamada[2,3]

[1] *Graduate School of Sciences and Technology for Innovation, Yamaguchi University, 2-16-1 Tokiwadai Ube Yamaguchi, 755-8611 JAPAN*
[2] *Graduates School of Sciences and Technology for Innovation, Yamaguchi University, 1677-1 Yoshida, Yamaguchi, 753-8511, JAPAN*
[3] *Research Center for Thermotolerant Microbial Resources, Yamaguchi University, 1677-1 Yoshida, Yamaguchi, 753-8511, JAPAN*
Email: izumi.k@yamaguchi-u.ac.jp, +81-(0)836-85-9663

*HIGHLIGHTS*

- Integration of fermentation and membrane dehydration is proposed.
- Acid-stable zeolite membranes were developed.
- Zeolite membranes were applied to concentrate bio-ethanol after fermentation.

**ABSTRACT**

Global warming is a challenge requiring international cooperation and collaboration between different research fields. One of the keys to tackle this challenge is to shift from the current fossil-fuel-based economy to a more sustainable economy. Bio-fuels and bio-chemicals are getting more attention, however, the energy consumed in a biomass conversion should be limited from a sustainability perspective.

We propose an on-site bio-conversion process by combining high-temperature fermentation and membrane separation. In this model, concentrated bio-ethanol is transported instead of raw biomass. The high-temperature fermentation does not require a water-cooling system, thus, reducing the energy requirement of the fermentation process. On the contrary, the concentration of bio-ethanol after fermentation is rather low e.g 5-10 % and requires a concentration process after the conversion. Membranes can offer an energy-efficient separation process compared to a conventional distillation. In addition, a membrane system has a simple design, allowing easier operation and maintenance.

Some types of zeolite membranes are commercially available and have been applied industrially to dry solvents. A-type zeolite membranes, the type of membranes commercially available, decomposed and lost their dehydration property when solutions obtained from fermentation were applied. We have developed new types of zeolite membranes, such as MOR and MFI membranes. These new membranes showed stable water-selectivity and successfully concentrated the bio-ethanol over 90%.

**Keywords**: membrane separation; zeolite; dehydration; bio-ethanol

**Acknowledgment**

The study was supported by e-Asia joint research program (e-Asia JRP).

IOP Conf. Series: Materials Science and Engineering **965** (2020) 011002 doi:10.1088/1757-899X/965/1/011002

The 1st International Conference on Engineering and Industrial Technology 2020 (ICEIT 2020)
11-13 September 2020, Thailand

# Introduction to Keynote Speaker

**Chin Kui CHENG**
College of Engineering
Universiti Malaysia Pahang

Dr. Cheng received his PhD in Chemical Engineering at The University of New South Wales (Australia). His PhD work is on syngas production from glycerol reforming. He is also an experienced scientist in the area of low temperature fuel cell, having worked in University of Alberta (Canada) for 2 years, developing PBI membrane for hydrocarbon conversion fuel cell. Dr. Cheng was previously a visiting researcher to Sheffield Hallam University (UK) in 2017 and also a visiting professor to the Prince of Songkla University (Thailand).

He is currently an Associate Professor in the Department of Chemical Engineering, College of Engineering. Besides his teaching role, Dr. Cheng leads the Reaction Engineering Group. His research fields are focused on the catalysis, reaction engineering and also waste treatment. Dr. Cheng has co-authored more than 140 journal articles in these fields, with an h-index of 27.

**SELECTED PUBLICATIONS**

[1] CK Cheng, SY Foo, AA Adesina. Glycerol Steam Reforming over Bimetallic Co–Ni/$Al_2O_3$. Industrial & engineering chemistry research 49 (21), 10804-10817. 2010
[2] SS Lam, RK Liew, CK Cheng, HA Chase. Catalytic microwave pyrolysis of waste engine oil using metallic pyrolysis char. Applied Catalysis B: Environmental 176, 601-617. 2015
[3] MMR Khan, TW Chuan, A Yousuf, MNK Chowdhury, CK Cheng. Schottky barrier and surface plasmonic resonance phenomena towards the photocatalytic reaction: Study of their mechanisms to enhance the photocatalytic activity. Catalysis Science & Technology. 2015
[4] $H_2$-rich synthesis gas production over Co/$Al_2O_3$ catalyst via glycerol steam reforming. CK Cheng, SY Foo, AA Adesina. Catalysis Communications 12 (4), 292-298. 2010
[5] RK Liew, WL Nam, MY Chong, XY Phang, MH Su, PNY Yek, NL Ma, ... Oil palm waste: an abundant and promising feedstock for microwave pyrolysis conversion into good quality biochar with potential multi-applications. Process Safety and Environmental Protection 115, 57-69. 2018
[6] BV Ayodele, MR Khan, CK Cheng. Catalytic performance of ceria-supported cobalt catalyst for CO-rich hydrogen production from dry reforming of methane. International Journal of Hydrogen Energy 41 (1), 198-207. 2016
[7] BV Ayodele, CK Cheng. Modelling and optimization of syngas production from methane dry reforming over ceria-supported cobalt catalyst using artificial neural networks and Box–Behnken design. Journal of Industrial and Engineering Chemistry 32, 246-258. 2015
[8] KW Siew, HC Lee, J Gimbun, SY Chin, MR Khan, YH Taufiq-Yap, ... Syngas production from glycerol-dry ($CO_2$) reforming over La-promoted Ni/Al2O3 catalyst. Renewable Energy 74, 441-447. 2015
[9] KH Ng, CH Lee, MR Khan, CK Cheng. Photocatalytic degradation of recalcitrant POME waste by using silver doped titania: Photokinetics and scavenging studies. Chemical Engineering Journal 286, 282-290. 2016
[10] KH Ng, CK Cheng .Photo-polishing of POME into $CH_4$-lean biogas over the UV-responsive ZnO photocatalyst. Chemical Engineering Journal 300, 127-138. 2016

IOP Conf. Series: Materials Science and Engineering **965** (2020) 011002 doi:10.1088/1757-899X/965/1/011002

The 1st International Conference on Engineering and Industrial Technology 2020 (ICEIT 2020)
11-13 September 2020, Thailand

***Keynote Speaker***

# Innovative Methods to Treat Palm Oil Mill Effluent

**Chin Kui Cheng**

*College of Engineering Universiti Malaysia Pahang*

ABSTRACT

Oil palm is the most cultivated commercial crop in Malaysia. In 2017 alone, Malaysia has produced 21 million metric tonnes of palm oil per annum from a total of about 450 operating mills. This accounts for 39% of world palm oil production and 44% of global palm oil exports. Every tonne of crude palm oil (CPO) produces approximately 2.5 to 3.5 tonnes of palm oil mill effluent (POME). This culminates in the total POME discharge of roughly 74 million metric tonnes. POME is a thick brownish, acidic, foul smelling colloidal wastewater. It is a complex mixture comprising of cellulosic materials, oils and fatty acids as well as lignin, pectin, carotene and phenols. Conventionally, it was treated using an open-ponding method. Nevertheless, this method requires creation of a series of lagoons; unfortunately, this consumes acreage of land which otherwise can be put into more productive usage. In order to treat POME more efficiently, various methods have been investigated such as adsorption, coagulation/flocculation, microbial fuel cells, ultrasonic-assisted membrane anaerobic system (UAMAS), up-flow anaerobic sludge fixed film (UASFF) reactor, modified anaerobic baffled bioreactor (MABB), up-flow anaerobic sludge blanket (UASB) reactors, expanded granular sludge bed (EGSB) reactors, membrane filtration, etc. Each aforementioned technologies exhibit their own unique strengths and also weaknesses. Herein, we explore the use of photocatalytic and also steam reforming methods, for the treatment of POME. The use of these two technologies were examined from the aspects of waste treatment, as well as their potential for producing useful gaseous products, i.e. $H_2$, $CH_4$. It is hope that the new knowledge gained from these innovative methods will spur more research activities in finding more effective sustainable ways to treat POME.

The 1st International Conference on Engineering and Industrial Technology 2020 (ICEIT 2020)
11-13 September 2020, Thailand

# Introduction to Keynote Speaker

**Dr. Mati HORPRATHUM**

*Keynote Speaker*

## Design and Engineering of Advanced Nanoscale Thin Films by Physical Vapor Deposition and Their Emerging Applications

Mati Horprathum

*Opto-Electrochemical Sensing Research Team, National Electronics and Computer Technology Center, Pathumthani 12120, Thailand*

ABSTRACT

Nanostructures have recently gained attention due to exhibiting large surface-to-volume ratios and unique properties, which attracted diverse ranges of applications. In this work, we present the emergence of engineering nanostructure-thin films growth via physical vapor deposition (PVD), glancing-angle deposition technique (GLAD) and combinatorial magnetron sputtering technique, developed at Opto-Electrochemical Sensing Research Team (OEC), NECTEC, Thailand, realized through the synergy of experiments, theory and potential in applications. Our proposed fabrication technique was well utilized for the preparations of the novel nanoscale thin films as alternative materials for sensor, optoelectronic and photovoltaic devices. We present morphological and nanostructural properties of various deposited materials and further explore advanced nanoscale thin films which enable controls of optical and electrical characteristics of these nanostructures, leading to improvements towards recent applications, i.e. plasmonic sensors, chemical gas sensors, smart windows, self-cleaning glasses, transparent conductive oxides (TCO), metallic glass materials and their future developments.

IOP Conf. Series: Materials Science and Engineering **965** (2020) 011003 doi:10.1088/1757-899X/965/1/011003

# Peer review declaration

All papers published in this volume of *IOP Conference Series: Materials Science and Engineering* have been peer reviewed through processes administered by the Editors. Reviews were conducted by expert referees to the professional and scientific standards expected of a proceedings journal published by IOP Publishing.

- **Type of peer review:** Double-blind
- **Conference submission management system:** http://iceitsu.org/Login/Login.aspx
- **Number of submissions received:** 46
- **Number of submissions sent for review:** 43
- **Number of submissions accepted:** 40
- **Acceptance Rate (Number of Submissions Accepted / Number of Submissions Received X 100):** 86.96%
- **Average number of reviews per paper:** 2
- **Total number of reviewers involved:** 39
- **Any additional info on review process:**
  The papers passed the preliminary review by the track chair will be reviewed again with the following guidelines:
  - Is the title and key words appropriate and relevant to its contents?
  - Is the abstract sufficiently informative?
  - Is the structure of the manuscript satisfactory?
  - Is the statement of objectives appropriate and achieved?
  - Is literature review appropriate and relevant? Does it address a sound motivation?
  - Are the description of research process and experimental procedure clearly written?
  - Are the results including all diagrams, figures, tables and captions presented clearly?
  - Are the conclusions supported by the evidence? Do they summarize the interesting findings of the research, including what have been learned and why it is useful?
  - Are the references update, relevant, and conform with academic standards?
  - Is the English correct and acceptable with clarity of expression, communication of ideas, readability and discussion of concepts?
  - Does the content of the work of sufficient scientific interest to justify its length?
- **Contact person for queries:**

IOP Conf. Series: Materials Science and Engineering **965** (2020) 011003 doi:10.1088/1757-899X/965/1/011003

Asst. Prof. Dr. Worapon Kiatkittipong (kiatkittipong_w@su.ac.th)
Silpakorn University

IOP Conf. Series: Materials Science and Engineering **965** (2020) 012001 doi:10.1088/1757-899X/965/1/012001

# The potential of attached growth of microalgae on solid surface for biomass and lipid production

**Sze Yu Chuah[1], Yaleeni Kanna Dasan[1,2], Yoke Wang Cheng[2], Jun Wei Lim[2,3], Yeek Chia Ho[4], Inn Shi Tan[5], Henry Chee Yew Foo[5], Peck Loo Kiew[6], Sim Siong Leong[7] and Man Kee Lam[1,2*]**

[1] Chemical Engineering Department, Universiti Teknologi PETRONAS, 32610 Seri Iskandar, Perak, Malaysia

[2] HICoE-Centre for Biofuel and Biochemical Research, Institute of Self-Sustainable Building, Universiti Teknologi PETRONAS, 32610 Seri Iskandar, Perak, Malaysia

[3] Fundamental and Applied Sciences Department, Universiti Teknologi PETRONAS, 32610 Seri Iskandar, Perak, Malaysia

[4] Civil and Environmental Engineering Department, Universiti Teknologi PETRONAS, 32610 Seri Iskandar, Perak, Malaysia

[5] Department of Chemical Engineering, Curtin University, Sarawak Campus CDT 250, 98009 Miri, Sarawak, Malaysia

[6] Department of Chemical and Petroleum Engineering, Faculty of Engineering, Technology & Built Environment, UCSI University, Jalan Mandarina Damai 1, 56000 Cheras, Kuala Lumpur, Malaysia

[7] Department of Petrochemical Engineering, Faculty of Engineering and Green Technology, Universiti Tunku Abdul Rahman, Kampar 31900, Perak, Malaysia

*E-mail: lam.mankee@utp.edu.my

**Abstract**. Microalgae cultivation is usually performed via suspended growth system; however, the subsequent harvesting technologies are either costly or energy intensive. Herein, attached growth of *Chlorella vulgaris* was investigated using low-cost supporting materials (polystyrene foam and cotton duct) to ease the microalgae harvesting process. Cotton duct served as a better supporting material than polystyrene foam as it promoted microalgae attachment. Evidently, the attached cultivation performed on cotton duct granted higher biomass yield (16.40 g/m$^2$) than the attached cultivation with polystyrene foam (11.70 g/m$^2$) after 14 days of cultivation. From the studied range of 0.1 – 0.3 v/v%, the optimal inoculum concentration for the attached cultivation of *C. vulgaris* on cotton duct was 0.3 v/v%. After 14 days of cultivation, the optimized attached cultivation of *C. vulgaris* was capable to produce a biomass yield of 18.2 g/m$^2$. The lipid content of dried microalgae from optimized attached cultivation (43 wt%) was higher than that of suspended cultivation (32.7 wt%).

## 1. Introduction

For the past decade, global energy demand underwent most conspicuous growth in 2018, with an increment of circa 2.3% [1]. Due to the upsurge of renewable energy, biofuel is often applauded as the potential surrogate of exhaustible fossil fuel [2]. Biofuel is readily produced from edible crops, crop wastes, woody parts of plants, and even algae [3]. Despite the potentiality of food-based biofuel, the food versus fuel controversy prompted the birth of third generation biofuel that derived from microalgae [4]. As compared to food crops, microalgae are better biofuel feedstock with faster growth rate, simpler cell structure (tolerate most of the environment), and higher lipid productivity [5]. Besides, microalgae-based biofuel is more environmental benign than fossil fuel in terms of lesser carbon footprint [6].

The commercialization of microalgae cultivation for biofuel production was eventually impeded by high production cost [7], which associated with cost-ineffective microalgae harvesting process (about 30 % of total cost [8]. Prevalently, suspended growth system was opted for microalgae cultivation whereby the microalgae dispersed freely over aqueous media [9]. Separation of microalgae from enormous amount of aqueous media could be performed using coagulation-flocculation, filtration, flotation, or centrifugation [10]. Aforesaid separation techniques are either costly or energy intensive. Lately, attached growth system was touted as a promising cultivation system to ease microalgae-media separation by forming biofilm on the surface of substrate. In relative to suspended growth system, the attached growth system necessitates lower water requirement, shorter hydraulic retention time, lower capital cost, and lesser monitoring [11]. Apart from high microalgae yield, the attached growth system allows the microalgae being concentrated in the form of biofilm, thereby guarantees a more effective microalgae separation.

To date, the microbial biofilm designs of attached growth system could be classified into three types, *viz.* permanently immersed biofilms, biofilms between liquid and gaseous phases, or permeated biofilms, depending on the biofilm arrangement and liquid medium supply [5]. For attached cultivation, environmental parameters that could influence biofilm growth include microalgae strains, supporting material for attachment, liquid medium flow rate, concentration of nutrients, and light intensity [12]. Previous studies demonstrated the potentiality of attached cultivation to replace suspended cultivation; nonetheless, this method had not been comprehensively investigated.

Herein, attached cultivation of *Chlorella vulgaris* microalgae was performed using cheap and readily available supporting materials, namely polystyrene foam and cotton duct. Despite diverse species of high lipid-bearing microalgae, *Chlorella vulgaris* was selected in this work owing to its small cell size and high tolerability to extreme environmental conditions [13]. After the screening of better supporting material (polystyrene foam or cotton duct), the attached cultivation of *Chlorella vulgaris* was further optimized for its biomass yield by varying inoculum concentration (0.1 – 0.3 v/v%). For optimization study, the reported biomass yields were biomass dry weights that quantified after 2 weeks of cultivation. Lastly, the lipid content of microalgae from optimal attached cultivation was assessed before compared with that from suspended cultivation.

## 2. Materials and methods

### *2.1. Stock culture preparation*

*Chlorella vulgaris* was sourced from Centre of Biofuel and Biochemical Research (CBBR), Universiti Teknologi PETRONAS. The microalgae were cultivated in Bold's Basal Medium (BBM). Every 1.0 L of BBM culture medium was made up of (a) 10 mL of $NaNO_3$ (2.5 g/L), $CaCl_2{\cdot}2H_2O$ (2.5 g/L), $MgSO_4{\cdot}7H_2O$ (7.5 g/L), $K_2HPO_4$ (7.5 g/L), $KH_2PO_4$ (17.5 g/L), NaCl (2.5 g/L); and (b) 1 mL of anhydrous EDTA (50 g/L), KOH (31 g/L), $FeSO_4{\cdot}7H_2O$ (4.98 g/L), $H_2SO_4$ (1 mL), $H_3BO_3$ (11.4 g/L), $ZnSO_4{\cdot}7H_2O$ (8.82 g/L), $MnCl_2{\cdot}4H_2O$ (1.44 g/L), $O_3$ (0.71 g/L), $CuSO_4{\cdot}5H_2O$ (1.57 g/L), $Co(NO_3)_2{\cdot}6H_2O$ (0.49 g/L). For the preparation of stock culture, 0.5 L of mother seed was inoculated into 1.0 L Duran bottle with an initial pH of 3.5, wherein the surrounding temperature was about 25 – 28 °C. During cultivation, compressed air was supplied to aerate the culture medium whereas cool-white fluorescent light (Philip TL-D 36W/865, light intensity of 60 – 70 μmol/m$^2$·s) was used to illuminate the culture medium [14].

### *2.2. Suspended cultivation*

By using optimum condition from previous study [13], suspended cultivation of *C. vulgaris* was carried out in a 5-L Duran bottle to serve as a control. Chicken manure compost was used as the nutrient to further grow *C. vulgaris* from seed culture. A 10 g of compost was dissolved in 0.6 L of tap water, and the mixture was stirred for 24 h. Centrifugation of mixture was then performed to separate insoluble solids from the compost solution. For suspended cultivation in sterilised bottle, a 5-L culture medium was prepared from 0.5 L of stock culture, 0.2 L of compost solution, and 4.3 L of tap water. The initial pH of culture medium was adjusted to 3 while the ambient temperature was around 25 – 28 °C. The culture medium was aerated and light-illuminated simultaneously, similar to the stock culture preparation. Tap water was added daily to compensate the loss of water from the culture medium through evaporation. Figure 1 shows the experimental set-up of suspended cultivation.

**Figure 1.** Experimental set-up of suspended cultivation.

### 2.3. *Attached cultivation*

Figure 2 illustrates two different experimental set-ups employed for attached cultivation of *C. vulgaris*. For a fair comparison with suspended cultivation, attached cultivation of *C. vulgaris* was performed using identical volumetric ratio of stock culture: compost solution: tap water (2.5: 1: 21.5). Under the illumination of cool-white fluorescent light, the culture medium was aerated with compressed air using air pump and air stones. The initial pH of culture medium was adjusted to 6, and the culture temperature was maintained in between 25 – 28 °C. Tap water was added daily to culture medium for compensation of evaporated water.

For screening of supporting materials, set-up Version 1.0 shown in Figure 2 (a) was used to ease the growth observing process. In this set-up, supporting materials (polystyrene foam and cotton duct) were cut into strips (surface area of 3 cm × 9 cm), which later placed in bioreactor. The bioreactors were modified from clear rectangular polypropylene containers, with size of 16.5 cm length × 11.5 cm width × 7.1 cm height. A total volume of 500 mL of culture medium (50 mL stock culture, 20 mL of compost solution, and 430 mL of tap water) was transferred to the bioreactor.

After the identification of suitable supporting material, set-up Version 2.0 shown in Figure 2 (b) was used for the optimization of attached cultivation for biomass yield. The aforesaid optimization was done by manipulating the inoculum concentration from 0.1 – 0.3 v/v%. The selected supporting material was cut into smaller pieces (surface area of 1 cm × 1 cm), and 300 pieces of supporting material were loaded into the bioreactor. Every piece of supporting material was numbered and weighed to facilitate the computation of biomass yield. The bioreactors were 1-L conical flasks instead of the polypropylene containers. For 0.1 v/v% inoculum concentration, the 1-L culture medium consisted of 100 mL of stock culture, 40 mL of compost solution, and 860 mL of tap water. The volume of compost solution was kept constant during the manipulation of inoculum concentration.

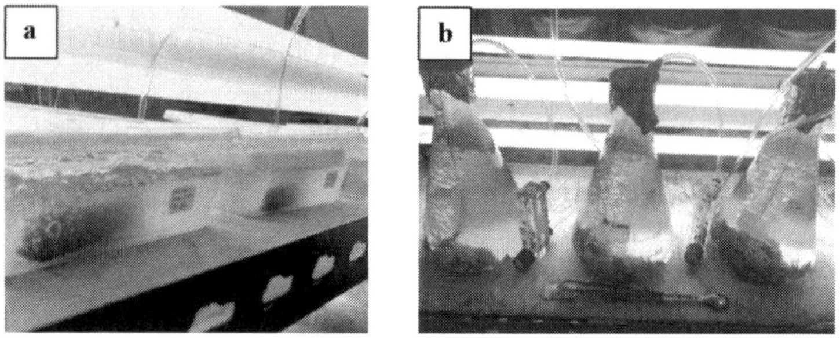


**Figure 2.** Experimental set-ups of attached cultivation: (a) set-up Version 1.0 (b) set-up Version 2.0.

*2.4. Measurement of suspended biomass concentration*
Several known volume of culture media were measured for their optical density at a specified wavenumber of 688 nm ($OD_{688}$) with the aid of UV-Vis spectrophotometer (Shimadzu UV-1280). Concurrently, these culture media were oven-dried at 100 °C to determine their suspended biomass concentration. From this preliminary work, the correlation between $OD_{688}$ and suspended biomass concentration was purposed as Eq. 1. Hence, the suspended biomass concentration of culture media could be estimated from $OD_{688}$ via Eq. 1.

$$\text{Biomass concentration (g/L)} = 0.5315(OD_{688}) + 0.0333;\ R^2 = 0.9721 \quad (1)$$

*2.5. Microalgae harvesting*
Different microalgae harvesting methods were used for suspended and attached cultivation, but aeration was stopped before any microalgae harvesting. For suspended cultivation, sedimentation method was applied wherein the microalgae were allowed to settle down naturally in 2 days' time. After settling, excess water was discarded, and the concentrated microalgae were oven-dried at 105 °C for 24 h. The dried microalgae were scrapped from the beaker and kept for further use. For attached cultivation, it was impractical to collect all the dried microalgae from supporting material by scrapping. Therefore, wet pieces of supporting material (microalgae and supporting material) were retrieved and oven-dried at 105 °C for 24 h. The biomass yield of attached cultivation was expressed as the weight of dried microalgae per unit surface area of supporting material ($g/m^2$).

*2.6. Microalgae lipid extraction*
The method of microalgae lipid extraction was adopted from the work of Bligh and Dyer [15]. Briefly, 0.2 g of dried microalgae (from suspended cultivation) and 50 pieces of dried supporting materials with microalgae (from attached cultivation) were transferred to pre-weighed beakers. A 60 mL of methanol-chloroform solution (volumetric ratio of 2:1) was poured into each beaker. Both mixtures were stirred at 400 rpm for 24 h in ambient temperature. The resulting mixture was filtered with SMITH filter paper (mean pore size of 20 – 25 μm) using a funnel. Then, the filtrate was air-purged at room temperature inside fume hood to evaporate the solvent before oven-dried at 105 °C for 24 h. Lastly, the dried microalgae lipid was weighed, and its lipid content was reported on the mass basis of dried microalgae.

## 3. Results and discussions

*3.1. Screening of supporting material*
In this study, polystyrene foam and cotton duct were purposed as low-cost supporting materials for attached cultivation of *C. vulgaris*. They were selected owing to their reusability and high availability [16]. Figure 3 depicts the transient profile of suspended biomass concentration for attached cultivation of *C. vulgaris* on polystyrene foam and cotton duct over 14 cultivation days. After 1 day of cultivation, the suspended biomass concentration dropped significantly, indicating the occurrence of *C. vulgaris*

microalgae attachment on supporting materials. The increment of suspended biomass concentration after day 1 eventually discloses that some microalgae detached from supporting materials as the initial adhesion was not firm. As compared with polystyrene foam, the use of cotton duct as supporting material recorded a lower suspended biomass concentration throughout 14 days of cultivation. A good supporting material should generate a low suspended biomass concentration as it favours the attached cultivation. Thus, cotton duct was preferred over polystyrene foam as a supporting material.

Figure 4 shows the biomass yield for attached cultivation of *C. vulgaris* on polystyrene foam and cotton duct on $10^{th}$ and $14^{th}$ cultivation day. The biomass yield of both attached cultivation increased from day 10 to 14, proving the microalgae growth in attached growth system. The attached cultivation of *C. vulgaris* performed with cotton duct granted a higher biomass yield (16.40 g/m$^2$) than that of polystyrene foam (10.05 g/m$^2$). This finding corroborates the microalgae attachment was more prevailed over cotton duct. Hence, cotton duct was employed for subsequent attached cultivation since it serves as a better supporting material as compared to polystyrene foam.

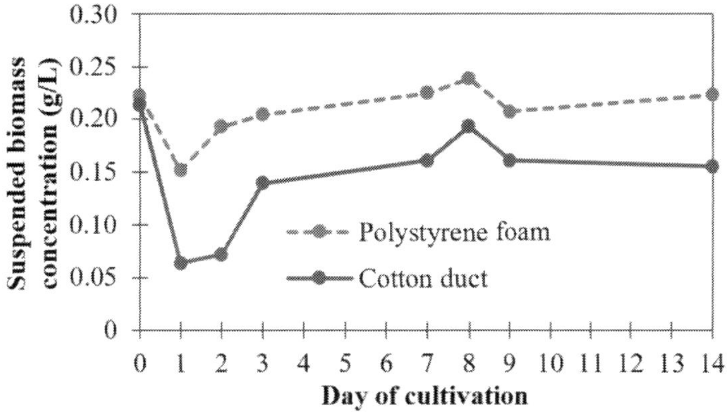


**Figure 3.** Transient profile of suspended biomass concentration for attached cultivation of *C. vulgaris* on polystyrene foam and cotton duct over 14 cultivation days.

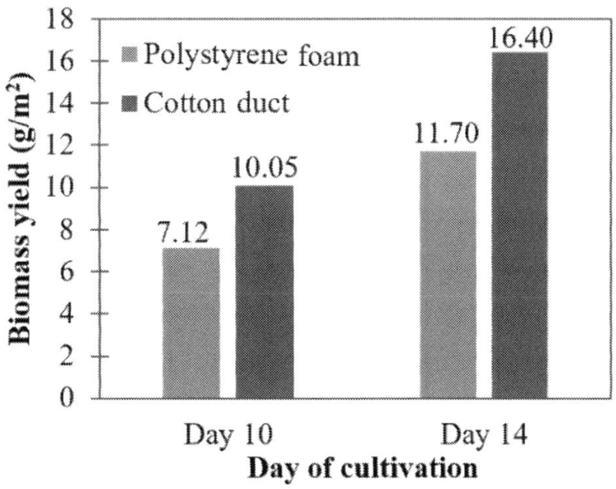


**Figure 4.** Biomass yield for attached cultivation of *C. vulgaris* on polystyrene foam and cotton duct on $10^{th}$ and $14^{th}$ cultivation day.

The possible reason for cotton duct outperformed polystyrene foam was the unique texture of cotton duct, which created protective clefts that shielded the microalgae biofilm from shear force in aerated liquid media [16]. By using cotton duct as supporting material, the biomass yield obtained in the current work (16.40 g/m$^2$) was comparable with that of previous research (16.20 g/m$^2$) conducted by Gross et al. [16]. Gross et al. [16] performed attached cultivation of *C. vulgaris* with a rotating algal biofilm system while this study examined attached cultivation of *C. vulgaris* with a simple 1-L conical flask. Surprisingly, the utilization of cotton duct gave similar biomass yield regardless of reactor configuration.

### *3.2. Inoculum concentration*

Inoculum concentration represents the concentration of stock culture in the cultivation medium. The attached cultivation of *C. vulgaris* on the cotton duct was optimized for its biomass yield by altering the inoculum concentration from 0.1 to 0.3 v/v%. According to Huang et al. [17], inoculum concentration could influence the light penetration to the biofilm, so optimal inoculum concentration should be used to maximize the microalgae growth. A dilute inoculum could result in fewer participating microalgae in photosynthesis whereas a concentrated inoculum might hinder the light transmission to the biofilm [18].

Figure 5 presents the influence of inoculum concentration on attached cultivation of *C. vulgaris* on cotton duct over 14 days of cultivation. The positive growth trend reveals that the microalgae could be cultivated via attached growth system. Figure 6 summarizes the final biomass yield for attached cultivation of *C. vulgaris* on cotton duct using different inoculum concentrations. In overall, the final biomass yield increased as the inoculum concentration increased from 0.1 – 0.3 v/v%.

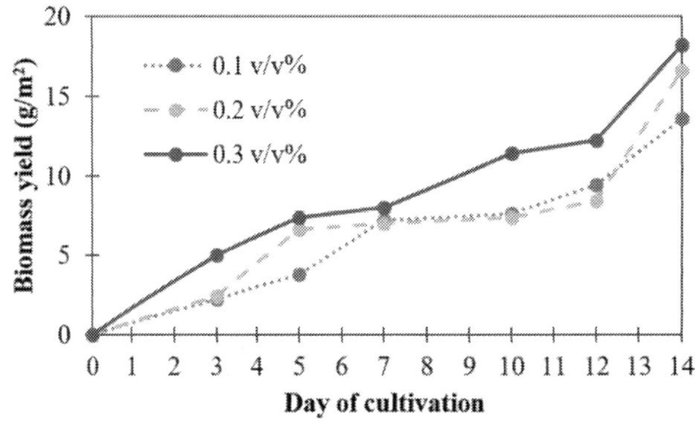


**Figure 5.** Influence of inoculum concentration on attached cultivation of *C. vulgaris* on cotton duct over 14 days of cultivation.

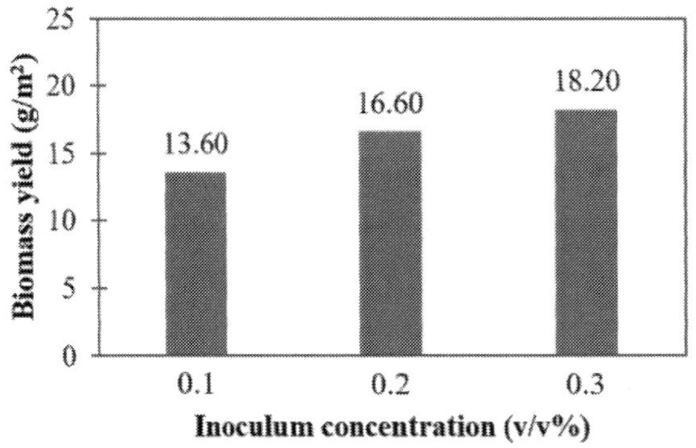


**Figure 6.** Final biomass yield for attached cultivation of *C. vulgaris* on cotton duct using different inoculum concentrations.

In present study, the biomass yield did not decline with greater inoculum concentration used, positing the cool-white fluorescent light was approachable by the microalgae biofilm for photosynthesis even at highest inoculum concentration of 0.3 v/v%. For inoculum concentration range of 0.1 – 0.3 v/v%, the increase of inoculum concentration enhanced the biofilm growth on cotton duct, which reduced the lengthy lag-phase and improved the survival rate of microalgae over other microbes [15]. Therefore, the optimal inoculum concentration for attached cultivation of *C. vulgaris* on cotton duct was 0.3 v/v% in this project.

### *3.3. Lipid content*

Lipid is the most precious part of microalgae since it could be converted into biodiesel, *viz.* a potential substitute for petroleum-based diesel [2]. The dried microalgae samples used for lipid extraction were collected from optimized suspended cultivation [13] and attached cultivation. Figure 7 compiles the lipid contents of dried microalgae from current work and previous studies. Under optimized conditions, the lipid contents of dried microalgae from suspended and attached growth systems were 32.7 wt% and 43 wt%, respectively. Therefore, the employment of attached cultivation on cotton duct could enhance the lipid content of *C. vulgaris* by *circa* 1.31 times. As compared to previous studies on attached cultivation, the lipid content of dried microalgae from optimized attached cultivation was nearly 3.64 – 4.83 times of literature lipid contents [11, 16]. This finding evinces the feasibility of optimization to maximize the lipid content of dried microalgae from attached cultivation. For cultivation of *C. vulgaris*, the attached growth system seems to be more effective than suspended growth system.

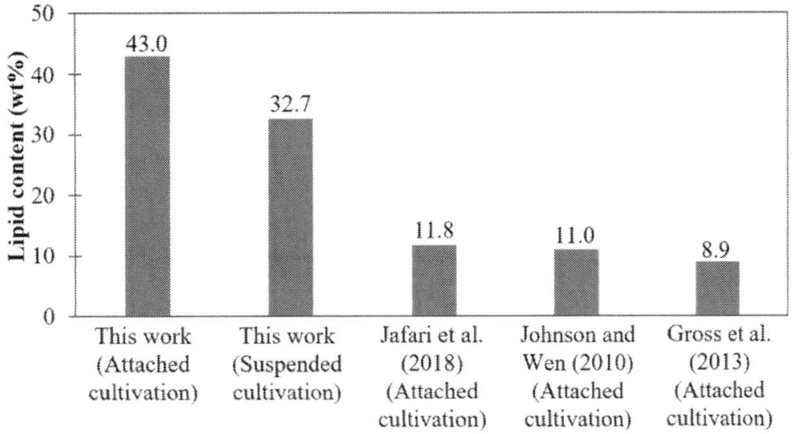


**Figure 7.** Lipid contents of dried microalgae from current work and previous studies

## 4. Conclusion

The present work investigated the potentiality of attached growth system to cultivate *Chlorella vulgaris* on low-cost supporting material like polystyrene foam and cotton duct. The declined suspended biomass concentration after 1 day of cultivation corroborated the microalgae attachment on supporting material. As compared to polystyrene foam, the cotton duct was more suitable to be used as supporting material since it attained lower suspended biomass concentration and higher biomass yield throughout 14 days of cultivation. For attached cultivation of *C. vulgaris* on cotton duct, the biomass yield was optimized by altering the inoculum concentration from 0.1 – 0.3 v/v%. The parallel trend between biomass yield and inoculum concentration suggested the light penetration was suffice for 0.1 – 0.3 v/v% inoculum concentration. In the specified range, the optimal inoculum concentration was 0.3 v/v%, which achieved highest biomass yield (18.2 $g/m^2$ on day 14). Herein, the lipid content of dried microalgae from optimized attached cultivation (43 wt%) was greater than that of suspended cultivation (32.7 wt%). Aforesaid finding shows the feasibility of attached cultivation for *C. vulgaris*, but further investigations should be conducted to scrutinize the effect of other environmental parameters on the biofilm growth.

**Acknowledgements**

The authors would like to acknowledge the financial support from Yayasan Universiti Teknologi PETRONAS (YUTP with cost center 0153AA-H46) and UCSI University Research Excellence &

Innovation Grant (REIG with grant no. REIG-FETBE-2020/019). Support from Ministry of Education Malaysia through HICoE award to CBBR is duly acknowledged (cost center 015MA0-052).

**References**

[1] Carr M and Hodges J 2019 *Global carbon emissions hit record in 2018 on higher fossil fuel consumption* (New York: Insurance Journal)
[2] Chisti Y 2007 Biodiesel from microalgae *Biotechnol. Adv.* **25** 294–306
[3] Khan M I, Shin J H and Kim J D 2018 The promising future of microalgae: current status, challenges, and optimization of a sustainable and renewable industry for biofuels, feed, and other products *Microb. Cell. Fact.* **17** 1–21
[4] Arshad M, Zia M A, Shah F A and Ahmad M 2018 An overview of biofuel *Perspectives on water usage for biofuels production* A Muhammad (Cham: Springer International Publishing) pp 1–37
[5] Mantzorou A and Ververidis F 2019 Microalgal biofilms: A further step over current microalgal cultivation techniques *Sci. Total. Environ.* **651** 3187–201
[6] Heo S and Choi J 2019 Potential and environmental impacts of liquid biofuel from agricultural residues in Thailand *Sustainability* **11** 1–14
[7] Choudhary P, Malik A and Pant K K 2017 Algal biofilm systems: An answer to algal biofuel dilemma *Algal biofuels: Recent advances and future prospects* S K Gupta *et al* (Cham: Springer International Publishing) pp 77–96
[8] Barros A I, Gonçalves A L, Simões M and Pires J C M 2015 Harvesting techniques applied to microalgae: A review *Renew. Sustain. Energy Rev.* **41** 1489–500
[9] Abo B O, Odey E A, Bakayoko M and Kalakodio L 2019 Microalgae to biofuels production: a review on cultivation, application and renewable energy *Rev. Environ. Health* **34** 91–9
[10] Singh G and Patidar S K 2018 Microalgae harvesting techniques: A review *J. Environ. Manage.* **217** 499–508
[11] Jafari N, Shafiee Alavijeh R, Abdolahnejad A, Farrokhzadeh H, Amin M M and Ebrahimi A 2018 An innovative approach to attached cultivation of *Chlorella vulgaris* using different materials *Environ. Sci. Pollut. Res. Int.* **25** 20097–105
[12] Zhuang L-L, Yu D, Zhang J, Liu F-F, Wu Y-H, Zhang T-Y, Dao G-H and Hu H-Y 2018 The characteristics and influencing factors of the attached microalgae cultivation: A review *Renew. Sustain. Energy Rev.* **94** 1110–9
[13] Tan X B, Lam M K, Uemura Y, Lim J W, Wong C Y, Ramli A, Kiew P L and Lee K T 2018 Semi-continuous cultivation of *Chlorella vulgaris* using chicken compost as nutrients source: Growth optimization study and fatty acid composition analysis *Energy Convers. Manage.* **164** 363–73
[14] Lam M K and Lee K T 2012 Potential of using organic fertilizer to cultivate *Chlorella vulgaris* for biodiesel production *Appl. Energy* **94** 303–8
[15] Bligh E G and Dyer W J 1959 A rapid method of total lipid extraction and purification *Can. J. Biochem. Phys.* **37** 911–7
[16] Gross M, Henry W, Michael C and Wen Z 2013 Development of a rotating algal biofilm growth system for attached microalgae growth with in situ biomass harvest *Bioresour. Technol.* **150** 195–201
[17] Huang Y, Xiong W, Liao Q, Fu Q, Xia A, Zhu X and Sun Y 2016 Comparison of *Chlorella vulgaris* biomass productivity cultivated in biofilm and suspension from the aspect of light transmission and microalgae affinity to carbon dioxide *Bioresour. Technol.* **222** 367–73
[18] Katarzyna L, Sai G and Singh O A 2015 Non-enclosure methods for non-suspended microalgae cultivation: Literature review and research needs *Renew. Sustain. Energy Rev.* **42** 1418–27

# Hydrothermal Treatment of Palm Oil Mill Effluent (POME) under Oxidative and Non-oxidative Conditions

**Zhan Sheng Lee, Chin Kui Cheng, Sim Yee Chin***

Faculty of Chemical and Process Engineering Technology, Universiti Malaysia Pahang, Lebuhraya Tun Razak, 26300 Gambang, Kuantan Pahang

chin@ump.edu.my

**Abstract**. The performances of oxidative (OHT) and non-oxidative (NOHT) subcritical hydrothermal treatments of palm oil mill effluent (POME) were investigated. The experiments were performed in a pressurised 500 mL-autoclave at different temperature (493 K – 533 K) and reaction time (2 h – 8 h). At 533 K and 8 h, the OHT reaction showed the highest removals of 5-day biochemical oxygen demand ($BOD_5$) and chemical oxygen demand (COD), recording 87.30% and 71.23% respectively, with the pH of liquid product attained 6.5 from an initial value of 3.5. The reduction of COD and $BOD_5$ in NOHT was lower than that in OHT, which were 61.43% and 68.02%, respectively. The mechanism of OHT reaction was via the free radical's pathway. In contrast, the organic compounds originally present in POME degraded into water-soluble products, accompanied by deoxygenation that consisted of decarboxylation and dehydration during the NOHT.

## 1. Introduction

The thriving palm oil industry in Malaysia has led to a massive production of viscous organic waste known as palm oil mill effluent (POME). The production of each tonne of fresh fruit bunches generates an approximately 0.66 tonne of POME in the palm oil milling process [1]. POME contains numerous suspended constituents, including the organelles, a number of nitrogenous compounds ranging from amino acids to proteins, various carbohydrates such as simple sugars and hemicellulose, a group of insignificant organic and mineral constituents, as well as free organic acids [2]. POME is acidic with an average pH of 4, attributed to organic acids. The typical 5-day biochemical oxygen demand ($BOD_5$) and chemical oxygen demand (COD) values of POME are 30,000 mg/L and 54,000 mg/L, respectively, which are higher than the other agricultural effluents [3].

Majority of the oil palm millers in Malaysia practise open ponding system for POME treatment. However, this treatment method needs long hydraulic retention time (up to 60 days) and large land bank for pond creation. Furthermore, the unpleasant odour as well as the complication in collecting biogas and sustaining the liquor distribution are the other drawbacks [4]. Over the years, different POME treatment systems have been evaluated, such as ultrafiltration membrane separation [5], steam reforming [6, 7], and advanced oxidation processes [8-10]. Nevertheless, these studies remained at laboratory-scale and were not investigated for their scalability for industrial application. This could be due to the technical impracticability and infeasible cost.

Subcritical hydrothermal treatment involves initiation of reaction at pressurised condition, without exceeding the critical pressure and temperature. This treatment method has been widely applied on wet biomass as it circumvents the drying pre-treatment. Some of the other benefits are the ease of process implementation, hazard-free reaction as well as the reasonable capital and operational costs due to mild operating conditions [11, 12]. Generally, hydrothermal reaction can be performed under either oxidative or inert (non-oxidative) atmosphere, where the former involves the addition of oxidant in the form of compressed air. Urrea et al. compared the reaction atmosphere on biopolymers in hydrothermal treatment of sludge [13]. They reported that OHT showed better results in terms of solubilisation, settleability and mineralisation in sludge treatment. Yousefifar et al. reported the study of OHT and NOHT on cellulose at different temperatures [14]. These authors concluded that non-oxidative process showed a significant role in solid degradation and solubilisation at temperature higher than 240°C. To our best knowledge, no studies of OHT and NOHT of POME have been done. POME contains a small portion of suspended solids that made up of lignocellulosic materials. Thus, POME could be treated following the method of lignocellulosic wastewater.

## 2. Materials and methods

### *2.1. POME sampling, storage and characteristics*

30-L of raw effluent from the end-of-pipe of a local milling facility was acquired. The temperature and pH value of the liquor were measured instantly, recorded readings of 360 K and 3.5, respectively. The average concentration of COD and $BOD_5$ of the fresh POME were 52,200 ± 800 mg/L and 18,020 ± 640 mg/L. The POME was preserved in a tight-sealed opaque container and stored in a chiller at 277 K. The POME was homogenised prior to hydrothermal reaction.

### *2.2. Subcritical hydrothermal POME treatment*

A pressure vessel (Model: Buchiglasuster Type 3E) of 1.0-L working capacity was used. 500 mL of the wastewater was stirred consistently at 500 rpm using an electromagnetic-driven propeller to maintain its homogeneity. The reactor temperature was varied between 493 K and 533 K with reaction time fixed at 2 h. Meanwhile, the influence of reaction time on POME was examined by varying the reaction time from 2 h to 8 h under a constant temperature of 533 K (highest temperature employed in the current study). All the operating parameters were identical for both NOHT and OHT, except the gas medium used to pressurise the reactor. For NOHT, 99.9995% purity nitrogen gas supplied by Alpha Gas Solution Malaysia was introduced into the vessel for purging and pressurising purpose prior to each experiment. The nitrogen gas was replaced by general industrial compressed air from Alpha Gas Solution Malaysia for OHT study. An initial pressure of 50 bar was fixed in the reactor, followed by gradual autogenous pressure build-up in concurrent with reaction temperature. The liquid product was sampled intermittently from the reactor for further analysis. All the experiment sets were triplicated for reproducibility assessment.

### *2.3. Liquid product analysis*

The major components in fresh POME was identified using a gas chromatography coupled with mass spectroscopy (GC-MS). 50 mL of POME was freeze-dried into solid powder using Labconco freeze dryer. The freeze-drying process was carried out at 193 K under a 0.14 mbar vacuum. The POME powder was transferred to an Agilent 7890A GC-MS via a single packed-bed column, namely Agilent J&W DB 5-ms (30 m × 0.25 mm ID, 0.25 μm df, Fused Silica). The MS detector and injector port of GC-MS were operated at 523 K while the operating temperature of GC oven was 573 K. The GC column temperature was raised from 323 K to 473 K at 8 K/min, and then to 573 K at 10 K/min. The compounds were identified using NIST mass spectral library. The liquid samples from various runs were subjected to pH, COD and $BOD_5$ analyses. A visual observation was done to compare the colour of all discharged products. A pH indicator strip from Merck was used to measure the pH value of POME and liquid products. For the COD analysis, two mL of the diluted fresh POME and liquid samples was carefully pipetted into the oxidation vials. A vial with distilled water was prepared as a baseline. A Hach DRB200 reactor was employed to digest the COD vials at 423 K for 2 h. The calculation of COD removal is shown in equation (1).

$$\text{COD removal (\%)} = (COD_i - COD_f)/COD_i \times 100 \qquad (1)$$

where $COD_i$ represents COD value of POME, ppm
$COD_f$ represents COD value of liquid discharge, ppm

The $BOD_5$ method was employed to measure the oxygen consumption by the microorganisms in the samples. One mL for each of the calcium chloride, magnesium sulphate, ferric chloride and phosphate buffer solutions were stirred thoroughly. The mixture was subsequently diluted to 1-L of dilution water. The dilution factors of the discharged products and raw POME were 100 and 1500 respectively. The diluted samples were shifted to a 300-mL incubation bottle. 1 N NaOH was added gradually to ensure that the pH was in the region of 6.5 to 7.5. The initial value of dissolved oxygen (DO) was recorded before depositing the incubation bottles in a BOD incubator. The incubation was done at 293 K for five consecutive days. The $BOD_5$ readings were obtained using equation (2). Equation (3) shows the calculation for $BOD_5$ removal percentage.

$$BOD_5 = (DO_i - DO_f) / V \qquad (2)$$

where $DO_i$ represents the initial DO value in ppm
$DO_f$ represents the DO value after five days in ppm
V represents the decimal volumetric fraction of sample used

$$BOD_5 \text{ removal (\%)} = (BOD_{5(i)} - BOD_{5(f)})/BOD_{5(i)} \times 100 \qquad (3)$$

where $BOD_{5(i)}$ represents the BOD value of POME, ppm
$BOD_{5(f)}$ represents the BOD value of liquid discharge, ppm

## 3. Results & discussion

### *3.1. Fresh POME Characterisation*

The fresh POME from the end-of-pipe of milling process was blackish brown and has unpleasant odour. Table 1 summarises the properties of raw POME in this study and in the past studies. The POME characteristics in this study were similar to literatures [6, 15, 16].

**Table 1.** The characteristics of fresh POME of the present study and literature works.

| Properties | Typical range [6, 15] | Present study [16] |
|---|---|---|
| Temperature (K) | 353-363 | 360 |
| COD (ppm) | 44,300-70,900 | 52,200 |
| $BOD_5$ (ppm) | 11,000-27,000 | 18,020 |
| pH | 3.6-4.3 | 3.5 |

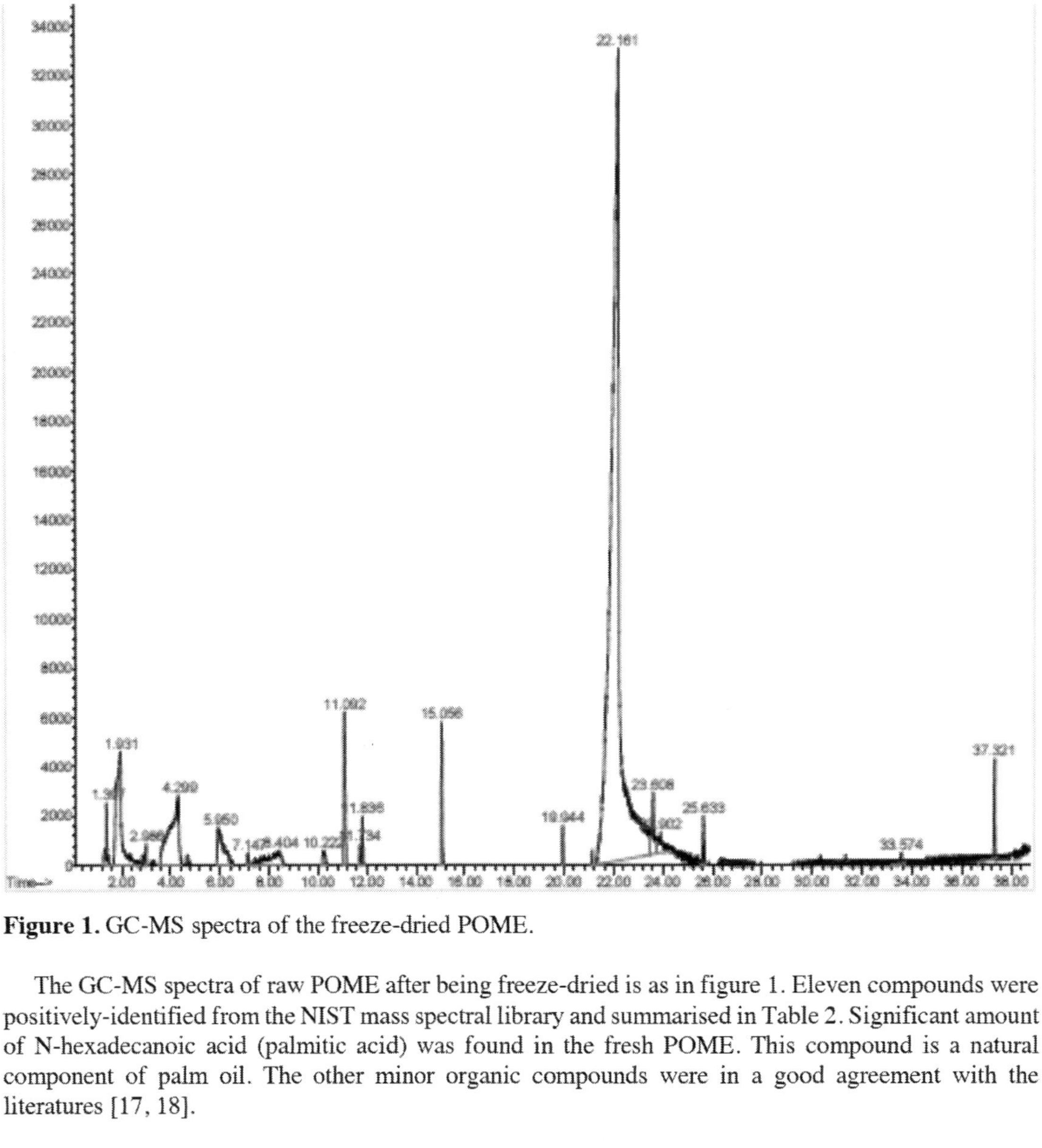


**Figure 1.** GC-MS spectra of the freeze-dried POME.

The GC-MS spectra of raw POME after being freeze-dried is as in figure 1. Eleven compounds were positively-identified from the NIST mass spectral library and summarised in Table 2. Significant amount of N-hexadecanoic acid (palmitic acid) was found in the fresh POME. This compound is a natural component of palm oil. The other minor organic compounds were in a good agreement with the literatures [17, 18].

**Table 2.** Identification of organic compounds in POME using GC-MS analysis.

| Retention time | Organics | Molecular formula | Molecular weight | $x_i$ (%) |
|---|---|---|---|---|
| **4.297** | Allyl ethyl ether | $C_9H_{10}O$ | 134.18 | 7.26 |
| **5.949** | 1-Buten-3-yne, 2-methyl- | $C_5H_6$ | 66.10 | 2.65 |
| **11.091** | 4-Hydroxy-2-methylacetophenone | $C_9H_{10}O_2$ | 150.17 | 2.42 |
| **11.735** | 1-Ethyl-2,2,6-trimethylcyclohexane | $C_{11}H_{22}$ | 154.292 | 0.17 |
| **11.835** | Ethyl (Z)-non-3-enyl carbonate | $C_{12}H_{22}O_3$ | 214.30 | 0.44 |
| **15.056** | Benzene, 1,4-dimethoxy-2-methyl- | $C_9H_{12}O_2$ | 152.1904 | 1.62 |
| **19.942** | 1-Propanol, 2-methyl- | $C_4H_{10}O$ | 74.122 | 0.28 |

| | | | | |
|---|---|---|---|---|
| **22.162** | n-Hexadecanoic acid | $C_{16}H_{32}O_2$ | 256.40 | 81.84 |
| **23.607** | 5-Octadecene, (E)- | $C_{18}H_{36}$ | 252.49 | 1.54 |
| **33.572** | Propylene oxide | $C_3H_6O$ | 58.079 | 0.42 |
| **37.319** | 1,4-Hexadiene, 3,3,5-trimethyl- | $C_9H_{16}$ | 124.223 | 1.35 |

### *3.2. Visual Appearance*

Figure 2 visualises the comparison among the raw POME and liquid products. The colour of liquid discharge of NOHT was darker than OHT. The presence of solid hydrochar (also found deposited on the reactor wall) may describe the dark colour of NOHT liquid product. The liquid discharge of NOHT exhibited a burnt charcoal smell while the OHT product was odourless. Therefore, the discolouration occurred in OHT could be due to the oxidation of tannins in POME.

**Figure 2.** (from left) Fresh POME, liquid product of NOHT at 533 K & 8 h and liquid product of OHT at 533 K & 8 h.

### *3.3. Effect of Hydrothermal Treatment Conditions*

The pH, COD and $BOD_5$ values for all the liquid products from both OHT and NOHT are tabulated in Table 3. The pH value of treatment effluent was higher than the fresh POME and showed a consistent increase with the elevation of reaction temperature and time. The However, the pH value indicated that the liquid discharge still exhibited a mild acidic nature. It could be caused by the remaining refractory organic acids in the discharged products [19]. The increase of pH supports the idea that the organic acids in POME showed further reactions as reaction temperature and time increased under hydrothermal treatment [20].

In NOHT, both $BOD_5$ and COD reduction increased gradually with the hydrothermal temperature. This was reported before by [21]. At 493 K, the final COD reading was 36400 ± 550 ppm while the final $BOD_5$ was 9178 ± 150 ppm. At 533 K, the COD reading was the lowest, at 21500 ± 300 ppm while the $BOD_5$ showed a mean value of 6763 ± 75 ppm. Based on the removal percentage shown in Table 3, prolonging the reaction time from 2 h to 8 h only marginally increased both the COD and $BOD_5$ reduction. The increments were only 8.88% and 4.46%. The highest COD and $BOD_5$ removal were achieved through NOHT at 533 K for 8 h at 61.43% and 68.02% respectively. The detailed reaction pathways were described in our previous work [14]. In NOHT, the reaction mechanism can be summarised as in a course of reactions, comprising of hydrolysis, decarboxylation and dehydration, that occur con-currently. Hydrolysis occurs in the beginning of the reaction to initiate the decomposition of POME into water-soluble intermediates [22]. The reaction is followed by decarboxylation and dehydration of the hydrolysed components that take place simultaneously.

For the OHT process, the COD and $BOD_5$ removal percentage were higher than NOHT under similar operating condition. The highest COD reduction was 71.23%, which was achieved at 533 K after an 8-

hour OHT duration. The superior COD and $BOD_5$ reduction were ascribed to the enhanced oxidation of organic compounds in the presence of air. The increase of temperature resulted a higher oxidation effectiveness, which is consistent with the literature finding [14]. The reactions that occurred during OHT is more complicated as it involves an additional series of oxidation reactions that are expected to take place in the organic content of POME. The chemical step involves the reaction between the supplied oxygen molecules or the oxidation intermediates and the organic compounds in POME. The organic compounds could be oxidised by a free-radical intermediate generated during the oxidation of another compound [23]. The formation of free radicals should be further investigated and supported with evidence, such as by using electron spin resonance or free radical inhabitation technique [24].

Despite the high COD and $BOD_5$ reduction, the final $BOD_5$ of the hydrothermally treated POME still did not comply the discharge regulatory (below 100 ppm). The increase in hydrothermal temperature did not improve dramatically the treatment performance. This may be owed to the refractory phenolic compounds present in POME. It was reported that POME has high lignin-to-cellulose ratio [25]. A catalytic reaction study is necessary in the future work to determine a suitable catalyst in achieving higher organic pollutants degradation. The heterogeneous catalytic reaction is potential to offer a higher versatility and greater efficiency in the treatment of pollutants with a promising economic benefit [26] Nonetheless, the degradability of the organic substances in POME by hydrothermal reaction is still remarkable compared to the existing open ponding treatment method [27].

**Table 3.** The analyses of pH, COD and $BOD_5$ of the end-of-pipe POME and all the liquid discharges.

| Process | Reaction temperature (K) | Reaction Time (h) | pH[a] | $BOD_5$[a] removal (%) | COD[a] removal (%) |
|---|---|---|---|---|---|
| **NOHT** | 493 | 2 | 5.5 | 49.07 | 30.27 |
| **NOHT** | 503 | 2 | 5.5 | 57.09 | 46.74 |
| **NOHT** | 513 | 2 | 5.8 | 59.92 | 48.85 |
| **NOHT** | 523 | 2 | 6.0 | 61.16 | 50.38 |
| **NOHT** | 533 | 2 | 6.3 | 62.47 | 58.81 |
| **NOHT** | 533 | 4 | 6.3 | 63.50 | 59.23 |
| **NOHT** | 533 | 6 | 6.5 | 66.21 | 61.00 |
| **NOHT** | 533 | 8 | 6.5 | 68.02 | 61.43 |
| **OHT** | 493 | 2 | 6.0 | 69.90 | 58.21 |
| **OHT** | 503 | 2 | 6.3 | 73.16 | 60.98 |
| **OHT** | 513 | 2 | 6.5 | 77.39 | 63.45 |
| **OHT** | 523 | 2 | 6.5 | 79.31 | 65.39 |
| **OHT** | 533 | 2 | 6.5 | 80.25 | 66.93 |
| **OHT** | 533 | 4 | 6.5 | 83.31 | 69.21 |
| **OHT** | 533 | 6 | 6.5 | 85.12 | 70.71 |
| **OHT** | 533 | 8 | 6.5 | 87.30 | 71.23 |

[a] Mean values from the triplicate measurement

## 4. Conclusion

Subcritical hydrothermal process is deemed a new pathway for a more effective POME treatment. The treatment performance was evaluated in the aspects of COD and $BOD_5$ removal, where OHT showed promising results compared to the NOHT. The best operating condition was determined at the temperature of 533 K and process time of 8 h under oxidative environment, with the highest COD (71.23%) and $BOD_5$ (87.30%) removal, respectively. In addition, the pH was 6.5 due to the elimination of carboxylic compounds. The liquid product from the treatment was comparable with the effluent via open ponding method. Post treatment of the hydrothermally treated liquid product is suggested to meet the local discharge standard. For NOHT, deoxygenation that comprises of both dehydration and decarboxylation is most likely the dominant post-hydrolysis process. In contrast, free radical pathway could be the major degradation process of the organics in OHT.

**Acknowledgments**
The authors appreciate Universiti Malaysia Pahang for supporting this project through UMP Flagship Strategic Leap 3 Grant (RDU172202) and UMP Short-Term Grant (RDU170325).

**References**

[1] Stichnothe, H. and F. Schuchardt, Life cycle assessment of two palm oil production systems. Biomass and Bioenergy, 2011. 35(9): p. 3976-3984.
[2] Wu, T.Y., et al., Pollution control technologies for the treatment of palm oil mill effluent (POME) through end-of-pipe processes. Journal of Environmental Management, 2010. 91(7): p. 1467-1490.
[3] Tabassum, S., Y. Zhang, and Z. Zhang, An integrated method for palm oil mill effluent (POME) treatment for achieving zero liquid discharge – A pilot study. Journal of Cleaner Production, 2015. 95: p. 148-155.
[4] Lee, Z.S., et al., Treatment technologies of palm oil mill effluent (POME) and olive mill wastewater (OMW): A brief review. Environmental Technology & Innovation, 2019: p. 100377.
[5] Ali, A.N.A., et al., Tackling colour issue of anaerobically-treated palm oil mill effluent using membrane technology. Journal of Water Process Engineering, 2015. 8: p. 221-226.
[6] Cheng, Y.W., et al., Syngas from catalytic steam reforming of palm oil mill effluent: An optimization study. International Journal of Hydrogen Energy, 2019. 44(18): p. 9220-9236.
[7] Cheng, Y.W., et al., Harnessing renewable hydrogen-rich syngas from valorization of palm oil mill effluent (POME) using steam reforming technique. Renewable Energy, 2019. 138: p. 1114-1126.
[8] Ng, K.H., et al., Photocatalytic degradation of recalcitrant POME waste by using silver doped titania: Photokinetics and scavenging studies. Chemical Engineering Journal, 2016. 286: p. 282-290.
[9] Ng, K.H. and C.K. Cheng, A novel photomineralization of POME over UV-responsive TiO2 photocatalyst: kinetics of POME degradation and gaseous product formations. RSC Advances, 2015. 5(65): p. 53100-53110.
[10] Saeed, M.O., et al., Treatment of POME using Fenton oxidation process: removal efficiency, optimization, and acidity condition. Desalination and Water Treatment, 2016. 57(50): p. 23750-23759.
[11] Abu Tayeh, H., et al., Subcritical hydrothermal pretreatment of olive mill solid waste for biofuel production. Bioresource Technology, 2016. 199: p. 164-172.
[12] Osada, M., et al., Low-Temperature Catalytic Gasification of Lignin and Cellulose with a Ruthenium Catalyst in Supercritical Water. Energy & Fuels, 2004. 18(2): p. 327-333.
[13] Urrea, J. L., García, M., Collado, S., Oulego, P., & Díaz, M. (2018). Sludge hydrothermal treatments. Oxidising atmosphere effects on biopolymers and physical properties. Journal of Environmental Management, 206, 284-290.
[14] Yousefifar, A., et al., Hydrothermal processing of cellulose: A comparison between oxidative and non-oxidative processes. Bioresource Technology, 2017. 226: p. 229-237.
[15] Ng, K.H., et al., A study into syngas production from catalytic steam reforming of palm oil mill effluent (POME): A new treatment approach. International Journal of Hydrogen Energy, 2018.
[16] Lee, Z.S., S.Y. Chin, and C.K. Cheng, An evaluation of subcritical hydrothermal treatment of end-of-pipe palm oil mill effluent. Heliyon, 2019. 5(6): p. e01792.
[17] Cheng, J., et al., Palm oil mill effluent treatment using a two-stage microbial fuel cells system integrated with immobilized biological aerated filters. Bioresource Technology, 2010. 101(8): p. 2729-2734.

[18] Cheng, Y.W., et al., Hydrogen-rich syngas production via steam reforming of palm oil mill effluent (POME) – A thermodynamics analysis. International Journal of Hydrogen Energy, 2018.
[19] Berge, N.D., et al., Hydrothermal Carbonization of Municipal Waste Streams. Environmental Science & Technology, 2011. 45(13): p. 5696-5703.
[20] Arauzo, P., Olszewski, M., & Kruse, A. (2018). Hydrothermal Carbonization Brewer's Spent Grains with the Focus on Improving the Degradation of the Feedstock. Energies, 11, 3226.
[21] Karagöz, S., et al., Low-Temperature Hydrothermal Treatment of Biomass: Effect of Reaction Parameters on Products and Boiling Point Distributions. Energy & Fuels, 2004. 18(1): p. 234-241.
[22] Williams, P.T. and J. Onwudili, Subcritical and Supercritical Water Gasification of Cellulose, Starch, Glucose, and Biomass Waste. Energy & Fuels, 2006. 20(3): p. 1259-1265.
[23] Bhargava, S.K., et al., Wet Oxidation and Catalytic Wet Oxidation. Industrial & Engineering Chemistry Research, 2006. 45(4): p. 1221-1258.
[24] Robert, R., et al., Intermediates in wet oxidation of cellulose: identification of hydroxyl radical and characterization of hydrogen peroxide. Water Research, 2002. 36(19): p. 4821-4829.
[25] Kang, S., et al., Hydrothermal conversion of lignin: A review. Renewable and Sustainable Energy Reviews, 2013. 27: p. 546-558.
[26] Arena, F., et al., Recent advances on wet air oxidation catalysts for treatment of industrial wastewaters. Inorganica Chimica Acta, 2015. 431: p. 101-109.
[27] Othman, M.R., et al., Treatment of effluents from palm oil mill process to achieve river water quality for reuse as recycled water in a zero emission system. Journal of Cleaner Production, 2014. 67: p. 58-61.

IOP Conf. Series: Materials Science and Engineering **965** (2020) 012003 doi:10.1088/1757-899X/965/1/012003

# Hydrogen sulfide reduction by conductive concrete with MFCs strain

**M Fukushima[1], K Promunuan[2], S O-Thong[2], T Suzuki[1], T Imai*[1]**

[1] Division of Environmental Engineering, Graduate School of Science and Technology for Innovation, Yamaguchi University, Japan
[2] Biotechnology Program, Faculty of Science, Thaksin University, Phatthalung, Thailand

*: imai@yamaguchi-u.ac.jp

**Abstract**. The objective of this study is to reduce the production of Hydrogen sulfide ($H_2S$) in sewer pipe by using conductive concrete and electricity-producing bacteria (EPB). This work was mainly focused on evaluating the effect of inoculation of EPB isolated from sewage sludge from municipal wastewater treatment plant for decreasing production of $H_2S$. The experimental results showed that inoculation of EPB could decrease the concentration of $H_2S$, indicating that produced $H_2S$ in an anaerobic sludge sediment layer (bottom of sewer pipe) was biologically oxidized by EPB. These results demonstrated that conductive concrete can provide an electron pathway from anaerobic sludge sediment in the bottom of sewer pipe to oxygen dissolved in surface water, and as a result the electron generated from hydrogen $H_2S$ oxidation in anaerobic environment was accepted by oxygen via conductive concrete.

## 1. Introduction

Hydrogen sulfide produced by sulfate-reducing bacteria (SRB) in an anaerobic sludge sediment layer causes corrosion of sewage pipes, shortening of service life, and the collapse of roads [1-5]. Therefore, a method of mitigating the activity of SRB by adding chemicals or aeration [6,7] has been proposed, but there is a problem that continuous maintenance cost is required. Besides, in recent years, the development of concrete with high resistance and approaches to increase the sulfuric acid resistance of sewer pipes such as lining with epoxy resin has been studied [8-10], but there are disadvantages such as higher manufacturing cost than ordinary Portland cement (OPC). Therefore, it is the current situation that it has not spread yet. The purpose of this study is to enable the oxidation/mitigation of sulfide (sulfide ion $S^{2-}$, hydrogen sulfide ion $HS^-$, and hydrogen sulfide $H_2S$) generated in the anaerobic sludge sediment layer in the sewer pipe, and the low cost and long durability. From the viewpoint of these properties, it is needed to develop the concrete pipes that contribute to the reduction of life cycle costs. In this study, we tried to verify the oxidation and mitigation effect of sulfide generated in the bottom of sewer pipe by the commercially available conductive concrete containing amorphous carbon as the conductive material (Figure 1. (a)). Hence, microbial fuel cells (MFCs) [11–15] are applied for the suppression of $H_2S$ production in sewage pipe as an alternative/useful method. MFCs are a device that uses electrically catalytic microorganisms to convert the organic or inorganic substance into electricity (Figure 1. (b)). Increasing interest has been given to MFCs due to its potential of generating alternative energy using organics contained in the wastewater. This study aimed to prevent the corrosion caused by $H_2S$ in sewage pipe with MFCs technology. The electricity-producing bacteria (EPB, it is same as MFCs strain) [16,17] are applied to oxidize from hydrogen sulfide to sulfur ($S^0$) by MFCs using the conductive concrete. By keeping a condition that the EPB can release electrons, the generation of $H_2S$ can be avoided (Figure 1.).

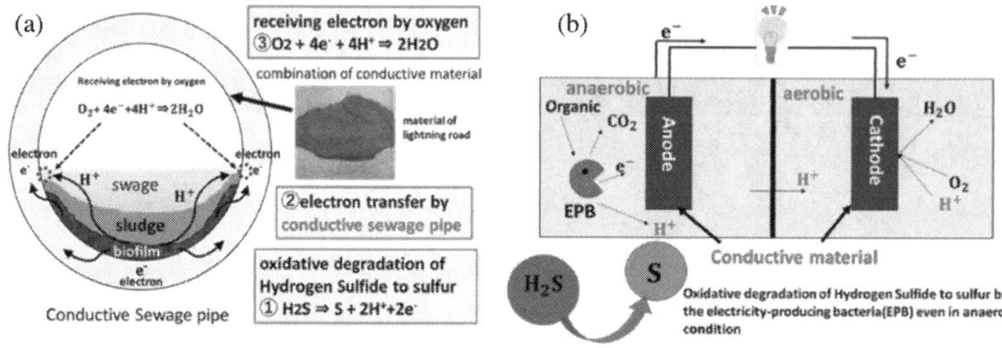


**Figure 1.** Oxidation and inhibition mechanism of sulfide by MFCs strain (EPB) assumed in this study.

## 2. Materials and Methods

Experiments were carried out in a room of which temperature was adjusted to 24°C ± 1°C. Unless otherwise stated, distilled water (SA-2100A•An type, Tokyo Rika Kikai Co., Ltd.) was used for adjusting the aqueous solution.

### *2.1. Preparation of conductive concrete and measurement of electrical resistivity*

The conductive concrete used in this study is Sun earth M5C (hereinafter refer to as Sun earth) manufactured by Sankosha Co., Ltd. Sun earth contains amorphous carbon produced from oil refining as a conductive material. As a result of the conductive carbon separation from the cement particles of Sun earth in gently running water, the amorphous carbon obtained from 100 g of Sun earth was about 50 g, and the maximum size was about 0.3 mm. The specific surface area measured by the BET method (Brunauer-Emmett-Teller method) using BELSORP-mini II (Microtrac Bell Co.) was 1.9 $m^2/g$. Although other conductive concrete such as containing conductive carbon fibres are commercially available instead of to Sun earth, Sun earth was selected in this study because it has a higher carbon content and higher conductivity than other conductive concretes. Regarding the preparation of the concrete specimen, Figure 2 shows the specimen prepared using ordinary Portland cement (OPC) and Sun earth. Specimens are classified into rod electrode specimens and cup-type specimens according to their shapes. The cup-shaped has a capacity of about 1.9 L. For the rod electrode specimens, both sets of 15ϕ×75 mm specimens were prepared. The experiments were conducted with these specimens connected via an external resistance of 100 Ω. The electrical resistivity of the conductive concrete was measured by a chemical impedance analyser (Hioki Denki Co., Ltd., IM3590) by the four-electrode method [18] used in this study. Since concrete contains a lot of electrolytes, the electrical resistivity changes with the period of application time. The electrical resistivity reported in this study is the value after 30 minutes, which was confirmed by preliminary experiments to be stable even in the low-frequency region where it takes a long time to converge to a constant value. Since it contains amorphous carbon as a conductive material, Sun earth has a higher water absorption rate than OPC. Therefore, regardless of the shape of the specimen, the water/cement ratio of OPC and Sun earth were set to 30% and 40%, respectively, to maintain the same workability (easy handling).

### *2.2. Differential culture of electricity-producing bacteria isolated from sewer sludge*

Excess sludge and digested sludge collected by the Ube City Eastern Wastewater Treatment Plant were separately cultivated using Miller's LB (Luria-Bertani) medium with $N_2$ gas purging at 30°C. Then, after extraction of DNA using the MORA-EXTRACT kit (Far East Pharmaceutical Co., Ltd.) and PCR operation by Touch Down method [19] using primer GC-341f and 518r [20] for amplification of about 200 bp 16S rRNA were analysed using DNA sequencer Ion S5 (Thermo Fisher Co., Ltd). Bacterial species identified using Windows free software BioEdit (SD Biosystems Co., Ltd.) and NCBI Blast data base. As a result, two species of *Morganella morganii* and *Proteus alimentorum* were identified from excess sludge, and two species of *Proteus mirabilis* and *Shigella sonnei* were identified from digested sludge with 99.7% or more homology (Table 1). It is suggested in the literature regarding microbial fuel cells that these are typical intestinal bacteria and these bacteria other than *Proteus alimentorum* are electricity-producing bacteria (EPB). Regarding *Proteus alimentorum*, it has been suggested in the literature that this bacterium is belonging to the same *Proteus* genus, including *Proteus mirabilis* identified in this study, are also EPB [21–24].

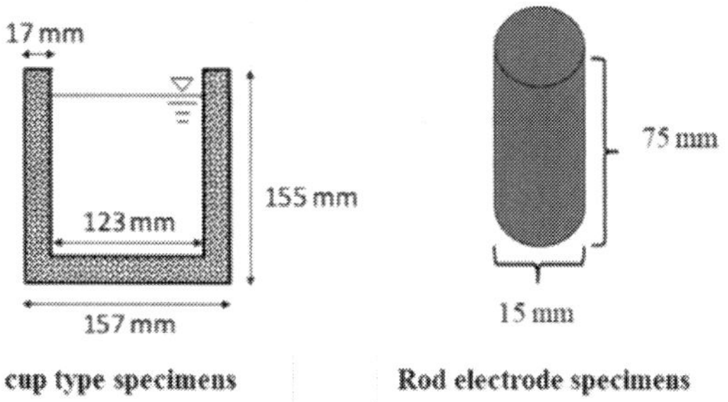


**Figure 2.** Two types of specimens made in this study.

Table 1. 16S rRNA gene sequence analysis of isolates

| Source | Isolate | Description |
|---|---|---|
| Aerobic sludge | M1 | *Morganella morganii* |
| | M2 | *Proteus alimentorum* |
| Anaerobic sludge | F1 | *Proteus mirabilis* |
| | F2 | *Shigella sonnei* |

## *2.3. Characterization of EPB with rod electrode*

Figure 3 shows the Schematic of device design, conductive carbon electrode (a), and Experimental configuration of the single-chamber MFC with air cathode. The purpose of this experiment is to measure voltage and reduce sulfide by inoculation of the EPB (MFCs strain) from aerobic sludge and anaerobic sludge (Table 1). The case of this experiment is MFC configuration and operation both anode and cathode of the microbial fuel cell were mixed electrode between cement and conductive (Figure 2). Table 2 shows the condition of this experiment.

The samples of Sun earth electrodes were prepared, each set of experiment contained digested sludge (SS concentration 17,160 mg/L, VSS concentration 12,800 mg/L) collected from the Ube City Eastern Wastewater Treatment Plant and Artificial wastewater (the composition is shown in Table 3) was mixed with 0.08 L and 0.62 L respectively. In addition, the above digested sludge, the other specimen was added with four types (as shown Table 2) of EPB isolated and cultured from excess sludge and digested sludge with each of concentration of approximately $10^{8}$ CFU/mL. The substrate of glucose and sodium sulfate was used the COD/$SO_4^{2-}$ ratio 0.67 and was added in the start-up step and loading substrate step. The COD of glucose was adjusted in the start-up step and the loading substrate step are 1,000 and 10,000 mg COD/L, respectively. The operation was controlled in 30 °C and measured voltage with data logger and sulfide every day. Sulfide was quantified by the methylene blue method according to the USEPA method 376.2. After confirming that sulfate was reduced/converted to hydrogen sulfide by SRB, glucose and magnesium sulfate were added.

Table 2. Condition of the experiment about inoculation of EPB

| Case | Condition of the experiment |
|---|---|
| A | Anaerobic sludge (AS) |
| B | AS+S1(M1+M2) |
| C | AS+S2 (F1+F2) |
| D | AS+S3 (M1+M2+S1+S2) |

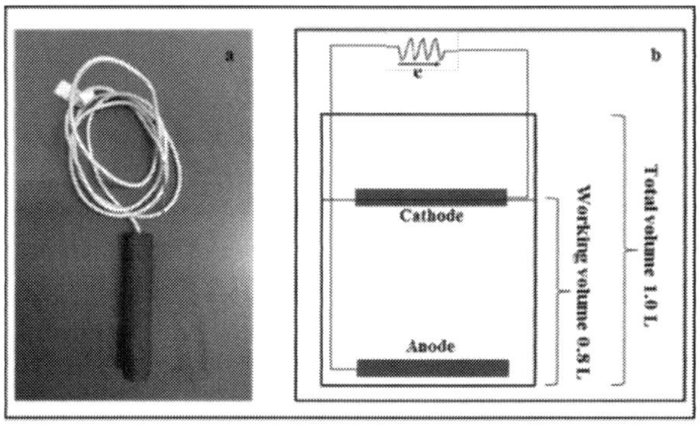


**Figure 3.** Schematic of device design, conductive carbon electrode (a), Experiment configuration (b)

Table 3. Composition of the substrates [25].

| | |
|---|---|
| $NaHCO_3$ | 2 g/L |
| $K_2HPO_4$ | 2 g/L |
| Yeast extract | 0.02 g/L |
| Glucose | 2 g/L |
| $(NH_4)_2HPO_4$ | 0.7 g/L |
| KCl | 0.75 g/L |
| $NH_4Cl$ | 0.85 g/L |
| $FeCl_3 \cdot 6H_2O$ | 0.42 g/L |
| $MgCl_2 \cdot 6H_2O$ | 0.81 g/L |
| $MgSO_4 \cdot 7H_2O$ | 0.25 g/L |
| $CoCl_2 \cdot 6H_2O$ | 0.18 g/L |
| $CaCl_2 \cdot 6H_2O$ | 0.15 g/L |

### *2.4. Application the EPB for reduction of sulfide with conductive concrete*

One sample of OPC and two samples of Sun earth were prepared, and each sample contained digested sludge (SS concentration 17,160 mg/L, VSS concentration 12,800 mg/L) collected from the Ube City Wastewater Treatment Plant and Artificial wastewater (the composition is shown in Table 3.) mixed with 0.14 L and 1.12 L, respectively (Table 4). In addition, the above digested sludge, the other specimen was added with four types of EPB (Table 1) separated and cultured from excess sludge and digested sludge so that each of them had a concentration of approximately $10^{8}$ CFU/mL and the experiment was started in the stationary state. After that, the pH, sulfate, and sulfide concentrations of the aqueous phase were measured periodically (The Method was same as 2.3).

Table 4. The condition of the experiment about inoculation of EPB

| Case | Condition of the experiment |
|---|---|
| A | OPC + anaerobic sludge AS) |
| B | Sun earth + AS |
| C | Sun earth + AS+S3 |

## 3. Results and Discussion

### *3.1. Measurement of electrical resistivity*

Figure 4. shows the results of measurement of the electrical resistivity of OPC and Sun earth. It was common that the electrical resistivity of both OPC and Sun earth tended to decrease with increasing power-frequency. However, at any power frequency, the electrical resistivity of Sun earth was significantly lower than that of OPC. From these results, it was suggested that the specimen prepared by using Sun earth had small electrical resistivity to the flow of electrons supplied by the oxidation of hydrogen sulfide and could easily move the electron from the anode to the cathode.

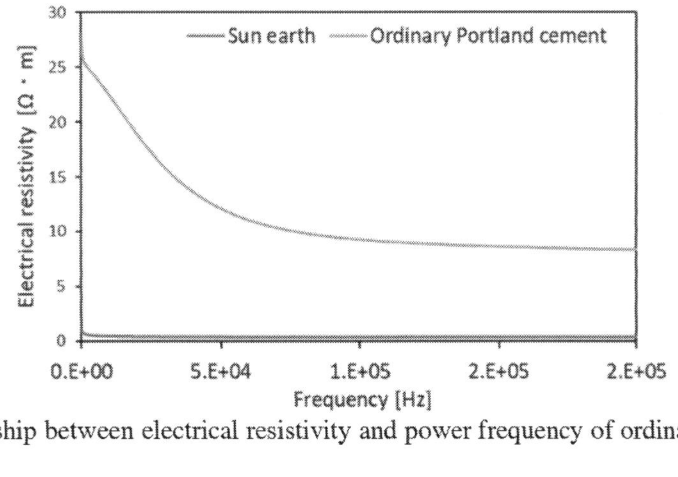


**Figure 4.** Relationship between electrical resistivity and power frequency of ordinary Portland cement and Sun earth.

### *3.2. Characterization of MFCs strain (EPB) with electrode*

Figure 5 shows the result of voltage for 10 days of experiment. "Start-up" means preliminary experiment and "Loading substrate" means main experiment. The result of experiment indicated the highest voltage by inoculation of S3. In Figure 6, the variation of sulfide for 10 days is shown. From this result, it was shown the effect of sulfide reduction by inoculation S2. As shown in Figure 7, it is revealed the comparison of sulfide generation rate of each case in this experiment. The result demonstrated that the effect of inoculation S2 could reduce more 60% than anaerobic sludge (AS) with no inoculation. From these results, it is suggested that the inoculum S2 (*Proteus mirabilis* + *Shigella sonnei*) is very effective for reduction of sulfide generation in sewer pipe.

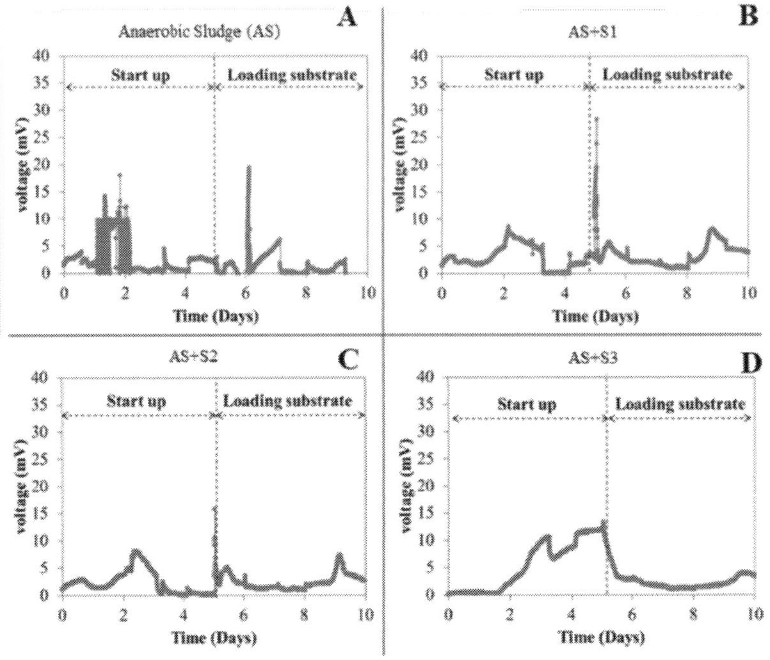


**Figure 5.** Electricity production from the rod electrode with a mixed culture in start-up and loading substrate step.

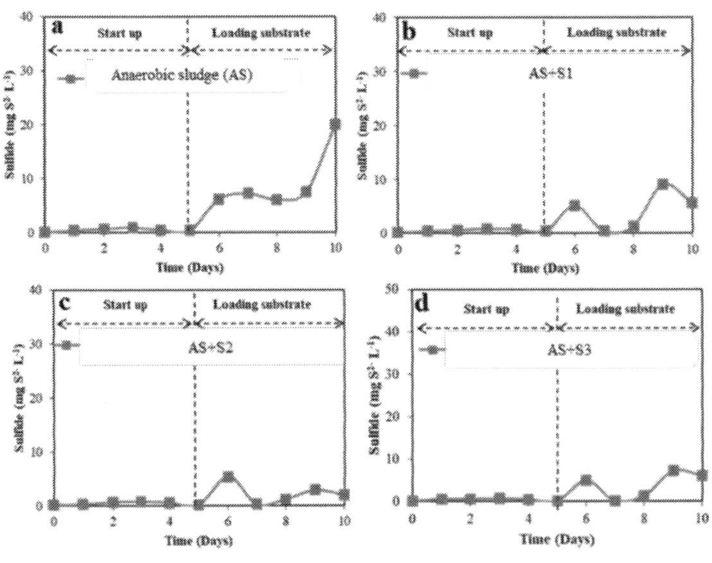


**Figure 6.** The reduction of sulfide by inoculation of electricity-producing bacteria

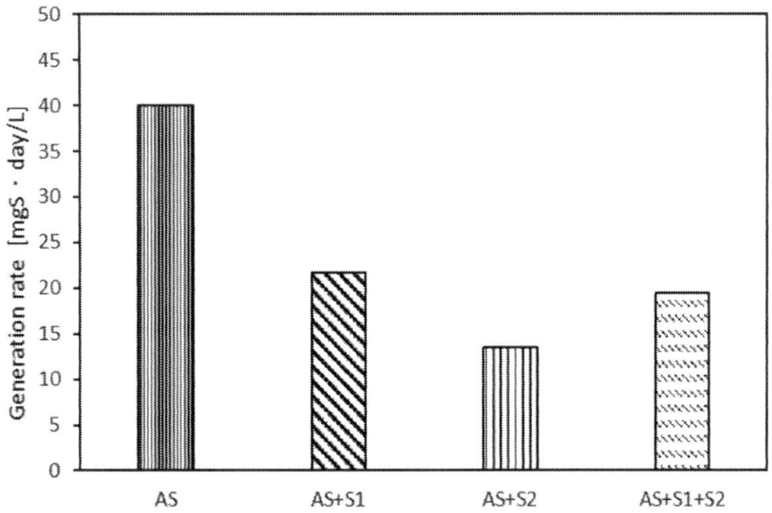


**Figure 7.** Comparison of the generation rate of sulfide in each experimental case

### *3.3. Application the MFCs strain (EPB) to reduction of sulfide with conductive concrete*

Figure 8 shows the results of evaluating the effect of the inoculation of MFCs strain (EPB) on the reduction of sulfide using cup-type specimens. First, focusing on the pH data, the pH decreased to about 4.5 and then increased to about 9 with the addition of glucose. This is the behaviour observed in the normal condition of anaerobic fermentation. It was occurred a decrease of pH by the formation of volatile fatty acid by acidogenic bacteria, and also occurred a rise in pH by the re-formation of alkali due to the volatilization of volatile fatty acid and/or the decomposition of volatile fatty acid and production of methane by methanogenic bacteria. Next, focusing on the concentration of sulfate, it significantly decreased within a few days in any case when magnesium sulfate was added, and this phenomenon did not depend on the type of concrete or the presence/absence of inoculum. Finally, focusing on the sulfide concentration in water, Sun earth showed lower sulfide concentration than that of OPC. As previously discussed in section *3.1*, this is due to Sun earth's higher conductivity than that of OPC. And when MFCs strain (EPB) were inoculated into Sun earth, the sulfide concentration decreased clearly. These results mean that 1) the use of conductive concrete and/or 2) inoculation of MFCs strain (EPB) can reduce sulfide generation. Therefore, to quantitatively evaluate this effect, Figure 9 shows the average amount of sulfide concentration generated during the 20 days experimental period. The reduction effect of sulfide was only 37% with Sun earth (without inoculation), whereas 69% of sulfide was reduced in the case of Sun earth (with inoculation), which is a combination of inoculation of MFCs strain (EPB) and Sun earth and EPB. These are favourable results demonstrating that biological oxidization of sulfide progresses and the reduction of sulfide thereby occurs. And, at the same time, it indicates the possibility of the acceleration of the biological oxidization/reduction of sulfide by MFCs strain (EPB) by developing more high-performance conductive concrete. Furthermore, the effect of sulfide reduction was obtained by inoculation of MFCs strain (EPB) even with OPC which has low conductivity. These results are shown that MFCs strain (EPB) can directly transfer electrons if they are provided an environment in which MFCs strain (EPB) can stably accumulate (i.e. provided conductive concrete as electron pathway).

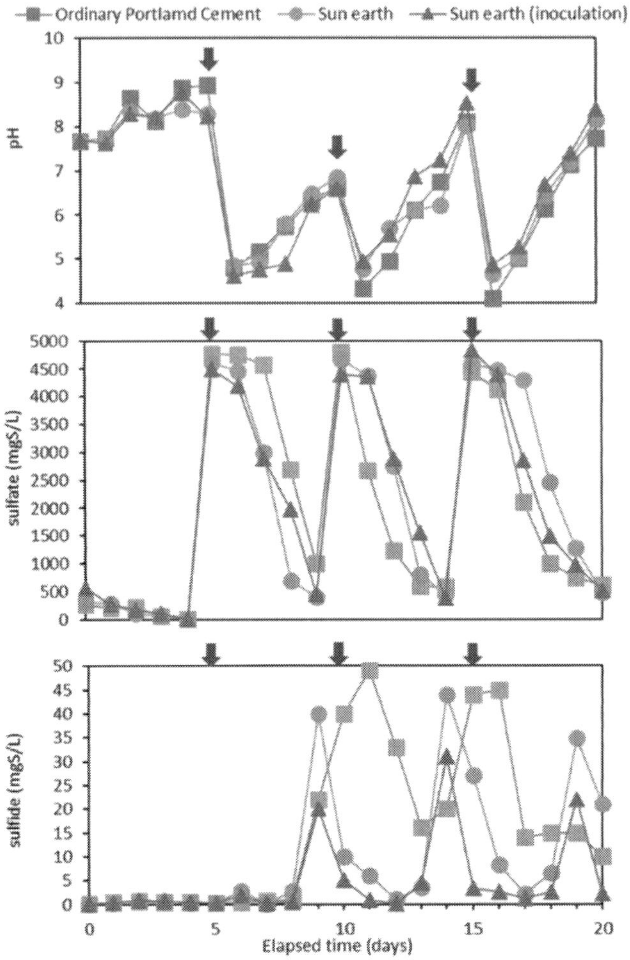


**Figure 8.** Results of the experiment conducted to evaluate the effect of inoculum on sulfide concentration. The arrows in the figure indicate the addition of glucose and magnesium sulfate. An initial concentration of glucose added was 1000 mg/L, but with 10000 mg/L on days 5, 10, and 15, and the concentration of magnesium sulfate was half of the glucose concentration

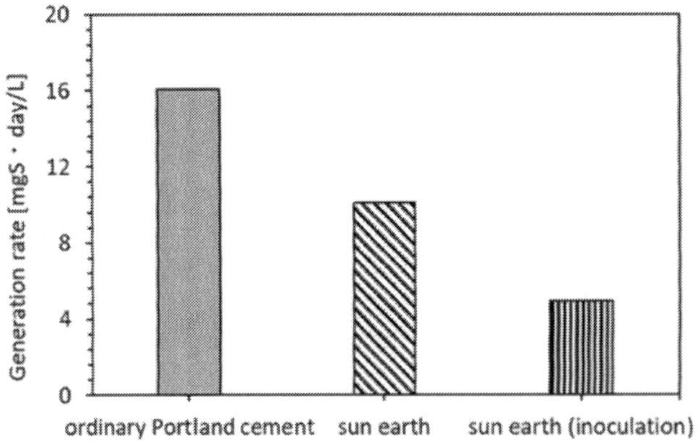


**Figure 9.** Comparison of sulfide concentration generated by ordinary Portland cement, Sun earth, and Sun earth (inoculation)

## 4. Conclusion

This study focused on conductive concrete as a new technology to mitigate hydrogen sulfide in the sewer pipe and aimed at demonstrating its effect. As the results of the experiments, conductive concrete served the electron transfer path and oxygen existing near the water surface could be used as electron acceptor. It made possible that hydrogen sulfide was oxidized and mitigated. Therefore, it was shown that conductive concrete was effective as a new technology for mitigating hydrogen sulfide even in an anaerobic environment. Additionally, it was confirmed that the sulfide was reduced by the inoculation of the EPB (MFCs strain), this result indicated the possibility of the acceleration of the biological oxidization/reduction of sulfide by the existence of EPB (MFCs strain) with conductive concrete. In the future, we plan to develop a new conductive concrete specialized for mitigating hydrogen sulfide and try to take root the EPB (MFCs strain) in the conductive concrete for a long period. In addition, the attempt to elucidate the mechanism of hydrogen oxidation and mitigation will be included.

## References

[1] Wu M, Wang T, Wu K and Kan L 2020 Microbiologically induced corrosion of concrete in sewer structures: A review of the mechanisms and phenomena *Constr. Build. Mater.* **239** 117813

[2] Li W, Zheng T, Ma Y and Liu J 2019 Current status and future prospects of sewer biofilms: Their structure, influencing factors, and substance transformations *Sci. Total Environ.* **695** 133815

[3] Shi X, Sang L, Wang X C and Jin P 2018 Pollutant exchange between sewage and sediment in urban sewer systems *Chem. Eng. J.* **351** 240–7

[4] Jiang G, Zhou M, Chiu T H, Sun X, Keller J and Bond P L 2016 Wastewater-Enhanced Microbial Corrosion of Concrete Sewers *Environ. Sci. Technol.* **50** 8084–92

[5] Jiang G, Keller J and Bond P L 2014 Determining the long-term effects of $H_2S$ concentration, relative humidity and air temperature on concrete sewer corrosion *Water Res.* **65** 157–69

[6] Lin H W, Lu Y, Ganigué R, Sharma K R, Rabaey K, Yuan Z and Pikaar I 2017 Simultaneous use of caustic and oxygen for efficient sulfide control in sewers *Sci. Total Environ.* **601–602** 776–83

[7] Zhang L, De Schryver P, De Gusseme B, De Muynck W, Boon N and Verstraete W 2008 Chemical and biological technologies for hydrogen sulfide emission control in sewer systems: A review *Water Res.* **42** 1–12

[8] Liu J, Zhang B, Qi W H, Deng Y G and Misra R D K 2020 Corrosion response of zinc phosphate conversion coating on steel fibers for concrete applications

[9] Xiao Z, Liu Y, Wang Y and Shi J 2020 TA/Fe(III) anti-chloride coating to protect concrete *J. Clean. Prod.* **259** 120922

[10] Jiang G, Sun J, Sharma K R and Yuan Z 2015 Corrosion and odor management in sewer systems *Curr. Opin. Biotechnol.* **33** 192–7

[11] Gorby Y A, Yanina S, McLean J S, Rosso K M, Moyles D, Dohnalkova A, Beveridge T J, Chang I S, Kim B H, Kim K S, Culley D E, Reed S B, Romine M F, Saffarini D A, Hill E A, Shi L, Elias D A, Kennedy D W, Pinchuk G, Watanabe K, Ishii S, Logan B, Nealson K H and Fredrickson J K 2006 Electrically conductive bacterial nanowires produced by Shewanella oneidensis strain MR-1 and other microorganisms *Proc. Natl. Acad. Sci. U. S. A.* **103** 11358–63

[12] Pfeffer C, Larsen S, Song J, Dong M, Besenbacher F, Meyer R L, Kjeldsen K U, Schreiber L, Gorby Y A, El-Naggar M Y, Leung K M, Schramm A, Risgaard-Petersen N and Nielsen L P 2012 Filamentous bacteria transport electrons over centimetre distances *Nature* **491** 218–21

[13] Rabaey K, Van De Sompel K, Maignien L, Boon N, Aelterman P, Clauwaert P, De Schamphelaire L, Pham H T, Vermeulen J, Verhaege M, Lens P and Verstraete W 2006 Microbial fuel cells for sulfide removal *Environ. Sci. Technol.* **40** 5218–24

[14] González T, Ureta-Zañartu M S, Marco J F and Vidal G 2019 Effect of Zeolite-Fe on graphite anode in electroactive biofilm development for application in microbial fuel cells *Appl. Surf. Sci.* **467–468** 851–9

[15] Lu D, Xing B, Liu Y, Wang Z, Xu X and Zhu L 2019 Enhanced production of short-chain fatty acids from waste activated sludge by addition of magnetite under suitable alkaline condition *Bioresour. Technol.* **289** 121713

[16] Vasieva O, Sorokin A, Szydlowski L and Goryanin I 2019 Do Microbial Fuel Cells have Antipathogenic Properties ? Computer Science & Systems Biology Do Microbial Fuel Cells have Antipathogenic Properties ? *J. Comput. Sci. Syst. Biol.* **12** 57–70

[17] Ng I S, Hsueh C C and Chen B Y 2017 Electron transport phenomena of electroactive bacteria in microbial fuel cells: a review of Proteus hauseri *Bioresour. Bioprocess.* **4**

[18] Dilushani P D R, Nawodani W R N and Bambaravanage T 2020 Soil Resistivity Analysis and Earth Electrode Resistance Determination **15** 26–35

[19] Takahashi S, Tomita J, Nishioka K, Hisada T and Nishijima M 2014 Development of a prokaryotic universal primer for simultaneous analysis of Bacteria and Archaea using next-generation sequencing *PLoS One* **9**

[20] Muyzer G, De Waal E C and Uitterlinden A G 1993 Profiling of complex microbial populations by denaturing gradient gel electrophoresis analysis of polymerase chain reaction-amplified genes coding for 16S rRNA *Appl. Environ. Microbiol.* **59** 695–700

[21] Gunasekaran G, Chongdar S, Naragoni S, Rodrigues P V. and Bobba R 2011 Microbial fuel cell constructed with micro-organisms isolated from industry effluent *Int. J. Hydrogen Energy* **36** 14914–22

[22] Rahimnejad M, Najafpour G D, Ghoreyshi A A, Shakeri M and Zare H 2011 Methylene blue as electron promoters in microbial fuel cell *Int. J. Hydrogen Energy* **36** 13335–41

[23] Patel R and Deb D 2018 Parametrized control-oriented mathematical model and adaptive backstepping control of a single chamber single population microbial fuel cell *J. Power Sources* **396** 599–605

[24] Choudhury P, Uday U S P, Mahata N, Nath Tiwari O, Narayan Ray R, Kanti Bandyopadhyay T and Bhunia B 2017 Performance improvement of microbial fuel cells for waste water treatment along with value addition: A review on past achievements and recent perspectives *Renew. Sustain. Energy Rev.* **79** 372–89

[25] Zhou W, Imai T, Ukita M, Sekine M and Higuchi T 2006 Triggering forces for anaerobic granulation in UASB reactors *Process Biochem.* **41** 36–43

IOP Conf. Series: Materials Science and Engineering **965** (2020) 012004 doi:10.1088/1757-899X/965/1/012004

# A simulation study of radiant chilled ceiling with dedicated outdoor air system for office buildings in Thailand

**Ashmin Aryal[1], Pipat Chaiwiwatworakul[1], Surapong Chirarattananon[1]**

[1] The Joint Graduate School of Energy and Environment, King Mongkuts's University of Technology Thonburi, 126 Prachauthit road, Bangmod, Thung Khru, Bangkok 10140

Corresponding author's e-mail: ashmin.1243@mail.kmutt.ac.th, asmin1243@gmail.com

**Abstract**. Radiant chilled ceiling (RCC) integrated with dedicated outdoor air system (DOAS) has been considered as a system offering building occupants a good thermal comfort with energy efficiency. This paper compares three different configurations of RCC and DOAS system in terms of cooling load, room condition and energy consumption. The results from TRNSYS simulation presented that all the configurations could maintain the indoor condition with the Predictive Mean Vote (PMV) value ±0.5 over the office hours 6:00-18:00. The mean radiant temperature of the room was found below the room air temperature which resulted from the low RCC surface temperature of around 19°C. No condensation occurred on the panel during the system operation. The study showed that the RCC with DOAS comprising of cooling coil (CC), round around coil (RC) and enthalpy wheel (EW) is the most energy saving configuration.

## 1. Introduction

Thailand is situated in a tropical zone where the climate is hot and humid all year-round. Diurnal ambient air temperature ranges 25°C-35°C with relative humidity 50%-70%. Natural environment cannot provide enough thermal comfort, thus artificial environment must be built by energy-intensive air conditioners. Air conditioners contribute to 70% of the total energy consumed in the building [1]. This fact corroborates the study of energy saving potential of the mechanical way to maintain the thermal environment. The conventional air-conditioning is energy intensive as stated. Thus, it is desirable to develop alternative air-conditioning approaches to thermal comfort at low energy expense.

An alternative solution to conventional air conditioning is radiant cooling system. It consumes less energy and it has been successfully implemented in moderate climatic conditions, but it is quite challenging in real tropical region like Thailand as its cooling performance is limited due to high moisture content in the air. Using radiant cooling system with natural ventilation under the tropical condition the supplied chilled water temperature to radiant cooling panel is limited to 24°C[2] due to condensation at the surface of the radiant panel. The solution to this problem of condensation is the use of Dedicated Outdoor Air System (DOAS) coupled with radiant chilled ceiling (RCC). Using this system, radiant panel is offloaded from the dehumidification task, alleviating the condensation problem. It allows the radiant system to handle sensible load and can provide better indoor air quality as well[3]. RCC has application in modern buildings, due to high thermal comfort, energy-saving potential, and advantage of integration with building design and retrofitting in old building[4].

IOP Conf. Series: Materials Science and Engineering **965** (2020) 012004 doi:10.1088/1757-899X/965/1/012004

The performance of RCC with DOAS will give an insight of its operation on the tropical climate and illustrate how it differs from the countries with milder climate. The objective of this paper is to assess the energy saving potential of different configuration of RCC with DOAS in an office building in Thailand and compare them with the help of mathematical simulation using TRNSYS.

## 2. Methodology

### *2.1. RCC with DOAS*

To harness the advantage of the energy saving, better thermal comfort and better vertical variation potential of RCC and to eliminate the prevailing disadvantage of its inability to handle latent load DOAS must be used. Thus, the system under investigation have two parallel units: Radiant unit (terminal unit) to take majority of the sensible load and the DOAS unit (auxiliary unit) to take the latent load and some part of the sensible load. Regarding the DOAS, various devices are combined to tackle the cooling load. Features of heat recovery like sensible wheel, enthalpy wheel, round around coil etc. are included with humidifier, dehumidifier and cooling coil for the performance enhancement of DOAS. DOAS when combined with different possible combination of the above-mentioned components have different energy performance under different climatic condition. In this study, three configurations of RCC and DOAS system were evaluated for their performance under the hot and humid climate of Thailand. The weather record from a meteorological station located at seven-story building of the school of bio resources and technology, in Bang Khun Tien Campus. King Mongkut's University of technology, Thonburi (latitude 14.5° N and 100.5° E) was used as input for the weather condition in TRNSYS.

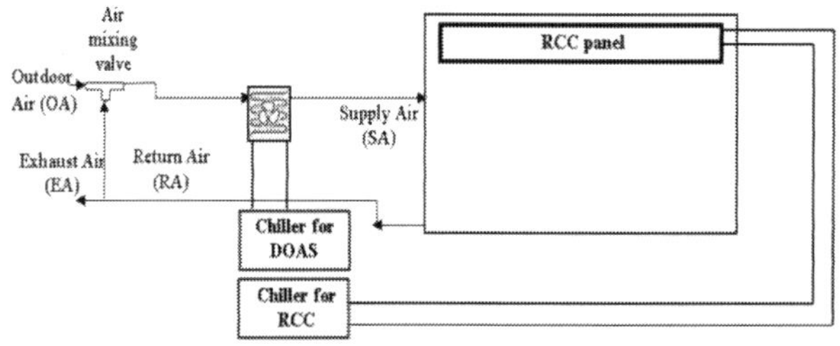


(a) RCC with DOAS (only Cooling Coil (CC))

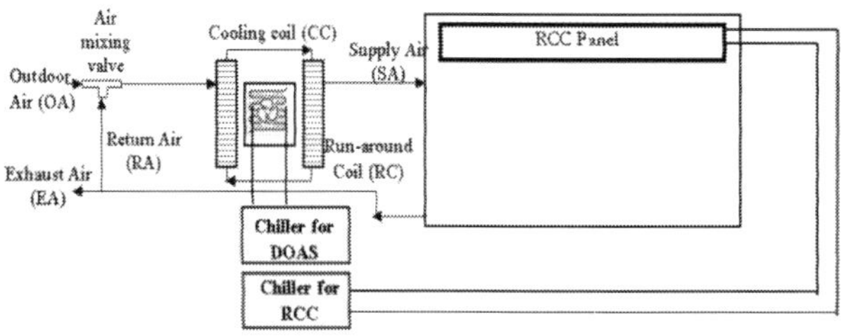


(b) RCC with DOAS (with cooling coil (CC) and round around coil (RC))

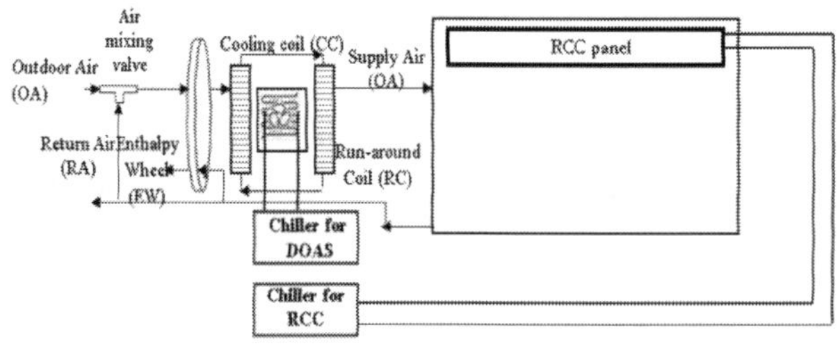


(c) RCC with DOAS (with cooling coil (CC) and round around coil (RC) and enthalpy wheel (EW))

**Figure 1.** The selected configurations of RCC and DOAS system

IOP Conf. Series: Materials Science and Engineering **965** (2020) 012004 doi:10.1088/1757-899X/965/1/012004

'Figure 1(a)' shows the diagram of the configuration 1, RCC with Cooling coil (CC) as DOAS, which reduces the temperature and humidity of the outside air (OA) and supply it to the radiant room (supply air: SA). The air inside the room will be exhausted (exhaust air, EA) in the same amount as OA supplied to the room. The return air(RA) is mixed in the air mixing valve with the OA. 'Figure 1(b)' is the diagram of configuration 2, which have an additional run-around coil (RC) than configuration 1, for dedicating the cooling coil to handle the latent heat by precooling the SA on the primary side of the RC before going to CC. The air then passes through the secondary side where it gets reheated after passing through the CC. Configuration 3, shown in 'Figure 1(c)', where enthalpy wheel (EW) is added to the components of configuration 2, which aids heat recovery from EA. The chilled water at 7°C is supplied to DOAS and 15°C (for configuration1 and 3), 15.2 °C (for configuration 2) is supplied to RCC by two different chillers as shown in 'Figure 1'. Chilled water at 15.2°C and 15°C are respectively supplied to configuration 2 and configuration 1 and 3 to ensure the comfort requirement is met without and condensation on the panel.

### *2.2. RCC with DOAS*
The building model in this study is a one storied square shaped office building prototype, width 20m × 20 m, height 3 m (area 400m$^2$). The building facades faces four cardinal directions (N, E, S and W) with fenestration on all the façade. Figure 2 shows the 3-d model of the building to be conditioned.

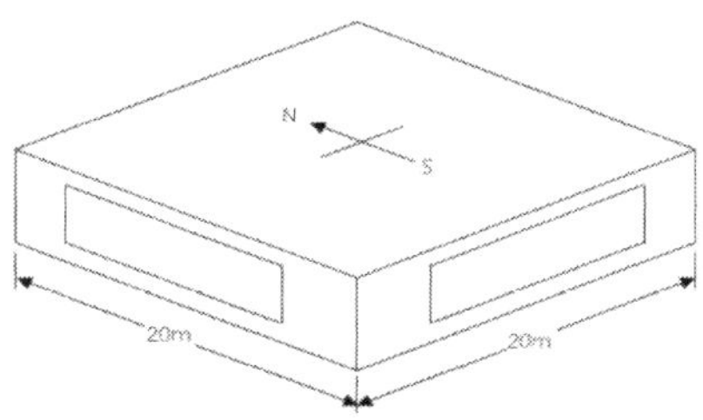


**Figure 2.** Building Model

The building is squared in shape (20×20×3 m) with fenestration on each facade. The opaque part of the wall consists of polyurethane insulation layer (0.02 m), plaster (0.01 m) brick (0.1 m) and plaster (0.01 m) with a total thickness of 0.14 m and the total heat transfer coefficient (U) is 1.008 W/m$^2$. K. The windows on the four walls of the building consist of clear glass of thickness 6 mm. The window to wall ratio is 0.4. The U value of the glass is 5.73 W/m$^2$. K and shading coefficient (SC) is 0.5. The roof is made up of 0.15 m concrete cement layer and 0.05 m polyurethane layer with U value of 0.519 W/m$^2$. K. The roof area is 400 m$^2$ on which there is RCC of 320 m$^2$ where the U value is 0.487 W/m$^2$. K. Chilled water is supplied to the RCC at the rate of 14,000 kg/hr. through a chiller at 15/15.2°C. The pipes for chilled water flow on the panel have 0.145 m spacing, the inside diameter of the pipe is 0.0012 m. The floor is made up of 0.15 m concrete cement layer with U value of 3.448 W/m$^2$. K.

### *2.3. Building usage*
It was assumed that the office was operated from 6:00-18:00. The power density of the lighting and the equipment was set as at 6 W/m$^2$ and 10 W/m$^2$, respectively. The building occupancy was assumed as 1 person for 10 m$^2$ floor area. The variation of occupancy, lighting and equipment power with time is shown in 'Figure 3'. The office is assumed to have a high air tightness with air infiltration of 0.3 ACH (210 cfm/floor) and ventilation rate calculated according to AHSRAE standard is 11.5 cfm/person. There are two period of operation of the system. The precooling period from 2:00-6:00 where the air is cooled with recirculated air from the room by the DOAS at 1300kJ/hr. The operational period where 1000kJ/hr of the fresh OA is mixed with the 300kJ/hr of RA in mixing valve, RCC is also operated in this time period. Air in both the period are passed through air mixing valve but during operational period the air is mixed at the valve while during the precooling period the OA is not supplied to the air mixing valve.

### 2.4. *Comfort parameters*

The evaluation of human comfort in TRNSYS is based on the Fanger's PMV model, with the clothing insulation of 0.5 clo, activity level of 1.2 met, wind speed 0.05 m/s. These parameters are inserted in the TRNSYS for simulation.

### 2.5. *TRNSYS Simulation*

TRNSYS 17[5] software was employed to simulate the operation of the radiant system with different DOAS configuration.

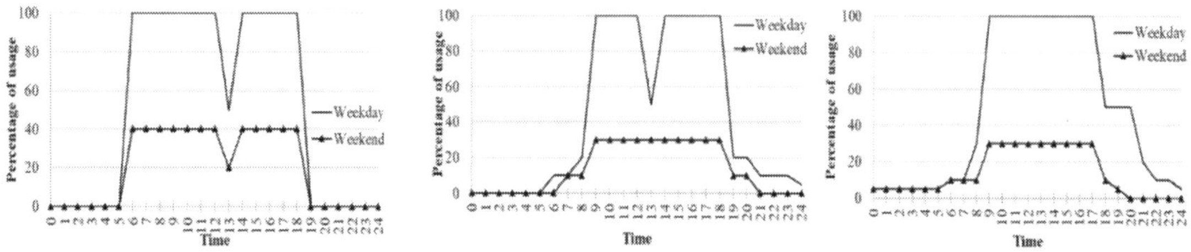


(a) Occupant density (b) Equipment density (c) Lighting power density

**Figure 3.** Characteristics of building usage

The simulation was run in a manner that it provided thermal comfort condition during the working hours by connecting different modules (which are mathematical codes) to make a complete system. Three configurations comprising RCC with DOAS were built in modules of respective configurations and were connected to building type 56 module in TRNSYS simulation studio. Type 56 module is facilitated with TRNBuild sub-program for assigning the different parameters for the building. RCC is contained in the type 56 module of TRNBuild sub-program. Type 9e facilitate the user to feed the user defined water temperature values of the chilled water to RCC created through a file. Type 9e also serves the purpose of reading the weather data from a data file created by the user via the detail weather data from the workstation. Type 752 is a module for cooling coil which uses bypass fraction approach for CC to solve for the outlet air temperature. The air flow rate into the CC is 1000 kg/hr. Type 689 is a module for RC that is responsible for transferring sensible heat from inlet air stream to secondary air stream. Type 667 is a module for EW where two air streams are passed near each other so that both energy and moisture will be transferred. Type 14 h is a module for a time dependent forcing function which has a behaviour characterised by a repeated pattern, used for scheduling the operation hours for the CC and the overall system. Type 33c is a module for Psychometrics (dry bulb temperature and humidity ratio known). When the dry bulb temperature and humidity ratio of a state is known it calls the TRNSYS psychrometric routine, returning dewpoint temperature, wet bulb temperature, relative humidity, absolute humidity and enthalpy. Type 99 is the radiation processor and weather data reader as well which takes input data, total radiation and horizontal radiation of one-hour interval on a horizontal surface. This component then interpolates the radiation data, calculates several quantities related to the position of the sun and estimates solar radiation on several surfaces of either fixed or variable orientation. Type 65 c is the graphic online plotter with output file graphics component, used for displaying selected system variables. It prints the given data, once per time step to a user defined external file. Unit descriptors (kJ/hr, kg/s, °C, etc.) are not printed to the output file. The indoor condition, chilled water and air condition and cooling load are generated and calculated from TRNSYS.

### 2.6. *Power consumption by the chiller*

A general reciprocating air-cooled chiller model[6][7]was used to calculate the instantaneous power consumption (kW). This model requires rated capacity, rated full-load power, cooling load, outside air dry-bulb temperature and chilled water supply temperature to calculate the instantaneous power consumption of the chiller.

## 3. Results and Discussion

### *3.1. Simulation for a hot and humid day*

'Figure 4 (a)' shows sample data of June 5 taken as a base/reference case to represent the hot and humid day. The total solar radiation and the diffuse solar radiation are relatively low and close showing a cloudy day and the air temperature is between 28-35°C., relative humidity 60-90 %. The radiation in each building façade is shown in 'Figure 4(b)' which is calculated by TRNSYS module type 16i. From 'Figure 4(b)' total solar radiation had the highest value of 689.5 W/m$^2$ at 16:00 hrs. The incident solar radiation on the west wall has highest value of 518.9 W/m$^2$ and the highest temperature was 33.92°C.

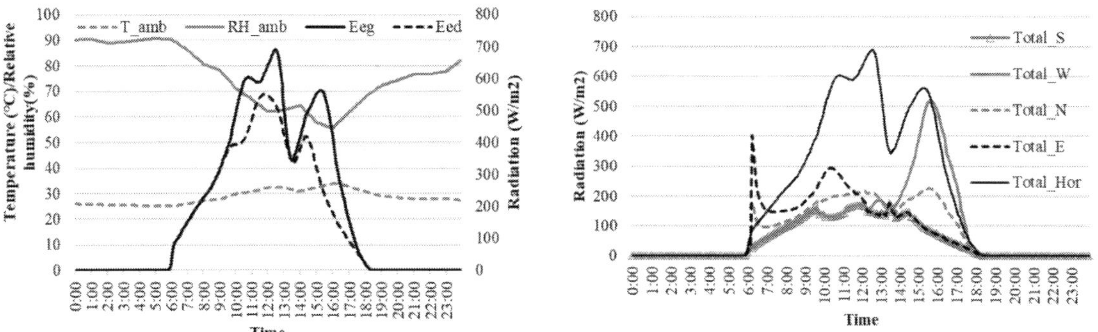


(a)The solar radiation and the ambient condition (b) Total radiation on each building façade

**Figure 4.** Weather and outdoor solar radiation

*3.1.1 Configuration 1.* The simulation results show the different condition of the air and chilled water used to cool the room and the indoor condition of the room as well. According to the 'Figure5(b)', the chilled water supply temperature (Tchs) is 15°C, which picks up the heat from the room and the resulting return chilled return water temperature (Tchr) is in the range of 16-16.9°C. with an increment of (1-1.9°C). Surface temperature of the RCC (Tpanel) is between 18-18.6°C during the air conditioning period of 6:00 to 18:00. For the DOAS. 'Figure 5(a)', the condition of the outside air is, temperature(Tamb) 25-33.92°C, humidity ratio (Wamb) 0.018-0.02 kgw/kgda which is mixed with the RA during the office hours and introduced directly during the precooling period results in air temperature (Tmix) and humidity ratio (Wmix) in the range of 24-32°C and 0.0093-0.0178 kgwv/kgda.The air is further reduced by the cooling coil (CC) to supply temperature (Tsup,ven ) 9°C and humidity ratio(Wsup,ven) 0.007 kgw/kgda. The air coming from the CC is mixed with the return air (RA) brings the room temperature at the range of 23-25°C with relative humidty 57.5-68.9%(humidity ratio around 0.0125 kgw/kgda). The dew point temperature (Tdp) in the room lies in between 13.4-17.6°C which is below the RCC panel temperature ensuring no occurrence of condensation as in 'figure 6'. Initial drop in Tdp in the room can be seen in the graph due to precooling period of the room where there is no internal cooling load. It can be seen from 'Figure 6' that the mean radiant temperature (Tmrt) and operative temperature (Top) is almost equal to the room air temperature (Trm) during the office operational hour which means the cooling in the room is dominated by RCC. During the initial preccoling period, Tmrt and Top are higher than Trm illustrating the dominant form of cooling is DOAS. Tmrt is higher than Trm from 6:00 a.m. to 9:35 a.m., the time taken by the radiant system to cool the walls.

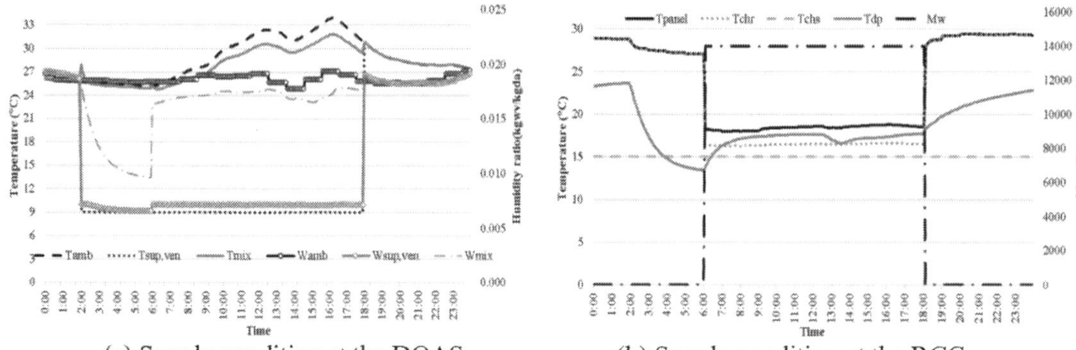


(a) Supply condition at the DOAS (b) Supply condition at the RCC

**Figure 5.** Supply condition of water and air in configuration 1

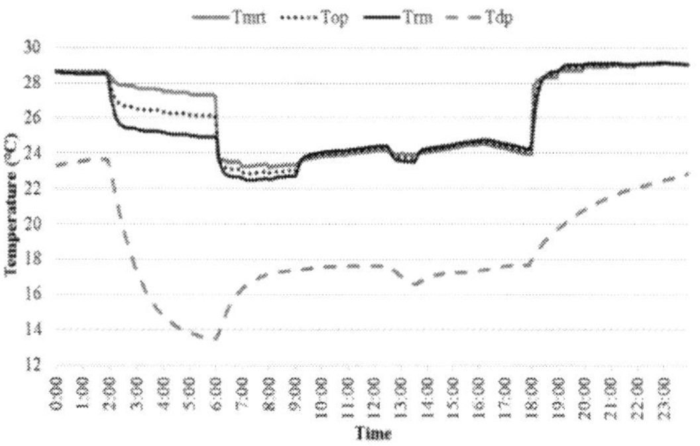


**Figure 6.** Indoor condition of configuration 1

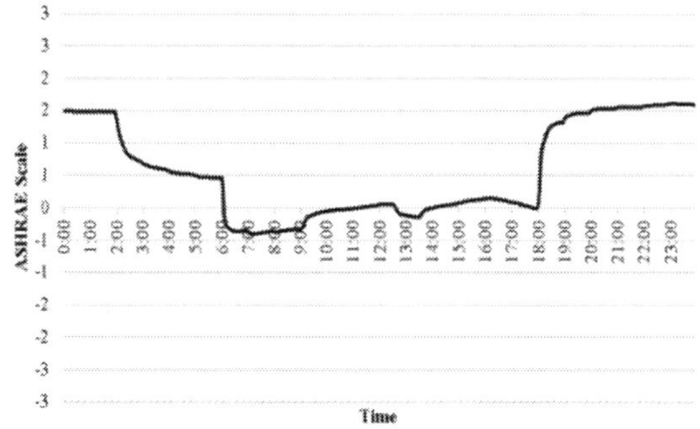


**Figure 7.** Comfort configuration of configuration 1

It can be seen from 'Figure 7' that the thermal sensation of the room represented by PMV is withtin comfort range of ± 0.5 as prescribed by ASHRAE 55 [8].

*3.1.2. Configuration 2.* 'Figure 8' to 'Figure 10' show the condition of the room air-conditioned by configuration 2. During the precooling period from 2:00-6:00 the air is recirculated from the room while during the operational period the hot and humid ambient air (Tamb ≈ 25-35 °C, Wamb ≈ 0.019 kgw/kgda ) is mixed with the RA(Tmix ≈ 25-32.4°C, Wmix ≈ 0.009-0.0173 kgw/kgda coil where the air is precooled at (Tpre ≈ 17.5-22.5°C, Wpre ≈ 0.017 kgw/kgda ) ('Figure8(a)'). The air temperature and humidity ratio after passing through the cooling coil is same as that of configuration 1 but the supply air temperature to the room is raised by reheating of the air on the secondary side of the run-around coil (Tre≈17-20.5°C). The Trm and Wrm are in the range of 23.9-27°C and it can be seen from 'Figure 8(a)' and 'Figure 9'. Tdp is in the range of 13.9-17.6°C and Tpanel is in the range of 18.5-19.4, ensuring no condensation. The dew point temperature of the room was brought down from 23.3°C to 13.9°C to avoid condensation in the RCC panel. The SA to the room for configuration 2 is higher than configuration 1, enabling the RCC to absorb more heat. The mean radiant temperature and the operative temperature are lower than the air temperature in configuration 2, as seen in 'Figure 6'. The difference between the Tmrt, Top and Trm in configuration 2 is more than that in the configuration 1.

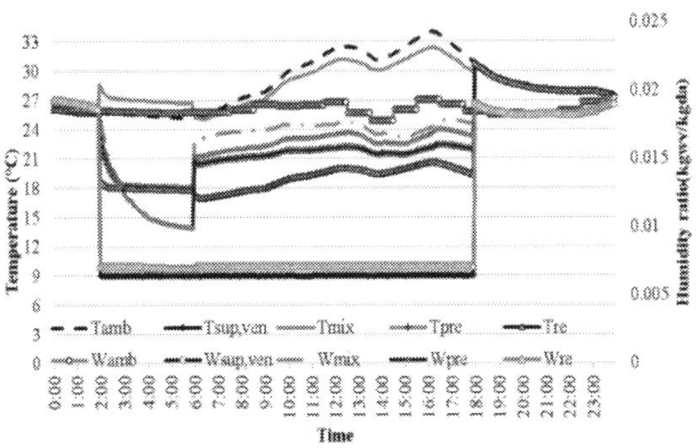


(a) Supply condition at the DOAS

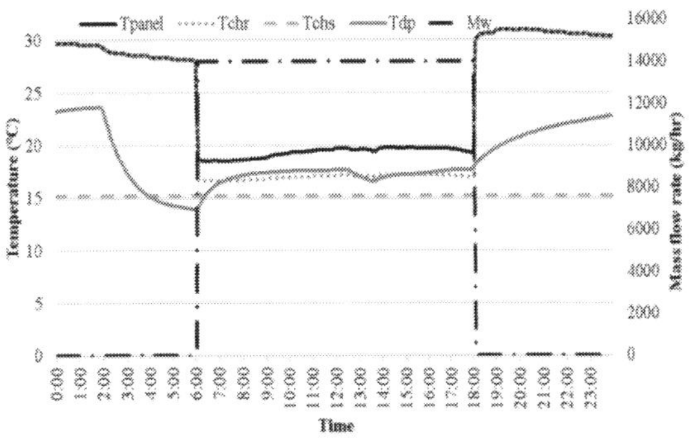


(b) Supply condition at the RCC

**Figure 8.** Supply condition of air and water on configuration 2

The PMV for configuration 2 lies nearly at ±0.5 ('Figure 10') but the PMV is higher than that of configuration 1. The panel temperature of RCC must be reduced to maintain the PMV of configuration 2 as same as configuration 1.

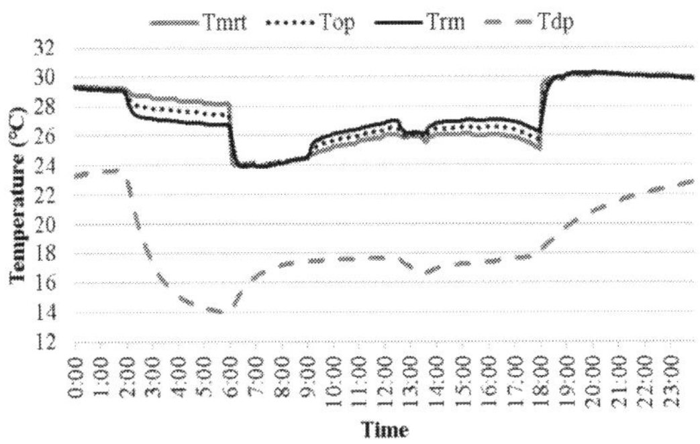


**Figure 9.** Indoor condition for configuration 2

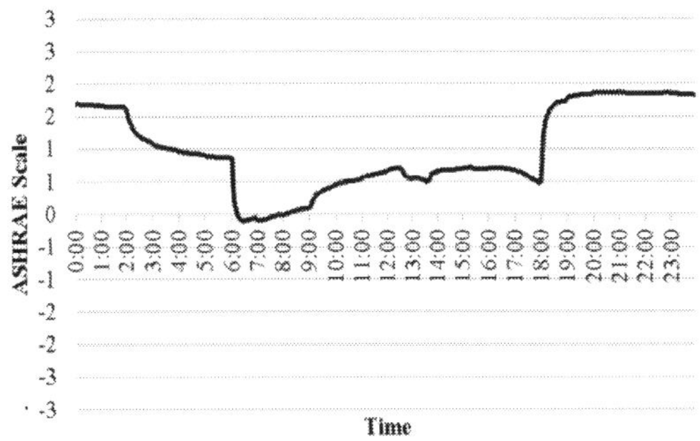


**Figure 10.** Comfort configuration for configuration 2

*3.1.3. Configuration 3.* 'Figure 11' shows the condition of the air and water at various states of the configuration 3. From 2:00-6:00 the air is recirculated from the room to mixing valve while during the operational period the hot and humid ambient air (Tamb ≈25-35 °C, Wamb ≈ 0.019 kgw/kgda ) is mixed with the RA(Tmix ≈ 23.7-32.4°C, Wmix ≈0.0163-0.0173) in the air mixing valve. The air from the mixing valve passes through the enthalpy wheel (Tew ≈ 23.93-29.3, Wew ≈ 0.01-0.017 kgwv/kgda) where it exchanges heat and moisture with the exhaust air and then passes via the primary side of the RC to the CC and then the secondary side of the RC and then eventually supplied to the room. The temperature of the mixed air is dropped with reduction in humidity (0.005kgw/kgda) after the air passes through EW. The air then flows through the primary side of the run-around coil with precooling temperature (Tpre) close to that of the configuration 2 with lowers humidity. The installation of the EW therefore has a huge contribution to reduction of humidity, but this will lead to reduction in the capacity of the run-around coil. The reheat temperature of configuration 3 is (16.4-19.1°C), lower than that of the configuration 2. Tpanel is in the range of (18.2-19.6°C) (as in 'Figure 12') while the dew point temperature in the room is in the range of (13.9-17.6°C) which means there is no condensation in the room. The dew point temperature of the air inside the room was brought down from 23.3°C to 13.9 °C during the precooling period. The mean radiant temperature is higher than the air temperature during the initial operation of the RCC panel from 6:00 to 8:10 and after this period the air temperature is higher than the mean radiant temperature and the operative temperature illustrating the room is cooled by the mode of radiation (as in 'Figure 13'). The PMV in configuration lies near to ± 0.5 (as in 'Figure 14')

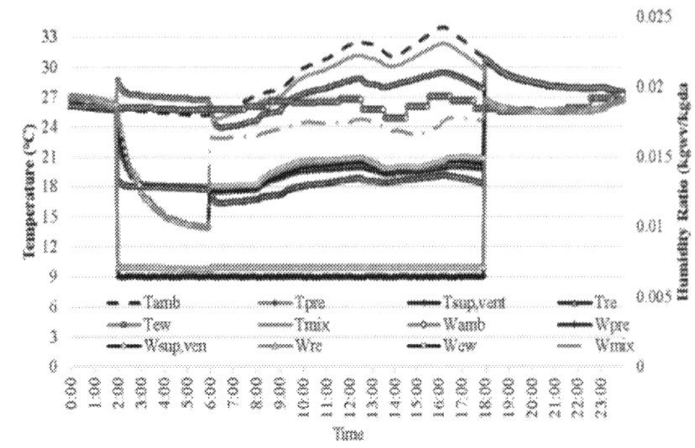


**Figure 11.** Supply condition at the DOAS in configuration 3

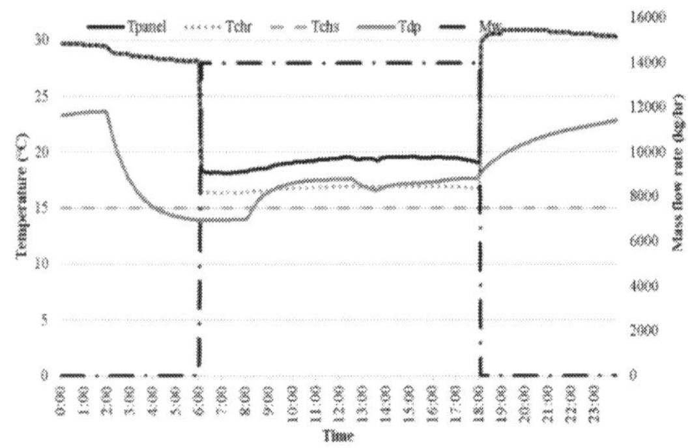


**Figure 12.** Supply condition at RCC in configuration 3

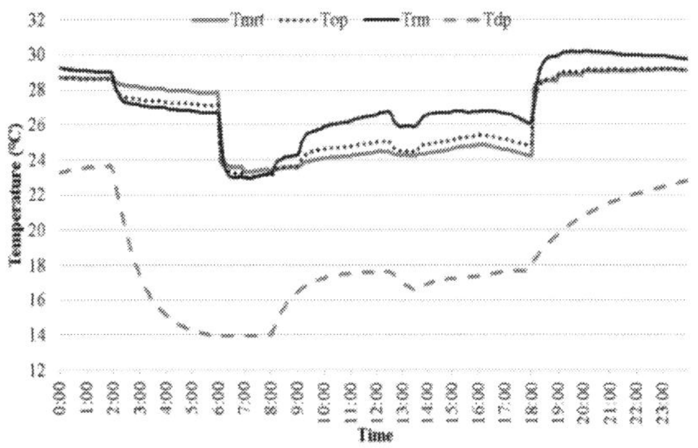


**Figure 13.** Indoor condition in configuration 3

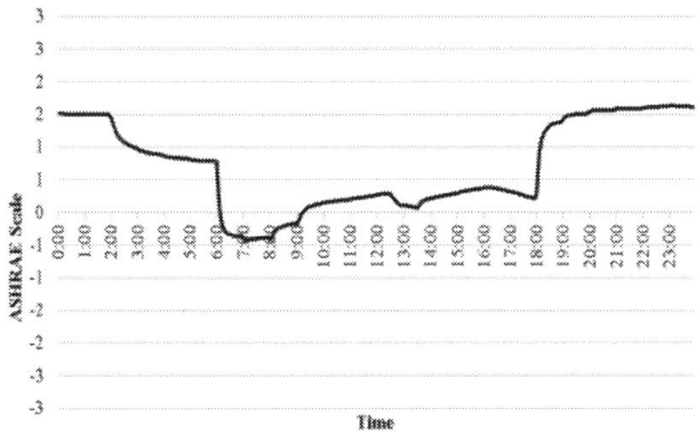


**Figure 14.** Comfort condition in configuration 3

### *3.2. Daily cooling Load Comparison*

For configuration 1, the total cooling load is between 8.7-48 kWth during working hours (8:00-17:00) ('Figure 15(a)'). Most of the load is taken by the RCC (21-20.2kWth) which accounts for 62% of the total load. The DOAS has cooling load of 8.7-18.2 kWth with latent heat slightly higher than sensible heat. Due to the air infiltration of 0.3 ACH/400 cfm, the DOAS system must be started at 2 am to bring down the dew point temperature in the room to safeguard against condensation.

'Figure 15 (b)' is the cooling load for configuration 2 (6-45.8 kWth) which is lower than configuration 1 (as 15.2°C, lower chilled water also enables some saving). However, in the case of configuration 2, more sensible heat is handled by the RCC (22.9-32 kWth) due to the reheating of the air on the secondary side of the round around coil after exiting the cooling coil. 'Figure 15 (c)' shows the reduction of load in configuration 3 as compared to the first two configuration. The cooling load of the system is in the range of 6-42.7 kWth. The load handled by RCC is in the range of (21.8-32.2 kWth) while the total cooling of DOAS is in the range of (6-11 kWth). Lower cooling load with this configuration is due to the heat recovery feature of the enthalpy wheel which contributes to lower CC load and the capability of the round around coil also dedicates the cooling coil to latent heat handling. The higher temperature entering the radiant room with low humidity ensures more sensible load handling by the radiant chilled ceiling. The sensible load handling capacity of the radiant chilled ceiling in configuration 3 is increased as the EW preconditioned the air and then it passes via the RC which dedicates the cooling coil to handle more latent load.

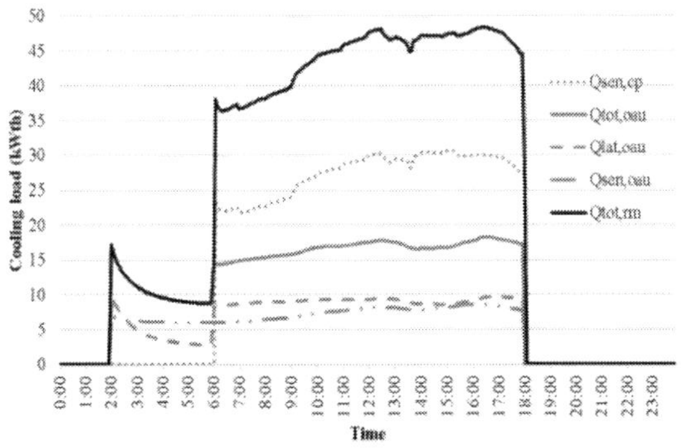


(a) Cooling load for configuration 1

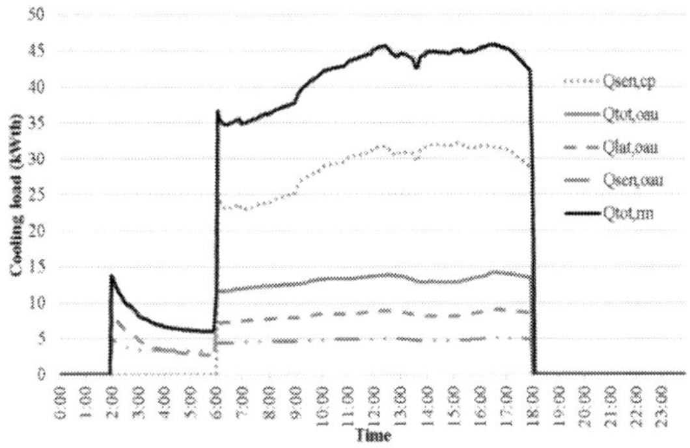


(b) Cooling load for configuration 2

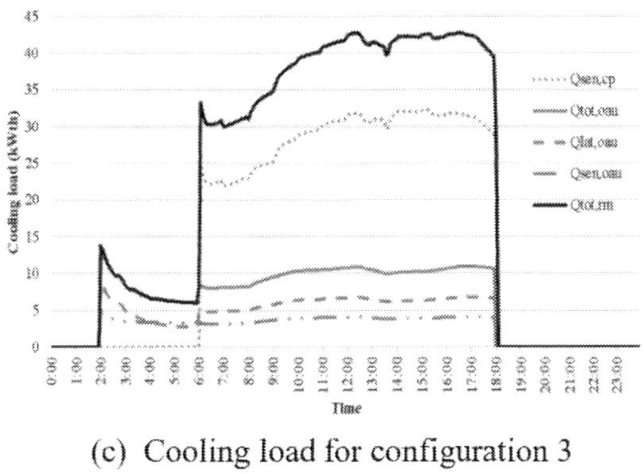


(c) Cooling load for configuration 3

**Figure 15.** Daily cooling load comparison

### *3.3. Monthly Cooling energy for the RCC and DOAS*

'Figure 16 (a), (b) and (c)' shows the monthly cooling load of three configurations of RCC and DOAS. The maximum cooling energy consumption is for the month of May with cooling energy of 17479.34 kWth for configuration 1, 16187.08 kWth for configuration 2 and 15018 kWth, for configuration 3. The minimum cooling energy consumption is for the month of February with cooling energy of 14175 kWth for configuration 1, 13026 kWth for configuration 2, 12339.21 kWth for configuration 3.

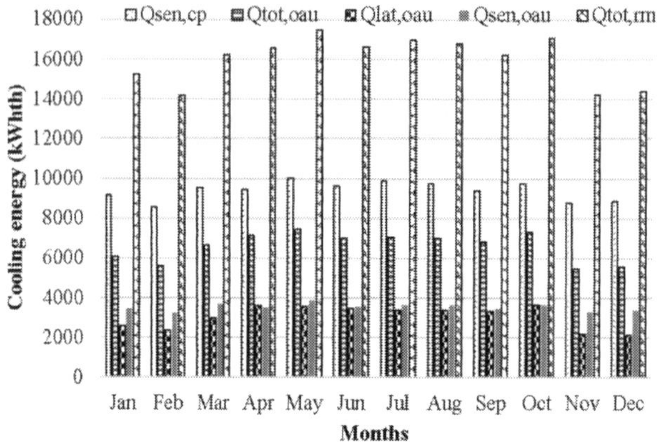


(a) Configuration 1

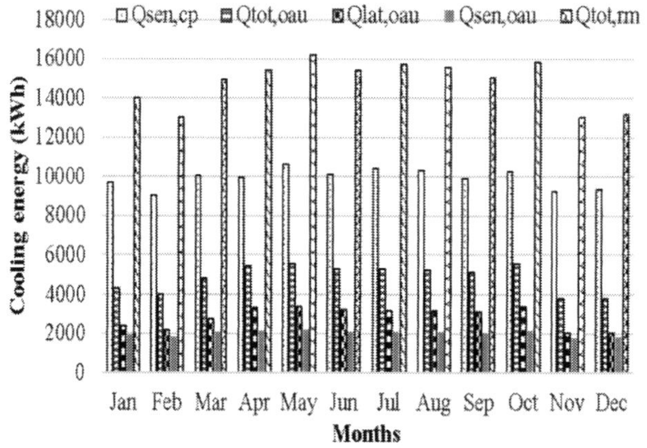


(b) Configuration 2

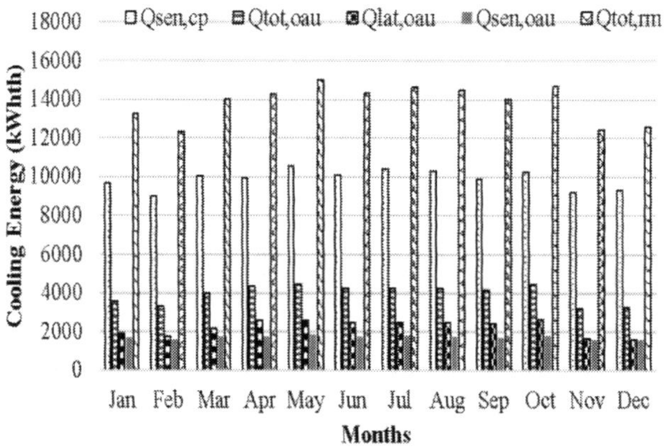


(c) Configuration 3

**Figure 16.** Monthly cooling load comparison of 3 configurations of RCC with DOAS.

*3.4. Annual simulation for the RCC and DOAS*

The annual cooling load of the three configuration of RCC with DOAS have been compared in this section. 'Figure 17' shows the simulation results for the whole year period. It can be observed that configuration 2 has 7.6% and configuration 3 has 13.4% less energy consumption than configuration 1. The lower energy consumption for configuration 2 is due to the ability of round around coil to dedicate the cooling coil to handle more latent load by precooling the air and the higher SA temperature due to heating of the supply air after the cooling coil ensures more sensible load handling from the RCC. The lowering of energy consumption in configuration 3 is due to the heat recovery feature of enthalpy wheel in conjunction with higher SA with less humidity. The share of cooling load handled by RCC panel is 58.8%, 67.1% and 71.5% in configuration 1, 2 and 3, respectively. The high load handling in configuration 2 and 3 is due to the higher air supply temperature provided by DOAS as compared to configuration 1.

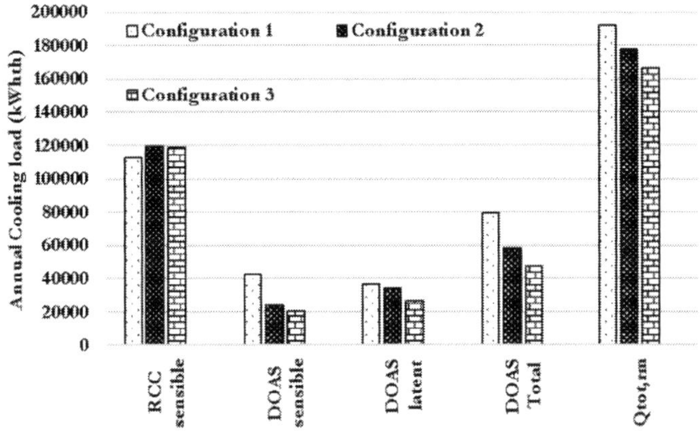


**Figure 17** Annual cooling load comparison

*3.5. Electricity Consumption analysis by the chiller*

With the help of the cooling load generated by TRNSYS and the reciprocating air-cooled chiller model[6] the electrical energy consumed by the chiller producing chilled water at 15/15.2°C for RCC and the chiller producing chilled water at 7°C for DOAS were calculated. The energy consumed by the chiller for RCC is about 58%,64% and 68% out of total energy consumed for running configuration 1,2 and 3, respectively. Less energy is devoted to chiller for configuration 2 as the chiller water temperature is 15.2°C. Configuration 2 and 3 save 8% and 13% electricity as compared to configuration 1. Configuration 3 saves energy by running the chiller dedicated to produce 15°C as major load. 'Table 1' shows the energy consumed by the two chillers that supply chilled water to the RCC and DOAS.

Table 1 Electricity consumption while operating the chiller with different configuration

| Type | Configuration | | |
|---|---|---|---|
| | 1 | 2 | 3 |
| RCC Chiller energy ($kW_{elec}$) | 29903 | 31483 | 31487 |
| DOAS chiller energy ($kW_{elec}$) | 23240 | 17480 | 14633 |
| System energy ($kW_{elec}$) | 53143 | 48963 | 46120 |

From the annual cooling load and annual energy consumption of the chiller, the Coefficient of performance (COP) of the respective chiller for different configuration is also calculated. The COP of the chiller for RCC and DOAS is 3.78 and 3.41 with difference of 0.34 for configuration 1, that for configuration 2 is 3.79and 3.33 with difference of 0.08 and COP of the chiller of RCC and DOAS for configuration 3 is 3.77 and 3.23 with difference of 0.54 respectively.

## 4. Conclusion

This paper reported the comparative study of RCC equipped 3 DOAS configuration in hot and humid climate. The dew point temperature is always lower than the RCC panel temperature for all the investigated configuration, ensuring no condensation. While the mean radiant temperature and the operative temperature are lower than the air temperature for all cases meaning cooling is dominated by the RCC. This shows that the RCC equipped with DOAS can be used in Thailand climate without any condensation. The annual simulation shows that configuration 2 has 7.6% and configuration 3 has 13.4% less annual cooling load than configuration 1. The RCC chiller for configuration 1,2 and 3 consume 58%,64% and 78% of the total energy consumed by the chillers for running the system. The Coefficient of performance (COP) of the chiller for RCC is higher than that of chiller for DOAS. Configuration 3 is the most energy efficient configuration for hot and humid climate of Thailand.

## Acknowledgement

The financial support from the National Science and Technology Development Agency (NSTDA), grant no. P-16-51993 through this research project is gratefully acknowledged.

## References

[1] A.-U.-W. Tantiwichien, "An Experimental and Simulated Study on Thermal Comfort," *Int. J. Eng. Technol.*, vol. 5, no. 2, pp. 177–180, 2013.
[2] P. Vangtook and S. Chirarattananon, "An experimental investigation of application of radiant cooling in hot humid climate," *Energy Build.*, vol. 38, no. 4, pp. 273–285, 2006.
[3] J. Jeong, "Simplified ceiling radiant cooling panel and enthalpy wheel models for dedicated outdoor ari system design," 2004.
[4] M. S. Shin, K. N. Rhee, S. H. Park, M. S. Yeo, and K. W. Kim, "Enhancement of cooling capacity through open-type installation of cooling radiant ceiling panel systems," *Build. Environ.*, vol. 148, pp. 417–432, 2019.
[5] Solar Energy Laboratory, "Trnsys 17," *Univ. Wisconsin-Madison*, pp. 1–29, 2012.
[6] J. W. Jeong, S. A. Mumma, and W. P. Bahnfleth, "Energy conservation benefits of a dedicated outdoor air system with parallel sensible cooling by ceiling radiant panels," *ASHRAE Trans.*, vol. 109 PART 2, pp. 627–636, 2003.
[7] California Energy Commission, "Alternative calculation method approval manual for the 2001 Energy Efficiency Standards for Nonresidential Buildings," 2001.
[8] ANSI/ASHRAE, "ANSI/ASHRAE 55:2017 Thermal Environmental Conditions for Human Occupancy," *Ashrae*, vol. 2017, p. 30, 2017.

IOP Conf. Series: Materials Science and Engineering **965** (2020) 012005 doi:10.1088/1757-899X/965/1/012005

# The influence of calcination temperature on (photo)electrochemical property of $TiO_2$ prepared by the sol-gel method

**Yingyot Poo-arporn[1]*, Kanokrat Luengprasert[2], Surangrat Tonlublao[1], Sirinuch Loiha[2], Anuchit Ruangvittayanon[1] and Rungtiva P. Poo-arporn[3]**

[1]Synchrotron Light Research Institute, 111 University Avenue, Nakhon Ratchasima, 30000, Thailand

[2]Materials Chemistry Research Center, Department of Chemistry, Faculty of Science, Khon Kaen University, Khon Kaen 40002, Thailand

[3]Biological Engineering Program, Faculty of Engineering, King Mongkut's University of Technology Thonburi, Bangkok, 10140

* yingyot@slri.or.th

**Abstract**. In this work, titanium dioxide ($TiO_2$) was successfully synthesized via the sol-gel method using Titanium (IV) isopropoxide as a precursor. The mixture was heated to 115°C to obtain the solid powder of TiO2. Ti K-edge XANES measurement was performed to determine its structure. Three pre-edge peaks at energy levels between 5460-5480 eV were clearly observed. These features correlated to 1s→d electronic transition in the distorted-octahedral structure of anatase $TiO_2$. The in situ XANES experiment at 700°C revealed no phase transformation from anatase to rutile. There was no characteristic of the rutile phase on the XRD pattern of $TiO_2$ calcined at 700°C. This result confirmed the stability of anatase. SEM images revealed the circular shape of $TiO_2$ particles with a diameter of around 50 nm. The highest (photo)electrochemical activity of $TiO_2$ was obtained at the calcination temperature of 300°C as investigated by a cyclic voltammetry.

## 1. Introduction

Photochemical processes in heterogeneous systems have been an area of great interest over the past three decades. In 1972, Fujishima and Honda discovered the photoelectrochemical properties of titanium dioxide ($TiO_2$) electrodes on the splitting of water under solar energy [1]. They also extended its application to the area of environmental photocatalysis including self-cleaning [2], photoinduced hydrophilicity [3], and antifogging [4]. More recently, many studies have been published on the photocatalytic applications of $TiO_2$ for the decomposition of organic compounds (pollutants) [5-8]. The important advantages of $TiO_2$ for pollutant treatment are: (i) the process occurs under ambient conditions; (ii) oxidation of the substrates to $CO_2$ is completed; (iii) the photocatalyst ($TiO_2$) is inexpensive and has a high turnover; (iv) $TiO_2$ can be supported on suitable reactor substrates, and (v) the process offers great potential as an industrial technology to detoxify wastewater.

The four polymorphs of $TiO_2$ are found in nature including anatase (tetragonal), brookite (orthorhombic), rutile (tetragonal), and $TiO_2$ (B) (monoclinic) [9]. Two additional high-pressure forms have been synthesized starting from rutile: TiO2(II) [10]. The structures of rutile and anatase can be discussed in terms of octahedral ($TiO_6$). Anatase can be regarded to be built up from octahedral connected by their vertices; in rutile, the edges are connected. $TiO_2$ can be prepared by the sol gel process. The sol gel process involves the evolution of inorganic networks through the formation of a colloidal suspension (sol) and gelation of the sol to form a network in a continuous liquid phase (gel) [11]. The precursors for synthesizing this colloid consist of a metal or metalloid element surrounded by various reactive ligands. Metal alkoxides are most popular because they are commercially available in high purity and react readily with water [12-14]. The encapsulated liquid can be removed from a gel by either evaporative drying or drying with supercritical drying. The resulting solid products are known as a xerogel and an aerogel, respectively. Heat treatment is necessary to burn off any residue organics or to oxidize the sample to a catalytically useful form. Heating is usually done in the presence of a reactive gas. The exposure of the sample to a high temperature over an extended period leads to sintering and consequently a decrease in surface area [15]. The process can also cause the material to crystallize into different structural forms. Thus, the physical characteristics of the product depend on parameters such as temperature, heating rate, time, and gaseous environment.

In this work, $TiO_2$ was prepared by sol-gel method. The characteristics of the $TiO_2$ were investigated by transmission electron spectroscopy (TEM), x-ray powder diffraction (XRD), x-ray absorption near edge structure (XANES) and cyclic voltammetry (CV).

## 2. Experimental

$TiO_2$ was prepared by the sol-gel method. Titanium (IV) isopropoxide was added to the mixture of 2-propanol and acetic acid under magnetic stirring. The suspension was stirred at room temperature for 20 min. Water was then added in the well mixed solution and continuously stirred for 60 min. The gel was then dried at 115°C to obtain the solid powder of $TiO_2$ ($TiO_2$-D). To study the influence of temperature on the $TiO_2$ structure, the samples were calcinated in air at 300°C, 500°C and 700°C for 1 hour.

To investigate the morphology of $TiO_2$, TEM analysis of the samples was performed using a Talos Thermoscientific operated at 200 keV. The powders were prepared for theTEM analysis by dropping the suspension of the powders in isopropanol on to a copper grid. The structural property of $TiO_2$ was characterized by x-ray powder diffraction (XRD) (Smartlab, Rigaku) and x-ray absorption near the edge structure (XANES) (BL2.2, Synchrotron Light Research Institute, Thailand). To record the XRD pattern, the detector scanned the peak position from the sample as a function of 2θ by starting at the 15° to 90° range and a scan speed of 2.42 deg/min. Ti K-edge XANSE spectra was recorded on a transmission mode. The energy dispersive monochromator, Si(111) single crystal and NMOS linear image sensor were employed. The measurement of diffuse reflectance spectra was performed on Lambda 950 UV/Vis/NIR spectrometer (PerkinElmer). The study of $TiO_2$ photocatalytic activity was obtained from cyclic voltammetry (CV). CV was performed in a three-electrode configuration with a Pt rod counter electrode and Ag/AgCl reference electrode. The cyclic voltammograms were recorded by the Autolab electrochemical workstation (PGSTAT302N, Metrohm). Sodium sulfate (1.0 M $Na_2SO_4$) was employed as a supporting electrolyte. For photocatalytic reaction, a 200 W Hg arc lamp (LOT) was used for excitation.

## 3. Results and discussion

The structural property of the prepared catalysts was investigated by XANES. The normalized Ti K-edge XANES spectra of standard anatase $TiO_2$ ($TiO_2$-A), rutile $TiO_2$ ($TiO_2$-R) and $TiO_2$ calcined at various temperatures are shown in Fig. 1. Three pre-edge peaks at energy levels between 4925-4950 eV of $TiO_2$-A indicated the electronic transition of electrons from 1s to 3d orbitals. Although this electron transition in Ti atoms was forbidden by selection rules, the mixing of d and p orbitals of Ti and O atoms in the $TiO_2$ system produced the pseudo-p character in the d orbital which consequently allows 1s→3d transition. The characteristics of these peaks have been reported by Farges et al. [17]. The quadrupole transition of the 1s electron to the $t_{2g}$ levels of the octahedron structure was denoted as A1. For dipole transition, $1s \rightarrow t_{2g}$ and $1s \rightarrow e_g$ were denoted as A2 and A3. After calcination, $TiO_2$-300 revealed three pre-edge peaks similar to the anatase. Although the calcination temperature was increased to 700°C, there was no indication of the rutile phase of $TiO_2$. This result indicated the stability of the anatase phase at this temperature.

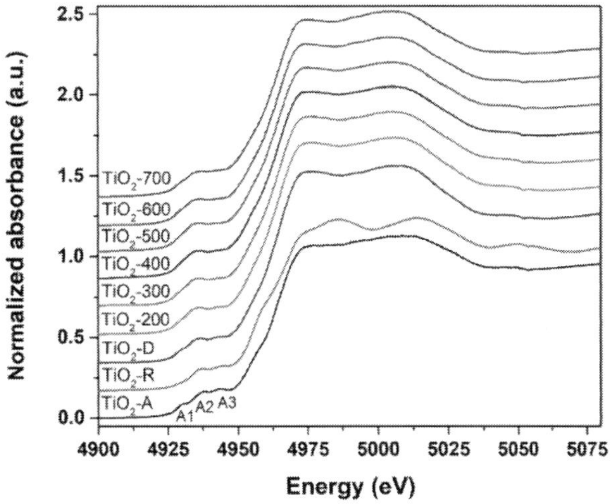


**Figure 1.** Normalized Ti K-edge XANES of anatase $TiO_2$ ($TiO_2$-A), rutile $TiO_2$ ($TiO_2$-R), non-calcined $TiO_2$ ($TiO_2$-D), and $TiO_2$ calcinated at 300°C, 400°C, 500°C, 600°C, 700°C, respectively

The structure of $TiO_2$ was further determined by means of x-ray diffraction spectroscopy. The diffraction patterns of $TiO_2$ calcined at 300°C, 500°C and 700°C are shown in Fig. 2. The prepared $TiO_2$ film showed the XRD patterns of anatase (JCPDS No. 83-2243). By calcination at 300°C, $TiO_2$-300 exhibited broad and not well-defined peaks of the anatase phase, suggesting a small grain size of $TiO_2$. The crystalline structure of anatase could be improved by increasing the calcination temperature. The sharper and higher intensity of XRD peaks of $TiO_2$-500 and $TiO_2$-700 could be clearly observed. The result indicated the improvement in the crystallinity of anatase $TiO_2$. It has been reported that the typical temperature required for the anatase to rutile phase transformation reported by Bastow and Withfield is ca. 600°C [16]. However, there was no indication of rutile phases in the XRD pattern although the calcination temperature was increased to 700°C.

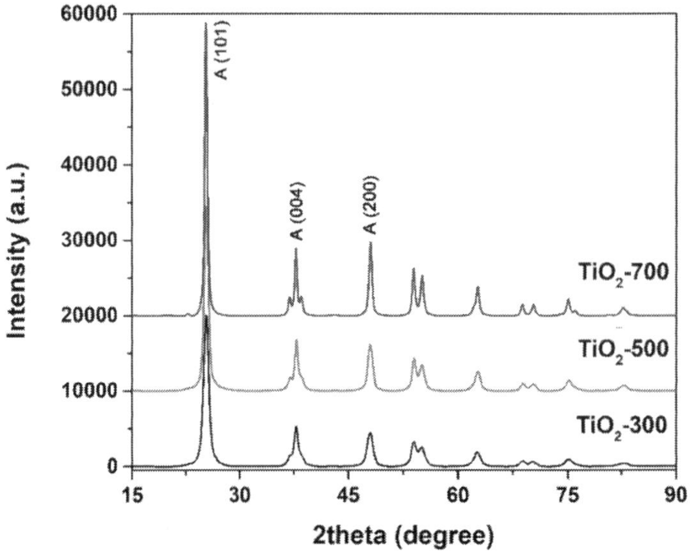


**Figure 2.** XRD patterns of $TiO_2$ calcinated at 300°C, 500°C and 700°C.

The morphology and size of the prepared $TiO_2$ were characterized by a transmission electron microscope. TEM images of $TiO_2$ calcined at 300°C, 500°C and 700°C are shown in Fig. 3. By the calcination at 300°C, as seen in Fig. 3A, the TEM image for $TiO_2$ displayed a sphere-like morphology with average particle-size of about 10 nm in diameter. The larger size of the $TiO_2$ particles was observed when the calcination temperature was increased to 500°C and 700°C. The thermal treatment affected the crystal growth, indicating the aggregation of $TiO_2$ nanoparticles when the temperature increases. The smaller size of the $TiO_2$-300 particles was also evident from the broader XRD peaks.

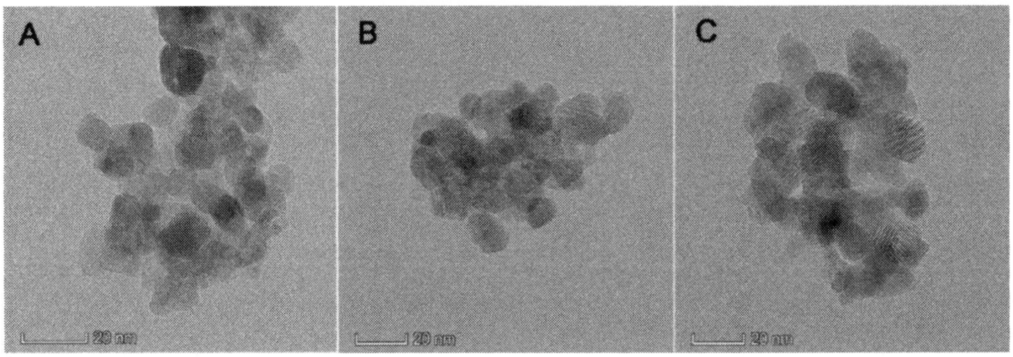


**Figure 3.** TEM images of $TiO_2$ calcinated at (A) 300°C, (B) 500°C and (C) 700°C.

Fig. 4(A) shows the diffuse reflectance spectra of the standard anatase and $TiO_2$ calcinated at 200°C and 700°C. The increasing of reflection was started around 350 nm due to the reduction of photon absorption. For $TiO_2$-200 and $TiO_2$-700, the absorption edge displayed a gradual blue shift. Further analysis of reflectance spectra was performed to determine the energy different between the lowest conduction band the highest valence (band gap). The calculation of band gap can be done by applying the Kubelka–Munk method which based on the following equation: $F(R) = \frac{(1-R)^2}{2R}$, where R is the reflectance and F(R) is proportional to the extinction coefficient (α)[18]. For anatase $TiO_2$, the estimated value of band gap was 3.27 eV (Fig. 4B). However, the smaller values were observed on $TiO_2$-200 (3.23 eV) and $TiO_2$-700 (3.25 eV).

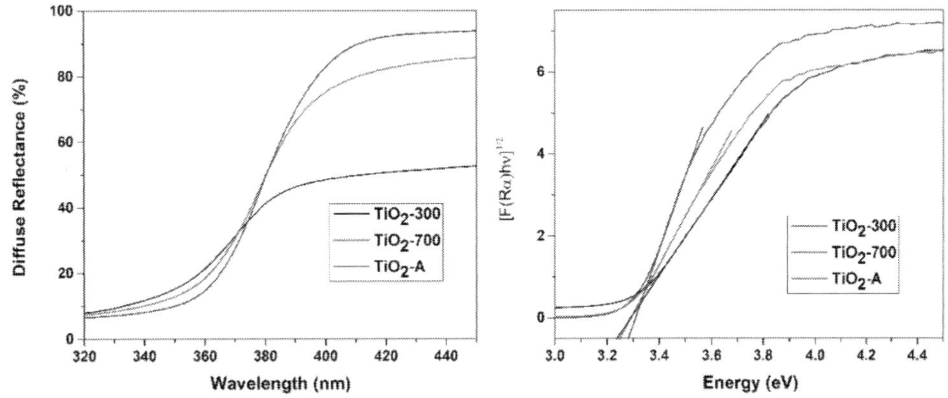


**Figure 4.** (A) UV-Vis diffuse reflectance spectra and (B) Kubelka-Munk plots and bandgap energy estimation for the standard anatase and $TiO_2$ calcinated at 200°C and 700°C.

The photoelectrochemical properties of the prepared $TiO_2$ samples were studied by the photoelectrochemical oxidation of ethanol in water. Photocurrent-voltage curves of $TiO_2$-300, $TiO_2$-500 and $TiO_2$-700 in 1.0 M $Na_2(SO_4)$ under chopped irradiation are shown in Fig. 5. When the illumination was exposed on the $TiO_2$ surface, an anodic polarization coursed in a rapidly rising anodic current (photocurrent). In contrast, this photocurrent immediately disappeared after the light exposure was shut off. For a reverse scan, the experiment was conducted in the dark to provide the background of a cathodic current. The irradiation of the $TiO_2$ electrode during cathodic polarization also exhibited a similar result. Chopped irradiation on the $TiO_2$-300 electrode yielded the average anodic photocurrent of approximately 2 $\mu$A. The photocurrent is due to the formation of electron-hole pairs; holes formed in the bulk which are forced to move to the surface ($TiO_2$/electrolyte interface), whereas electrons are moving towards the ITO/$TiO_2$ interface. Under irradiation, the generated photoelectrons may be trapped by the $TiO_2$ surface groups (TiOH). Once the applied potential is positive for the electron trap potential, a flow of electrons from the trap band to the conducting substrate occurs and an anodic current is generated.

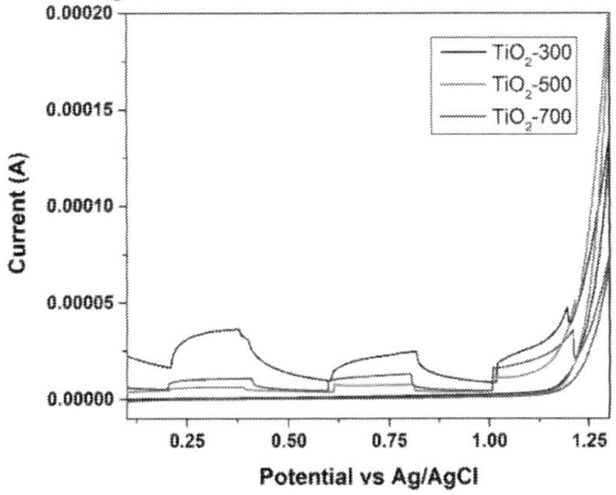


**Figure 5.** Photocurrent-voltages curve of $TiO_2$ calcinated at 300°C, 500°C and 700°C in aqueous 1.0 M $Na_2(SO_4)$ under chopped irradiation.

Since calcination temperatures affect the structural property of $TiO_2$, the photoelectrochemical properties of $TiO_2$ with different calcination temperatures were investigated. The anodic photocurrent of

$TiO_2$-300 was higher than those of $TiO_2$-500 and $TiO_2$-700. The results suggested that the photocurrent of $TiO_2$ depends on the crystallinity of anatase. As mentioned before, the generated photoelectrons can be trapped by the $TiO_2$ surface groups (TiOH) [19]. Presumably, the photoelectrons of $TiO_2$-300 were trapped by TiOH on the surface, resulting in the delay of the photocurrent. It is clearly seen that $TiO_2$-500 and $TiO_2$-700, with a small amount of TiOH, exhibited a very rapid rise/decay of the photocurrent. Kavan et al. [20] reported the photocurrent of single crystal anatase without any delay of the photocurrent. However, the photocurrent obtained from $TiO_2$-700 was reduced when the calcination temperature was raised to 700°C. The drop in photocurrent might be due to the aggregation of crystalline anatase during the thermal treatment, resulting in the enhancement of charge-transfer resistance of $TiO_2$.

## 4. Conclusions

The titanium dioxide photocatalyst was successfully prepared by the sol gel method. The average particle size of 10 nm of $TiO_2$-300 was obtained. TEM images revealed the aggregation of $TiO_2$ particles when the calcination temperature was increased to 700°C. XRD patterns as well as the Fe K-edge XANES spectrum confirmed the formation of the anatase phase. There was no indication of a rutile structure in any samples. The photocurrent-voltage curves clearly revealed the influence of anatase crystallinity on the $TiO_2$ photocurrent.

## 5. Acknowledgement

This work was supported financially by the Synchrotron Light Research Institute, Thailand, as the subsidy fund of fiscal year 2019 (grant number SLRI2019-S09).

## 6. References

[1] A. Fujishima, and K. Honda, "Electrochemical photolysis of water at a semiconductor electrode," Nature, vol. 238, pp. 37-38, 1972.

[2] M. Lettieri, D. Colangiuli, M. Masieri, and A. Calia, "Field performances of nanosized $TiO_2$ coated limestone for a self-cleaning building surface in an urban environment," Building and Environment, vol. 147, pp. 506-516, 2019.

[3] H. -M. Kim, D. Kim, and B. Kim, "Photoinduced hydrophilicity of $TiO_2/WO_3$ double layer films," Surface and Coatings Technology, vol. 271, pp. 18-21, 2015.

[4] C. Garlisi and G. Palmisano, "Radiation-free superhydrophilic and antifogging properties of e-beam evaporated $TiO_2$ films on glass," Applied Surface Science, vol. 420, pp. 83-93, 2017.

[5] J. Kim and B. -K. Lee, "Enhanced photocatalytic decomposition of VOCs by visible-driven photocatalyst combined Cu-$TiO_2$ and activated carbon fiber," Process Safety and Environmental Protection, vol. 119, pp. 164-17, 2018.

[6] O. Monfort, T. Roch, M. Gregor, L. Satrapinskyy, D. Raptis, P. Lianos, and G. Plesch, "Photooxidative properties of various $BiVO_4/TiO_2$ layered composite films and study of their photocatalytic mechanism in pollutant degradation," Journal of Environmental Chemical Engineering, vol. 5, pp. 5143-5149, 2017.

[7] S. Feizpoor, A. Habibi-Yangjeh, and K. Yubuta, "Integration of carbon dots and polyaniline with $TiO_2$ nanoparticles: Substantially enhanced photocatalytic activity to removal various pollutants under visible light," Journal of Photochemistry and Photobiology A: Chemistry, vol. 367, pp. 94-104, 2018.

[8] Y. Lee, S. Lee, H. Sil K., J. T. Moon, J. B. Joo, and I. Choi, "Multifunctional and recyclable $TiO_2$ hybrid sponges for efficient sorption, detection, and photocatalytic decomposition of organic pollutants," Journal of Industrial and Engineering Chemistry, vol. 73, pp. 328-335, 2019.

[9] U. Diebold, "The surface science of titanium dioxide," Surface Science Reports, vol. 48, pp. 53-229, 2003.

[10] P. Y. Simons and F. Dachille, "The structure of $TiO_2$II, a high-pressure phase of TiO2," Acta

Crystallographica, vol. 23, pp. 334-336, 1967.

[11] D. P. Debecker, V. Hulea, and P. H. Mutin, "Mesoporous mixed oxide catalysts via non-hydrolytic sol–gel: A review," Applied Catalysis A: General, vol. 451, pp. 192-206, 2013.

[12] M. Yang, W. Liu, C. Jiang, S. He, Y. Xie, and Z. Wang, "Fabrication of superhydrophobic cotton fabric with fluorinated $TiO_2$ sol by a green and one-step sol-gel process," Carbohydrate Polymers, vol. 197, pp. 75-82, 2018.

[13] F. Huang, B. Motealleh, W. Zheng, M. T. Janish, C. B. Carter, and C. J. Cornelius, "Electrospinning amorphous $SiO_2$-$TiO_2$ and $TiO_2$ nanofibers using sol-gel chemistry and its thermal conversion into anatase and rutile," Ceramics International, vol. 44, pp. 4577-4585, 2018.

[14] M. Crişan, N. Drăgan, D. Crişan, A. Ianculescu, I. Niţoi, P. Oancea, L. Todan, C. Stan, and N. Stănică, "The effects of Fe, Co and Ni dopants on $TiO_2$ structure of sol-gel nanopowders used as photocatalysts for environmental protection: A comparative study," Ceramics International, vol. 42, pp. 3088-3095, 2016.

[15] D. Valdespino, G. Rojas-George, J. Muñoz-Saldaña, B. M. Murguía, and M. P. Cruz-Jáuregui, "Controlling micro-porous size in $TiO_2$ pellets processed by sol-gel and rapid liquid phase sintering," Ceramics International, vol. 45, pp. 14510-14516, 2019.

[16] F. Farges, "Ti K-edge XANES studies of Ti coordination and disorder in oxide compounds: Comparison between theory and experiment," Physical Review B, vol. 56, pp. 1809-1819, 1997.

[17] T. J. Bastow and H. J. Whitfield, "47,49Ti NMR: Evolution of crystalline $TiO_2$ from the gel state," Chemistry of Materials, vol. 11, pp. 3518-3520, 1999.

[18] R. Lopez, R. Gomez, "Band-gap energy estimation from diffuse reflectance measurements on sol–gel and commercial $TiO_2$: a comparative study" Jounal of Sol-Gel Science and Technology, vol. 61, pp. 1–7, 2012.

[19] D. Jiang, H. Zhao, S. Zhang, and R. John, "Characterization of photoelectrocatalytic processes at nanoporous $TiO_2$ film electrodes: photocatalytic oxidation of glucose," Journal of Physical Chemistry B, vol. 107, pp. 12774-12780, 2003.

[20] L. Kavan, M. Gra1tzel, S. E. Gilbert, C. Klemenz, and H. J. Scheel, "Electrochemical and photoelectrochemical investigation of single-crystal anatase," Journal of American Chemical Society, vol. 118, pp. 6716-6723, 1996.

# Monitoring system for 5-kW Solar pumping system

**Sivapong Phetsong[1], Patiwat Somasri[1], Teerasak Hudakorn[1], and Nantawatana Weerayuth[2]**

1 Department of Mechanical Engineering, Silpakorn University, Nakhon Pathom, Thailand

2 Department of Mechanical Engineering, Ubonratchathani University, Ubonratchathani, Thailand

E-mail: phetsong_s@su.ac.th

**Abstract**. Solar pumping system has been used extensively in agriculture and irrigation system. It is an old pumping system replacement because the old pumping system requires more maintenance and replacement. This article is reported on the study and development of the solar pumping system and the remote monitoring system. The system consists of the 5-kW solar panel and the 2.2kW submersible pump with a solar pump controller. It has been installed in the rural area which cannot easily accessible from the city. The monitoring system has been developed by integrating the electrical power modules, water measurement modules, weather measuring modules, and main controller modules. All of them communicate over the RS485 bus. The main controller unit integrates with software, gather all data from every modules and display on touch screen. It also makes data post-processing and records it on a USB drive and prepare data protocol for sending via communication modules. The system can work properly but it still need some further development in the future.

**Keywords:** Solar, pumping system, remote monitoring, Solar pumping system, monitoring system.

## 1. Introduction

Solar pumping systems have been used extensively in agriculture. They have to be used as an irrigation system for both agriculture and normal living in rural areas. The system consists of the solar panels (or PV), submerge pumps, system controller/inverter, and water tank. Generally, the solar pumping system should be required any service/maintenance as least as possible. In the past, the pumping system used the gasoline/diesel engine or even the natural gas engine which required periodic maintenance and also make some noise, air pollution and also unpredictable fuel cost. When comparing it with the solar pumping system, the solar pumping system does not need any fuel and also requires only a few service/maintenance because it has only a few mechanical parts.

The use of solar pumping systems in agriculture and rural areas will reduce system maintenance and also has a longer service life than the pumping system which using the engine. The system has fewer mechanical parts that required lower maintenance.

The solar pump system in the present day does not require batteries. We do not have to worry about the battery lifetime or the problem of battery due to the bad weather. The solar pumping system also supports integration between solar electricity systems with AC power in the building system which increases the capability of the system and reduces discontinuity problems [1][2].

Because the pumping systems installation areas are in agricultural areas or remote areas. The remote monitoring systems are essential. Administrators can inspect the work council without access to inspect the area. The remote monitoring system can notify problems in advance when it is necessary to be maintained which makes the work planning much more effective in advance. Besides, remote data collection from each pumping system can also lead to the analytic and planning system as well by analyzing data from several pumping system sites.

In this article, we present a system for recording the performance of solar pumping systems, consisting of various sensors such as voltage, current, power, temperature/humidity, flow rate and water quantity. The system also includes data display and export to communication devices suitable for use in the other application as well.

## 2. Background and Theory

In general, the solar power system can be divided into 2 main groups, the first group do not connect to external power sources. The system converts solar energy into the electrical energy and store into the battery, as shown in fig 1. Then the system supplies electrical power from the battery and solar system into the system. This system relies on batteries to accumulate energy during electricity generation. It is possible to use continuously during the night and in the low sunlight for a long time. The second group produces electricity and connects to the existing electrical systems which reduce the use of electrical energy in buildings from electricity production by the solar system as shown in figure 2. This system does not need a backup battery because all electrical energy can be supply to the grid and there is no need to collect energy in the battery.

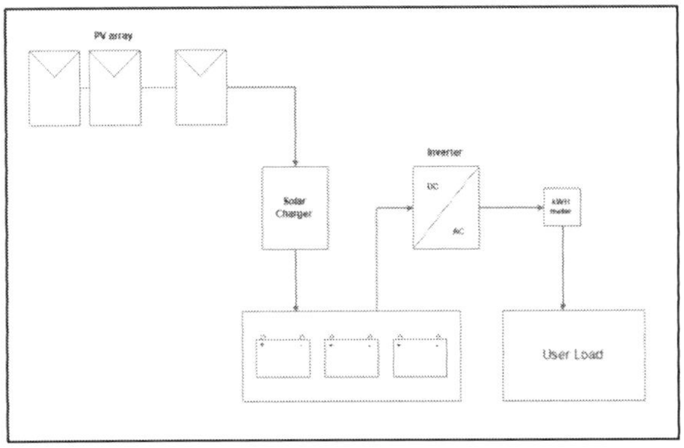


**Figure 1.** Off-Grid or Stand Along Solar System

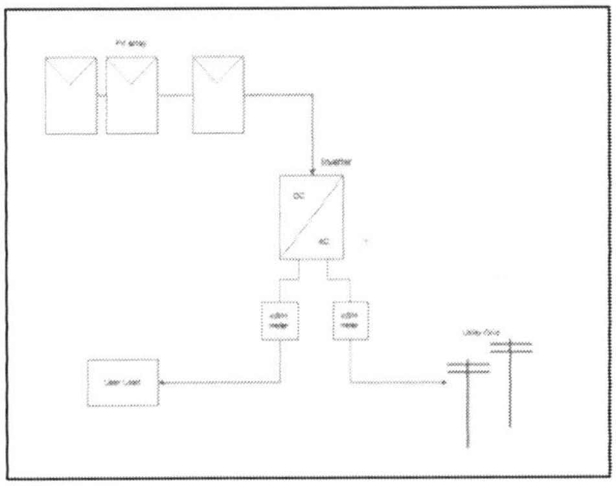

**Figure 2.** On-Grid Solar System.

For the solar pumping system, the advantages of both systems are applied together, as shown in figure 3. the system does not need batteries and only works when there is sufficient light. The energy will be stored in the form of a water head by pumping it to a higher place instead, figure 4 Therefore, it does not need batteries anymore. most systems are based on industrial electric motor control systems, so it can control and adjust power independently, the same as the industrial electric motors or water pumps. Some systems also work with AC power supplies to enhance system performance in the insufficient sunlight situation or continuously work in the night.

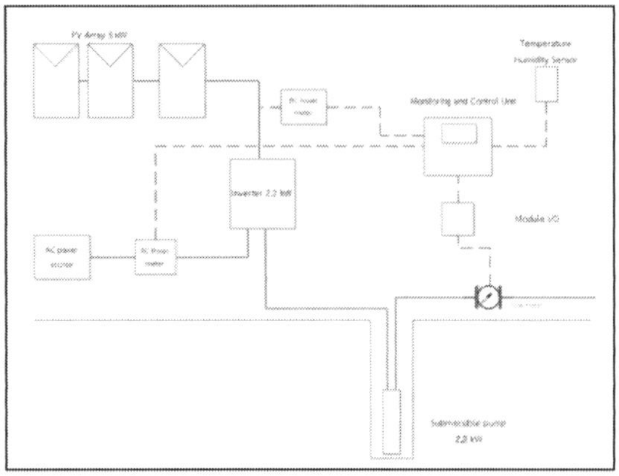

**Figure 3.** Solar pumping system

**Figure 4.** Water tank in Solar pump site

The remote monitoring system consists of a set of measuring equipment. Various signals will be collected and transformed into a form that can be easily processed. (Fig 5.) For the solar pumping system, the key variable is the measurement of solar power which be converted to electrical power in a period of time. The second thing is the output of the system that is the water pump power in the form of electrical energy. We can also measure the volume of water which reflects the overall efficiency values. The system may equip with additional sensors to increase the ability to monitor and control the operation of the system.

## 3. Method and Experiment

The solar pumping system with a remote monitoring system in this site consists of a 2.2 kW submersible pump with a 2.2 kW inverter that supports both AC and solar power. It works with a 5-kilowatts of solar panel using 250 watts' panel in 10 panels series per string and 2 string in parallel. The system setup the inverter in MPPT (Maximum Power Point Tracking) mode for the highest solar efficiency [3]. There are 3 level sensors in the water tank to control the operation of the pump and the flowmeter to measure the rate of the total water flow. These sensors have been required to calculate the system efficiency and also used to measure the parameter which is required in the remote monitoring system because the pump controller does not have any communication function support [4][5].

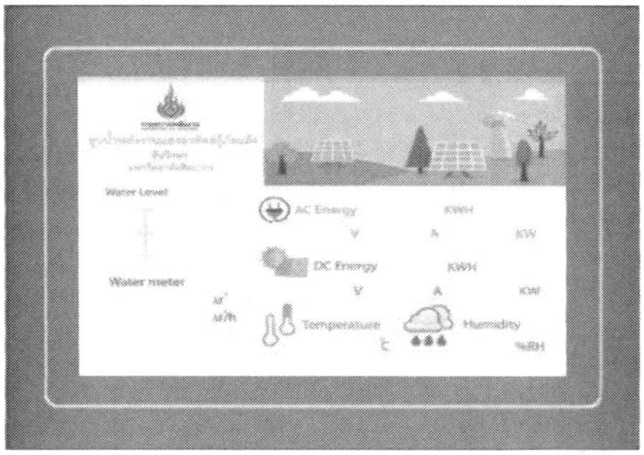

**Figure 5.** Monitoring system display example

In the monitoring system module, the system has been designed to measure several electrical quantities and measure the amount of water. By separating into independent modules in design, the measuring parameter will be processed in the different modules and communicate with each other over the RS485 bus. All modules can be controlled by the main processor in the main modules [6]. With this design, we can separate the system into several modules, 1. Solar power measurement module, 2. AC Power measurement module, 3. water level and flow module, 4. Weather monitoring module, and 5. Main control equipment and display. (Fig. 6) In the operation, the main control device is responsible for communication to extract data from various measuring modules and display it on the screen. Including data processing Then saved to the memory installed on the machine and also prepare data for communicating to external devices via additional communication modules. Other module measured parameters related to their modules and calculate basic data as needed.

The solar power measurement module measures the electrical energy from the solar panel. Since the system has 10 PV in series, the system voltage is between 300-400 Volts. The signal conditioning module has been required to use. By using voltage divider to lower voltage range and shunt-resistor integrate into the system, we can measure voltage, current, and power of the system, as shown in figure 7. Furthermore, when using the time integration, we can record the cumulative electrical energy that has been developed from the solar panel. All parameters will be calculated and stored in the internal register, to wait for the main control module to gather data from itself [7].

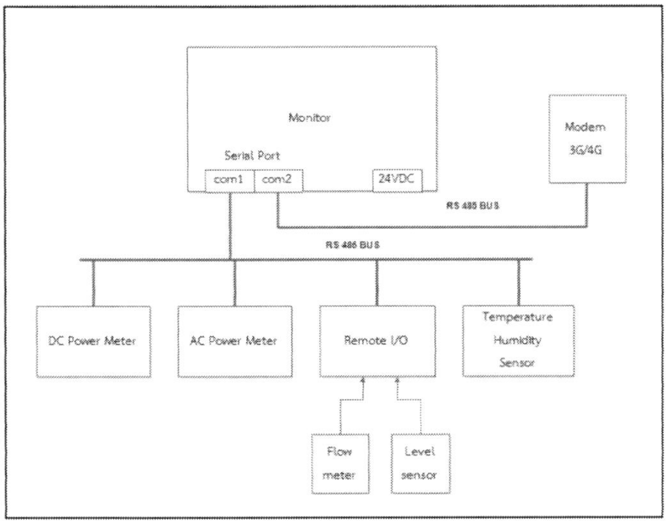


**Figure 6.** System Diagram

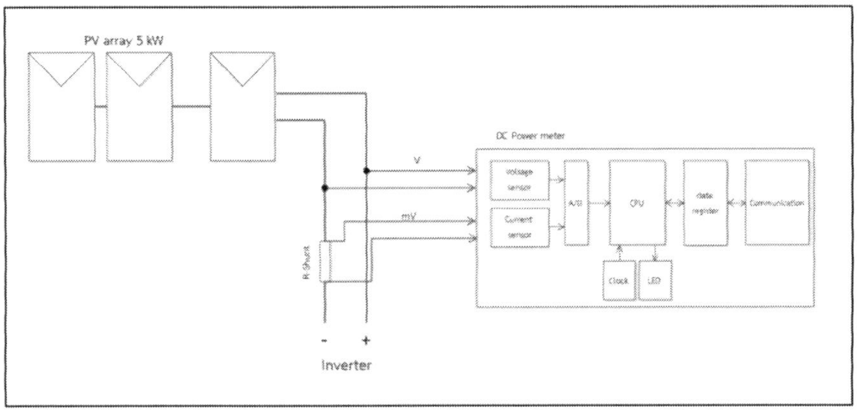


**Figure 7.** Solar Power measurement

The AC Power measurement modules use the ready-made instruments which support one channel single-phase AC power measurement which includes voltage, current, power, total energy. The meter has integrated with a signal conditioning circuit and also integrates the current transformer. All parameters stored in the internal register and ready for the main controller. In the water level and flow measurement, the system uses the electrical conductivity between the two electrode bars to check the water level. It also uses a pulse-rate flow meter installed in the water pipe which can send a pulse signal to the measuring device. (Fig.8) The device can read the status and using the counting circuit to calculate the water volume and calculate the flow rate. In addition to the other sensors, the system measures the temperature and humidity for reference, which makes it able to estimate the environment that is sunny, cloudy or rainy. Users can check with these data when needed if they founded the electric generation or water flow rate seem to be abnormal.

**Figure 8.** Flow meter with pulse output

The main controller unit integrates with a 7-inch ready-made display. The device supports RS485 connection, so it can connect to all other devices in the system. The software is developed and integrated into the device so that the system can display various information in real-time. It can calculate additional parameters other than the measurement data. All measurements and calculated data will be recorded to the flash memory and ready to be used in the future. The system also prepares the data format available to transmit to an external communication module that can send out the data to the remote site or control center.

```
{api_key,name,temp,hum,acpower,acV,acA,ackWH,dcpower,dcV,dcA,dckWH,flowTotal,flowRate,Level,timestamp}

{RFVdyrs17g,Pumpwater01,36.3,49.70,17,231.20,0.14,0.71,274.9,2.57,46.3,11.95,3.6,10,7/17/201812:00:41 PM}
```

**Figure 9.** Data Protocol and sample

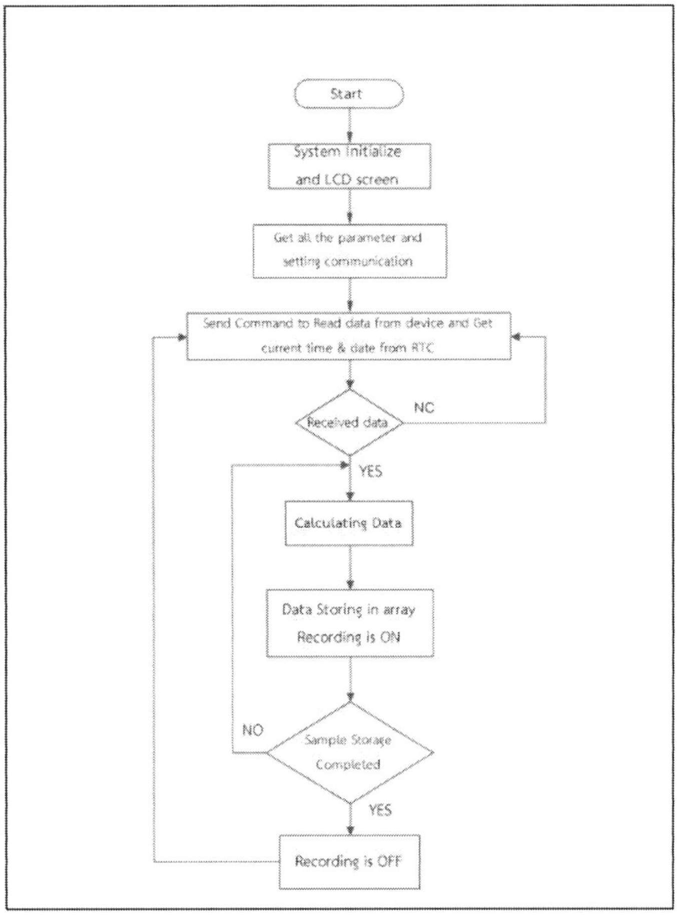


**Figure 10**. Flowchart

## 4. Results

From the monitoring system we designed, when all the equipment has been integrated, the system can operate by using the external power supply which is supplied separately to the other parts of the system. The system can work continuously even in the night time. (Fig 11, Fig 12) The Screen display all information from the system include the system status. Users can also monitor and control all functions directly from the device.

The data received to the system will be shown on screen. It shows only the current data and accumulated data. To display historical data or display in a graph, users have to use another software to process the recorded data to create a graph as well. (Fig 13) The system can create weather data, Solar energy, Electrical power production data, And the water data. [8]

Furthermore, When the communication equipment has been integrated into the system The system can transfer all data to the remote server(offline) or to the cloud. (Fig 14) The communication system can be used for both general long-distance wireless communication systems or through the network of various communication systems as well.

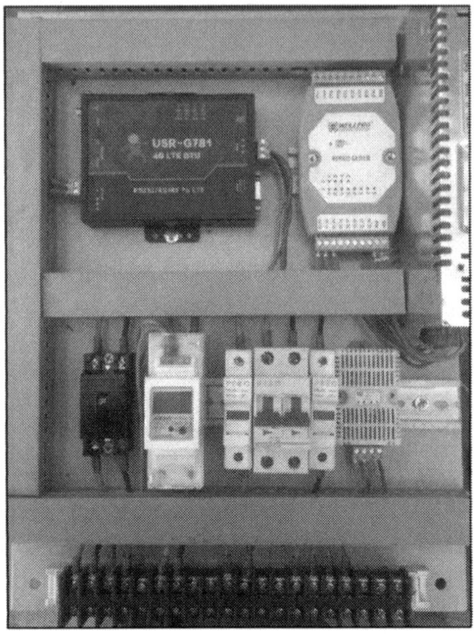


**Figure 11.** Modules in enclosure

**Figure 12.** System is running

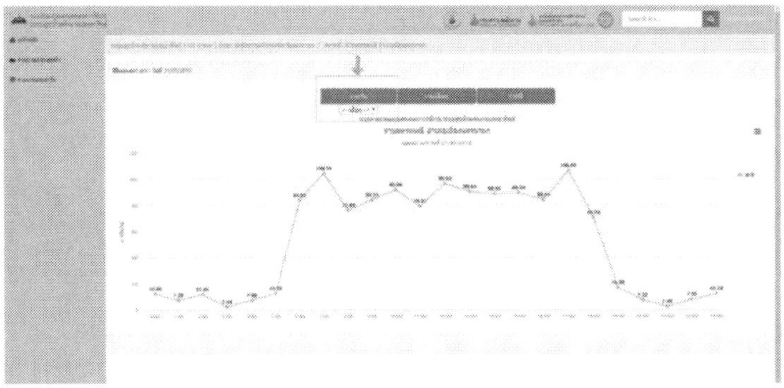

**Figure 13.** System can create historical data

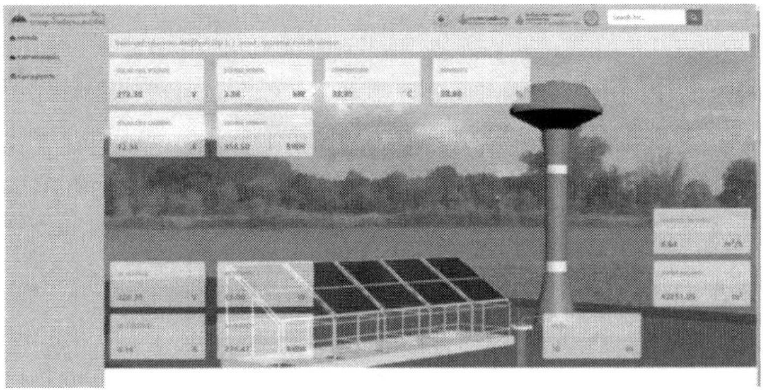

**Figure 14.** Remote monitoring on the cloud

## 5. Discussion and Conclusion

After installing the monitoring system, the user can easily check the system status and parameter and not necessary to measure data by themselves. The system can monitor all modules and create an alarm if the system is not working properly. When integrating with communication devices, the system can report information to the user directly. Users can plan service/maintenance while they are in office.

The current work, which is under development, is the construction of the data infrastructure including data and system security systems. Due to the device is far away, the system security must be the first priority. [9] Moreover, when there are more install sites, the number of data increases. The system infrastructure and the management system will be the next step. Also, access control should be the thing that is unavoidable.

## References

[1] S.S. Chandel, M. Nagaraju Naik and Rahul Chandel, "Review of Solar photovoltaic water pumping system technology for irrigation and community drinking water supplies", Renewable and Sustainable Energy Reviews, vol. 49, pp. 1084-1099, 2015.

[2] Mansur Aliyu et al., "A review of solar-powered water pumping systems", Renewable and Sustainable Energy Reviews, vol. 87, pp. 61-76, 2018.

[3] Abdelmalek Mokeddem, et al., "Performance of a directly-coupled PV water pumping system", Energy conversion and management, vol. 52, pp. 3089-3095, 2011.

[4] M. Benghanem, K.O. Daffallah and A. Almonhammedi, "Estimation of dialy flow rate of photovoltaic water pumping systems using solar radiation data", Results in Physics, vol. 8, pp. 949-954, 2018.

[5] Arunendra K. Tiwari, Vilas R. Kalamkar, "Performance investigation of solar water pumping system using helical pump under the outdoor condition of Nagpur, India", Renewable Energy vol. 97, pp. 737-745, 2016.

[6] Yasin Kabalci et al., "Design and implementation of a solar plant and irrigation system with remote monitoring and remote control infrastructures", Solar Energy, vol. 139, pp.506-517, 2016.

[7] Soumia Benlabna, Nallapaneni Manoj Kumar and Ali Tahri, "Data acquisition system: On the solar photovoltaic module and weather parameters monitoring", Procedia Computer Science, vol. 132, pp. 873-879, 2018.

[8] Amit Kumar Rohit et al., "Virtual lab based real-time data acquisition, measurement and monitoring platform for solar photovoltaic module", Resource-Efficient Technologies, vol. 3, pp. 446-451, 2017.

[9] Argyrios Samourkasidis, Evangelia Papoutsoglou and Ioannis N. Athanasiadis, "A template framework for environmental timeseries data acquisition", Environmental Modelling & Software, vol. 117, pp. 237-249, 2019.

# Effect of Catalyst-Assisted Hydrothermal/Organosolv Process for Fractionation of Parawood

**Araya Srisuwansakda[1], Penjit Srinophakun[2], Verawat Champreda[3], Nopparat Suriyachai[3]***

[1]Interdisciplinary of Sustainable Energy and Resources Engineering, Faculty of Graduate School, Kasetsart University, Bangkok 10900, Thailand

[2]Department of Chemical Engineering, Faculty of Engineering, Kasetsart University, Bangkok 10900, Thailand

[3]National Center for Genetic Engineering and Biotechnology (BIOTEC), National Science and Technology Development Agency (NSTDA), Khlong Luang, Pathum Thani 12120, Thailand

*E-mail: nopparat.sur@ncr.nstda.co.th

**Abstract.** Fractionation of lignocellulosic components is a pre-requisite in biorefineries in order to separate the main compositions of raw material for further maximum utilization to biofuels, chemicals, and materials. Generally, a single step process is often insufficient to assure full availability of polymeric sugars from raw material. Therefore, the purpose of this study is to investigate two-step fractionation comprising hydrothermal and organosolv processes for the efficient separation of lignocellulosic biomass with a promising recovered yield in terms of quality and quantity. In this study, the effects of acids and bases concentration were studied in the two-step fractionation process, since they play different important roles in separation process. It was observed that the optimal condition of hydrothermal process was performed at 180°C for 60 min in the presence of 3.3%(w/w) of $H_2SO_4$ followed by organosolv process at 200°C for 60 min with 5%(w/w) of NaOH, resulting in the efficiency of hemicellulose and lignin removal of 87.96% and 46.39%, respectively. In addition, the high recovery of hemicellulose as pentose sugars were observed. The remaining cellulose was obtained with 80.22%. These results demonstrated the successive hydrothermal and organosolv process for fractionation of Parawood.

## 1. Introduction

Currently, energy needs are the driving force for exploring new energy resources. Biorefinery is a facility that integrates biomass conversion processes and equipment to produce fuels, power, materials and chemicals from agricultural residues. A biorefinery takes advantage on valorization of various components from biomass feedstock for value-added products. Agricultural biomass is not only considered as an interesting raw material for production of alternative energy but also a carbon-neutral raw material, therefore it does not cause global warming.

Lignocellulosic biomass is one of the most abundant carbon sources, and the most often used raw material for bioenergy production, since it is highly available and renewable. The use of lignocellulosic biomass, which includes energy crops, agro-residues, and forest waste from industry, have gained worldwide attention in recent years, as it is cheap and available [1]. However, Lignocellulosic materials have a complex structure consisting of cellulose, hemicellulose, and lignin, resulting in a high recalcitrant for degradation and resist on enzymatic digestion [2][3]. An efficient fractionation approach is thus required for separation of these biopolymers for subsequent valorization to different products.

Parawood is a light-colored medium-density tropical hardwood obtained from the Para rubber tree (*Hevea brasiliensis*). Parawood is one of the most important economic trees in Thailand due to its availability as one of the main plantation crops in Asia [4]. The constituent biopolymers can be separated for further conversion to fuels, chemicals, and materials in biorefineries.

Fractionation of lignocellulosic components represent an initial step for value-added utilization of the individual biopolymers in integrated biorefineries. The fractionation process entails disruption of lignocellulose biopolymeric matrix and subsequent separation of its components into isolated fractions according to selectivity of the reaction media and process conditions [5]. Several techniques have been developed for biomass fractionation including various chemical (acid, alkaline, and oxidation) and thermochemical methods (steam explosion, autohydrolysis, and organosolv) or their combinations in multi-stage processes [6].

A two-step process comprising hydrothermal and organosolv processes are used for fractionation of lignocellulosic components from Parawood. Hydrothermal process using liquid hot water (LHW) is a physico-chemical pretreatment method that are used for specific separation of hemicellulose content and has been widely applied to enhance cellulose digestibility of various types of agricultural residues [7]. Hydrothermal process has several advantages over other leading pretreatment methods (e.g. diluted acid, alkali, and steam explosion) in that it has no chemical requirement and produces less inhibitory byproducts such as HMF and furfural [8]. The superheated liquid water auto-ionizes into hydronium ions, which act as a promoter for cleavage of ester bonds of acetyl side chains of hemicelluloses resulting in the formation of acetic acid [9]. The in situ generated acid then autocatalyzes hydrolysis of the hemicellulose and alteration of lignin structure, leading to increased enzyme accessibility to the cellulose fibers. Apart from Hydrothermal process, several homogeneous acids have been used as external catalysts in hydrothermal process of biomass. Dilute sulfuric are normally used to enhance hemicellulose solubilization from various agricultural residues [10-12]. However, the lignin that cannot be removed via hydrothermal process is partly rearranged on the surface of the lignocellulosic biomass exhibiting an inhibitory effect on downstream enzymatic hydrolysis [13].

Organosolv process has attracted a research interest, as it proposes an effective method to remove lignin with the use of organic solvents. The benefits of organosolv process is the isolation of high-quality lignin and high-purity lignin-free cellulose [14,15]. The lignin recovered is sulfur free, while the organic solvents used (ethanol) can easily be recovered which is a significant advantage for small scale biorefinery plants [16]. The addition of an organic solvent allows for better mass transfer and the dissolution of lignin [17], reducing its recondensation on the external surface area of the pulp [18]. Typically, Organosolv process are catalytically assisted with bases, such as NaOH. Sodium hydroxide can enhance lignin removal and lead to biomass swelling by solvation and saponification with minimal hydrolysis of the polysaccharide fractions [10-12].

However, using single step process, it is often difficult to achieve the appropriate biomass decomposition, thus the above-mentioned methods are combined to increase efficiency the separation of lignocellulosic biomass with a promising recovered yield in terms of quality and quantity [19,20]. For these reasons, there is significant research interest in investigating the best fractionation method for lignocellulosic materials. Sequential hydrothermal process for solubilization of hemicellulose and organosolv process for the removal of lignin and the production of high purity pulps have been published [15,21,22]. These studies investigated the effect of the different pretreatment techniques on the physical and chemical properties of the pulps, together with the saccharification effect of the residual solid.

In order to explore the use of external catalyst concentration on fractionation process, this study reports the use of $H_2SO_4$ as an acid catalyst for hydrothermal process and the use of NaOH as a base catalyst for organosolv process of Parawood. The effect of the catalyst concentration for improving hemicellulose/lignin solubilization was tested by measuring sugar recovery and yield in the liquid fraction, saccharified solid faction and lignin yield. An efficient two-step process will be optimized on reaction conditions based on process parameters to achieve high fractionation efficiency with a high quality of derived products.

## 2. Materials and methods

### *2.1. Materials*

Parawood was provided from Asia Biomass Co., Ltd., Thailand. The biomass was physically reduced using a cutting mill and sieved to the particle size of 0.5-1 mm, and then dried at 70°C for 24 hours. The main composition of Parawood contained 32.06 wt% cellulose, 23.76 wt% hemicellulose and 31.28 wt% lignin with minor contents of 5.10 wt% ash and 7.80 wt% extractive, according to the standard NREL method [23].

### *2.2. Methods*

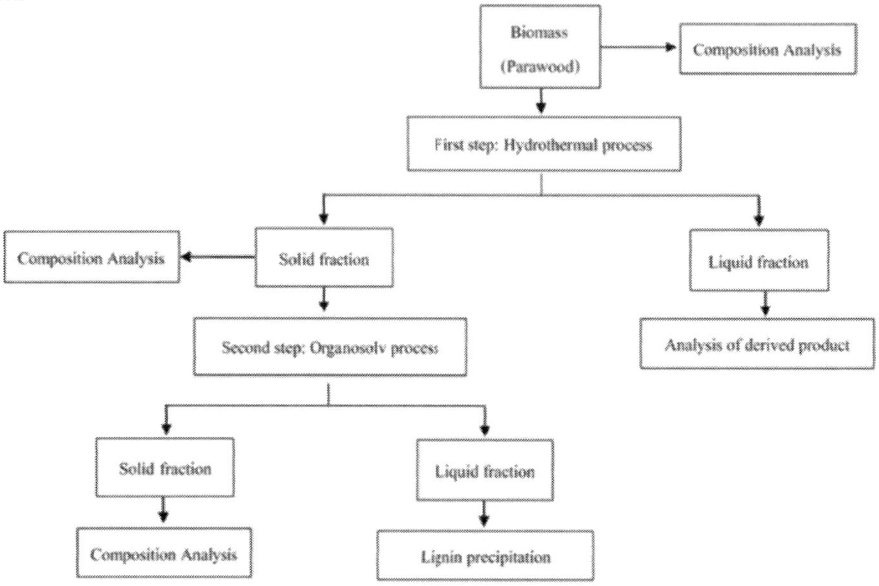


**Figure 1**. Process flow chart of the two-step fractionation.

*2.2.1. Hydrothermal process.* The process was performed in a 600 ml stainless-steel high-pressure reactor (Parr Instrument Company). Biomass loading was 10% with a working volume of 100 ml. The first stage was maintained at a constant temperature of 180°C for 60 min in the absence and presence of sulfuric acid ($H_2SO_4$) by varying concentrations (based on %w/w of biomass loading) and the initial pressure was adjusted at 20 bars by nitrogen. After LHW process, the solid fraction was separated by filtration by filter paper using a Buchner funnel and washing with deionized water until pH was neutral. The obtained solid was then dried at 70°C for further fractionation in the second stage. Chemical compositions of the pretreated solids were analyzed using the standard NREL method. The liquid fraction was collected for analysis of sugar and inhibitory by-products by High Performance Liquid Chromatography (HPLC).

*2.2.2. Organosolv process.* The remaining solid fraction from hydrothermal process (180°C, 3.3% $H_2SO_4$, 60min) was used as material in organosolv process. The organic solvent was an aqueous solution of ethanol (95% v/v). The parameters of the second stage were maintained at 200°C for 60 min. The process was performed at different sodium hydroxide (NaOH) concentration (based on %w/w of biomass loading). After organosolv process, the solid fraction was separated from the liquid fraction by filtration. In liquid phase, the isolated lignin was separated by precipitation technique. All obtained fractions were collected for further analysis.

*2.2.3. Lignin precipitation.* According to the practical procedure for precipitation of lignin, DI water was added with 3 time based on the initial volume of liquid fraction, and then the pH was adjusted equal to 2 by sulfuric acid. The supernatant was collected at room temperature for 24 h. After precipitation was completed, lignin was separated by filtration by filter paper using a Buchner funnel and dried for further utilization.

*2.2.4. Analytical methods.* The composition of raw material and pretreated biomass, including cellulose, hemicellulose and lignin was determined according to National Renewable Energy Laboratory (NREL) as standard protocols. The profiles of sugars and contaminated products (e.g. lactic acid, formic acid, acetic acid, levulinic acid, HMF and furfural) in the liquid fractions were analyzed by high performance liquid chromatograph using an Aminex HPX-87H column (Bio-Rad, Hercules, CA, USA) operating at 65°C with 5 mM $H_2SO_4$ as the mobile phase with the flow rate of 0.5 mL/min. The percentage of biomass composition (cellulose, hemicellulose and lignin) was determined based on the remaining contents in the remaining solid compared to their respective contents in the raw material. The calculations of remaining cellulose, hemicellulose and lignin removal were described according to the following equations, respectively.

$$\%\ \text{Remaining cellulose} = \frac{C_F}{C_I} \times 100 \tag{1}$$

$$\%\ \text{Hemicellulose removal} = \frac{H_I - H_F}{H_I} \times 100 \tag{2}$$

$$\%\ \text{Lignin removal} = \frac{L_I - L_F}{L_I} \times 100 \tag{3}$$

Where, Remaining cellulose is the percentage of remaining cellulose after fractionation process compared to initial cellulose content (%); $C_F$ is the total amount of remaining cellulose content in solid fraction after the process (g); $C_I$ is the total amount of cellulose in the raw material (g) and hemicellulose removal efficiency is the percentage of hemicellulose removal after fractionation process compared to initial hemicellulose content (%); $H_I$ is the total amount of hemicellulose in the raw material (g); $H_F$ is the total amount of remaining hemicellulose content in solid fraction after the process (g) and lignin removal efficiency is the percentage of lignin removal after fractionation process compared to initial lignin content (%); $L_I$ is the total amount of lignin in the raw material (g); $L_F$ is the total amount of remaining lignin content in solid fraction after the process (g).

The $C_5$ sugars recovery and $C_5$ sugars and lignin yield were calculated as the percentage of the obtained product based on content in the raw material (dry weight basis).

$$\text{Product recovery} = \frac{\text{Grams of the product obtained}}{\text{Grams of the product in raw material} \times \text{Percentage of the product removal}} \times 100 \tag{4}$$

$$\text{Product yield} = \frac{\text{Grams of the product obtained}}{\text{Grams of the product in raw material}} \times 100 \tag{5}$$

## 3. Results and discussion

### *3.1. Hydrothermal process*

The effect of acid catalyst concentration in hydrothermal process for fractionation of parawood was studied by varying $H_2SO_4$ concentration (without acid catalyst, 0.6%w/w, 3.3%w/w, and 12.5%w/w). The operating condition was kept constant at 180°C for 60 min. The compositional changes in solids fraction after hydrothermal process are summarized in Table 1. Acidic condition plays an important role in separation of hemicellulose content. It was found that standard hydrothermal process showed particularly removal of hemicellulose fraction. According to Figure 2, the presence of $H_2SO_4$-catalyzed led to higher removal of hemicellulose than non-catalytic process. The highest hemicellulose removal was obtained at 12.5% $H_2SO_4$ with 94.95%. This indicates the increasing of acid concentration led to an increasing trend on solubilization of hemicellulose. For further purpose of this step, it was proven that the hydrothermal process demonstrated a less effect on the fraction of lignin and cellulose. However, the severe condition in fractionation could affect the degradation of derived sugars as the results of products in liquid fraction.

**Table 1.** The composition of remaining solid after hydrothermal process at 180°C for 60 min under different acid concentration

| Run | Process condition: $H_2SO_4$ concentration (% w/w) | Cellulose (%)[a] | Hemicellulose (%)[a] | Lignin (%)[a] | Ash (%)[a] |
|---|---|---|---|---|---|
| 1 | without acid catalyst | 31.28 | 6.19 | 28.02 | 1.00 |
| 2 | 0.6 | 31.46 | 5.36 | 25.48 | 1.93 |
| 3 | 3.3 | 31.47 | 3.45 | 25.27 | 1.83 |
| 4 | 12.5 | 29.23 | 1.20 | 23.33 | 1.50 |

[a] % Dry weight basis

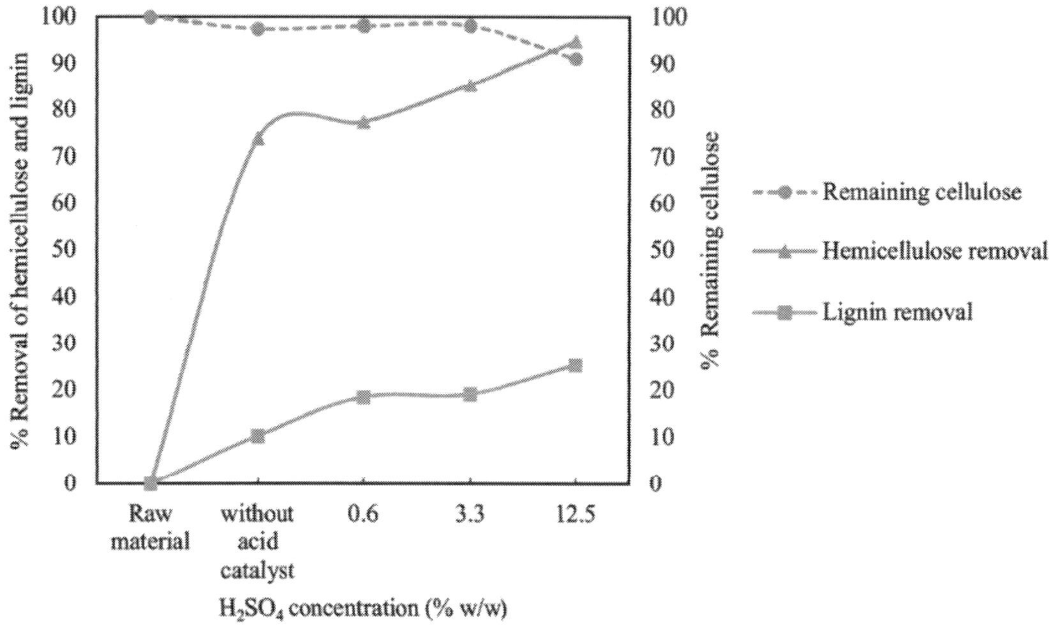


**Figure 2.** The efficiency of hydrothermal process on the response of hemicellulose removal, lignin removal and remaining cellulose under different acid concentration at 180°C for 60 min.

Tables 2 and 3 show the product profiles in liquid fraction after hydrothermal process at 180°C for 60 min under acidic condition. The product was analyzed in form of derived sugars and contaminated products. Hydrothermal in the presence of acid catalyst led to extensive solubilization and hydrolysis of the hemicelluloses into sugars and by-products. Increasing of $H_2SO_4$ concentration led to increase $C_5$ sugars concentration. This result demonstrated that $C_5$ recovery was substantial declined due to the degradation of sugars into by-products. In contrast, addition of the acid concentration from 0.6% to 3.3% in hydrothermal process led to increased $C_5$ sugars yield compared to standard condition. Further increasing the acid concentration above 12.5% led to decreased $C_5$ yield due to the marked increased formation of side-products. Generally, furfural is released from degradation of pentose sugars, while hydroxymethylfurfural (HMF) is generated from degradation of hexose sugars. Particularly, the high amount of HMF and furfural were found under the harsh condition. Thus, the suitable condition of hydrothermal process was at 180°C for 60 min with the presence of 3.3 %w/w $H_2SO_4$, resulting in almost complete of hemicellulose removal with the highest yield of $C_5$ sugars.

**Table 2.** The sugar profiles in liquid fraction after hydrothermal process at 180°C for 60 min under different acid concentration

| Run | $H_2SO_4$ concentration (% w/w) | Concentration (g/L) | | | | | | Total $C_5$ Sugar (g) | $C_5$ recovery (%) | $C_5$ yield (%) |
|---|---|---|---|---|---|---|---|---|---|---|
| | | $C_6$ sugars | | $C_5$ sugars | | | | | | |
| | | Glucose | | Xylose | | Arabinose | | | | |
| | | Mono mer | Oligo mer | Mono mer | Oligo mer | Mono mer | Oligo mer | | | |
| 1 | without acid catalyst | 0.24 | 1.61 | 2.29 | 8.42 | 0.46 | 0.52 | 1.17 | 58.54 | 43.29 |
| 2 | 0.6 | 0.62 | 1.09 | 3.93 | 7.36 | 0.55 | 0.00 | 1.18 | 56.59 | 43.83 |
| 3 | 3.3 | 1.71 | 0.28 | 9.04 | 0.93 | 2.44 | 0.00 | 1.24 | 53.78 | 45.97 |
| 4 | 12.5 | 2.05 | 0.16 | 9.23 | 0.20 | 1.18 | 0.00 | 1.06 | 41.39 | 39.30 |

**Table 3.** The contaminated compounds in liquid fraction after hydrothermal process at 180°C for 60 min under different acid concentration (con.)

| Run | $H_2SO_4$ concentration (% w/w) | Concentration (g/L) | | | | | |
|---|---|---|---|---|---|---|---|
| | | Lactic acid | Formic acid | Acetic acid | Levulinic acid | HMF | FF |
| 1 | without acid catalyst | 1.13 | 1.75 | 8.98 | 0.00 | 0.13 | 0.78 |
| 2 | 0.6 | 1.46 | 1.54 | 7.24 | 1.33 | 0.26 | 0.98 |
| 3 | 3.3 | 3.28 | 2.59 | 9.34 | 0.00 | 0.79 | 1.74 |
| 4 | 12.5 | 4.88 | 3.01 | 12.28 | 0.01 | 1.11 | 2.19 |

### *3.2. Organosolv process*

The second step fractionation experiments of Parawood were continually conducted. The remaining solid fraction from hydrothermal process (180°C, for 60 min with 3.3%w/w $H_2SO_4$) was used as material in organosolv process. The process was conducted by varying NaOH concentration (without basic catalyst, 5%w/w, 10%w/w). The operating conditions were kept constant at 200°C for 60 min. The compositional changes in solids fraction after two step process are summarized in Table 4. Since basic condition plays an important role in separation of lignin fraction. It was found that organosolv process in the presence of base catalyst could efficiently remove lignin fraction. According to Figure 3, the presence of NaOH-catalyzed led to higher removal of lignin than non-catalytic process. The highest lignin removal was obtained at 10% NaOH with 56.91%, which corresponded to the highest lignin yield (23.23%). However, a high NaOH concentration (10%) showed marked decrease of cellulose fraction due to the degradation of cellulose into contaminated products. According to the purpose of fractionation process, the high recovered yield of three fractions (cellulose, hemicellulose, and lignin) are required. This indicates that the optimal condition would be considered at the middle point in order to obtain the high yield of all desired fraction. Thus, the suitable condition of organosolv process was at 200°C for 60 min with the presence of 5%w/w NaOH, resulted in 46.39% of lignin removal and 19.25% of lignin yield.

**Table 4.** The composition of remaining solid after sequential organosolv process at 200°C for 60 min under different base concentration

| Run | Process conditions NaOH concentration (% w/w) | Cellulose (%)[a] | Hemicellulose (%)[a] | Lignin (%)[a] | Ash (%)[a] | Lignin precipitation (g)[b] | Lignin yield (%) |
|---|---|---|---|---|---|---|---|
| 1 | without basic catalyst | 28.03 | 3.28 | 18.21 | 1.90 | 0.5271 | 16.85 |
| 2 | 5 | 25.72 | 2.86 | 16.77 | 1.50 | 0.6020 | 19.25 |
| 3 | 10 | 21.52 | 2.65 | 13.48 | 2.40 | 0.7265 | 23.23 |

[a] % Dry weight basis
[b] Dry weight after precipitation process

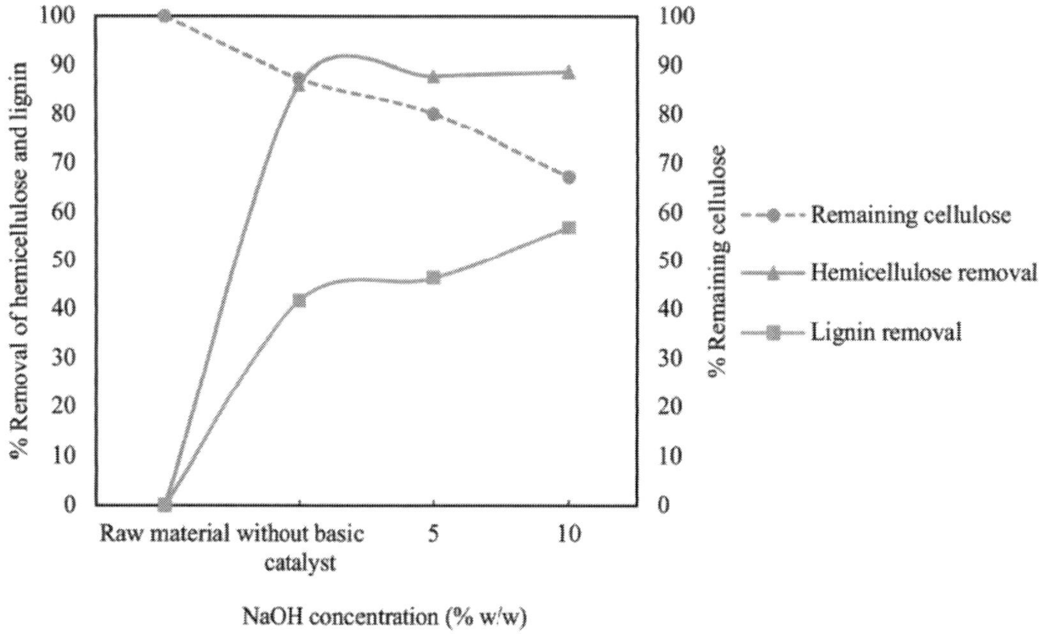


**Figure 3.** The efficiency of sequential organosolv process on the response of hemicellulose removal, lignin removal and remaining cellulose under different base concentration at 200°C for 60 min

## 4. Conclusion

An integrated system of two step fractionation was found to have a significant impact on specific separation of lignocellulosic biomass. The specific separation of both hemicellulose and lignin in two step process was optimized. The suitable condition was observed as follows; first step: the presence of 3.3% $H_2SO_4$ at 180°C for 60 min, second step: the presence of 5% NaOH at 200°C for 60 min. The work demonstrated a sequential process for the efficient separation of Parawood in integrated biorefinery.

## Acknowledgements

The author would like to thank Thailand Advanced Institute of Science and Technology and Tokyo Institute of Technology (TAIST-Tokyo Tech) for funding, Kasetsart University for suggestions. This work was supported by Integrative Biorefinery Laboratory (IBL), National Science and Technology Development Agency (NSTDA).

## References

[1] Mood, S.H., Golfeshan, A.H., Tabatabaei, M., Jouzani, G.S., Najafi, G.H., Gholami, M., Ardjmand, M. *Lignocellulosic biomass to bioethanol, a comprehensive review with a focus on pretreatment. Renew. Sust. Energy Rev.* 2013, **27**, 77–93.

[2] Liu H., Zhang Y.X., Hou T., Chen X., Gao C., Han L., Xiao W. *Mechanical deconstruction of corn stover as an entry process to facilitate the microwave-assisted production of ethyl levulinate. Fuel Process. Technol.* 2018, **174**:53–60. doi: 10.1016/j.fuproc.2018.02.011.

[3] Jönsson L.J., Martín C. *Pretreatment of lignocellulose: Formation of inhibitory by-products and strategies for minimizing their effects. Bioresour. Technol.* 2016, **199**:103–112. doi: 10.1016/j.biortech.2015.10.009.

[4] MAF Mazlan, Y Uemura, N Osman, S Yusup, *Fast pyrolysis of hardwood residues using a fixed bed drop-type pyrolyzer*, Ene Conv Manag **98** (2015) 208–214.

[5] Huang HJ, Ramaswamy S, Tschirner UW, Ramaraob BV. *A review of separation technologies in current and future biorefineries.* Sep Purif Technol 2008, **62**:1-21.

[6] Hu, F., Ragauskas, A., 2012. *Pretreatment and lignocellulosic chemistry. Bioenerg. Res.* **5**, 1043–1066.

[7] A. Kruse, E. Dinjus, *Hot compressed water as reaction medium and reactant properties and synthesis reactions, J. Supercrit. Fluids* **39** (2007) 362–380

[8] T.N. Ang, G.C. Ngoh, A.S.M. Chua, *Comparative study of various pretreatment reagents on rice husk and structural changes assessment of the optimized pretreated rice husk, Bioresour. Technol.* **135** (2013) 116–119.

[9] C.C. Teo, S.N. Tan, J.W.H. Yong, C.S. Hew, E.S. Ong, *Pressurized hot water extraction (PHWE), J. Chromatogr. A* **1217** (2010) 2484–2494.

[10] B. Yang, C.E. Wyman, *Pretreatment: the key to unlocking low cost cellulosic ethanol, Biofuels Bioprod. Biorefin.* **2** (2007) 26–40.

[11] J. Xu, J.J. Cheng, *Pretreatment of switchgrass for sugar production with the combination of sodium hydroxide and lime, Bioresour. Technol.* **102** (2011) 3861–3868.

[12] X. Li, T.H. Kim, *Low-liquid pretreatment of corn stover with aqueous ammonia, Bioresour. Technol.* **102** (2011) 4779–4786

[13] Kristensen, J.B.; Thygesen, L.G.; Felby, C.; Jorgensen, H.; Elder, T. *Cell wall structural changes in wheat straw pretreated for bioethanol production. Biotechnol. Biofuels* 2008, **1**, 5.

[14] Zhang, K.; Pei, Z.; Wang, D. *Organic solvent pretreatment of lignocellulosic biomass for biofuels and biochemicals: A review. Bioresour. Technol.* 2016, **199**, 21–33.

[15] Huijgen, W.J.J.; Smit, A.T.; de Wild, P.J.; den Uil, H. *Fractionation of wheat straw by prehydrolysis, organosolv delignification and enzymatic hydrolysis for production of sugars and lignin. Bioresour. Technol.* 2012, **114**, 389–398

[16] Vallejos, M.E.; Zambon, M.D.; Area, M.C.; da Silva Curvelo, A.A. *Low liquid-solid ratio fractionation of sugarcane bagasse by hot water autohydrolysis and organosolv delignification. Ind. Crops Prod.* 2015, **65**, 349–353.

[17] Oliet, M.; García, J.; Rodríguez, F.; Gilarrranz, M. *Solvent effects in autocatalyzed alcohol–water pulping. Chem. Eng. J.* **2002**, *87*, 157–162.

[18] Oliet, M.; Rodríguez, F.; García, J.; Gilarranz, M.A. *The effect of autocatalyzed ethanol pulping on lignin characteristics. J. Wood Chem. Technol.* **2001**, *21*, 81–95.

[19] Chen, H.; Liu, J.; Chang, X.; Chen, D.; Xue, Y.; Liu, P.; Li, H.; Han, S. *A review on the pretreatment of lignocellulose for high-value chemicals. Fuel Process. Technol.* 2017, **160**, 196–206.

[20] Solarte-Toro, J.C.; Romero-García, J.M.; Martínez-Patiño, J.C.; Ruiz-Ramos, E.; Castro-Galiano, E.; Cardona-Alzate, C.A. *Acid pretreatment of lignocellulosic biomass for energy vectors production: A review focused on operational conditions and techno-economic assessment for bioethanol production. Renew. Sust. Energy Rev.* 2019, **107**, 587–601.

[21] Garrote, G.; Falqué, E.; Domínguez, H.; Parajó, J.C. *Autohydrolysis of agricultural residues: Study of reaction byproducts. Bioresour. Technol.* 2007, **98**, 1951–1957.

[22] Amendola, D.; De Faveri, D.M.; Egües, I.; Serrano, L.; Labidi, J.; Spigno, G. *Autohydrolysis and organosolv process for recovery of hemicelluloses, phenolic compounds and lignin from grape stalks. Bioresour. Technol.* 2012, **107**, 267–274.

[23] Sluiter, A., Hames, B., Ruiz, R., Scarlata, C., Sluiter, J., Templeton, D., & Crocker, D. (2004). *Laboratory analytical procedure (LAP).* Golden: National Renewable Energy Laboratory (NREL).

IOP Conf. Series: Materials Science and Engineering **965** (2020) 012008 doi:10.1088/1757-899X/965/1/012008

# Efficiency of protein extraction from palm kernel cake via different chemical extraction methods

**Namfon Chaipet[1], Marisa Raita[2], Wanwipa Siriwatwechakul[1*], Verawat Champreda[2]**

[1]Sirindhorn International Institute of Technology, Thammasat University - Rangsit Campus, 99 Moo 18, on Phahonyothin Road, Khlong Luang, Pathum Thani 12120, Thailand.

[2]National Center for Genetic Engineering and Biotechnology - Integrative Biorefinery Laboratory (IBL) 113 Thailand Science Park, Phahonyothin Road, Khlong Nueng, Khlong Luang, Pathum Thani 12120 Thailand

*Email: wanwipa@siit.tu.ac.th

**Abstract**. Protein isolation from defatted palm kernel cake (DPKC) was investigated under 3 conditions, including different chemical reagents (alkaline, saline, acidic), concentrations in the range of 0.01- 0.3M and reaction time (0.5-4 h). The reaction contained 1 g biomass in 50 mL of solutions, with mixing speed 250 rpm, and the extraction occurred at 35°C for 2 hrs. Extractable protein of 35.06%, 12.40% and 4.34% based on available protein in the raw material were obtained in aqueous of 0.25M NaOH, 0.01M NaCl and 0.1M HCl. Afterward, DPKC was isolated with 0.25M NaOH for 0.5, 1, 2, 3 and 4 hrs, achieving the protein concentration of 24.16%, 28.99%, 32.72%, 34.39%, and 34.84%, respectively. The protein isolation process resulted in 48.32- 60.57 %w/w and 7.27-34.60 %w/w removals of protein and ash, according to analysis of the solid residues. The oil extraction process mostly effected the composition of lignin, which can remove lignin up to 54.71% (w/w).

## 1. Introduction

Oil palm is a major economic crop in Malaysia, Indonesia, and Thailand and a main source for producing cooking oil and biodiesel. Generally, the palm oil industry provides in crude palm oil of 22% and palm residues of 78% based on a weight basis [1]. Several by-products including empty palm fruit bunch (EPFB), palm fiber (PF), palm kernel shell (PKS) and palm kernel cake (PKC) are main palm wastes in industrial palm oil processing. PKC is palm residue obtained from cracking and pressing of palm nuts. It can be used for producing animal feed supplement in ruminants, swine, and poultry due to its high protein and fiber content, approximated at 15-21% and 50% respectively [2, 3]. The variety of essential amino acids from PKC such as Arginine, Valine, Methionine, Cysteine, Lysine, and others have been reported in previous works [4], leading to the possibility of its use as food supplements in infants, preschool children, and adults [5]. Moreover, the hydrolyzed protein from PKC showed a high value of antioxidant activities [6].

Many researches have reported protein isolation from PKC using physical and chemical treatment such as alkaline, salt-alkaline, and enzymatic methods [6-8]. The chemical method such as alkali treatment leads to a simple and low cost process. However, the method provides lower efficiency, generates of unwanted products, and produces wastewater treatment. Arifin et al 2009 had reported the use of alkaline and salt treatment in protein extraction of PKC [9]. It was found that the isolated protein concentration from alkaline treatment based on 0.03 M NaOH at 35°C for 2 hr with the ratio of 30:1 liquid/solid showed in 38.51% (g/g) of biomass. While, protein concentration of saline treatment was 25.81% (g/g) contained in 0.2 M NaCl at 35°C for 2 hr with the ratio of 60:1 liquid/solid [9]. The increase of reaction temperature at 80 °C extracted PKC protein by 0.03 M NaOH for 4 h resulted in higher soluble protein with 56.6% as reported by Choi and Mohamad 2019. The obtained protein had potential for using in cosmetic products [10]. Besides, a study on the synergistic method of protein extraction using alkaline and enzymatic treatment found that papain gave the extractable protein with the highest antioxidant activity compared to the other protease enzymes (Alcalase, Chymotrypsin, Pepsin, Trypsin, Flavourzyme, and Bromelain). The extracted protein can be considered for its potential use in food industries [6]. In addition to alkaline, salt alkaline and enzyme methods, some studies investigated extraction of protein from plants using acidic solution. Eromosele et al 2007 studied the effects of pH in extracting protein from African yam bean. The protein was extracted with distilled water, adjusted to pH 1-12 with HCl and NaOH, and extracted at room temperature (ca. 30°C) for 2 hr [11]. The extraction resulted in protein concentration of 14.8% and 18.4 % at pH 2 and pH 10, respectively [11].

This work studied the efficiency of protein extraction from PKC using different chemical reagents (saline, acidic, and alkaline solution), reagent concentration, and reaction time. The efficiency of protein extraction was evaluated based the soluble protein by Bradford method. In addition, the effect of oil and protein extraction process was evaluated in form of % removal and % soluble liquid product recovery (by-product).

## 2. Materials and methods

### *2.1. Material*
Palm kernel cake (PKC) obtained from Suksomboon Palm Oil, Co. Ltd. (Chonburi, Thailand). The biomass was physically grinded by a cutting mill (Retsch ZM2000, Haan, Germany), and sieved (mesh #14-40). The milled PKC was dried at 70 °C for 3 days, and stored in plastic bag at room temperature before use. All analytical grade chemicals were provided from major chemical suppliers (Sigma-Aldrich, Fluka, and Merck).

### *2.2. Oil extraction*
The PKC oil extraction was performed in a Soxhlet extractor using hexane as a solvent with the ratio of 1:35 (w/v) for 6 h. After that, the mixture of hexane and oil was distilled by a rotary evaporator (USA N-1200BV-W). The isolated oil was dried at 100±3°C, until the weight remained unchanged following the AOAC method (2014) [12]. The percentage of fat removal was calculated according to **Equation 1**. Whereas, the defatted solid sample was dried at 70 °C for 3 days, and then weighed before further use.

$$\text{Fat removal (\%w/w)} = \frac{\text{Weight of extracted oil (g)}}{\text{Initial weight of oil in sample (g)}} \times 100 \qquad (1)$$

### *2.3. Protein extraction*
The defatted PKC was extracted by different aqueous media that were alkali (NaOH), salt (NaCl pH 9), and acid (HCl) under the ratio of solid and liquid (1:50 (w/v)) at 35°C for 2 h, and stirred at 250 rpm on multi-position hot plate stirrer (CHINCAN, SP200-2T, China). The reaction was performed by different aqueous media concentrations (0.01-0.3 M) and mixing times (0.5-4 h). The slurry from protein extraction was separated by filtered through Whatman paper No.4. The solid residue after protein extraction was washed by distilled water until neutral, and then dried 70°C for 3 days before analyzing the biomass composition [13]. The concentration of soluble protein obtained from liquid phase was

analyzed by Bradford assay using bovine serum albumin as a standard [14]. The protein concentration was determined according to **Equation 2**. The concentration of the extracted protein was based on percentage of protein recovered in the soluble phase compared to the available protein in the raw material.

$$\text{Protein concentration (\% g/g protein)} = \left(\frac{\text{Weight of total protein in supernatant (gram)}}{\text{Weight of total protein in PKC (gram)}}\right) \times 100 \quad (2)$$

*2.4. Biomass composition*

Chemical biomass compositions (cellulose, hemicelluloses, lignin and ash) were analysed by the National Renewable Energy Laboratory (NREL) method [8].The biomass samples (0.3 g) was added to 3 mL of 72% (v/v) $H_2SO_4$, and then vortexed for 2 h. Afterward, the sample mixture was added to 84 mL of distilled water, and sterilized at 121°C for 1 h. The liquid phase was adjusted to pH 7 with $CaCO_3$ powder before analysing the sugar profiles and acid soluble lignin. The sugar profile was analysed by high-performance liquid chromatography (SPD-M10A DAD, Shimadzu, Japan) equipped with a refractive index detector using an Aminex HPX-87H column (Bio-Rad, Hercules, CA, USA) operating at 65°C with 5 mM H2SO4 as the mobile phase at a flow rate of 0.5 ml/min. The acid soluble lignin was determined by UV-Visible spectrometer at 240 nm. Whereas, the solid residue was filtered through crucible por. 5, and then burnt in the furnace at 575°C for 3 h for analysing the acid insoluble lignin. The 0.3 g of biomass sample was burnt in the furnace at 575°C for 3 h for analysing ash content. The efficiency of the operated conditions was determined in term of the percentage of pulp yield, hemicellulose removal, and lignin removal according to the **Equation 3-5**, respectively.

$$\text{Pulp yield (\%w/w)} = \frac{\text{Final weight after pretreated PKC (g)}}{\text{Initial weight of raw PKC (g)}} \times 100 \quad (3)$$

$$\text{Hemicellulose removal (\%w/w)} = \left[\frac{H_I-(H_F \times \%\text{ pulp yield})}{H_I}\right] \times 100 \quad (4)$$

Where hemicellulose removal is the percentage of hemicelluloses removed after protein extraction compared with the initial hemicellulose content; $H_I$ is the percentage of initial hemicellulose content; $H_F$ is the percentage of remaining hemicellulose content in the final solid recovery.

$$\text{Lignin removal (\%w/w)} = \left[\frac{L_I-(L_F \times \%\text{ pulp yield})}{L_I}\right] \times 100 \quad (5)$$

Where lignin removal is the percentage of lignin removed after protein extraction compared with the initial lignin content; $L_I$ is the percentage of initial lignin content; $L_F$ is the percentage of remaining lignin content in the final solid recovery.

The oil content was extracted with light petroleum for 6 h in a Soxhlet extractor according to AOAC method. The extracted fat was weighed, and calculated as a percentage of the sample [12]. The protein content of PKC sample was determined according to Kjeldahl method (protein conversion factor of 6.25 following the method of AOAC (2000) [15]. The biomass sample (1.5 g) was mixed with 20 mL of 98% v/v sulfuric acid, and added to 14 g of $K_2SO_4$ and 1.6 g of $CuSO_4 \cdot 5H_2O$. The sample was digested at 420°C until the color changed, and then added 100 mL of distilled water in the mixture. After that, it was distilled, and 100 mL of NaOH was added during process. The $NH_3$ gas from distillation process was trapped in boric acid contained in 1 mg of methyl red, 1 mg of bromocresol green, and 1 mL of ethanol. Later, the boric acid solution was titrated with 0.1 N HCl to the endpoint. The protein content was determined according to **Equation 6-7.**

$$\text{Kjeldahl nitrogen, } \% = \frac{(V_s - V_b) \times M \times 14.01}{W \times 10 \times A} \tag{6}$$

$$\text{Protein content (\%w/w)} = \%\text{Kjeldahl nitrogen} \times F \times P \tag{7}$$

Where $V_S$ = volume (mL) of standardized acid used to titrate a test
$V_b$ = volume (mL) of standardized acid used to titrate reagent blank
M = molarity of standard HCl
14.01 = atomic weight of N
W = weight (g) of test portion or standard
10 = factor to convert mg/g to percent
A = the accuracy of procedure and equipment as 0.8244
F = factor to convert N to protein (PKC conversion factor of 6.25)
P = Pulp yield (%)

The protein content ($P_F$) of biomass can be used to calculate the efficiency of protein removal of each condition. The equation was expressed as:

$$\text{Protein removal (\%w/w)} = \left(\frac{P_I - P_F}{P_I}\right) \times 100 \tag{8}$$

Where protein removal is the percentage of protein removed after protein isolation process compared with the initial protein content; $P_I$ is the percentage of initial protein content; $P_F$ is the percentage of protein content in the final solid recovery.

## 3. Results and discussion

### *3.1. Raw PKC composition*

The chemical composition of raw PKC is shown in **Figure 1**. The result showed that the PKC sample contained 46.43% carbohydrate by weight, which were divided into 7.04% cellulose and 39.39% hemicellulose. The other major components included 20.03% lignin, 17.37 % crude protein, 12.65 % of oil/fat, and 4.02 % of ash on a dry weight basis. Normally, the protein content of PKC was in a range of 14.5-19.6% (w/w), carbohydrates 50-62% (w/w) and fat content depending on the extraction methods and the type of palm [4].

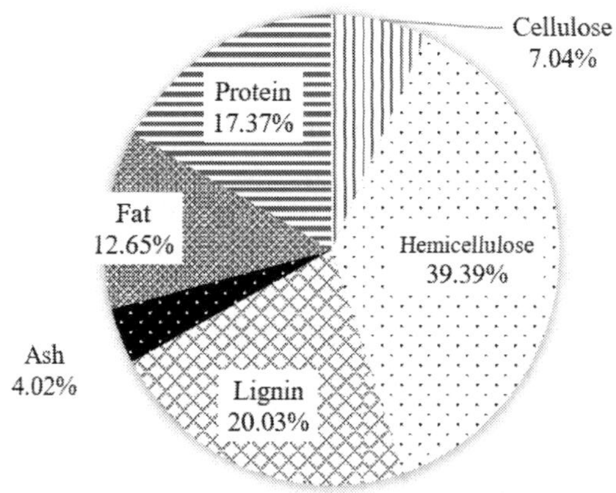


**Figure 1**. Proximate compositions of the raw material on a dry weight basis

### *3.2. Oil extraction*

The PKC oil was extracted by hexane, and the result showed that the pulp yield after defatted PKC was 81.71%, leading to 90.04 % (w/w) fat removal (**Table 2 and 3**). After oil extraction, the composition of defatted PKC composed of cellulose 6.99%, hemicellulose 38.35%, lignin 9.07%, ash 3.67%, fat 1.26% and protein 17.11% by weight (**Table 2**). The result showed that the process affected the composition of oil and lignin, resulting in lignin removal up to 54.71% (w/w). Chang et al. 2014 extracted oil from PKC with hexane at room temperature (24°C) for 6 h. The defatted PKC still contained oil in pulp up to 7.87% (w/w), but the research did not report in %protein yield from alkaline extraction because the process was preparing protein isolates to hydrolyse with enzymes [16]. However, Lau et al 2015 reported that extracting protein from mesocarps of palm oil was challenging because of high level of interfering compounds such as oil, polysaccharide and phenolics. They found that oil removing could improve the efficiency of protein extraction [17].

### *3.3. Protein extraction*

*3.3.1. The effect of chemical reagent types and concentrations.* The effect of chemical reagent types (NaOH, NaCl, and HCl) and concentrations (0.01- 0.3 M) on protein extraction was studied. The investigation included the ratio of solid and liquid (1:50, w:v) at 35°C for 2 h with mixing speed 250 rpm, as shown in **Figure 2**. An increase in NaOH concentration from 0.01 to 0.3 M led to an increase in the amount of the extracted protein from 5.88% to 35.06%. Particularly, the extraction with 0.25 M NaOH achieved the highest extraction with 35.06% of protein extracted compared to the other chemical reagents. Increasing NaCl concentration from 0.01 to 0.3 M with pH adjusted to 9 resulted in the decline in the protein extraction (6.01-12.40%). The isolated protein extracted by HCl solution showed the lowest protein content with 3.05-4.34%.

The highest protein isolated by saline and acid treatment was 12.40 and 4.34% with 0.01 M of NaCl and 0.1 M of HCl, respectively. The use of NaOH in protein extraction provided disruption of the biomass structure and increased the protein solubility. The high concentration of NaOH helps to break down hydrogen bond of polypeptide. Thereby, the surface change of protein would increase, and then the protein would have more ability to dissolve in water. The result of protein extracted from PKC by NaOH treatment is comparable to that of previous works [5]. The addition of 1.5% sodium hexametaphosphate–assisted (SHMP) in alkaline treatment adjusted to pH 10 led to protein yield at 28.37% contained at 50°C for 1 h [5]. Besides, the protein concentration extracted from different feedstocks such as red pepper seed [18], watermelon seed [19], quinoa seed [20], Africa yam bean [11], cottonseed [21] and rice bran [22] using alkaline process was the range of 12-84%.

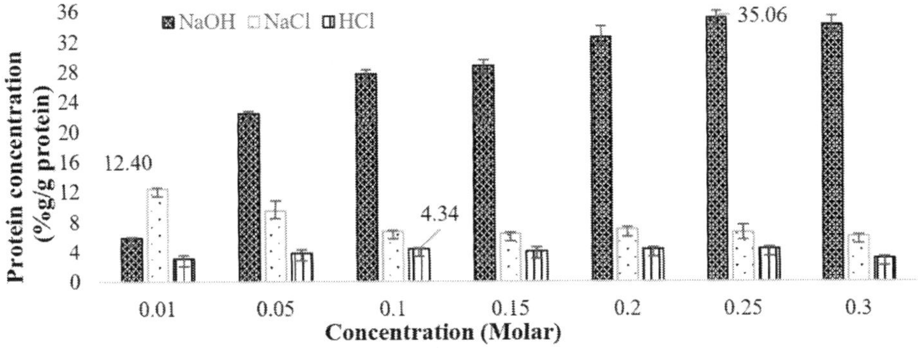


**Figure 2**. Effect of chemical reagents and concentrations on the protein concentration

The analysis of Pearson Product Movement Correlation was calculated by using 2-tailed (significant at the 0.05 level) to compare the effects of chemical reagents and concentrations on protein concentration (**Table 1**). It found that increase of soluble protein concentration extracted by NaOH concentration significantly ($P<0.05$), and showed a positive correlation to 86.2%. While, rising of NaCl concentration is significant ($P<0.05$), and showed a negative correlation to 80.5%. Increasing of HCl concentration is no significant ($P>0.05$) corresponded to the lowest efficiency for extracting protein from PKC.

**Table 1.** Analysis of Pearson Product Movement Correlation in different chemical reagents and concentrations on protein content.

| Correlations | | NaOH | NaCl | HCl |
|---|---|---|---|---|
| | Pearson Correlation | .862* | -.805* | .137 |
| Concentration | Sig. (2-tailed) | .013 | .029 | .770 |
| | N | 21 | 21 | 21 |

*3.3.2 The effect of reaction time.* The protein extraction from defatted PKC was investigated in using 0.25 M NaOH with the ratio of solid to liquid of 1:50 (w/v) at 35°C for different reaction times (0.5- 4 h) with mixing speed 250 rpm. The increase in reaction time from 0.5 to 4 h led to an increase of protein concentration in the range of 24.16-34.84%. However, the extracted protein after 2 h increased slightly (**Figure 3**). Therefore, the reaction time at 2 h was optimum for protein isolation from PKC. An increase of protein yield was consistent with rising extraction time, which is similar to the previous works. Shen et al 2008 reported that extractable protein from tea using alkaline method was performed in the range of 1-6 h, leading to increase the protein yield from 12 to 38%; however, after 4 h, the protein yield showed marginal increase [23].

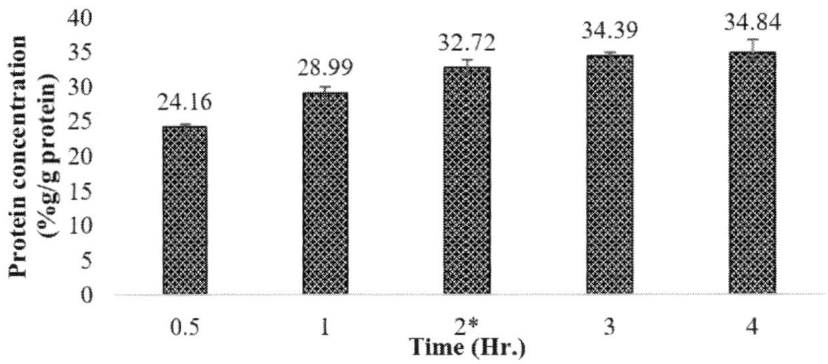


**Figure 3.** Effect of reaction time on protein concentration

*3.3.3. Biomass compositions and soluble liquid products.* The biomass composition from PKC after oil and protein extraction was analyzed by NREL and Kjeldahl method as listed in **Table 2**. In oil extraction, hexane was showed to efficiently reduce the fat and lignin content from 12.65% to 1.26% and 20.03% to 9.07%, respectively. These were equivalent to the fat and lignin removal of 90.04% and 54.71%, respectively (**Table 3**). The protein, ash, cellulose, and hemicellulose were not significantly removed from the defatting process. Later, the extractable protein with 0.25 M NaOH at 0.5-4 h presented the reduction of protein content in the solid residues to the range of 6.49-8.59% compared to the raw PKC (17.37%) corresponding to the efficiency of protein removal with 48.32-60.57%. While, the composition of cellulose, hemicellulose, lignin, and fat composition had slightly affected in this stage of protein extraction process at different reaction times, except the efficiency of ash removal showed in the range of 7.27-34.60%.

In the liquid phase, it found that the recovery of cellulose, hemicellulose and soluble protein achieved in the range of 1.63-5.38% (g/g cellulose), 0.88-4.12% (g/g hemicellulose), and 24.16-34.84% (g/g protein), respectively after protein extraction process in different reaction times at 0.5, 1, 2, 3 and 4 h (**Table 4**). This result showed the efficiency of product recovery. However, the loss of cellulose, hemicellulose, and protein content was still observed. The loss of cellulose and hemicellulose under the protein extraction condition using 0.25 M NaOH at 35 °C might be due to their degraded to by-products e.g. hydroxymethylfurfural (HMF), furfural, acetic acid, and formic acid according to previous works [24].

**Table 2.** PKC compositions after oil and protein extraction

| Process | Time (h) | Pulp yield (%) | Composition in the solid phase (%w/w) | | | | | | |
|---|---|---|---|---|---|---|---|---|---|
| | | | Cellulose | Hemicellulose | Lignin | Ash | Fat | Protein | Other |
| Raw material | - | - | 7.04 | 39.39 | 20.03 | 4.02 | 12.65 | 17.37 | - |
| Oil extraction | 6 | 81.71 | 6.99 | 38.35 | 9.07 | 3.67 | 1.26 | 17.11 | 5.26 |
| Protein extraction | 0.5 | 67.91 | 6.73 | 37.89 | 8.44 | 3.38 | Nf | 8.59 | 2.89 |
| | 1 | 67.24 | 6.40 | 37.62 | 8.86 | 3.22 | Nf | 8.08 | 3.07 |
| | 2 | 65.41 | 6.29 | 37.35 | 8.79 | 2.84 | Nf | 7.91 | 2.22 |
| | 3 | 64.00 | 6.39 | 36.53 | 8.63 | 2.64 | Nf | 7.69 | 2.12 |
| | 4 | 61.70 | 6.26 | 36.02 | 8.74 | 2.28 | Nf | 6.49 | 1.91 |

Nf; no information, *other: oil, wax, starch and other

**Table 3.** Efficiency of PKC composition after oil and protein extraction

| Process | Time (h) | %Efficiency in the solid phase | | | | | |
|---|---|---|---|---|---|---|---|
| | | Cellulose removal | Hemicellulose removal | Lignin removal | Ash removal | Fat removal | Protein removal |
| Oil extraction | 6 | 0.76 | 2.63 | 54.71 | 8.69 | 90.04 | 1.50 |
| Protein extraction | 0.5 | 3.61 | 1.18 | 3.13 | 7.27 | Nf | 48.32 |
| | 1 | 8.32 | 1.87 | 1.04 | 11.23 | Nf | 51.30 |
| | 2 | 9.86 | 2.54 | 1.42 | 20.63 | Nf | 52.27 |
| | 3 | 8.54 | 4.63 | 2.21 | 25.57 | Nf | 53.56 |
| | 4 | 10.31 | 5.92 | 1.63 | 34.60 | Nf | 60.57 |

Nf; no information, *other: oil, wax, starch and other

**Table 4.** The soluble products obtained from protein extraction process

| Process | Time (h) | Composition in the liquid phase (g/100g of initial composition) | | |
|---|---|---|---|---|
| | | Cellulose | Hemicellulose | Protein content |
| Oil extraction | 6 | - | - | - |
| Protein extraction | 0.5 | 1.63 | 0.88 | 24.16 |
| | 1 | 2.27 | 1.67 | 28.99 |
| | 2 | 3.47 | 1.84 | 32.72 |
| | 3 | 4.56 | 3.86 | 34.39 |
| | 4 | 5.38 | 4.12 | 34.84 |

## 4. Conclusion

Alkaline (NaOH) was the best solution for isolating protein from palm kernel cake (PKC). Increasing solution concentration influenced the concentration of soluble protein except when acidic solution (HCl) used. This research suggested the time for protein extraction from PKC as 2 h, resulted in a high yield of protein and had lower by-products. From the study suggested extracting protein from PKC with 0.25M NaOH, but the concentration was higher than other research, which usually perform at 0.03-0.15 M of NaOH concentration. Thereby, my research had studied the effect of high concentration in form of by-products and composition of biomass after treatment. For functional properties of protein isolates such as nutrition value and purity will be further analyzed.

**Acknowledgements**
This research obtained financial supports from Thailand Advanced Institute of Science and Technology-Tokyo Institute of Technology (TAIST-TokyoTech) and Sirindhorn International Institute of Technology, Thammasat University. In addition, we had received facilities of laboratory equipment, chemical, location and assistance from Integrative Biorefinery Laboratory (IBL), BIOTEC (NSTDA). The authors thank everyone involved on this occasion.

**References**
[1] Abdullah, N. and F. Sulaiman, *The oil palm wastes in Malaysia.* Biomass now-sustainable growth and use, 2013. **1**(3): p. 75-93.
[2] Aspar, H.M., *Malaysian palm kernel cake as animal feed.* Notes, 2001. **1**(366): p. 360.
[3] Saeed, M., et al., *Use of mannan-oligosaccharides (MOS) as a feed additive in poultry nutrition.* J. World Poult. Res, 2017. **7**: p. 94-103.
[4] Alimon, A., *The nutritive value of palm kernel cake for animal feed.* Palm Oil Dev, 2004. **40**(1): p. 12-14.
[5] Chee, K.-L. and M.-K. Ayob, *Optimization of hexametaphosphate-assisted extraction and functional characterization of palm kernel cake protein.* Food science and technology international, 2013. **19**(2): p. 109-122.
[6] Zarei, M., et al., *Production of defatted palm kernel cake protein hydrolysate as a valuable source of natural antioxidants.* International Journal of Molecular Sciences, 2012. **13**(7): p. 8097-8111.
[7] Mi, S., et al., *The synergism of hot water pretreatment and enzymatic hydrolysis in depolymerization of lignocellulosic content of palm kernel cake.* Journal of Molecular Catalysis B: Enzymatic, 2016. **134**: p. 37-42.
[8] Klein-Marcuschamer, D., et al., *The challenge of enzyme cost in the production of lignocellulosic biofuels.* Biotechnology and bioengineering, 2012. **109**(4): p. 1083-1087.
[9] Arifin, B., et al., *Protein extraction from palm kernel meal.* Journal of Applied Sciences, 2009. **9**(17): p. 2996-3004.
[10] Choi, W., et al., *Antioxidant properties of crude and cold ethanol precipitated protein from plam kernel cake (PKC) as potential cosmeceutical agent.* Journal of oil palm research, 2019. **31**(1): p. 159-164.
[11] Eromosele, C., et al., *Extractability of African yam bean (Sphenostylis stenocarpa) protein in acid, salt and alkaline aqueous media.* Food Hydrocolloids, 2008. **22**(8): p. 1622-1628.
[12] Sanderson, P., *A new method of analysis of feeding stuffs for the determination of crude oils and fats*, in *Recent advances in animal nutrition.* 1986, Butterworths, London, UK. p. 77-81.
[13] Sluiter, A., et al., *Determination of structural carbohydrates and lignin in biomass.* Laboratory analytical procedure, 2008. **1617**: p. 1-16.
[14] Kruger, N.J., *The Bradford method for protein quantitation*, in *The protein protocols handbook.* 2009, Springer. p. 17-24.

IOP Conf. Series: Materials Science and Engineering **965** (2020) 012008 doi:10.1088/1757-899X/965/1/012008


[15] Thiex, N.J., et al., *Determination of crude protein in animal feed, forage, grain, and oilseeds by using block digestion with a copper catalyst and steam distillation into boric acid: collaborative study.* Journal of AOAC International, 2002. **85**(2): p. 309-317.

[16] Chang, S.K., et al., *Biochemical characterisation of the soluble proteins, protein isolates and hydrolysates from oil palm (Elaeis guineensis) kernel.* Food Bioscience, 2014. **7**: p. 1-10.

[17] Lau, B.Y.C., et al., *Method developments to extract proteins from oil palm chromoplast for proteomic analysis.* SpringerPlus, 2015. **4**(1): p. 1-13.

[18] Firatligil-Durmus, E. and O. Evranuz, *Response surface methodology for protein extraction optimization of red pepper seed (Capsicum frutescens).* LWT-Food Science and Technology, 2010. **43**(2): p. 226-231.

[19] Wani, A.A., et al., *Effect of temperature, alkali concentration, mixing time and meal/solvent ratio on the extraction of watermelon seed proteins—a response surface approach.* Biosystems engineering, 2006. **94**(1): p. 67-73.

[20] Aluko, R. and E. Monu, *Functional and bioactive properties of quinoa seed protein hydrolysates.* Journal of Food Science, 2003. **68**(4): p. 1254-1258.

[21] Zhang, B., et al., *Alkaline extraction method of cottonseed protein isolate.* Mod Appl Sci, 2009. **3**(3): p. 77-82.

[22] Gadalkar, S.M., P.R. Gogate, and V.K. Rathod, *Recovery of Proteins from Rice Mill Industry Waste (Rice Bran) Using Alkaline or NaCl-Assisted Alkaline Extraction Processes.* Journal of Food Process Engineering, 2017. **40**(3): p. e12430.

[23] Shen, L., et al., *Studies on tea protein extraction using alkaline and enzyme methods.* Food Chemistry, 2008. **107**(2): p. 929-938.

[24] Jönsson, L.J., B. Alriksson, and N.-O. Nilvebrant, *Bioconversion of lignocellulose: inhibitors and detoxification.* Biotechnology for biofuels, 2013. **6**(1): p. 16.

IOP Conf. Series: Materials Science and Engineering **965** (2020) 012009 doi:10.1088/1757-899X/965/1/012009

# Extraction of lutein from marigold flower using solvent extraction

**P Kasemwattanarot[1], S Chuachang[1], P J Seepheng[1] and P Ramakul[1, 2]**

[1] Department of Chemical Engineering, Faculty of Engineering and Industrial Technology, Silpakorn University, Nakhon Pathom 73000, Thailand

[2] Corresponding author. E-mail address: ramakul_p@su.ac.th

**Abstract**. The solvent extraction of lutein from marigold flowers was studied. The influence of three variables namely, the extraction temperature, ratio of the marigold flowers in the solvent, and ratio of ethanol in the solvent were investigated by a central composite design (CCD). The relationship between the yield and these variables was obtained by a regression model. The optimum condition was also determined. The highest yield for lutein (91.4%) was predicted by the regression model, which showed good agreement with the experimental results.

## 1. Introduction

Lutein is a yellow pigment that is synthesized from plants and can be found in green leafy vegetables in high quantities; such as, spinach, yellow carrots, and kale. Lutein is a member of the carotenoid group, which is used as a food additive in order to adjust the flavor and color. Lutein also provides benefits for human health because of its strong antioxidant properties, and can help maintain normal vision for humans [1,2]. Additionally, lutein significantly inhibits breast cancer cell growth [3]. Dwyer et al., 201 [4]ameliorating cardiovascular diseases. Lutein is also an accessory pigment by absorbing light energy in a blue light region and changing it into chlorophyll. Furthermore, carotenoids containing lutein are involved in photoprotection; that is, eliminating photooxidative damage because of over-illumination [5].

Marigold flower is a high source of lutein. As such, for extracting lutein from marigold flowers, the solvent extraction method is essential because lutein is bound with a light-harvesting complex (LHC). However, the extraction of supercritical carbon dioxide is not effective for the extraction of lutein from marigold flowers, as it is a polar molecule. Polar organic solvents; such as, ethyl ether, methanol, and ethanol can increase the extraction yield of lutein. Yen et al. [6] and Macias-Sanchez et al. [7] also reported that the extraction of lutein from *Scenedesmus almeriensis* by supercritical fluid extraction (SFE) provided a low yield of lutein, but a high yield of beta-carotene (less polar). In addition, Ruen-Ngam et al. [8] extracted lutein from *Chlorella vulgaris* by solvent extraction that resulted in a good efficiency of 53% of lutein recovery and 43% of selectivity with the lutein content of 7.9 g/kg dry weight. Therefore, the extraction technique from marigold should be a promising method for the extraction of lutein.

In this study, ethanol was selected as a cosolvent because it was a safe solvent with a low environmental impact. The mixture of water and ethanol was used to extract lutein from marigold flowers by a one-step extraction process. The samples were analysed by an ultraviolet–visible (UV-Vis) spectrophotometer. Although the product (lutein) was still in the solvent, this method was reliable because it was simple, rapid and economical. The method achieved the best balance between the extraction performance and the complicated apparatus. The influences of the composition of ethanol, solvent/solid ratio, and pH of the solvent were investigated. The optimum point was determined by a central composite design CCD.

## 2. Experiment

*2.1 Materials, Chemical and Apparatus*

Marigold flowers were collected from a shrine in Thailand. They were milled and sieved through a 200-mesh sieve (the size was in a range between 62-74 µm) for use in all experiments. The analytical standard of lutein was purchased from Sigma® (Germany). The main solvent was double-deionized water (Milli-Q; Millipore, Billerica MA, USA), with 18.2 $M\Omega \cdot cm^{-1}$ conductivity, and the ethanol (Labscan®, Ireland) was of an analytical grade. The concentration of lutein was determined by a UV-Vis spectrophotometer at 470 nm [9] by using a calibration curve between the absorbance and lutein concentration standard with a linearity of 0.9991.

*2.2 Procedure*

The procedure is shown in Figure 1. A total of 20 experiments were run in triplicate using 100 g of marigold flowers placed in a beaker. After that, 300 ml of a mixture between water and ethanol was added. The samples were mixed using a magnetic stirrer with a stirring speed of 100 rpm for 30 minutes to achieve the equilibrium. Finally, the marigold flowers part was removed by a 0.25 µm filter (Double-Ring; Hangzhou, China), and the extracted solutions were sampled into the UV-Vis spectrophotometer.

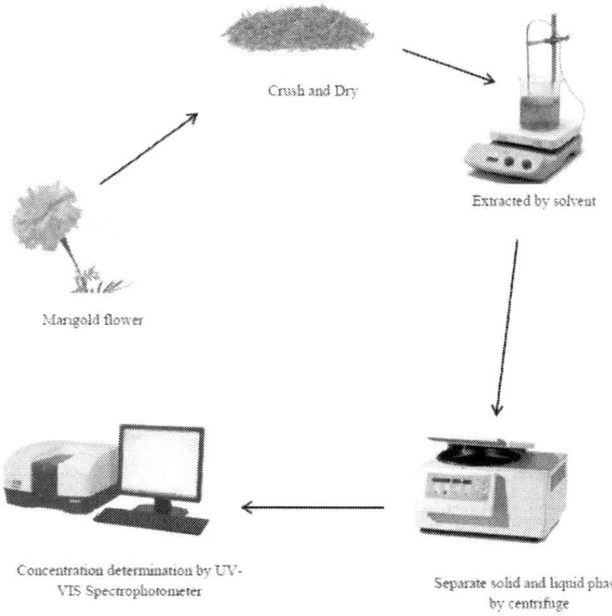


Fig. 1. Extraction procedure.

*2.3 Experimental Design*

The extraction temperature, solvent/solid ratio, composition of the ethanol in the water were varied between 41.6-58.4°C, 0.5-1.5, and 23.2-56.8% (v/v), respectively. A CCD [10] was performed to determine the influences of these factors on the extraction yield. The results were compiled on five levels: $-\alpha$, 1, 0, +1, and $+\alpha$, where $\alpha = 2^{n/4}$ and n was the number of the variables. Therefore, in this research, the $\alpha$-value for four variables was 1.68. The coded values and actual values are shown in Table 1, and there were 20 conditions of the experiment that are tabulated in Table 2.

Table 1. Actual levels and CCD code levels.

| Variable | Unit | Variable Level | | | | |
|---|---|---|---|---|---|---|
| | | **-1.68** | **-1** | **0** | **1** | **1.68** |
| Temperature (A) | °C | 41.6 | 45.0 | 50.0 | 55.0 | 58.4 |
| Solid/solvent ratio (B) | g/100 mL | 0.5 | 0.7 | 1.0 | 1.3 | 1.5 |
| Composition of the ethanol in the water (C) | %(v/v) | 23.2 | 30.0 | 40.0 | 50.0 | 56.8 |

Table 2. Experimental conditions by CCD and the experimental results.

| **Exp. no.** | **Extraction Temperature (°C) (A)** | **Solid/solvent ratio g/ 100 ml (B)** | **Composition of the ethanol in the water (%) (C)** | **%Yield** |
|---|---|---|---|---|
| 1 | 45.0 | 0.7 | 30.0 | 63.21 |
| 2 | 55.0 | 0.7 | 30.0 | 63.90 |
| 3 | 45.0 | 1.3 | 30.0 | 58.90 |
| 4 | 55.0 | 1.3 | 30.0 | 68.83 |
| 5 | 45.0 | 0.7 | 50.0 | 65.37 |
| 6 | 55.0 | 0.7 | 50.0 | 85.00 |
| 7 | 45.0 | 1.3 | 50.0 | 61.77 |
| 8 | 55.0 | 1.3 | 50.0 | 89.30 |
| 9 | 41.6 | 1 | 40.0 | 71.69 |
| 10 | 58.4 | 1 | 40.0 | 74.00 |
| 11 | 50.0 | 0.5 | 40.0 | 77.61 |
| 12 | 50.0 | 1.5 | 40.0 | 64.78 |
| 13 | 50.0 | 1 | 23.2 | 52.51 |
| 14 | 50.0 | 1 | 56.8 | 69.10 |
| 15 | 50.0 | 1 | 40.0 | 77.37 |
| 16 | 50.0 | 1 | 40.0 | 75.40 |
| 17 | 50.0 | 1 | 40.0 | 75.93 |
| 18 | 50.0 | 1 | 40.0 | 76.02 |
| 19 | 50.0 | 1 | 40.0 | 75.69 |
| 20 | 50.0 | 1 | 40.0 | 76.37 |

*2.4 Lutein Content and % Yield Determination*

Firstly, the total amount of lutein in the marigold flowers had to be determined. The marigold flowers (100.0 g) were soaked by a solvent (water + ethanol) for 20 minutes to achieve the equilibrium. When the extraction was completed, the solvent was gathered and stored in a bottle, while the marigold flowers were taken to be extracted again by using the virgin solvent. The extraction was done 10 times to be certain that the concentration of the lutein in the solvent was zero. Finally, the total lutein content in the marigold flowers was determined by summing up the lutein content in all 10 solvent beakers. It was found that the total content of the lutein in 100.0 g of marigold flowers was 19.04 mg. This ratio of the value of 19.04 mg/100g was used as the initial content of lutein for all the experiment samples. The extraction yield is defined in Eq.(1) 1 as a ratio between the weight of the lutein extracted into the solvent ( $w_{c,ex}$ ) and the initial content of lutein in the marigold flowers ( $w_{c,i}$). For example, when 100.0 g of marigold flowers was extracted, the initial content of the lutein was 19.04 mg.

$$\%Yield = \frac{W_{c,ex}}{W_{c,i}} \times 100 \qquad (1)$$

## 3. Results and Discussion

From the experimental results by the central composite design (CCD), the effects of the temperature ($X_1$), solid/solvent ratio($X_2$), and the composition of the ethanol in the water ($X_3$) were investigated (Figures 2-4).

*3.1 Effect of the Temperature*

As can be seen in Figure 2, the yield of the lutein gradually increased as the extraction temperature was increased from 41.6°C to 58.4°C because a high temperature could reduce the viscosity and increase the diffusivity of the solvent into the marigold flowers' cells. Simultaneously, this enhanced the solubility of the lutein in the cells, which resulted in a higher rate of dissolution into the solvent. However, when the temperature was above 55°C, the yield started to decrease due to the disintegration of the solid matrix, which prevented the lutein molecules from penetrating and dissolving into the solvent [11].

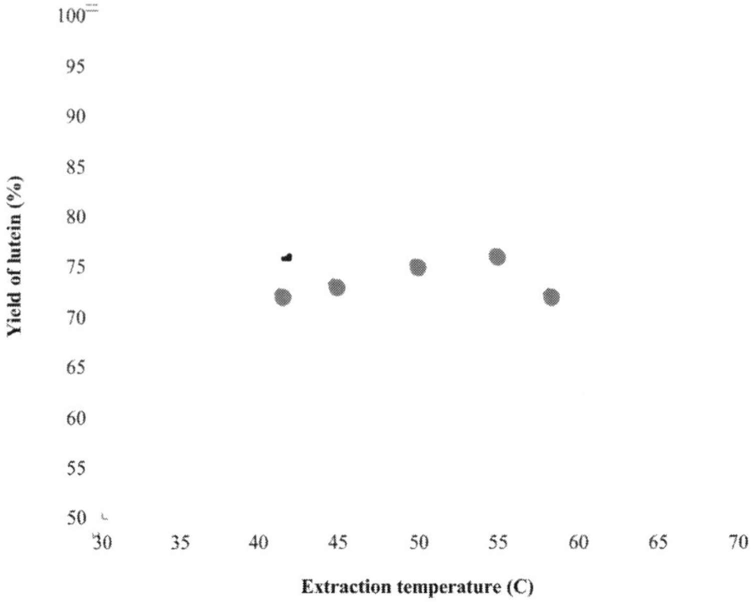


Fig. 2. Plot of the effect of the temperature on the %yield of lutein.

*3.2 Effect of the Solid/Solvent Ratio*

As shown in Figure 3, the yield was lower with an increasing solid/solvent ratio. It was noted that the solid/solvent ratio increased by decreasing the volume of the solvent. This could be explained by the mass transfer in Eq.(2).

$$J = k'(C_s - C_B) \quad (2)$$

Where $J$ is the flux of the lutein transferred from the solid matrix (mg/cm$^2$-s), $k'$ is the mass transfer coefficient (cm/s), and $C_s$ and $C_B$(mg/cm$^3$) are the concentrations of lutein in the solvent that are very near the solid matrix and in the bulk solvent, respectively. From the experimental result, when the solvent volume was decreased, the ratio of the solid part (marigold flower) rose, and there was a lower difference between $C_s$ and $C_B$. This caused a lower mass transfer because when the volume of the solvent was decreased, $C_B$ increased, and as can be seen from Eq.(2), this resulted in a lower flux.

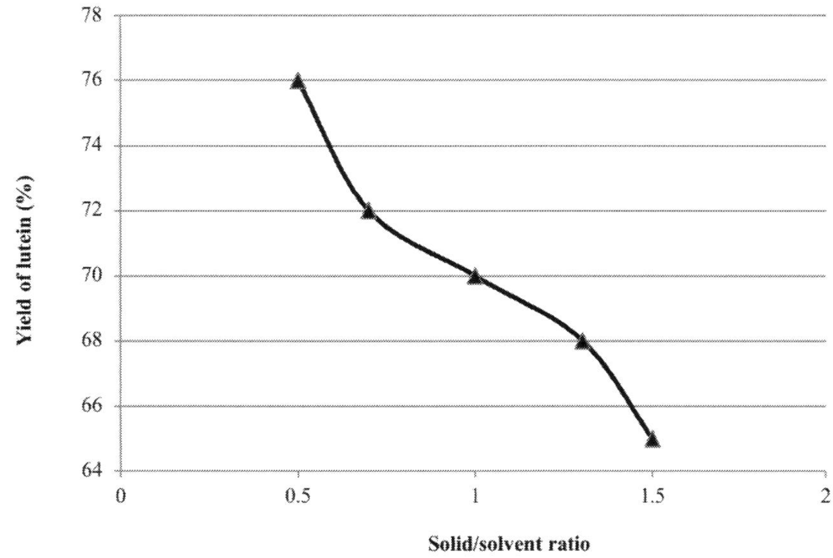


Fig. 3. Plot of the effect of the solid/solvent ratio on the %yield of the lutein.

*3.3 Effect of the Composition of the Ethanol in the Water*

It can be observed in Figure 4 that the yield of the lutein enhanced when the composition of the ethanol in the water increased. This could be described by the fact that ethanol could penetrate the cell walls. Therefore, this made a higher contact area between the solvent and the solids. The solvent could then easily gain access to leach the lutein out and dissolve it [11].

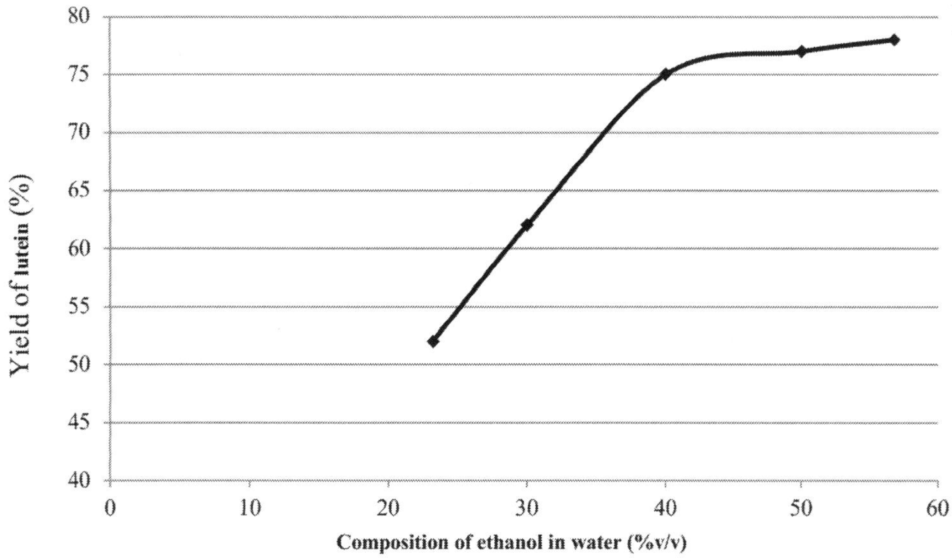

Fig. 4. Plot of the effect of the composition of the ethanol in the water on the %yield of lutein.

### *3.4 Regression Model and Statistical Analysis*

A second-order polynomial regression model was employed for modeling the yield of the lutein. The experimental results were statistically analysed using Minitab 17 statistical software (Minitab Inc., State College, PA, USA).

As a result, the quadratic regression equation for the independent and dependent variables could be expressed by the correlation coefficient ($R^2$) of 0.92 as follows:

$$Y = 76.07 + 4.52A - 1.48B + 5.45C - 0.80A^2 - 1.38B^2 - 5.05C^2 \quad (3)$$

The value of the coefficients characterized the magnitude of the effect. The sign (+,-) indicated the way that this would respond to the %yield of the lutein. From the ANOVA, the P-values of B, $A^2$, $B^2$, AB and BC were higher than 0.05. This inferred that these effects were not significant with a 95% confidence [10]. Therefore, the B, $A^2$, $B^2$, AB and BC terms were omitted from Eq.(3)., and the regression equation was shortened with the $R^2$ of 0.85 as follows:

$$Y = 76.07 + 4.52A + 5.45C - 5.05C^2 + 4.57AC \quad (4)$$

### *3.5 Optimisation and Model Validation*

The significant factors; namely, extraction temperature (A) and the composition of the ethanol in the water (C), were optimized using a response surface. Figure 5 shows the 3-D response surface plot, which was created by Eq. (4). From the observation to the response surface plot, the extraction temperature of 55°C (A=1.00) and composition of the ethanol in the water of 56.8% (C=1.68) were selected to be the optimal operating conditions. The highest yield of 91.48% was predicted by Eq.(4).

In order to validate the optimized conditions, the extraction temperature (A) and solid/solvent ratio were set at the optimized conditions of 55°C and 56.8%, respectively, in the various amounts for the initial mass of marigold flowers of 150 g, 200 g, and 250 g, respectively. The results are shown in Figure 5 and indicate that the yields were approximately 90% when the initial mass of marigold flowers was increased, which demonstrated that the optimized conditions predicted by the CCD and Eq.(4).were satisfactory.

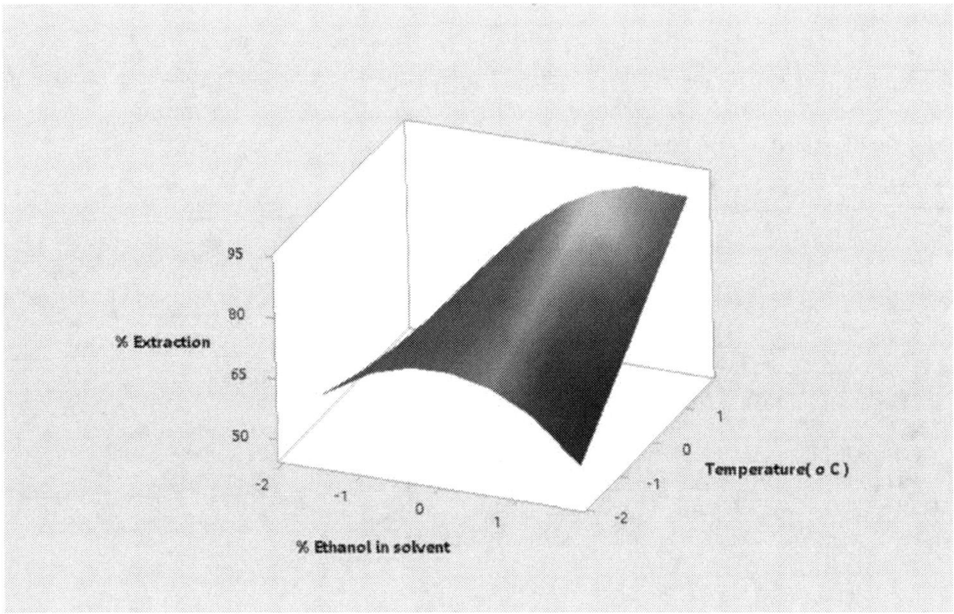


Fig. 5 3-D response surface plot of the effects of temperature and ethanol composition on the %yield of lutein.

Table 3. Yield of lutein in the different initial masses of marigold flowers.

| Exp. no. | Extraction Temperature (°C) | Composition of the Ethanol in water (%) | Initial Mass of Marigold Flowers (g) | Yield of Lutein (%) |
|---|---|---|---|---|
| 1 | 55.0 | 56.80 | 100 | 90.2 |
| 2 | 55.0 | 56.80 | 150 | 90.5 |
| 3 | 55.0 | 56.80 | 200 | 91.4 |
| 4 | 55.0 | 56.80 | 250 | 90.5 |

## 4. Conclusion

A simple, rapid, and economical solvent comprising ethanol/water could extract lutein from marigold flowers. High yields of lutein were obtained. The extraction temperature of 54°C and the composition of the ethanol in the water of 56.8%(v/v) were the optimal conditions, and the yield of 91.4% was predicted by the regression equation. The validity of the equation was evaluated by being compared to the experimental data, which resulted in achieving a good agreement.

## 5. Acknowledgment

The authors greatly appreciate the financial support by the Silpakorn University Research, Innovation and Creative Fund. Thanks is also given to the Department of Chemical Engineering, Faculty of Engineering and Industrial Technology, Silpakorn University, Thailand, for the chemical and apparatus support.

## 6. References

[1] Perry A, Rasmussen H, Johnson E J, 2009 *J. Food Compos. Anal.* **22** 9

[3] Barbara D A, William W A 2002 *Science.* **298** 2149

[3] Gong X, Smith J R, Swanson H M, Rubin L P, 2018 *Molecules.* **23**, 905

[4] Dwyer J H, Navab M, Dwyer K M, Hassan K, Sun P, Shircore A, Hama-Levy S, Hough G, Wang X, Drake T, Merz C N, Fogelman A M 2001. *Circulation* 103 2922

[5] Frank H A, Cogdell R J, 1996 *Photochem. Photobiol.* **63** 257

[6] Yen H W, Chiang, W C, Sun C H 2012 *J. Taiwan Inst. Chem. Eng.* **43** 53

[7] Macias-Sanchez M D, Fernandez-Sevilla J M, Fernandez F G A, Garcia M C C, Grima E M, 2010 *Food Chem.* **123** 928.

[8] Ruen-Ngam D, Shotipruk A, Pavasant P, Machmudah S, Goto M 2012. *Chem. Eng. Technol.* 35 255

[9] Goula A M, Ververi M and Adamopoulou A 2017 *Ultrason. Sonochem.* **34** 821

[10] Montgomery D C 2013 *Design and analysis of Experiment.* (John Wiley and Sons Inc.)

[11] Calvo M M, Dado D, Santa-Maria G 2007. *Eur. Food. Res. Technol.*, **224** 567

# Coaction of bio-sorption and bio-filtration for the remediation of domestic and agricultural wastewater contaminated with heavy metal.

**T Runguphan and J Kitpichai**
Department of Biotechnology, Faculty of Engineering and Industrial Technology, Silpakorn University, Sanam Chandra Palace Campus, 6 Rajamankha Nai Rd, Chedi, Muang, Nakhon Pathom, Thailand.

Corresponding address (Email: runguphan_t@silpakorn.edu, trunguphan@gmail.com)

**Abstract**. Most heavy metals are well known for their toxicity and carcinogenic properties and when disposed-off through wastewater from anthropological activities poses a serious threat to the human population and other living organisms of the receiving body. The presence of heavy metals also hampers biodegrading potential for microorganism to remediate wastewater due to theirs toxic-inhibition. In this study we combine the coaction of bio-sorption and bio-filtration with aim to improve the remediation efficiency of the Copper ($Cu^{2+}$) contaminated domestic and agricultural wastewater samples in Nakhon Pathom Province, Thailand. Rice husk, agricultural waste from rice farming abundant in the area, was used as natural bio-sorbent material and tested to identify its optimum adsorption conditions. The adsorbed wastewater after sorption was fed through bio-filtration tank (sand-filled and bio-augmented with *Pseudomonas stutzeri* DMST 28410) and underwent bio-filtration regimes to complete the coaction remediation. Results from the present study showed rice husk grounded to the size of 106$\mu$m powder at optimal sorption conditions (pH 7, dosage 10g/L, mixing rate 150rpm., sorption time 90min., initial $Cu^{2+}$ concentration 12mg/L) was effective and removed more than 40% of the metal concentration from the wastewater samples. For domestic wastewater, coaction with bio-filtration further improved the copper removal efficiency up to 96% when treated for 180min. Poultry farm wastewater yielded similar copper removal results, with sorption accounted for about 36% and coaction with bio-filtration totalling up to 97% copper removal. Wastewater parameter, Chemical Oxygen Demand (COD) from both sources after bio-sorption exhibited rises in concentration. This poses the need to further explore other pre-treatment methods for rice husk for it to fully be utilized as bio-sorbent in the removal of copper and other heavy metals from wastewater.

## 1. Introduction

Water demand generates wastewater, the characteristics of which depend on the end use of water [1]. The discharge of these wastewater into the environment, in addition to those discharged naturally, cause contamination of the natural water systems making it unsuitable for use. Ideally the wastewater requires to be treated prior to discharge into the receiving water bodies [2].

Contamination of ground and surface water by different organic and inorganic pollutants are a major factors of environmental problems. The generally uncontrolled nature of wastewater collection causes different sources of contaminants to be mixed in the process. Often one would find the wastewater to contain organic contaminations causing high Chemically Oxygen Demand (COD) and also inorganic contaminants like toxic chemicals or heavy metal, further causing high toxicity in the water [3]. An example being agricultural wastewater with intensive farming and heavy use of chemical herbicides, pesticides and insecticides. Effluent with high COD and high metal ions if discharged into the natural water streams without the proper treatment processes could risk contaminating the groundwater.

Different remediation processes such as chemical oxidation, coagulation, solvent attraction, membrane permeation, and adsorption have previously been employed to remove pollutants in the wastewater [4], [5], [6]. These treatment processes however often involve high material and operating costs. The inhibitory effects of the heavy metal ion in wastewater contaminated with copper further complicated the treatment process [7]. It is the ability of damaging or suppressing microbial organelles by reducing bacterial membrane permeability and damaging nucleic acids and as the result, the presence of contaminant copper significantly impact the effectiveness of the conventional wastewater treatment [8]. This presents a significant challenge for the in-situ bioremediation of wastewater at contaminated sites.

Conventional bioremediation of the wastewater mainly uses naturally occurring microorganisms and other aspects of the natural environment to degrade the wastewater of its contaminants [9]. The remediation is deemed as an important approach to the environmentally and economically sustainable water management. The degrading potential thus the remediation, however, is hampered by the presence of the toxic heavy metal. Possible counter measure to the problem could be to add a remediate system in between wastewater discharge (high COD and high metal ions) and the treated effluent water (low COD and low metal ions) prior to the conventional biological wastewater treatment (bio-filtration) unit. An additional unit aims to reduce the concentration, hence its inhibitory effect, of the metal through adsorption process so the less toxic wastewater could be treated more effective biologically. The treatment system, utilizes metal adsorption unit, is to work in conjunction with biological filtration unit. These two in-between processes: - bio-sorption (using bio-sorbent materials) to reduce the toxicity of copper and bio-filtration to further remove COD summing up the term coaction in the title of this research.

The conversion of byproduct of natural materials especially from agricultural wastes into bio-sorbent material presents an attractive, environmental friendly and economical viable alternative to other sorbent materials. Rice husk, which is the low value waste from rice farming, highly abundant in Nakhon Pathom area makes them a good source of cheap raw material as natural adsorbent. Rice husk contains about 20% silica has been reported as a good adsorbent for many metals such as lead, cadmium, selenium, copper, zinc, and mercury [10].

Major components of rice husk which may contribute for the sorption of metal ions are carbon and silica. Silica composes of silicates in which each oxygen atom is shared between two adjacent tetrahedrons. The ionic nature of the Si-O bond due to the large different their electro-negativity and benefits chemical adsorption of the metal ions. The material physical characteristics especially the bulk density of about and solid density of 0.73 g/ml 1.5g/ml respectively [11], with high porosity and large surface area per unit gram of material make them a good natural source of both physical and chemical adsorbent.

The primary objective of this study was to combine the coaction of bio-sorption and bio-filtration with aim to improve the remediation efficiency of the Copper ($Cu^{2+}$) contaminated domestic and agricultural wastewater samples in Nakhon Pathom Province, Thailand. Evaluation of how various sorption parameters such as pH, agitation, reaction time, sorbent dosage and initial sorbate concentration that could influence the biodegradation potential, both the COD removal and metal reduction, of the wastewater from contaminated sites were also investigated.

## 2. Materials

Investigation on the coaction of bio-sorption and bio-filtration for the remediation of domestic and agricultural wastewater contaminated with copper was set up at the Department of Biotechnology Laboratory, Faculty of Engineering and Industrial Technology, Silpakorn University, Sanam Chandra

Palace Campus. The scope was set to focus on domestic and agricultural discharges at low to moderate strength COD concentrations. The sources of wastewater were selected and obtained within the vicinity areas as follow,

- Cafeteria discharge: moderate strength COD of around 500-600mg/L for the domestic wastewater samples.
- Poultry farm discharge: low strength COD of around 100-200mg/L for the agricultural wastewater samples.

Rice husk as bio-sorbent material was sourced locally and prepared (washed/oven dried and sterilized) before any modification to be used in the bio-sorption process through size reduction. Contaminant copper was prepared from Copper (II) Sulphate ($CuSo_4.5H_2O$) solutions with the simulated concentration of 10mg/L as the initial wastewater copper contamination.

## 3. Methodology

### *3.1. Bio-sorption*

Parametric studies were conducted to determine the optimum adsorption parameters. These included:-

- Size (unmodified, grounded to 106$\mu$m, 150$\mu$m, 212$\mu$m respectively) - tested at mixing/agitation speed of 150rpm. and 60minutes reaction (sorption) time.
- An initial 10g/L bio-sorbent dosage then varied from 0-20g/L at 2.5g/L increments.
- An initial 10mg/L $Cu^{2+}$ contamination concentration then varied from 0-20mg/L at 2mg/L increments.
- Reaction time from 0-180mintues at 30mintues increments.

### *3.2. Bio-filtration*

The copper tolerance bacteria used in the study was *Pseudomonas stutzeri* DMST 28410 obtained from the Department of Medical Sciences, Thailand. Cells were harvested in the exponential growth phase with the colony forming unit per millilitre (CFU/ml) measured against McFarland standards. For this study, the bacterial concentration was prepared so its visibility was matched to Scale 1 on the McFarland standard scales in which the number of cells of $300x10^6$ CFU/ml could be approximated. Bacterial dosage was a constant 10% volume by volume through all column experiments (figure 1).

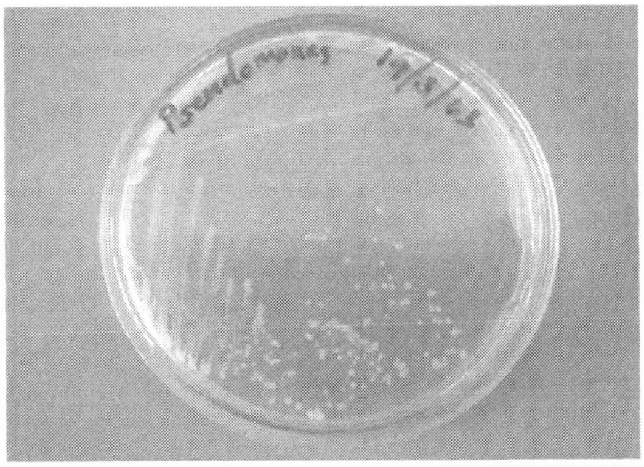

**Figure 1.** *Pseudomonas stutzeri* DMST 28410.

All filtration experiments were conducted in glass columns at three different points as follow,

- Top section (wastewater inlet): 2cm. diameter, thickness 0.2cm.
- Bottom section (treated water outlet): 0.5cm. diameter, thickness 0.15cm.
- Columns height: 30cm.

The packed column was filled with autoclaved quartz sand with particle size <0.45 $\mu$m (washed with deionized water and heating in an oven at 110°C for 4 hours to remove any traces impurities.

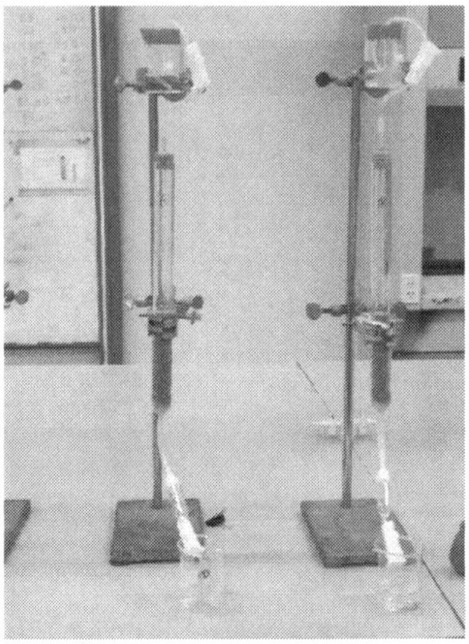

**Figure 2.** Set-up of apparatus for sand-filled bio-augmented columns.

The bio-filtration process was performed with 3 retention stages as follow,

1) flow-through: the control valves at column inlet and outlet were set to open fully so that the incoming wastewater flowed through column without being retained.
2) 3 hours retention/contact time: the inlet control valve open to let the incoming wastewater in while the outlet control valve at bottom, of the column remain shut to achieve retention time of 3 hours.
3) 6 hours retention/contact time: Similar to the 3hours retention period but with outlet valve remained close for 6 hours to achieve the desired retention period.

The effluent water coming out from the bio-filtration column were measured for their standard water quality parameters, conducted in triplicate for the averaging value determination.

### *3.3. Measurement of Standard water quality parameters*

#### *3.3.1. pH and wastewater temperature.*

The pH and the temperature of the wastewater samples (both raw and treated) were measured periodically in accordance with the treatment processes.

#### *3.3.2. Chemical Oxygen Demand (COD)*

The COD levels (measured in milligram per litre of wastewater discharge) for the initial untreated raw wastewater and the treated were measured before and after the remediated experiments (both bio-sorption and bio-filtration) following APHA, 1998, Standard Methods for the Examination of Water and Wastewater (Methods: 5220 B. Opened Reflux Titrimetric Method).

#### *3.3.3. Copper contamination concentration*

A quantitative analysis of copper in a soluble copper salt was performed by complexometric standard titration method.

## 4. Results and discussion

### *4.1. Control condition*

Cafeteria initial COD concentration was 524mg/L and with the stimulated concentration of copper contamination yielded $Cu^{2+}$ concentration of 10.9mg/L. The higher value than the added stimulated concentration could be attributed to the small traces amount of copper present in the wastewater prior to the addition. The COD value for poultry farm discharge was 128mg/L with $Cu^{2+}$ approximately 10mg/L. These initial COD and copper contamination in the wastewater remained unchanged for the non-treated samples (Control condition) and were above the standard limits set out by the governing body, Pollution Control Department of Thailand, of 60mg/L. Discharging wastewater without any treatments neither in the form of bio-sorption, bio-filtration nor both resulted in wastewater not passing water regulatory limits and risk damaging the natural water conditions.

*4.2. Bio-sorption*

The optimum adsorption conditions using rice husk as bio-sorbent tested on cafeteria discharge wastewater (initial COD of 524mg/L) can be concluded as follow,

- grounded to powder with 106$\mu$m. particle size
- at 150rpm mixing rate
- with 90min sorption time
- with 12mg/L initial $Cu^{2+}$ concentration
- at 10g/L sorbent

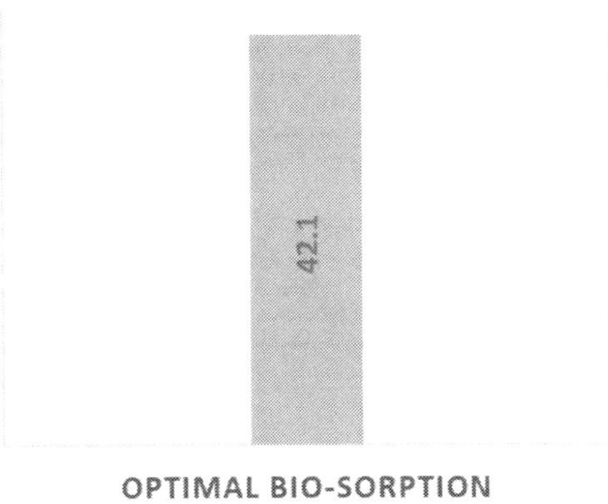


**Figure 3.** %Copper removal: Bio-sorption.

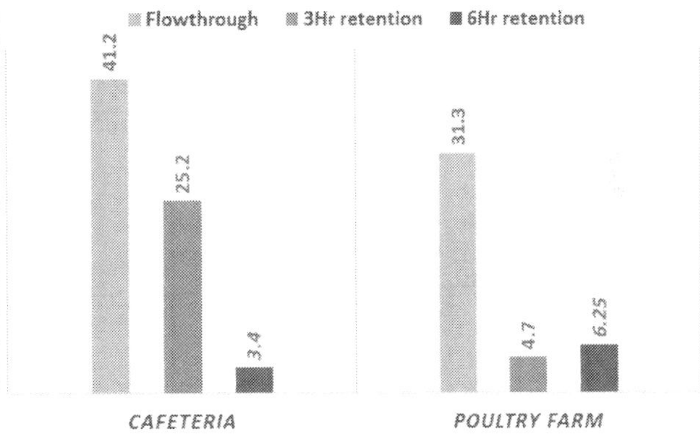


**Figurer 4.** %COD removal: Bio-filtration.

These conditions resulted in 42.1 % of the copper contaminants being effectively removed from 12mg/L to 6.95mg/L as seen in figure 3. Rice husk with minimal modification to the actual agricultural waste material i.e. simply washed, dried, and grounded to small powder could substantially reduce almost half the initial concentration of copper in the wastewater.

*4.3. Bio-filtration*

Tests for the influence of bio-filtration on the wastewater were conducted on both wastewater sources, cafeteria and pouty farm discharges with no copper contamination. Results showed that the flow-through system was most effective in removing the COD (both moderate and low strength) from the two wastewater sources. Flow-through system achieved the highest %COD removal with 41.2% (figure 4) for cafeteria discharge and 31.3% for poultry farm, respectively. Bio-filtration through sand-filled column bio-augmented with degrading bacterium *Pseudomonas stutzeri* DMST 28410 proves sufficient in removing nearly half COD contamination from the wastewater in relatively short retention time.

*4.4. Coaction*

*4.4.1. Copper removal*

IOP Conf. Series: Materials Science and Engineering **965** (2020) 012010 doi:10.1088/1757-899X/965/1/012010

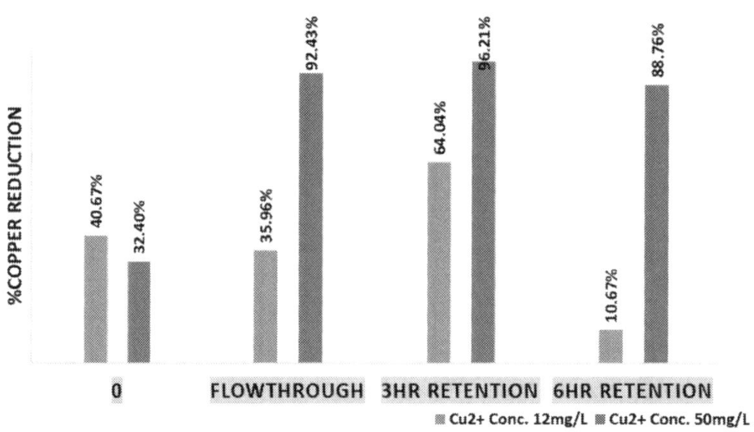


**Figure 5.** %Copper removal through coaction: cafeteria discharge.

Wastewater samples were treated under coaction of bio-sorption and bio-filtration to determine their copper removal efficiency. Copper contamination in the wastewater were at the concentration of 12mg/L and 50mg/L. According to 12mg/L was the optimum concentration from conditions optimization in Section 4.2, there was also reported from previous research that *Pseudomonas stutzeri* bacteria was able to withstand at this concentration and still yielded considerable copper reduction. Figure 5. shows coaction resulted in higher copper reduction efficiency with improved copper removal at 50mg/L initial $Cu^{2+}$ concentration compare to the less contaminated concentration.

When considering retention periods, flow-through, 3hr or 6hr retention yielded no significant improvement in the copper removal. The maximum copper reduction of 96.2%obtained with the 3hr retention.

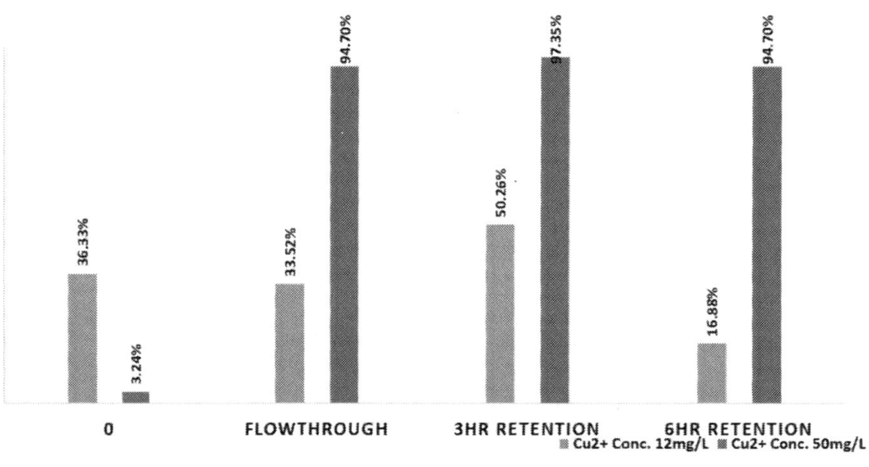


**Figure 6.** %Copper removal through coaction: poultry farm discharge.

Similar conclusion could be drawn from agricultural sample from poultry farm discharge where coaction resulted in better copper reduction efficiency. Improvement in copper reduction at 50mg/L $Cu^{2+}$ initial concentration also was shown in figure 6. Filtration retention parameters i.e. flow-through, 3hr or 6hr yielded no significant improvement in the copper removal with maximum copper reduction 97.4% obtained at 3hr retention.

*4.4.2. COD concentration*

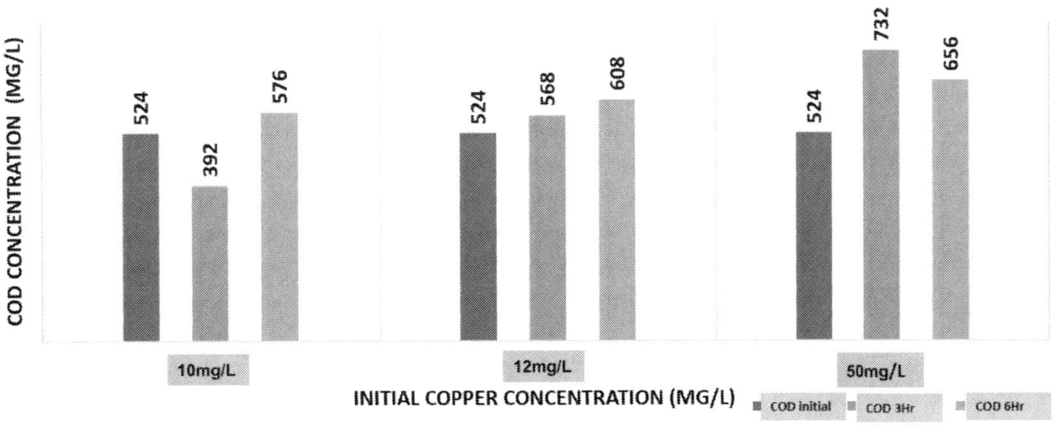


**Figure 7.** COD concentration through coaction: cafeteria discharge.

Wastewater parameter, Chemical Oxygen Demand (COD) from both sources after bio-sorption exhibited slight rises in concentration as seen in figure 7 and figure 8. The wastewater COD from cafeteria discharge in figure 7 was around 524mg/L initially, with 10mg/L stimulated $Cu^{2+}$ contamination, the value first dropped down to 392mg/L before raised up to 576mg/L for 3hr and 6hr retention times, respectively. This represents an increase in COD concentration of around 10% from the initial COD. Cafeteria discharge with the initial $Cu^{2+}$ concentration of 12mg/L and 50mg/L yielded similar increase in COD concentration, 16% and 40% for 12mg/L and 50mg/L $Cu^{2+}$, respectively. The rises in COD concentration could be attributed to the fact that during adsorption with rice husk bio-sorbent, some of the organic contents of the fine powder rice husk was leached out, darker colour adsorbed water discharged out of bio-sorption process, as seen from visual inspection, causing the rises in the COD concentration.

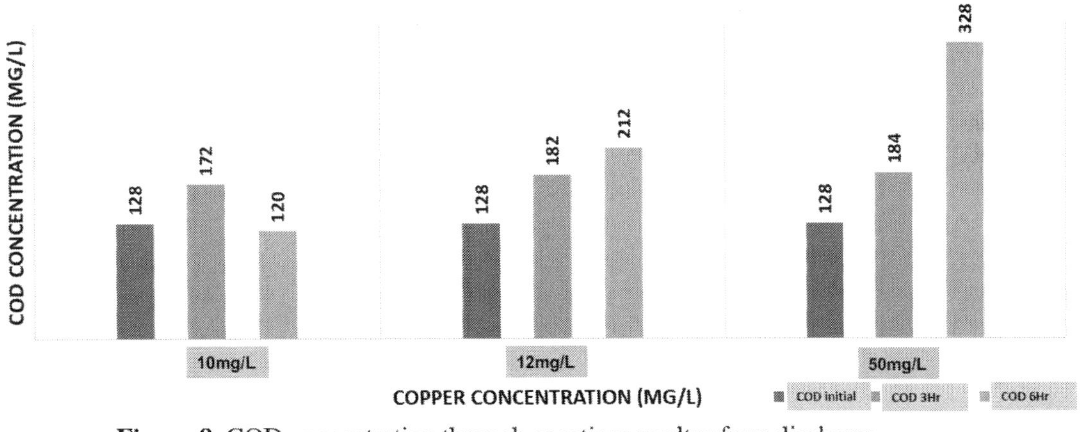


**Figure 8.** COD concentration through coaction: poultry farm discharge.

Similar results also appear with wastewater from poultry farm discharge, presented in figure 8. The initial COD concentration of 128mg/L recorded a raise in values for all three-copper contamination concentrations (10,12 and 50mg/L). Higher copper contamination, $Cu^{2+}$ of 50mg/L posted the highest increase in COD of up to 50%, the same as from the moderated strength COD of cafeteria discharge.

These increases in COD concentration after bio-sorption with the treated rice husk poses the need to further explore alternative pre-treatment methods for it to fully be utilized as bio-sorbent in the removal of copper and other heavy metal from wastewater. Modification methods to the raw agricultural waste materials for example through strong acidic and alkali pre-treatment have been reported to further improved sorption properties [12][13] may help lessen the wastewater COD discharging out from the biosorption process. Further investigation into thermal modification [14] for example microwaving and carbonisation[15] present possible pre-treatment methods that could potentially be integrated in bio-sorbent preparation with the aim to fully be utilized as bio-sorbent in the removal of copper and other heavy metals from wastewater.

## 5. Conclusion

Treated rice husk (grounded powder) is an effective bio-sorbent material for removing heavy metal ions $Cu^{2+}$ from domestic and agricultural wastewater contaminated with copper. Coaction of bio-sorption and bio-filtration through bio-augmentation of *Pseudomonas stutzeri* DMST 28410 further enhanced the removal of copper metal ions in both domestic and agricultural wastewater samples. This could be attributed the process of further bio-sorption through metal uptake and/or biomineralization by microbial cell of the bacteria. An increase of wastewater COD concentration after bio-sorption by rice husk poses the need to further explore alternative pre-treatment methods or additional measures (modification to the bio-sorbent) to fully utilize its bio-sorbent material prospect.

## References

[1] Grady C P, Daigger G T and Love N G 2011 *Biological wastewater treatment* (New York: CRC Press)

[2] Bromley D W, Taylor D C and Parker D E 2018 Water Reform and Economic Development: Institutional Aspects of Water Management in the Developing Countries* *Econ. Water Resour.* 443-65 https://doi.org/10.4324/9781351159289-28

[3] Zhong H, Wang H, Liu X, Liu C, Liu G, Tian Y and Chen Y 2017 Degradation and characteristic changes of organic matter in sewage sludge using Vermi-biofilter system *Chemosphere* **180** 57-64 https://doi.org/10.1016/j.chemosphere.2017.03.121

[4] Latifoglu A and Gürol M D 2000 Advanced Oxidation Processes in Water Treatment *Chemical Water and Wastewater Treatment* **6** 137-43 https://doi:10.1007/978-3-642-59791-6_13

[5] Zhang S and Zhao Z 2010 Coagulation-Sedimentation Study of Petrochemical Wastewater for Advanced Treatment *Int. Conf. on E-Product E-Service and E-Entertainment* https://doi:10.1109/iceee.2010.5660446

[6] Janyasuthiwong S 2020 Adsorption Technology for Metal Contaminated Wastewater Treatment *Metal Removal and Recovery from Mining Wastewater and E-Waste Leachate* 33-58 https://doi:10.1201/9780138747527-3

[7] V G V and Misra A K 2016 Copper contaminated wastewater – An evaluation of bioremedial options *I. Blt. Env.* **27**(1) 84-95 https://doi:10.1177/1420326x16669397

[8] Zinicovscaia I 2016 Conventional Methods of Wastewater Treatment *Cyanobacteria for Bioremediation of Wastewaters* 17-25 https://doi:10.1007/978-3-319-26751-7_3

[9] Kaushal S and Singh J S 2017 Wastewater Impact on Human Health and Microorganism-Mediated Remediation and Treatment Through Technologies *Agro-Environmental Sustainability* 235-50 https://doi:10.1007/978-3-319-49727-3_12

[10] Shamsollahi Z and Partovinia A 2019 Recent advances on pollutants removal by rice husk as a bio-based adsorbent: A critical review *J. Environ. Manage.* **246** 314-23 https://doi:10.1016/j.jenvman.2019.05.145

[11] Zou Y and Yang T 2019 Rice Husk, Rice Husk Ash and Their Applications *Rice Bran and Rice Bran Oil* 207-46 https://doi:10.1016/b978-0-12-812828-2.00009-3

[12] Wongjunda J and Saueprasea P 2010 Biosorption of Chromium (VI) Using Rice Husk Ash and Modified Rice Husk Ash *Environ. Res.* **4**(3) 244-50 https://doi:10.3923/erj.2010.244.250

[13] Shukla S K 2014 Metal Decontamination From Chemically Modified Rice Husk Film *Adv. Mater. Letrs.* **5**(6) 352-55 https://doi:10.5185/amlett.2014.1018

[14] Razavi Z, Mirghaffari N and Rezaei B 2014 Performance Comparison of Raw and Thermal

IOP Conf. Series: Materials Science and Engineering **965** (2020) 012010 doi:10.1088/1757-899X/965/1/012010

Modified Rice Husk for Decontamination of Oil Polluted Water *CLEAN - Soil, Air, Water,* **43**(2) 182-90 https://doi:10.1002/clen.201300753

[15] Mishra A, Taing K, Hall M W and Shinogi Y 2017 Effects of Rice Husk and Rice Husk Charcoal on Soil Physicochemical Properties, Rice Growth and Yield *Agri. Sci.* **08**(09) 1014-32 https://doi:10.4236/as.2017.89074

# Anatomical and physical properties of tendril perversion of *Coccinia grandis* (L.) Voigt

**T Prombanchong[1], A Suriyawong[2], P Srisongkram[3] and N Suwanpayak[*3]**

[1] Division of Science, Faculty of Science and Technology, Rajamangala University of Technology Srivijaya, Nakhon Si Thammarat Saiyai Campas, Nakhon Si Thammarat, 80110, Thailand
[2] Department of Engineering, King Mongkut's Institute of Technology Ladkrabang, Prince of Chumphon Campus, Chumphon 86160, Thailand
[3] Department of General Science, King Mongkut's Institute of Technology Ladkrabang, Prince of Chumphon Campus, Chumphon 86160, Thailand

Email: ksnathap@gmail.com

**Abstract**
The structure of biomaterial is interest in biomimetic and biomechanics research. The physical properties, mechanical, characteristic of twisting and bending of plant tendrils are also thought-provoking to study. In this study, we proposed an innovative biomaterial spring device for holding an object and providing support. The device uses *Coccinia grandis* (L.) Voigt tendril as inspired by the relations of left-hand (LH) and right-hand (RH) climbing tendrils with a single branch. To study the morphology and anatomy, testing stress, length, number of the helix of the tendril which relation of mimic string material *Coccinia grandis* (L.) Voigt tendril and helical metal which have two helixes, three helixes and four helixes were tested by using food texture analysis. The anatomical showed many cell types and slight differences between straight and coiled tendrils. The result showed the tendril of two helixes has the highest stress and four helixes the least. Also, tensile testing showed that the number of helixes does not affect the stress of the tendril models.

**Keywords:** Tendril, anatomy, biomimetic, *Coccinia grandis* (L.) Voigt

## 1. Introduction
Biomimetic is an important interdisciplinary field which has many research areas for instance engineering, chemistry and biology, including being applied in the synthesis of the material innovation process. The biological structures are composited of many cell types and substance supporting cell structure such as the fibre cell and lignin, having amazing properties the characteristic of biomaterial encourage study [1]. In the 15th century, Leonardo da Vinci designed the wings of a glider [2] which was inspired by the bat's wings although his design was not successful. But, it inspired the next generation of inventors to develop the idea continuously until the first airplane was successfully flown in the 19th century by the Wright brothers, both of whom were inspired by the flight of birds. Biomimetic science and technology have been developed in widely different devices in various fields to work together as a system. Subsequently, this science needs a multi-disciplined team, such as computer science, electronics engineering, mechanical engineering, physics, chemistry and biology all come together to solve problems. Nature has many hidden secrets materials, phenomena and complex

materials. Scientists use two methods to mimic nature [1, 3], the first method begins with the problem to be solved and analysis it by searching for living entities that have similar problems and try to imitate the creature's methodolgy. The second method is based on the discovery of natural data until the knowledge and understanding of intelligent adaptation of the mechanism of life which then becomes the inspiration to develop new technology. Biomimetics can be divided into three types [4]: (i) mechanism-driven biomimetics is to solve engineering problems by seeking examples in nature to study and solve the problems, such as for instance the lotus effect that tries to find materials that clean themselves [5]. (ii) Organism-driven biomimetics is the study of living organisms to search for distinctive properties and using those features to create new products such as velcro. (iii) Integrative biomimetics is the guideline that combines both the approaches mentioned above, such as studying the movement of cockroaches. One classic example the lotus effect was successfully studied by a researcher and applied in the field of self-cleaning surface [6] attracting interest for various applications such as solar cells, textiles, anti-freezing and anti-snow surfaces, filters, microfluidics, medical devices, blood vessel replacements by fabricating flexible hybrid inorganic nanowires decorated microfibers [7].

Tendril is interested in natural products which are the part of the trunk, flower, leaf or branch of a plant that changes into a spiral style for support [8]. The tendril has a long, twisting, spiral-like springs to make it flexible. When the wind blows back and forth wherein ivy plants, which are of interest to researchers in various fields, the tendril of each plant provide highly effective support using biology material. The spiral shape of the climbing tendril supports very effectively and can be explained by studying similar homoclinic and heteroclinic orbits in the three-dimensional configuration problem of a cylinder constrained rod [9]. The gourd's tendril is positioned opposite the petiole, and it does not grow directly from the axillary but the axillary bud. An anatomical structure compared to natural plants is gourd flowering, whether male or female, occurs at the position of the axillary near the tendril, including where the gourd divides. The new branches are separate from the axillary positions and close to the tendril. Therefore, there is a hypothesis that the tendril of the gourd may emerge from flowers, inflorescences, branches and leaves [10].

This research aimed to study anatomy and physical properties of *Coccinia grandis* (L.) Voigt tendril at Chumphon Province, southern Thailand. The relationship between magnitude stress and extraction length, size, number of perversion of the tendril as biomaterial springs, for holding an object with flexibility at a certain distance depending on the size of the tendril, to support trees or branches in the desired direction.The results obtained are compared with the tendril model made from metal wire, then designed to use a tendril simulation to create innovation by using adhesion with support for flexible strength.

## 2. Methodology

In this study, we use the tendril of *Coccinia grandis* (L.) Voigt at Chumphon province, southern Thai land. *Coccinia grandis* (L.) Voigt tendrils are natural products which are the part of the trunk but acting as branches that separate from the trunk. The harvested samples are as shown in Figure 1(a), where the tendril has two helixes and Figure 1(b) where it has three helixes. The growth of the tendril can be divided into four steps, as showed in Figure 1(c). The first step in the tendril growth is growth in a straight line to find near support to twist before becoming a helix (Step 1). After the tendril achieves the support, the tendril gradually becomes helical at both the tendril launching and end point due to growth, the distance between the source of the tendril and the support is constant (Step 2-3). The tendril helical has the characteristic of a helical spring, competitive and flexible, enabling it to withstand the wind (Step 4). The primary deformation of the tendril structure is formed by bending and twisting when it becomes a helical structure like a helical spring.

The characteristics of *Coccinia grandis* (L.) Voigt tendril is shown in Figure 1 (a)-(b) after already making contact and spiraling around the support, which has two helixes and three helixes, respectively. In this study, we found the tendril had two helixes showed the most stress and four helixes the least stress. Figure 1 (a)-(b) shows the geometrical relations of left-hand (LH) and right-hand (RH) climbing towel gourd tendrils on the same plant branch. Through understanding the twist deformation features of innovative three-dimensional elastic chiral mechanical metamaterials, an innovative multi-functional tetrachical cylindrical tube is proposed, which can generate cylindrical tube twist deformation upon compression loading conditions [11].

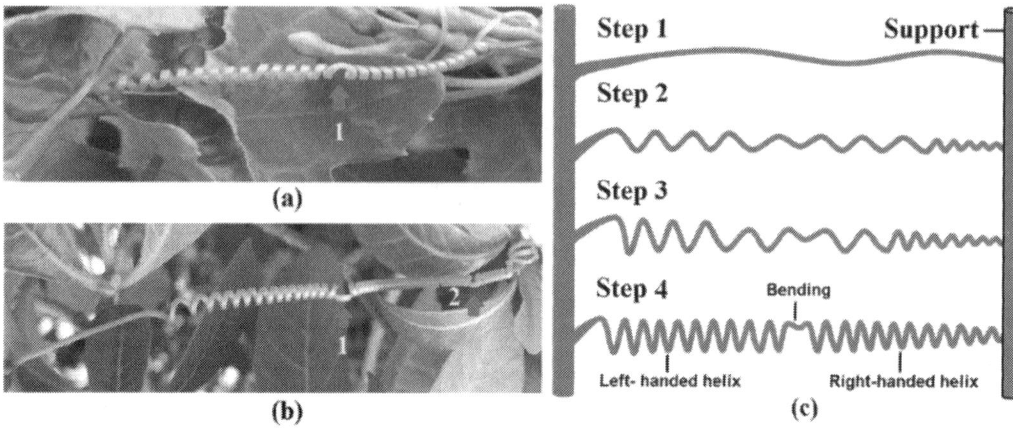


**Figure 1.** The tendril of characteristics of *Coccinia Grandis* (L.) Voigt in Chumphon Province, southern Thailand, (a) tendril has two helixes and (b) tendril has three helixes and (c) steps of tendril growth.

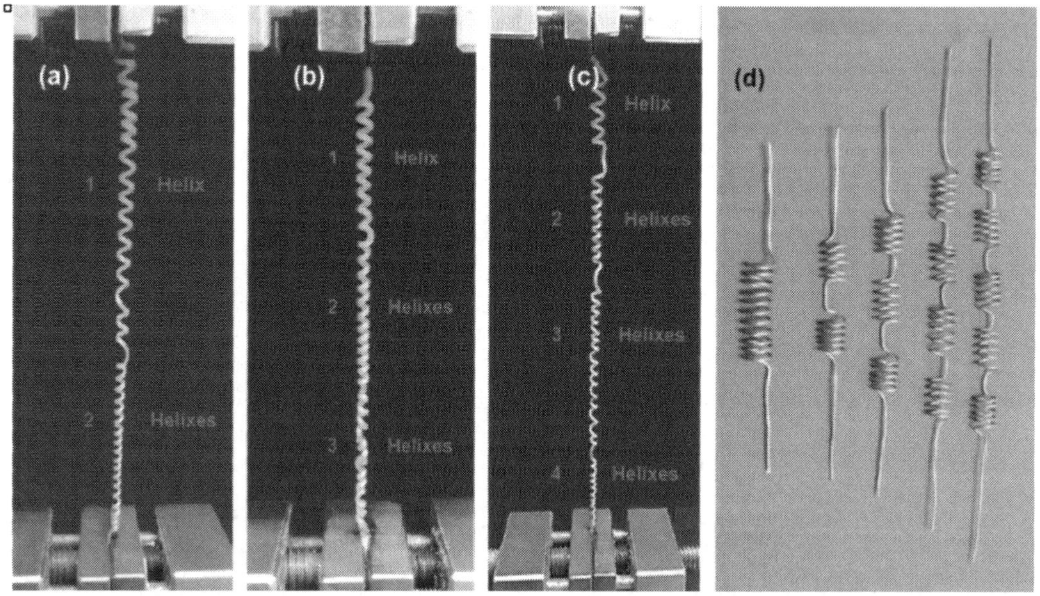


**Figure 2.** Under tensile testing of the tendril and helical metal (a) tendril with 2 helixes (b) tendril with three helixes, (c) tendril with four helixes and (d) tendril models using an aluminium coil that have one to five helixes.

The *Coccinia grandis* (L.) Voigt tendril was tested by using a texture analyzer in the tension mode. Figure 2(a)-(c) shows the tendrils have one bend (two helixes), two bend (three helixes) and three bend (four helixes), respectively. The tendril that has one bend and has two helical tendrils that are right-handed (RH) the helix rotates clockwise (CW) and left-handed (LH) the helix rotates counterclockwise (CCW). Figure 2(d) is a model to mimic the tendril by winding an aluminium coil into spring and turning it to find the mechanical properties of the model as tendrils, including comparing the mechanical properties with the natural the *Coccinia grandis* (L.) Voigt tendril properties. To observe the *Coccinia grandis* (L.) Voigt anatomy, the trunk, straight and coil tendril were cross-sectioned by using the free hand technique [12]. The small pieces were put into a plate containing distilled water. The thinnest part was selected and then stained with 0.01% v/v. Safranin-O and dissolved in distilled water for 1 min. The stained section was gently transferred to a slide, two drops of water were added and then the section was covered with a coverslip. The slide was observed and the image captured under a light microscopic.

## 3. Results and discussion

The experimental testing results of the *Coccinia grandis* (L.) Voigt tendril shown in Figure 3-5. The results are divided into two parts: the first part, showed the morphological structure of the internal structure of the cross-section of the tendril is shown in Figure 3, secondly mechanical testing of tendrils and tendril models are shown in Figure 4.

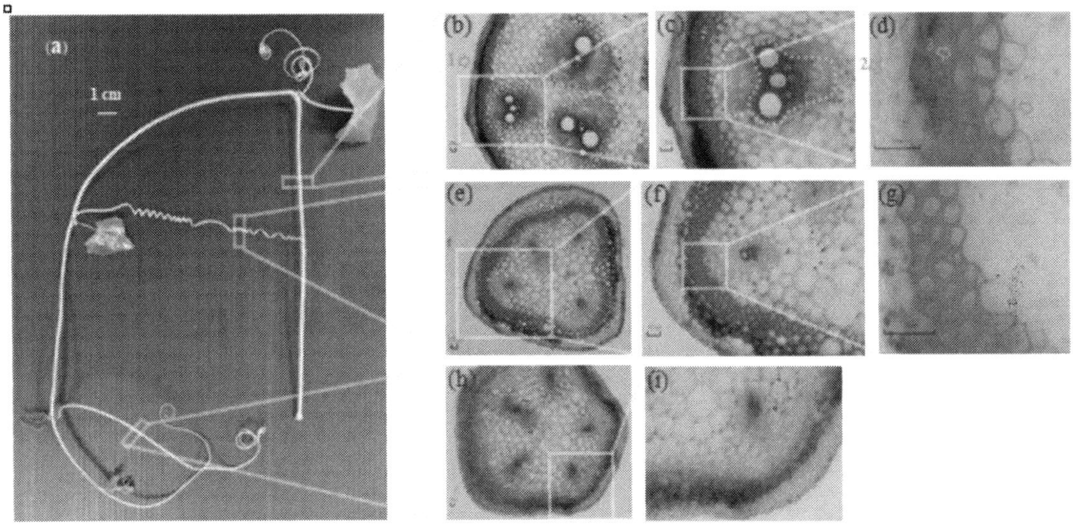


**Figure 3.** The morphological and anatomical of the trunk straight and coiled *Coccinia grandis* (L.) Voigt, (a) photograph representing the morphological of the straight trunk and coiled tendril, (b)-(d) the trunk under light microscopic, magnifying power 4x 10x and 40x, respectively, (e)-(g) the coiled tendril under magnifying power 4x, 10x and 40x, respectively and (h)-(i) the straight tendril under magnifying power 4x and 10x, respectively.

The *Coccinia grandis (L.)* Voigt belongs to the Curcubitaceae family. Similar to plants in the same family, it has tendrils that support the trunk to find sunlight or take the plant to the numerous environmental niches [11]. The photograph in Figure 3 shows the morphological of the trunk, straight and coiled tendril was used to study anatomical. The result representing many cell types are shown in Figure 3(b)-(i) identified with the blue arrows 1-6 which are the epidermis, collenchyma, fibre, parenchyma, xylem and phloem, respectively. The epidermis collenchyma and parenchyma are primary cell wall structure whereas the xylem phloem is a set of 4 cell types; some dead (haveing a secondary cell wall) and some alive (not accumulating a secondary cell wall). The specify cell type xylem and phloem have lignin achromatize by safranin-O. The fibre cell (red colour) with the blue arrow number 3 was not founded in the straight or immature tendril is shown in Figure 3(e)-(i). The cross-section revealed that the fibre cells found in the trunk were symmetrical whereas the mature tendril was asymmetric, as shown in Figure 3(b) and (e). The fibre cell of the tendril representing three layers was slightly similar to the wild cucumber tendril representing two layers which were ribbon [13]. The twist of the tendril involved with the lignified gelatinous fibre cells (g-cells) provided a supporting structure via the cell wall lignification water flux and stiffness of the cellulose micro-fibrils [14]. The differential lignification of 2 layers result in cell contraction and coil via the hydrophobic of lignin [13].

The mechanical properties were determined by using a tensile analyzer to measure force, mechanical extension and stress stretching at 1-3 times, affecting the strength by preventing impact and sedimentation. The create a climber model was created by using aluminium coil with a diameter of 0.56 ± 0.01 to study, its properties, including the comparison of straight wire as shown in Figure 4. The experiment compared the mechanical properties of the tendril model made from metal wire with four samples that are two helixes, three helixes, four helixes and linear metal wire. The results showed that in the tendril models, the helix numbers is not have different effects, as shown in Figure 4(b), but were different from the linear wire.

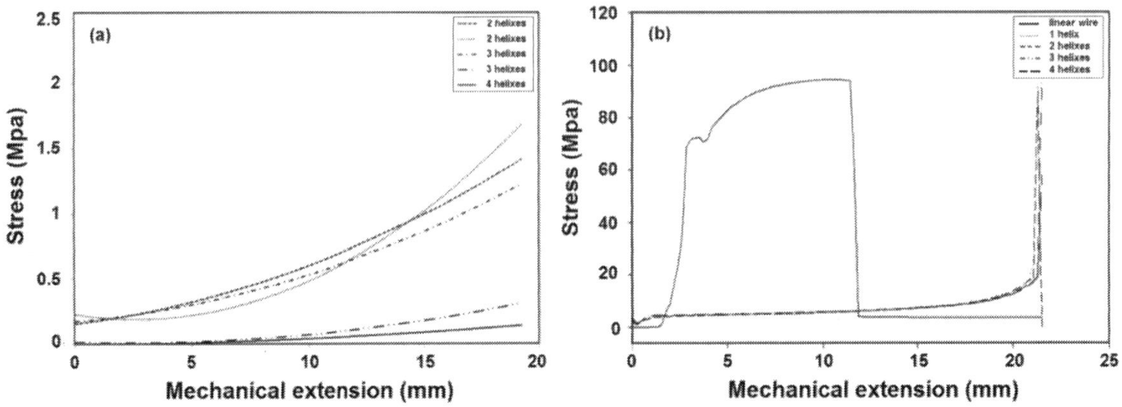


**Figure 4.** The stress of (a) the *Coccinia grandis* (L.) Voigt tendril having two helixes, three helixes and four helixes and (b) the stress of the tendril metal model having two to five helixes.

Figure 4 (b) shows the relationship between stress and mechanical extensions from the root. The results show that the number of helixes is much less stressed compared to the tendril. The smaller helixes mean that the number of helixes may be due to support a defined distance. With too many tendrils, the tendril is longer so it can be attach to the support. The handgrip is less secure when it has a smaller number of helixes because they can hold support a short distance. In this study, the tendril spring model can be applied like biomimetics to develop a spring tension to hold maintain a support for setup camping as shown in Figures 5(a) and (b).

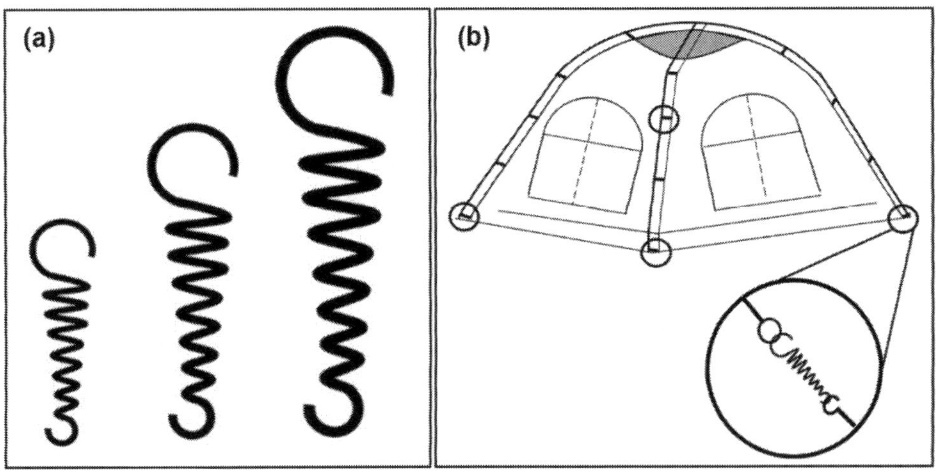


**Figure 5.** (a) a designing of a tendril spring model can be applied for the elastic holding of the support and (b) a useful application of the tendril spring model.

## 4. Conclusions

A natural product is a miracle product having interesting biomaterials by changing the physical characteristics according to the living environment. In this study, the experimental results of the tendril inversion affected the tensile force and physiological of the trunk, straight and coiled tendril of *Coccinia grandis* (L.) Voigt. The experimental results showed the strength of the tendrils is not as hypothesized but that the amount of bending affects the strength of the spring tendril. On the other hand, the amount of bending varies the tendril's weakness also increases, the same as reported by the metal models that the amount of the bending does not affect the strength. The tendril is very flexible as a natural spring, which is the adaptation of plants that they can hold and be wind resistant or other forces that can cause the trunk to fall off, this is a one of the biomimetic to create a new innovative device production.

## Acknowledgements

The authors would like to thank King Mongkut's Institute of Technology Ladkrabang Prince of Chumphon Campus for providing the financial support and thank the Department of General Science for supporting the research laboratory.

## References

[1] Plessis A D, Broeckhoven C, Yadroitsava I, Yadroitsev I, Hands C H, Kunju R and Bhate D 2019 *Addit. Manuf.* **27** 408
[2] Richardson P L *Notes Rec* **73**(3)
[3] Helms M, Swaroop S V and Goel A K 2009 *Design Studies* **30** 606
[4] Feng J, Zhang W, Liu C, Guo M and Zhang C 2019 *Sci. Rep.* **9** 5051
[5] Teodorescu M 2014 *J. Text.* **2014** 154184
[6] Zhang M, Feng S, Wang L and Zheng Y 2016 *Biotribology* **5** 31
[7] Chen R, Wan Y, Wu W, Yang C *et al.* 2019 *Mater. Des.* **162** 246
[8] Ensikat H J, Ditsche-Kuru P, Neinhuis C and Barthlott W 2011 *Beilstein J. Nanotechnol.* **2** 152-161.
[9] Malabarba J, Reichelt M, Pasquali G and Mithöfer A 2019 *J. Plant Growth Regul.* **38** 39–45
[10] Sousa-Baena M S, Sinha N R, Hernandes-Lopes J and Lohmann L G 2018 *Front. Plant Sci.* **9**
[11] Wu W, Geng L, Niu Y, Qi D, Cui X and Fang D 2018 *Extreme Mech. Lett.* **20** 104
[12] Immanuel S M and Nanditha V 2019 *J Pharmacogn. Phytochem.* **8**(3) 4296
[13] Isnard S and Silk,WK. 2009 *Am. J. Bot.* **96**(7) 1205

# Zinc electrodeposition of mild steel in the presence of ginger, pomegranate and celery extracts

**Roland T. Loto[1*], Cleophas Loto[1], and Marvellous Akinyele[1]**
[1]Department of Mechanical Engineering, Covenant University, Ogun state, Nigeria

Email: tolu.loto@gmail.com

**Abstract.** Zn electrodeposition on low carbon steel (LCS) within an acid bath composed of $ZnCl_2$, $H_3BO_3$ and KCl solution at predetermined quantities of ginger, celery and pomegranate extracts with respect to plating time was evaluated. Results obtained showed the plant extracts significantly influenced Zn electrodeposition on LCS with respect to concentration. Ginger extracts induced the highest Zn electrodeposited with values between 0.0867 g and 0.1219 g. Comparison of corrosion resistance between the non-plated and Zn electrodeposited LCS (with and without extract additives) were determined with potentiodynamic polarization method. Results showed ginger, pomegranate and celery extracts enhanced the corrosion resistance of Zn electrodeposited LCS with highest corrosion rate values of 2.48 mm/y, 2.78 mm/y and 1.95 mm/y compared to 7.96 mm/y for non-plated LCS and 6.58 mm/y for Zn electrodeposited LCS without plant extracts.

## 1. Introduction

Electrochemical corrosion occurs due to redox reaction mechanisms between the substrate materials (metals and non-metals) and their environment. This phenomenon results in variable but significant degradation in their metallurgical, mechanical and physical properties (Popoola *et al.*,2013; Loto, 2017). The outcome of corrosion degradation ranges from mechanical failures to plant shutdown, hitches in refineries operations, breakdown during chemical processing, leakages in boiler operations etc. (Sastri, 2015). Carbon steel has exceptional mechanical properties; which gives it versatile application in nearly all industries (Lgaz *et al.*, 2018). However, the weak corrosion resistance of carbon steels limits their application in industrial operating conditions due to the presence of corrosive anions. Electrodeposition has very useful application in corrosion control of carbon steels (Basile *et al.*, 2010). The underlying aspect of electrodeposition involves strong adherence of a corrosion resistant coating of metal such as Zn, Cu etc. onto the carbon steel which prevents corrosive anions from getting to it (O'Connor *et al.*, 2018; Bicelli *et al.*, 2008). Coatings have extensive application in underground pipelines, shipping, automotive, building, jewelry etc. to obtain a finer material surface (Peabody, 2001; Makhlouf, 2011). Zn coating on carbon steels enhances its corrosion resistance by forming a protective covering over the steel when in contact with air (Yuan *et al.*, 2017). Electrodeposition of zinc from ionic liquids has been under studied due to their sustainable properties (Pereira *et al.*, 2017). Organic additives are used to improve the aesthetic qualities of Zn electrodeposition, growth of deposits, deposit structure as well as operating performance and energy efficiency (Abbott *et al.*, 2011; Feng *et al.*, 2017; Kavitha *et al.*, 2006). Organic compounds have been employed as corrosion inhibitors (Verma *et al.*, 2018; Loto, 2018). Research has documented the use of plant extracts in the electroplating of zinc on metals and this. However, application of plant extracts in the electroplating of zinc on metals and their alloys is still evolving (Loto *et al.*, 1992; Loto *et al.*, 2017). This research aims to study the effect of natural extracts of celery ginger and pomegranate on the zinc electrodeposition on mild steel at specific concentrations and plating times.

## 2. Material and methods

Low carbon steel (LCS) plate was machined to dimensions of 100 mm (length) x 10 mm (breadth) x 1.5 mm (thickness). Extracts of ginger, celery and pomegranate fruits were obtained by individually chopping the fruits into miniature quantities. They were subsequently pressed and squeezed to remove solid particles. Zn plate was machined to dimensions of 100 mm (length) x 10 mm (breadth) x 0.1 mm (thickness). The electrodeposition bath consists of $ZnCl_2$, $H_3BO_3$ and KCl. The acid chloride electrolyte was prepared by the dissolution of anhydrous $ZnCl_2$ (71g), crystalline KCl (207g) and $H_3BO_3$ (35g) in 1 L of distilled $H_2O$. The prepared plant extracts added to the acid chloride electrolyte in volumetric concentrations shown below in Table 1. Electrodeposition was performed on LCS by immersing it in the electrolyte baths with different plant extracts and plating time. LCS was made the cathode while the Zn plate was the anode. The plating time was limited to 15 min and 18 min. Potentiodynamic polarization of electroplated LCS was performed Digi-Ivy potentiostat. Plots were obtained at scan rate of 0.0015 V/s between potentials of −1.5 V to +1.5 V. Corrosion current density $J_{cr}$ (A/cm$^2$) and corrosion potential $E_{cr}$ (V) values were obtained Tafel extrapolation. Corrosion current $I_{cr}$ (A) was obtained from the intercept between the anodic-cathodic polarization plots.

**Table 1.** Concentration of plant extracts in bath solution

| **Plant Extract Additives** | Ginger | Celery | Pomegranate |
|---|---|---|---|
| | 0.08 | 0.08 | 0.08 |
| **Volumetric Concentration (L/L)** | 0.12 | 0.12 | 0.12 |
| | 0.16 | 0.16 | 0.16 |

## 3. Results and discussion

### *3.1 Electrodeposition results*

Data for Zn electrodeposition on LCS are shown in Tables 2 and 3. Data obtained in the absence of the plant extracts are depicted in Table 2 while Table 3 presents the data at specific concentrations of celery, ginger and pomegranate extracts. Plant extracts significantly influenced the amount of Zn electrodeposited on LCS. 0.0558g and 0.00621g of Zn (Table 2) deposited on LCS without plant extracts. These values are significantly lower in most cases than the values obtained in Table 4 with the addition of extracts of celery, ginger and pomegranate at different plating times. Ginger extracts enhanced the highest weight gain with more Zn deposited on LCS at concentration of 0.08 L – 0.16 L. Results obtained for weight gained in the presence of pomegranate were generally the lowest. Pomegranate performed poorly until concentration of 0.16L where the weight gain is 0.0723g and 0.1614g. These values are higher than the values obtained in Table 2.

**Table 2.** Electrodeposition parameters without plant extracts

| **Sample Identification** | **Extract Concentration (l/l)** | **Time (min)** | **Weight Gained (mg)** |
|---|---|---|---|
| NA-15 | - | 15 | 0.0558 |
| NA-18 | - | 18 | 0.0621 |

**Table 3.** Electrodeposition parameters with celery, ginger and pomegranate extracts

| | | **Celery** | | **Ginger** | | **Pomegranate** | |
|---|---|---|---|---|---|---|---|
| **Extract Conc. (l/l)** | **Time (min)** | **Sample Identification** | **Weight Gained (mg)** | **Sample Identification** | **Weight Gained (mg)** | **Sample Identification** | **Weight Gained (mg)** |
| 0.08 | 15 | C-4-15 | 0.0763 | G-4-15 | 0.0867 | P-4-15 | 0.0582 |
| 0.08 | 18 | C-4-18 | 0.0883 | G-4-18 | 0.1115 | P-4-18 | 0.0602 |
| 0.12 | 15 | C-6-15 | 0.0995 | G-6-15 | 0.1096 | P-6-15 | 0.0664 |
| 0.12 | 18 | C-6-18 | 0.1011 | G-6-18 | 0.1109 | P-6-18 | 0.062 |
| 0.16 | 15 | C-8-15 | 0.1041 | G-8-15 | 0.1178 | P-8-15 | 0.0723 |
| 0.16 | 18 | C-8-18 | 0.1152 | G-8-18 | 0.1219 | P-8-18 | 0.1614 |

*3.2 Corrosion polarization studies*

Potentiodynamic polarization test was performed for the non-plated and Zn electrodeposited LCS at predetermined concentrations of the extracts and plating time (15 min and 18 min). Data obtained from the test are depicted in Table 4. Corrosion rate of Zn electrodeposited LCS without extract additive stands at 7.96 mm/y with corresponding corrosion current density of 6.76 $10^{-4}$ A/cm$^2$. The corrosion rate value results from the accelerated deterioration of Fe substrate metal of LCS into the electrolyte. Electrodeposition of Zn to LCS decreased the corrosion rate value to 6.58 mm/y as shown in Table 4 while the polarization resistance value increased to 39.3 Ω. This change in value only signifies slight improvement, but is insufficient to protect LCS. The presence of ginger extract in the electrodeposition process (4 ml and 15 mins plating time) reduced the corrosion rate of Zn electrodeposited LCS to 4.40 mm/y. At 6 ml, the corrosion resistance of LCS reduced further to 2.48 mm/y at polarization resistance of 104 Ω. Pomegranate extract at 4 ml and 6 ml gave corrosion rate value of 3.73 mm/y and 2.78 mm/y which slightly varies to the values obtained with ginger extract. The presence of celery extract at 4 ml and plating time of 15 mins gave corrosion rate of 3.75 mm/y which is quite similar to the values with ginger and pomegranate extract. Increase in plating time to 18 min significantly decreased the corrosion rate value of LCS to 1.95 mm/y.

**Table 4**. Potentiodynamic polarization data for non-plated and Zn electrodeposited LCS with respect to extract concentration and plating time

| **Control and No Additive** | | | | | | |
|---|---|---|---|---|---|---|
| **Sample** | **Extract Conc. (l/l)** | **Corrosion Rate (mm/y)** | **Corrosion Current (A)** | **Corrosion Current Density (A/cm$^2$)** | **Corrosion Potential (V)** | **Polarization Resistance (Ω)** |
| Control | - | 7.96 | 7.91E-04 | 6.76E-04 | -0.641 | 32.5 |
| NA-15 | - | 6.58 | 6.55E-04 | 5.59E-04 | -0.645 | 39.3 |
| **Ginger Additive** | | | | | | |
| **Sample** | **Extract Conc. (l/l)** | **Corrosion Rate (mm/y)** | **Corrosion Current (A)** | **Corrosion Current Density (A/cm$^2$)** | **Corrosion Potential (V)** | **Polarization Resistance (Ω)** |
| G-4-15 | 0.08 | 4.40 | 4.38E-04 | 3.74E-04 | -0.610 | 58.7 |
| G-6-15 | 0.12 | 2.48 | 2.47E-04 | 2.11E-04 | -0.780 | 104 |
| **Pomegranate Additive** | | | | | | |
| **Sample** | **Extract Conc. (l/l)** | **Corrosion Rate (mm/y)** | **Corrosion Current (A)** | **Corrosion Current Density (A/cm$^2$)** | **Corrosion Potential (V)** | **Polarization Resistance (Ω)** |
| P-4-15 | 0.08 | 3.73 | 3.71E-04 | 3.17E-04 | -0.295 | 69.3 |
| P-6-15 | 0.12 | 2.78 | 2.76E-04 | 2.36E-04 | -0.353 | 93 |
| **Celery Additive** | | | | | | |
| **Sample** | **Extract Conc. (l/l)** | **Corrosion Rate (mm/y)** | **Corrosion Current (A)** | **Corrosion Current Density (A/cm$^2$)** | **Corrosion Potential (V)** | **Polarization Resistance (Ω)** |
| C-4-15 | 0.08 | 3.75 | 3.72E-04 | 3.18E-04 | -0.693 | 69 |
| C-4-18 | 0.08 | 1.95 | 1.94E-04 | 1.66E-04 | -0.689 | 132 |

**Key: X-Y-Z**. X: Additive Name (No Additive, Celery, Ginger & Pomegranate); Y: Additive Amount (4 ml, 6 ml & 8 ml); Z: Plating Time (15 mins, 18 mins).

## 4. Conclusion

Extracts of ginger, pomegranate and celery improved the amount of Zn electrodeposited on LCS in the admixed solution of $ZnCl_2$, $H_3BO_3$ and KCl solution with respect to plating time and concentration. Corrosion rate from potentiodynamic polarization method confirmed the enhanced corrosion resistance of Zn electrodeposited steel samples in the presence of plant extracts. The corrosion resistance of the low carbon steel was observed to be significantly higher than the resistance of the non-plated steel and Zn electrodeposited steel without the extracts. Corrosion resistance increased with increase in concentration of the plant extracts and plating time. Ginger extract depicted the lowest corrosion rate compared to the non-plated steel which displayed weak corrosion resistance.

**Acknowledgement**
The author is grateful to Covenant University Ota, Ogun State, Nigeria for the sponsorship and provision of research facilities for this project.

**References**

1. Popoola LT, Grema AS, Latinwo GK, Gutti B and Balogun AS 2013 Corrosion problems during oil and gas production and its mitigation *Int. J. Ind. Chem.* **4** 35. https://doi.org/10.1186/2228-5547-4-35.
2. Loto RT 2017 Study of the corrosion behaviour of S32101 duplex and 410 martensitic stainless steel for application in oil refinery distillation systems *J. Mater. Res. Technol.* 6(3) 203–212.
3. Sastri VS 2015 Consequences of corrosion, In Challenges in corrosion, John Wiley & Sons, Inc. Hoboken, NJ. https://doi.org/10.1002/9781119069638.ch5.
4. Lgaz H, Chung I, Albayati MR, Chaouiki A, Salghi R and Mohamed SK 2018 Improved corrosion resistance of mild steel in acidic solution by hydrazone derivatives: An experimental and computational study *Arabian J. Chem.* **13(1)** 2934-2954. https://doi.org/10.1016/j.arabjc.2018.08.004.
5. Basile F, Benito P, Fornasari G, Monti M, Scavetta E, Vaccari A and Tonelli D 2010 A novel electrochemical route for the catalytic coating of metallic supports *Stud. Surf. Sci. Catal.* **175** 51-58.
6. O'Connor MP, Coulthard RM and Plata DL 2018 Electrochemical deposition for the separation and recovery of metals using carbon nanotube-enabled filters *Environ. Sci-Wat. Res.* **4** 58-66.
7. Bicelli LP, Bozzini B, Mele C and D'Urzo L 2008 A Review of nanostructural aspects of metal electrodeposition *Int. J. Electrochem. Sci.* **3** 356 – 408.
8. Peabody AW 2001 Control of pipeline corrosion, 2nd Ed., NACE International: The Corrosion Society, Houston, Texas.
9. Makhlouf ASM 2011 Current and advanced coating technologies for industrial applications, In book: Nanocoatings and ultra-thin-films: Technologies and applications, Woodhead Publishing Limited, Cambridge, UK. http://doi.org/10.1533/9780857094902.1.3.
10. Yuan L, Ding Z, Liu S, Shu W and He Y 2017. Effects of additives on zinc electrodeposition from alkaline zincate solution *T. Nonferr. Metal Soc.* 27(7) 1656–1664.
11. Pereira NM, Pereira CM, Araújo JP and Silva AP 2017 Zinc electrodeposition from deep eutectic solvent containing organic additives *J. Electroanal. Chem.* **801** 545–551.
12. Abbott AP, Barron JC, Frisch G, Ryder KS and Silva AF 2011 The effect of additives on zinc electrodeposition from deep eutectic solvents *Electrochim. Acta* **56(14)** 5272–5279.
13. Feng L, Sun X, Yao S, Liu C, Xing W and Zhang J 2014 Electrocatalysts and catalyst layers for oxygen reduction reaction, In Rotating electrode methods and oxygen reduction reaction, in Rotating electrode methods and oxygen reduction electrocatalysts, **175** 69-76. https://doi.org/10.1016/B978-0-444-63278-4.00003-3.
14. Kavitha B, Santhosh P, Renukadevi M, Kalpana A, Shakkthivel P, Vasudevan T, 2006 Role of organic additives on zinc plating *Surf. Coat. Tech.* **201(6)** 3438–3442.
15. Verma DK, Khan F, Bahadur I, Salman M, Quraishi MA, Vermac C and Ebenso EE 2018 Inhibition performance of glycine max, cuscuta reflexa and spirogyra extracts for mild steel dissolution in acidic medium: Density functional theory and experimental studies *Results Phys.* **10** 665–674.
16. Loto RT 2018 Surface coverage and corrosion inhibition effect of rosmarinus officinalis and zinc oxide on the electrochemical performance of low carbon steel in dilute acid solutions *Results Phys.* **8** 172-179
17. Loto CA, Olofinjana A and Loto RT 2014 Effect of manihot esculenta c. leaf extract additive on the zinc electroplating on mild steel in acid chloride solution *Int. J. Elect. Sci.* **9** 3746 – 3759.
18. Loto CA, Olefjord I and Mattsson H 1992 Surface effects of organic additives on the electrodeposition of zinc on mild *Corros. Prev. Control J.* **39** 142-149.

# Corrosion resistance study of UNS A96110 aluminum alloy in chloride-sulphate solution

**Roland T. Loto**[1*]
[1]Department of Mechanical Engineering, Covenant University, Ogun state, Nigeria

Email: tolu.loto@gmail.com

**Abstract.** Electrochemical analysis of the corrosion resistance of UNS A96110 aluminum alloy (UN6110) in neutral chloride media at 0.5%, 1%, 1.5%, 2%, 2.5%, 3% and 3.5% NaCl concentration, and 3.5% NaCl/sulphuric acid solution at concentrations of 0.00625 M and 0.1 M by potentiodynamic polarization method. UN6110 exhibited lower corrosion in the acid chloride solution compared to the results obtained from the neutral chloride media. The lowest corrosion rate (0.097 mm/y) was obtained from the neutral chloride solution at 0.5% NaCl corresponding to corrosion current density of 8.78 x $10^{-6}$ A/$cm^2$ and polarization resistance of 2927 Ω while the highest value 0.344 mm/y was obtained at 3.5% NaCl concentration. Polarization plot showed significant variation at the anodic portion of the polarization plots with visible potential transients on the passivation portion of the plot. The aluminium alloy exhibited passivation reaction mechanisms due to formation and breakdown of the inert protective oxide. Variation of the cathodic polarization plots shows activation controlled reduction reactions. In sulphate-chloride solution, corrosion rate UN6110 significantly increased to 15.855 mm/y at 0.00625 M and 20.017 mm/y 0.1 M. Optical images obtained from the neutral chloride solution showed the presence of large micro corrosion pits and grooves due to the electrochemical action of chlorides. This differs from the images obtain from the acid chloride solution which showed general surface deterioration couple with well-defined micro and macro corrosion pits.

## 1. Introduction

Aluminum has important industrial applications due to its exceptional toughness, corrosion resistance, and proportion of strength to weight, formability, good weldability, recyclability and impact resistance [1, 2]. The mechanical and physical properties attributed to aluminium can be modified to meet the specific demands of application conditions. The sole purpose of aluminium application in the automotive industry is to reduce vehicle mass without compromising specific strength, design and fuel efficiency requirements [3]. The corrosion resistance of aluminium is due to the formation of $Al_2O_3$ on the metal surface which self-heals when slightly abraded or placed in mildly corrosive environments [4]. The presence of alloying elements within the aluminium substrate matrix is responsible for its extensive structural application. However, the alloying element cause discontinuities in the aluminium matrix and creates heterogeneous metallurgical structure which results in the formation of microscopic voids, flaws and in most cases impurities. This increases the vulnerability of the alloy to localized corrosion deterioration [5-9]. In environments with threshold chloride concentrations, the passive film on aluminium alloy tends to breakdown exposing the steel to continuous corrosion deterioration [10]. Marine environments with its extensive structural architecture are highly corrosive to aluminium alloys due to the presence of chlorides, sulphates [11, 12]. Proper documentation of the threshold corrosion resistance of aluminium alloys is important for material selection purposes. UN6110 aluminum is a 6000-series aluminium alloy with silicon as the main alloy addition. They are extensively applied in automobile vehicle due to their high strength, stiffness to weight ratio, formability and weldability. This article focuses on the corrosion resistance of UN 6110 aluminium alloy in neutral chloride and sulphate contaminated artificial seawater.

## 2. Material and methods

UNS A96110 aluminium alloys (UN6110) procured disposable vehicular part was analysed with PhenomWorld scanning electron microscope (Model No. MVE0224651193) at the Materials Characterization Laboratory, Covenant University, Ota, Ogun State, Nigeria. The elemental composition of 0.5% Fe, 0.22% Cu, 0.15% Mn, 0.043% Mg, 0.2% Zn, 0.1%, 1.5% Si, 0.052% trace elements with the balance being Al was determined. UN6110 sample was cut, sectioned and placed in pre-hardened resin mounts with surface areas of 1 $cm^2$. The exposed surface underwent metallographic preparation with coarse emery papers (120, 240, 400, 600, 800, 1000, 1500, 2000, 2500 and 3000 grits), and 6 μm diamond polishing solution before cleaning with distilled $H_2O$ and $C_3H_6O$. Standard grade NaCl procured from Qualikems, India was formulated in volumetric concentrations of 0.5%, 1%, 1.5%, 2%, 2.5%, 3% and 3.5% NaCl in 200 mL of distilled $H_2O$. Standard grade $H_2SO_4$ at 98% purity was formulated in molar concentration of 0.00625 M, 0.0125 M, 0.025 M, 0.05 M and 0.1 in neutral chloride solution (3.5% NaCl concentration). Polarization test of UN6110 was performed with Digi-Ivy potentiostat. The potentiostat configuration consists of triple electrode system (mounted UN6110), Ag/AgCl reference electrode and Pt wire counter electrode. The electrodes were immersed in 200 mL of the neutral chloride and acid chloride media a cell container interfaced with the potentiostat and computer. Plots were obtained at scan rate of 0.0015 V/s between potentials of -1.75 V to +0.5 V. Corrosion current density, $C_J$ ($A/cm^2$) and corrosion potential, $C_P$ (V) were obtained from the plots by Tafel extrapolation method. Corrosion rate $C_R$ (mm/y) was calculated from the equation below;

$$C_R = \frac{0.00327 \times C_D \times E_Q}{D} \quad (1)$$

$E_Q$ is the equivalent weight (g) of UN6110, 0.00327 is a corrosion rate constant and $D$ is the aluminum alloy density (g).

## 3. Results and discussion

### *3.1 Electrodeposition results*

Results from potentiodynamic polarization test for UN6110 in the neutral chloride and acid chloride electrolyte are presented in Tables 1 and 2. Fig. 1 and 2 shows the corresponding polarization curves for UN6110 corrosion in neutral and acid chloride solution. The rate of deterioration of UN6110 in the acid chloride media was significantly higher compared to the neutral chloride solution from observation of the corrosion rate results. This is as a result of the oxidation of alloy surface by the combined effect of sulphates and chloride anions in the acid chloride solution compared to the effect of chloride anion from the neutral chloride solution. Corrosion rate results from both solutions increased with concentration due to the increased oxidation strength of the electrolyte. The corrosion rate obtained at 0.5% chloride concentration is 0.097 mm/y at corrosion current density of 2.29 x $10^{-5}$ $A/cm^2$ and polarization resistance of 1121 Ω. At 3.5% chloride concentration, the rate of corrosion has increased to 0.344 mm/y. The cathodic polarization curves in Fig. 1 are closely aligned with similar slopes which signifies the $O_2$ reduction and $H_2$ evolution mechanism are under activation control with respect to increase in chloride concentration. However, significant variations are visible on the anodic polarization curves signifying the deteriorating effect of the chloride anions. Miniature current transients are occurred on the anodic polarization curves at 0.5% and 1% chloride concentration resulting from the formation, breakdown and reformation of the passive film on UN6110. After 1% chloride concentration the passivated region of the anodic polarization curves visibly reduced as a result of collapse of the protective film coupled with increase in corrosion rate. This is also evident on the anodic polarization curves. The corrosion potential values shown in Table 1 differ with chloride concentration. At 0.5% chloride concentration, the corrosion potential initiated at -1.011 V and increased (with respect to concentration) to -1.047 V at 3.5% chloride concentration signifying the strong effect of cathodic reaction mechanism under activation control as earlier mentioned.

The polarization curves in Fig. 2 significantly differ from the curves in Fig. 1 due to the presence of suphates in the electrolyte solution. The passivated region present in Fig. 1 iss completely absent (excluding 0M sulphate solution) due to destruction of the passive film on the metallic alloy. The combined effect of sulphates and chlorides caused breakages on the protective film at rates faster than its ability to reform; hence the resulting accelerated corrosion compared to the values obtained in the neutral chloride solution. The corrosion rate results obtained increased with increase in sulphate concentration. At 0.00625 M concentration the corrosion rate of UN6110 to 15.855 mm/y (corrosion current density of 1.44 x $10^{-3}$ A/cm$^2$ and 4.46 x $10^{-4}$ A/cm$^2$, and polarization resistance of 18.60 Ω). The corrosion potential for UN6110 shifts from -0.677 V to -0.657 V due to dominant anodic reaction mechanism related to accelerated oxidation of UN6110 surface and increased corrosion though visible changes in cathodic polarization slope with respect tosulphate concentration shows cathodic reaction mechanisms counterbalanced the anodic counterparts.

**Table 1.** Polarization data of UN6110 corrosion in neutral chloride solution from 0.5% to 3.5% chloride concentration

| Sample | NaCl Conc. (%) | Corrosion Rate, $C_r$ (mm/y) | Corrosion Current, $C_i$ (A) | Corrosion Current Density, $C_j$ (A/cm$^2$) | Corrosion Potential, $C_p$ (V) | Polarization Resistance, $R_p$ (Ω) | Cathodic Tafel Slope, $B_c$ (V/dec) | Anodic Tafel Slope, $B_a$ (V/dec) |
|---|---|---|---|---|---|---|---|---|
| A | 0.5 | 0.097 | 8.78E-06 | 8.78E-06 | -1.011 | 2927.00 | -10.590 | -5.024 |
| B | 1 | 0.128 | 1.17E-05 | 1.17E-05 | -1.010 | 2203.00 | -10.630 | 3.915 |
| C | 1.5 | 0.284 | 2.58E-05 | 2.58E-05 | -1.043 | 1223.00 | -8.889 | 3.479 |
| D | 2 | 0.318 | 2.88E-05 | 2.88E-05 | -1.048 | 897.20 | -10.010 | 2.923 |
| E | 2.5 | 0.319 | 2.90E-05 | 2.90E-05 | -1.031 | 893.30 | -8.593 | 1.958 |
| F | 3 | 0.326 | 2.97E-05 | 2.97E-05 | -1.045 | 864.11 | -9.321 | 1.796 |
| G | 3.5 | 0.344 | 3.12E-05 | 3.12E-05 | -1.047 | 823.2 | -9.202 | 1.786 |

**Table 2.** Polarization data for UN6110 corrosion in neutral chloride solution (3.5% NaCl concentration) at 0 M to 0.1 M sulphate concentration

| Sample | $H_2SO_4$ Conc. (%) | Corrosion Rate, $C_R$ (mm/y) | Corrosion Current, $C_I$ (A) | Corrosion Current Density, $C_J$ (A/cm$^2$) | Corrosion Potential, $C_P$ (V) | Polarization Resistance, $R_p$ (Ω) | Cathodic Tafel Slope, $B_c$ (V/dec) | Anodic Tafel Slope, $B_a$ (V/dec) |
|---|---|---|---|---|---|---|---|---|
| A | 0 | 0.344 | 3.12E-05 | 3.12E-05 | -1.047 | 823.20 | -9.202 | 1.786 |
| B | 0.00625 | 15.855 | 1.44E-03 | 1.44E-03 | -0.677 | 18.60 | -3.315 | 5.635 |
| C | 0.0125 | 13.693 | 1.24E-03 | 1.24E-03 | -0.653 | 20.78 | -3.774 | 1.419 |
| D | 0.025 | 13.148 | 1.19E-03 | 1.19E-03 | -0.652 | 23.24 | -5.805 | 4.272 |
| E | 0.05 | 13.432 | 1.22E-03 | 1.22E-03 | -0.662 | 20.04 | -6.274 | 5.597 |
| F | 0.1 | 20.017 | 1.82E-03 | 1.82E-03 | -0.657 | 14.13 | -6.323 | 2.647 |

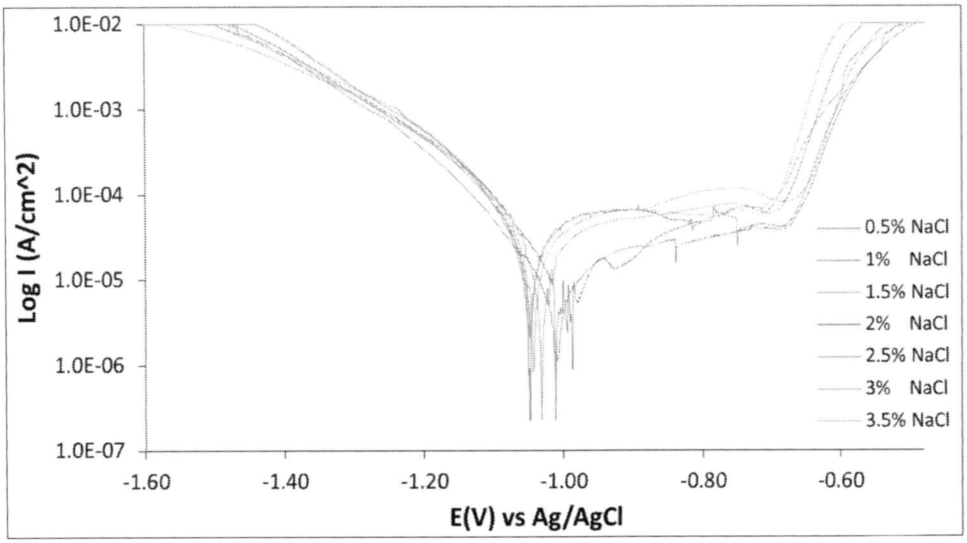


**Fig. 1** Potentiodynamic polarization plots of Al6110 corrosion in neutral chloride solution at 0.5% to 3.5% NaCl concentration

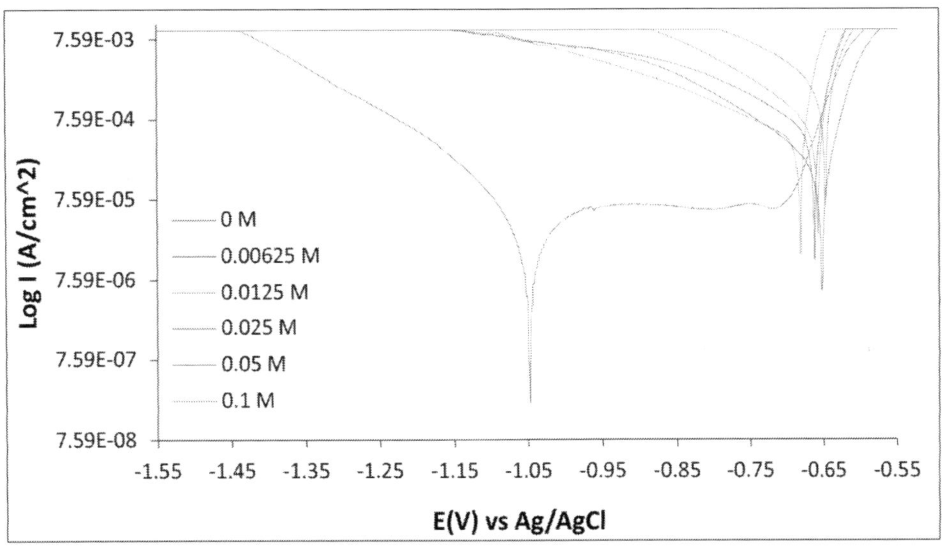


**Fig. 2** Potentiodynamic polarization plots of Al6110 corrosion in acid chloride solution at 0 M to 0.1 M sulphate concentration

*3.2 Optical microscopy studies*

Optical images for UN6110 corrosion in neutral and acid chloride solution are shown Fig. 3 to Fig. 5(b) at mag. x100. Fig. 3 shows the optical image before corrosion. Fig. 4(a) and (b) shows the optical images of UN6110 after corrosion from 0.5% and 3.5% chloride solution. Fig. 4(a) and (b) significantly contrast Fig. 3 due to the debilitating effect of chlorides on the alloy surface. The image in Fig. 4(a) shows extensive surface deterioration with micro-pits resulting from corrosion. Comparing the image to Fig. 4(b), the extent of corrosion damage is much higher with enlarged pits and connecting grooves due to higher degree of surface oxidation resulting increase chloride concentration. The images shown in Fig. 5(a) and (b) significantly differs from Fig. 4(a) and (b) due to differences in the constituents of the electrolyte where corrosion occurred. The deterioration mode in Fig. 5(a) is that of general surface deterioration at 0.00625 M sulphate concentration. However, at 0.1M sulphate concentration well defined micro and macro corrosion pits are visible on the alloy surface.

**Fig. 3** Optical image of UN6110 before corrosion

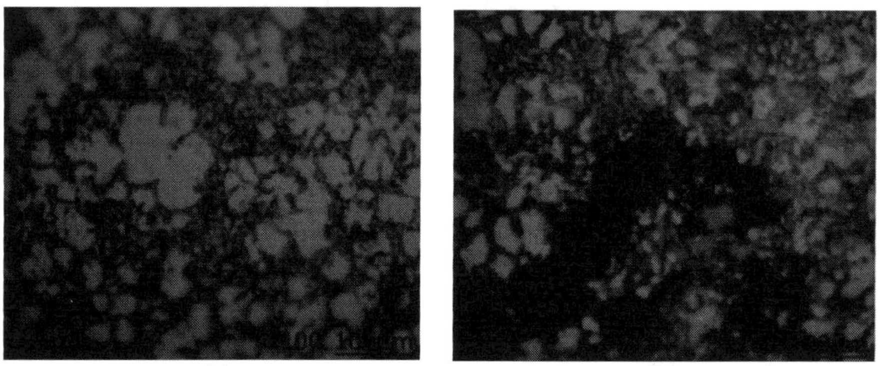

(a) (b)

**Fig. 4** Optical images of UN6110 after corrosion in neutral chloride solution (a) at 0.5% chloride concentration and (b) at 3.5% chloride concentration

(a) (b)

**Fig. 5** Optical images of UN6110 after corrosion in acid chloride solution (a) at 0.00625M sulphate concentration and (b) at 0.1M sulphate concentration

## 4. Conclusion

Electrochemical studies were performed on UN6110 aluminium alloy in neutral chloride and sulphate contaminated artificial seawater solution to assess its corrosion resistance. The corrosion resistance of the alloy in the acid chloride media was significantly lower compared to values obtained in the neutral chloride media. Corrosion rate increased with increase in chloride and sulphate concentration. Passivated regions on the polarization plots from the neutral chloride solution shows localized deterioration of the alloy dominated the corrosion reaction mechanism compared to the plots from the acid chloride media were passivation completely disappeared do to extensive anodic dissolution over the aluminium surface. Images from optical microscopy showed the extent and mode of corrosion damage on UN6110 from both solutions differ from each other.

**Acknowledgement**
The author is grateful to Covenant University Ota, Ogun State, Nigeria for the sponsorship and provision of research facilities for this project.

**References**

1. Birbilis NT, Muster TH, Buchheit RG 2011 Corrosion of Aluminum Alloys. Corrosion Mechanisms in Theory and Practice, Third Edition, CRC Press, pp. 705-736.
2. Zhang Z, Yang X, Zhang J, Zhou G, Xu X, Zou B 2011 Effect of welding parameters on microstructure and mechanical properties of friction stir spot welded 5052 aluminum alloy. *Mater. Des.* **32**:4461-4470.
3. Long RS, Boettcher E, Crawford D 2017 Current and future uses of aluminum in the automotive industry. *JOM.* **69(12)** 2635–2639.
4. Davis JR 1999 Corrosion of aluminum and aluminum alloys, Materials Park, OH, ASM International.
5. Loto RT, Phillip Babalola P 2018 Effect of alumina nano-particle size and weight content on the corrosion resistance of AA1070 aluminum in chloride/sulphate solution. *Results Phys.* **10**:731–737.
6. Loto RT 2019 Investigation of the influence of SiC content and particle size variation on the corrosion resistance of Al-SiC matrix composite in neutral chloride solution. *Int. J. Adv. Manuf. Tech.* **101(9–12)**:2407–2413. https://doi.org/10.1007/s00170-018-3137-9.
7. Loto RT, Babalola P 2019 Corrosion resistance of low SiC particle variation at low weight content on 1060 aluminum matrix composite in sulfate-contaminated seawater *Results Phys.* **13:**102241.
8. Nam ND, Phung VD, Thuy PTP, Dao VA, Kim SH, Yi JS 2019 Corrosion behaviours of hot-extruded Al-xMg alloys. *J. Mater. Res. Techn.* https://doi.org/10.1016/j.jmrt.2019.08.047.
9. Rana RS, Purohit R, Das S 2012 Reviews on the influences of alloying elements on the microstructure and mechanical properties of aluminum alloys and aluminum alloy composites. *IJSRP.* 2(6):1-7.
10. Natishana PM, O'Grady WE 2014 Chloride ion interactions with oxide-covered aluminum leading to pitting corrosion: A Review. *J. Electrochem. Soc.* **161(9)**:C421-C432.
11. Hou X, Gao L, Cui Z, Yin J 2018 Corrosion and Protection of Metal in the Seawater Desalination *IOP Conf Ser Earth Environ Sci* **108** 022037.
12. Fang HHP, Wu KK, Yeong CLY 1990 Corrosion of construction metals under simulated acid rain/fog conditions with high salinity *Water Air Soil Pollut* **53(3–4)** 315–325.
13. Koroleva EV, Thompson GE, Hollrigl G, Bloeck M 1999 Surface morphological changes of aluminium alloys in alkaline solution: Effect of second phase material *Corros Sci* **41(8)** 1475–1495.
14. Moon S-M, Pyun S-I 1997 The corrosion of pure aluminium during cathodic polarization in aqueous solutions *Corros Sci* **39(2)** 399–408.

IOP Conf. Series: Materials Science and Engineering **965** (2020) 012014 doi:10.1088/1757-899X/965/1/012014

# The Investigation of $SiO_2$ structure obtained from the combustion of rice husk

**Yingyot Poo-arporn[1]*, Samrerng Duangnil[1], Denchay Bamrungkoh[1], Supan Boonsuya[1], Surangrat Tonlublao[2], Dechmongkhon Kaewsuwan[2], Narong Chanlek[1], Hideki Nakajima[1], Anuchit Ruangvittayanon[1] and Rungtiva P. Poo-arporn[3]**

[1]Synchrotron Light Research Institute, 111 University Avenue, Nakhon Ratchasima, 30000, Thailand

[2] School of Physics, Institute of Science, Suranaree University of Technology, 111 University Avenue, Muang District, Nakhon Ratchasima 30000, Thailand

[3]Biological Engineering Program, Faculty of Engineering, King Mongkut's University of Technology Thonburi, Bangkok, 10140, Thailand

* yingyot@slri.or.th

**Abstract**. In this work, $SiO_2$ was obtained from the combustion of rice husk. The influence of temperature on its structure and morphology was investigated. At low combustion temperatures, the rice husk ash appeared to be black due to an insufficient temperature for the oxidation process. When the temperature was raised to 1000°C, the color changed to a pale pink. The elemental composition of the rice husk ash was determined by means of x-ray photoelectron spectroscopy. The survey analysis of the XPS spectra revealed the existence of Si, O, C, Ca P, K and N. The elemental fraction of Si was increased as a function of temperature. However, the N 1s XPS signal can be collected from the rice husk after the combustion at 300°C only. Si L edge XANES spectra were recorded to determine the structure of the $SiO_2$. The results suggested the similarity between the $SiO_2$ structure of the rice husk ash and the standard $SiO_2$. The microstructure of the rice husk ash was studied by SEM. It was clearly seen that the pores were created on the rice husk ash. By increasing the pyrolysis temperature, the cellulosic material was removed and consequently produced channels.

## 1. Introduction

Rice husk (RH) is a byproduct of the rice milling process. The outer covering of the rice grain is an agricultural waste material in all rice growing countries [1]. RH can be used as a fuel for power plants [2], a raw material in the production of activated carbon [3] and as the feed stock of industrial chemicals based on silica and silicon compounds [4]. RH is composed of hemicellulose, cellulose, and lignin totaling about 75 – 90% [5] with small amounts of Si and other trace elements such as Fe, Mn, Ca, Na, K, and P [6]. Rice husk ash (RHA) can be obtained from the combustion of RH under controlled conditions [7]. On burning, cellulose and lignin are removed, leaving behind silica ash. The controlled temperature and environment of burning yields better quality of rice husk ash in terms of its particle size. RHA consists mainly of silica ($SiO_2$) (>90%) [8]. The content of impurities in RHA is influenced by geographic location, fertilizers, and agrochemicals. The structure of silica in RHA depends on the temperature and duration of combustion [9]. The amorphous silica can be produced by burning RH at temperatures lower than 700°C under oxidizing conditions [10]. Silica is a raw material that can be used in many applications such as ceramics, agriculture, and medicine [11-13]. Many authors reported the fabrication of silica from RH to increase the value of agricultural waste and to reduce the disposal and pollution problems [14-19].

In this work, silica was fabricated by combustion under an oxygen atmosphere at various temperatures. The characteristics of the silica were investigated by scanning electron spectroscopy (SEM), x-ray photoelectron spectroscopy (XPS) and x-ray absorption near edge structure (XANES).

## 2. Experimental

Rice husk ash was prepared by the combustion of rice husk (Surin, Thailand) under an $O_2$ (99.99%) at a flow rate of 50 ml/min and ambient pressure for 1 hour. The combustion temperatures were varied from 300-1000°C at a heating rate of 5 °C/min with the interval of 100°C and denoted as RHO300, RHO400, RHO500, RHO600, RHO700, RHO800, RHO900 and RHO1000, respectively. The morphology of the RHA was studied by SEM QUANTA 450, FEI. XPS and XANES were performed to investigate the electronic properties and structural information. XPS (PHI VersaProbe II) was carried out at the SUT-NANOTEC-SLRI Joint Facility, BL5.3: SUT-NANOTEC-SLRI, Synchrotron Light Research Institute. The excitation energy was 1486 eV using an Al K$\alpha$ source. Si $L_3$-edge XANES was performed at the BL3.2a: PES beamline of the Synchrotron Light Research Institute, Thailand.

## 3. Results and discussion

The burning of RH under an $O_2$ atmosphere at different temperatures caused the different appearance in the colors of the final ashes (Fig. 1). At 300°C, the color of the RHA was black, suggesting an insufficient temperature for the combustion process. Its elemental compositions were dominated by carbon. The firing of RH at 400-600°C caused changes in the RHA colour from black to white grey. This means that the residual carbon content of RHA obtained in this temperature range was low. The pink color appeared when the temperature was raised to 700°C. At 1000°C, the color of the RHA was pinkish.

**Figure 1.** Optical images of rice husk ash obtained from the combustion of rice husk under an $O_2$ atmosphere at 300°C, 400°C, 500°C, 600°C, 700°C, 800°C, 900°C and 1000°C.

The microstructures of the RHO300, RHO500 and RHO1000 were characterized from a scanning electron micrograph (Fig. 2). The morphological structures of all RHAs have a predominantly rough and undulating surface texture. The major parts of all RHAs retained their original shape. The dense membrane structures were observed on both the inside and outside surfaces without the presence of micropores. The cross-section of RHA revealed an interlayer which consisted of a crisscross mesh of chips between the two surfaces. This inside structure was arranged in a loose honeycombed configuration with a high number of holes. The size of holes on the RHA interlayers were varied ranging from several nanometers to several microns. The presence of these pores correlated to its high surface area when it is ground to powder.

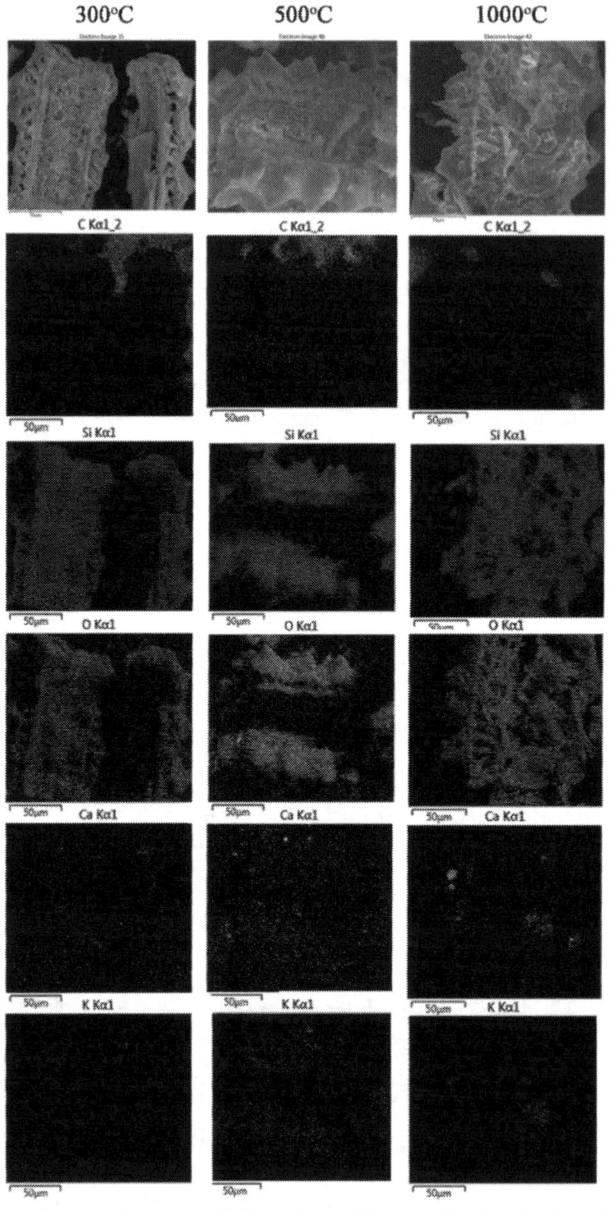


**Figure 2.** The illustration of SEM images of the selected RHA obtained after the combustion of RH under an $O_2$ atmosphere at 300°C, 500°C 1000°C, respectively, and the correspondence EDS spectrum.

The elemental distribution of RHA was further studied by EDS analysis. It could be seen that the internal reticular porous structure was a carbon-rich layer while silicon was packed on the outer surface. The EDS analysis also explored the existing trace elements in all RHAs. The trace elements consisted of K, P, and Ca. At low combustion temperatures (RHO300), dispersal among these elements was observed. Their agglomeration occurred when the temperature was increased, whereby the RHO1000 exhibited the greatest agglomeration. Calcium is an essential element required for structure role in the membrane and cell wall of plants [20]. For RHO300, the remained carbon due to uncompleted combustion of rice husk maintained the high dispersion ability of calcium. However, the lower carbon content was observed on RHO500 and RHO1000. Thus, the remained calcium was favored to agglomerate in RHA at high temperature.

X-ray photoelectron spectroscopy was conducted to examine the chemical composition of carbon compounds in the RHAs (Fig. 3). The survey scan explored the contribution of Si, O, C, K, P and Ca atoms in all samples. For RHO300, the elemental relative ratio of Si:O:C:K:P:Ca was 12.30:44.30:38.15:1.10:0.55:0.62. When the combustion temperature was increased to 500°C, the relative concentration of Si atoms was increased which contrasted with the reduction of carbon. This result suggested the transformation of solid phase carbon to the gas phase carbon dioxide. However, although the temperature was increased to 1000°C, the relative concentration of carbon atoms was not changed significantly. The existing carbon in the RHA confirmed the appearance of a white grey color. As shown in Fig. 3A, the Si 2p XPS spectra of RHA after calcination at all temperatures revealed a similar peak position at 103.0 eV due to the electron emission from $SiO_2$. The result agreed well with the O 1S XPS spectra that exhibited the photoelectron peak due to the O-Si bond at around 532.5 eV. Among the trace elements, potassium exhibited the highest atomic concentration. The K 2p XPS peaks appeared at 293.2 eV and 296 eV and corresponded to the $K^{+}$ in the $K_2O$ compound.

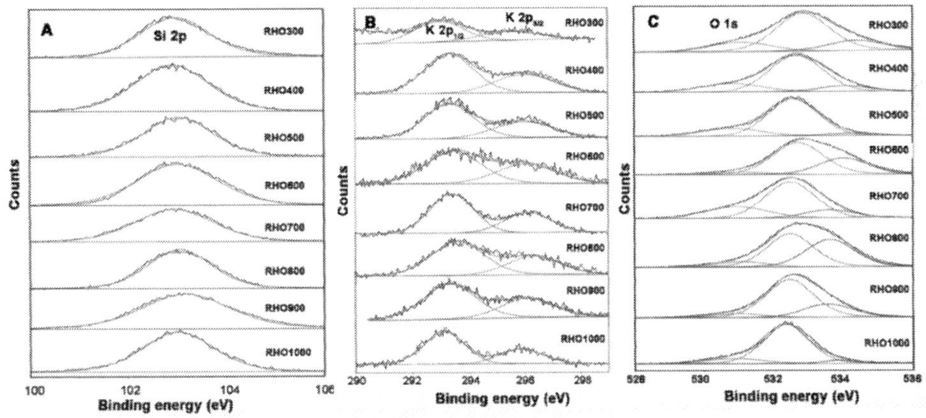


**Figure 3.** The Si 2p (A), K 2p (B) and O 1s XPS spectra rice husk ash after combustion at various temperatures under an $O_2$ atmosphere.

X-ray absorption spectroscopy was employed to investigate the structure and composition of the RHA samples. XANES spectra were recorded at the energy range of 110-120 eV to examine the photoabsorption characteristics of the Si atom (Fig. 4). Normalized Si L-edge XANES of standard $SiO_2$ revealed the main absorption peaks at around 105.7 eV, 106.3 eV and 108.1 eV. Si L-edge XANES spectroscopy involves the electronic transition of Si 2p states. The first two well resolved peaks are transitions to the first unoccupied 3s-like states, from the spin-orbit splitting of Si 2p ($2p_{3/2}$ and $2p_{1/2}$). The peak at 108.1 eV was due to a 2p to 3d transition, which features a broad peak (due to multiple scattering). It has been reported that the splitting and relative peak ratios of the first two peaks relate to the long-range order of Si–O tetrahedral coordination. For the well crystalline phase, the increase in shared oxygen atoms led to better resolved splitting with an increase in the ratio between the intensity of the first and second peaks [20]. The comparison of the XANES spectrum between standard $SiO_2$ and RHAs showed the similar features for RHA and standard $SiO_2$. The results suggested the formation of the $SiO_2$ compound in RHA. However, as described above, the rice husk ash after combustion at 300-

700°C was in a short-range order coordination. A combustion temperature of 1000°C was necessary to transform the $SiO_2$ from the amorphous to the crystalline phase.

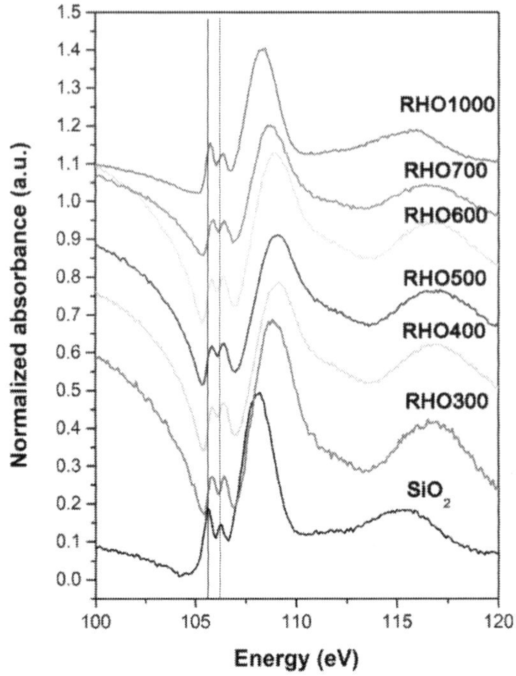


**Figure 4.** Normalized Si L-edge XANES of standard $SiO_2$ and rice husk ash after combustion at temperature 300°C, 400°C, 500°C, 600°C, 700°C and 1000°C.

## 4. Conclusions

Rice husk ash can be prepared by firing under an oxidizing atmosphere. The combustion temperatures of 300-1000oC caused changes in RHA color from black to pinkish. The morphological structure of all RHAs has a predominantly rough and undulating surface texture. After combustion, the major parts of all RHAs retained their original shape. The dense membrane structures were observed on both the inside and outside surfaces without the presence of micropores. The cross-section of RHA revealed an interlayer which consisted of a crisscross mesh of chips between the two surfaces. Using EDS analysis, it could be seen that the internal reticular porous structure was a carbon-rich layer while silicon was packed on the outer surface which agreed well with the XPS result. The rice husk ash after combustion at 300-700°C was in a short-range order coordination. A combustion temperature of 1000°C was necessary to transform the $SiO_2$ from the amorphous to the crystalline phase.

## 5. Acknowledgement

This work was supported financially by the Synchrotron Light Research Institute, Thailand, as the subsidy fund of fiscal year 2019 (grant number SLRI2019-S09).

## 6. References

[1] Z. Shamsollahi and A. Partovinia, "Recent advances on pollutants removal by rice husk as a bio-based adsorbent: A critical review," Journal of Environmental Management, vol. 246, pp. 314-323, 2019.
[2] R. Pode, "Potential applications of rice husk ash waste from rice husk biomass powerplant,"

Renewable and Sustainable Energy Reviews, vol. 53, pp. 1468-1485, 2016.

[3] D. Kalderis, D. Koutoulakis, P. Paraskeva, E. Diamadopoulos, E. Otal, J. O. del Valle and C. Fernandez-Pereira, "Adsorption of polluting substances on activated carbons prepared from rice husk and sugarcane bagasse," Chemical Engineering Journal, vol. 144, pp. 42-50, 2008.

[4] C. Real, M. D. Alcala and J. M. Criado, "Preparation of silica from rice husk," Journal of the American Ceramic Society, vol. 79, pp. 2012-2016, 1996.

[5] A. Rahman and J. lsmail, "Preparation and characterization of a spherical gel from a low-cost material," Journal of Materials Chemistry, vol. 3, pp. 931-934, 1993.

[6] H. Moayedi, B. Aghel, M. M. Abdullahi, H. Nguyen and A. S. A Rashid, "Applications of Rice Husk Ash as Green and Sustainable Biomass," Journal of Cleaner Production, vol. 273, pp. 117851, 2019.

[7] S. Zhang, S. Zhu, H. Zhang, X. Liu and H. Zhang, "Evaluation of pyrolysis behavior and products properties of rice husk after combined pretreatment of washing and torrefaction," Biomass and Bioenergy, vol. 127, pp. 105293, (2019).

[8] SK S. Hossain, L. Mathur and P. K. Roy, "Rice husk/rice husk ash as an alternative source of silica in ceramics: A review," Journal of Asian Ceramic Societies, vol. 6, pp. 299-313, 2018.

[9] S. R. Almeida, C. Elicker, B. M. Vieira, T. H. Cabral, A. F. Silva, P. J. S. Filho, C. W. Raubach, C. A. Hartwig, M. F. Mesko, M. L. Moreira and Sergio Cava, "Black SiO2 nanoparticles obtained by pyrolysis of rice husk," Dyes and Pigments, vol. 164, pp. 272-278, 2019.

[10] J. A. S. Costa and C. M. Paranhos, "Systematic evaluation of amorphous silica production from rice husk ashes," Journal of Cleaner Production, vol. 192, pp. 688-697, 2018.

[11] Z. Yanga, Z Zhao, J. Yu and Z. Ren, "Preparation of silica ceramic cores by the preceramic pyrolysis technology using silicone resin as precursor and binder," Materials Chemistry and Physics, vol. 223, pp. 676-682, 2019.

[12] R. Si, Y. Han, D. Wu, F. Qiao, L. Bai, Z. Wang and H. Yan, "Ionic liquid-organic-functionalized ordered mesoporous silica-integrated dispersive solid-phase extraction for determination of plant growth regulators in fresh Panax ginseng," Talanta, vol. 207, pp. 120247, 2020.

[13] V. Mamaeva, C. Sahlgren and M. Lindén, "Mesoporous silica nanoparticles in medicine-Recent advances," Advanced Drug Delivery Reviews, vol. 65, pp. 689-702, 2013.

[14] R. A. Bakar, R. Yahya and S. N. Gan, "Production of high purity amorphous silica from rice husk," Procedia Chemistry, vol. 19, pp. 189 – 195, 2016.

[15] S. K. Hubadillah, M. H. D. Othman, A. F. Ismaila, M. A. Rahman, J. Jaafara, Y. Iwamoto, S. Honda, M. I. H. M. Dzahirc and M. Z. M. Yusop, "Fabrication of low cost, green silica based ceramic hollow fibre membrane prepared from waste rice husk for water filtration application," Ceramics International, vol. 44, pp. 10498–10509, 2018.

[16] S. Sankar, S. K. Sharma, N. Kaur, B. Lee, D. Y. Kim, S. Lee, H. Jung, "Biogenerated silica nanoparticles synthesized from sticky, red, and brown rice husk ashes by chemical method," Ceramics International, vol. 42, pp. 4875-4885, 2016.

[17] W. Wang, H. Chen, J. Fang, and M. Lai, "Large-scale preparation of rice-husk-derived mesoporous SiO2@TiO2 as efficient and promising photocatalysts for organic contaminants degradation," Applied Surface Science, vol. 467–468, pp. 1187-1194, 2019.

[18] J. Cui, F. Cheng, J. Lin, J. Yang, K. Jiang, Z. Wen, and J. Sun, "High surface area C/SiO2 composites from rice husks as a high-performance anode for lithium ion batteries," Powder Technology, vol. 311, pp. 1-8, 2017.

[19] J. Kong, S. Gao, Y. Liu, X. Jin, D. Wei, S. Jiang, K. Ye, J. Wang, P. Xing, and X. Luo, "Recycling of carbonized rice husk for producing high purity silicon by the combination of electric arc smelting and slag refining," Journal of Hazardous Materials, vol. 380, pp. 120827, 2019.

[20] Philip J. White, "Calcium channels in higher plants," Biochimica et Biophysica Acta, vol. 1465, pp. 171-189, 2000.

[21] S. A. Shaw, D. Peak and M. J. Hendry, "Investigation of acidic dissolution of mixed clays between pH 1.0 and –3.0 using Si and Al X-ray absorption near edge structure," Geochimica et

Cosmochimica Acta, vol. 73, pp. 4151–4165,2009.

IOP Conf. Series: Materials Science and Engineering **965** (2020) 012015 doi:10.1088/1757-899X/965/1/012015

# The preparation of composites between polyaniline-silver (PANI-Ag) via interfacial polymerization

**W Panloetparnich[1], V Loryuenyong[1] and A Buasri[1,*]**

[1]Department of Materials Science and Engineering, Faculty of Engineering and Industrial Technology, Silpakorn University, Nakhon Pathom, 73000, Thailand

*Corresponding author: achanai130@gmail.com

**Abstract.** In this study, polyaniline-silver (PANI-Ag) composites between aniline monomer and Ag particles in silver nitrate were prepared via interfacial polymerization method at room temperature. The objective was to study the effects of mass ratios of aniline monomer to Ag particles in silver nitrate (1:0.5, 1:1, 1:1.5, 1:2 and 1:2.5) on the electrochemical properties. The characterization techniques included FTIR, XRD, SEM and EDAX. FTIR and XRD results presented the success of the synthesis of the composites and crystallization of the PANI-Ag composites due to the presence of silver. SEM and EDAX techniques indicated that polyaniline nanofibers were successfully covered on the Ag particles. The CV results showed the higher specific capacitance with increasing amounts of Ag particles.

## 1. Introduction

Nowadays, devices for storing electrical energy which are batteries and capacitors are widely used in electrical industries. However, the process of charge-discharge electricity in batteries take quit a long time and limiting the usage in many applications. Therefore, the researchers are currently interested in studying and developing the capacitors to be more capacitance and electrical energy storage. Capacitors can charge-discharge electricity very quickly than batteries. Generally, the electrodes are produced from carbon materials (carbon nanotubes (CNTs)), transition metal oxide (NiO, Co3O4, MnO2, TiO2, SnO2 and CuO) [1-7] and conducting polymers are simple chemical and electrochemical synthesis, economical price and high environmental stability.

Among conductive polymer, polyaniline (PANI) has gained more attention to produce electrode due to there are 4 oxidation states (leucoemeraldine, emeraldine base, emeraldine salt and pernigraniline). It has high specific capacitance pseudocapacitor. Although conductive state of PANI is only green protonated emeraldine. On the other hand, PANI has lower conductivity than metal. Thus, the researchers try to improve electrical properties of PANI via metal doping method. Silver particle (Ag) is transition metal, there are high electrical properties and high surface area. After electrode synthesis, Ag particle can reduce electrical resistivity or charge transfer resistance. [8]

Therefore, different methods have been used to improve capacitors properties. In the Interfacial polymerization technique, the polymerization is occurred in immiscible organic/aqueous biphasic system. This technique no add surfactant, simply synthesize and variety of organic solvent. Recently, Mahesh D. Bedre et al. prepared PANI-Ag nanocomposites via employing interfacial polymerization using ammonium persulphate as an oxidizing agent. Resulting in the reduction of $Ag^{+}$ to $Ag^{0}$ and to conducting emeraldine salt.[9] Ahmad Abdolahi et al. synthesized PANI-Ag composites with core-shell structure by interfacial polymerization method.[10] Kuk Ro Yoon et al. synthesized polypyrrole-silver (PPy-Ag) nanocomposites by aqueous polymerization (PPy-Ag-Aq) and interfacial polymerization (PPy-Ag-IF). The silver particles were dispersed in PPy-Ag-IF better than in PPy-Ag-Aq.[11]

In this research, the PANI-Ag composites use addition of Ag in PANI to reduce electrical resistance and to enhance the electrochemical properties. The transition metal particles are widely used due to the presence of unique magnetic, optical and catalytic properties. The electrodes was based on the current composite materials were prepared via interfacial polymerization by dispersing silver particles and aniline monomer in the water phase and inorganic phase, respectively. The objectives of this work is to study morphology and electrochemical properties of PANI-Ag composites.[10]

## 2. Experimental

### *2.1. Materials*

Aniline monomer (LOBA CHEMIE PVT.LTD) purchased from India, silver nitrate (POCH) purchased from Poland, ammonium persulphate (Ajax Finechem Pty.Ltd) purchased from Poland Australia, chloroform (ACl Lacscan) purchased from Thailand and sulfuric acid (EMSURE GO made in Germany).

### *2.2. Synthesis of polyaniline-silver composites*

The synthesis of composite materials between polyaniline and Ag particles from interfacial polymerization by preparing solutions with 2 systems. The first system is organic phase by put 0.36 ml of aniline monomer in 20 ml of chloroform and the second system is silver nitrate in distilled water by mixed 0.23 g of ammonium persulfate with 0.2 g of silver nitrate in10 ml of sulphuric acid (1 M) after that mixed and stirred solution each system for 30 minutes. The mixing 2 systems, pour the second system into the first system and leave for interfacial polymerization for 24 hours then filtered and washed with acetone and distilled water. It will turn into a precipitate and colorless solution. Finally dried for 48 hours. PANI-Ag composites with different mass ratio of Ag particles in silver nitrate (1:0.5, 1:1, 1:1.5, 1:2 and 1:2.5).

### *2.3. Characterizations*

PANI-Ag composites were characterized by fourier transform infrared spectroscopy (FTIR) (NICOLET, Impact, United States), x-ray diffraction (XRD) (SHIMADZU, Lab X, XRD 6100, Japan), scanning electron microscopy (SEM) (MIRA3 TESCAN, Australia) of composites sputtered coated with Pd/Au.

### *2.4. Preparation and characterization of the electrode*

The electrochemical analysis of PANI-Ag composites was carried out through the standard three-electrode cell configuration. The counter electrode was platinum wire, the reference electrode was Ag/AgCl and the working electrode was prepared by mixing 80 %w/w of the PANI-Ag composites, 10 %w/w of carbon black (Vulcan XC 72R) and 10 %w/w of polyvinylidene fluoride. The mixture was dissolved in 1-methyl-2-pyrrolidinone (NMP) solvent, then drop coated onto the glassy carbon electrode (GCE). The prepared GCE working electrode was dried at 80 °C for 24 hours to eliminate the solvent.[11] The CV test using DY2311 Potential (Digi-Ivy, Inc, United States) and the electrolyte was 1M $H_2SO_4$ solution and all the electrochemical analyses were tested at room temperature.

## 3. Results and discussion

### *3.1. Fourier transform infrared spectroscopy*

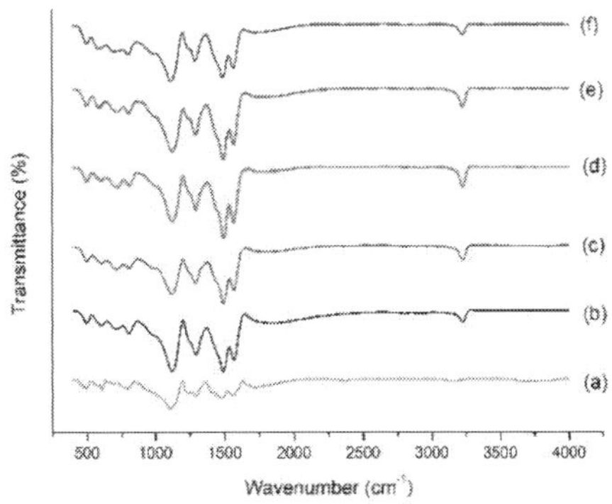


**Figure 1.** FTIR spectra of the PANI-Ag composites (a) Pure PANI, (b) PANI-Ag0.5, (c) PANI-Ag1.0, (d) PANI-Ag1.5, (e) PANI-Ag2.0 and (f) PANI-Ag2.5.

The FTIR spectra of pure PANI and PANI-Ag composites were recorded and shown in Figure 1. Pure PANI and PANI-Ag composites has the main peaks at 1108.98 $cm^{-1}$,1478.74 $cm^{-1}$ and 1559.57 $cm^{-1}$ associated with C=N and C=C stretching of quinoid and benzenoid rings in the polyaniline structure. The peak at about 3228.53 $cm^{-1}$ is O-H stretching from $HNO_3$ and not presented in pure PANI. Based on FTIR results, it could be confirmed that polyaniline was successfully synthesized due to the characteristic peaks corresponding to emeraldine salt (ES) is 3138.82 $cm^{-1}$, 1559.57 $cm^{-1}$, 1478.74 $cm^{-1}$, 1294.30 $cm^{-1}$ and 1238.45 $cm^{-1}$.[9,10]

### *3.2. X-ray diffraction*

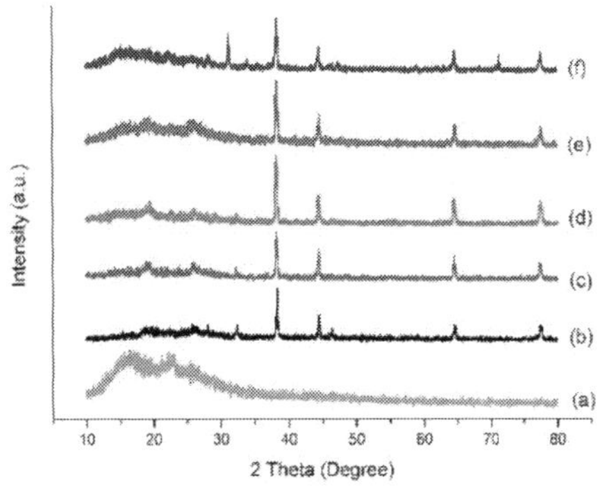


**Figure 2.** XRD patterns of the PANI-Ag composites (a) Pure PANI, (b) PANI-Ag0.5, (c) PANI-Ag1.0, (d) PANI-Ag1.5, (e) PANI-Ag2.0 and (f) PANI-Ag2.5.

The XRD patterns of PANI-Ag composites are presented in Figure 2. The Figure 2a showed only broad peak at about 2θ value of 10-35°. This is the characteristic peak of PANI corresponded to amorphous nature of PANI. The results. In the PANI-Ag composites indicated the appearance of the FCC structure of

IOP Conf. Series: Materials Science and Engineering **965** (2020) 012015 doi:10.1088/1757-899X/965/1/012015

Ag particles, as indicated by the (1 1 1), (2 0 0), (2 2 0) and (3 1 1) diffraction peaks (2θ) at 38.1°, 44.2°, 64.3° and 77.5°, respectively. As a result, All peak of PANI-Ag composite match with the **JCPDS file 04-0783** of Ag. [9]

*3.3. Scanning electron microscopy*

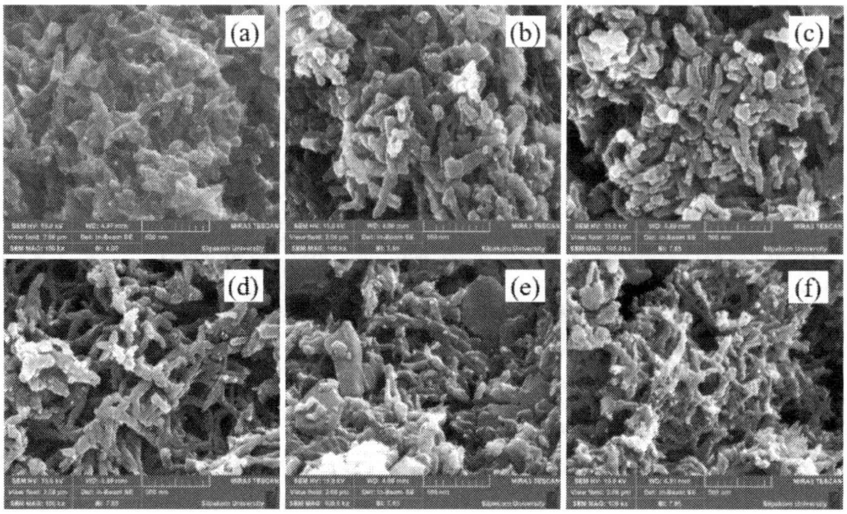


**Figure 3.** SEM images of the PANI-Ag composites (a) Pure PANI, (b) PANI-Ag0.5, (c) PANI-Ag1.0, (d) PANI-Ag1.5, (e) PANI-Ag2.0 and (f) PANI-Ag2.5.

The morphology of PANI-Ag composites is shown in Figure 3. Normally, three stages are proposed in PANI synthesis: 1) nucleation, 2) initial growth, 3) secondary growth. Figure 3a, in interfacial polymerization form only PANI nanofiber in initial growth stage. It is proposed the secondary growth stage is suppressed due to the secondary growth stage leads to PANI in micro-size scale. [8] From Figure 3b-f shown the formation of polyaniline nanofibers with surface roughness. However, increasing Ag contents found more gap between polyaniline nanofibers. From Figure 3e, PANI-Ag2.0 exhibited the formation of polyaniline nanofibers and sheets. From Figure 3f, PANI-Ag2.5 found the smallest polyaniline nanofibers compared to other Ag contents. Thus, SEM images indicated the interfacial polymerization caused the formation of aniline monomer on Ag particles that act as the growth centers and Ag particles are covered with a layer of polyaniline nanofibers.[9,10]

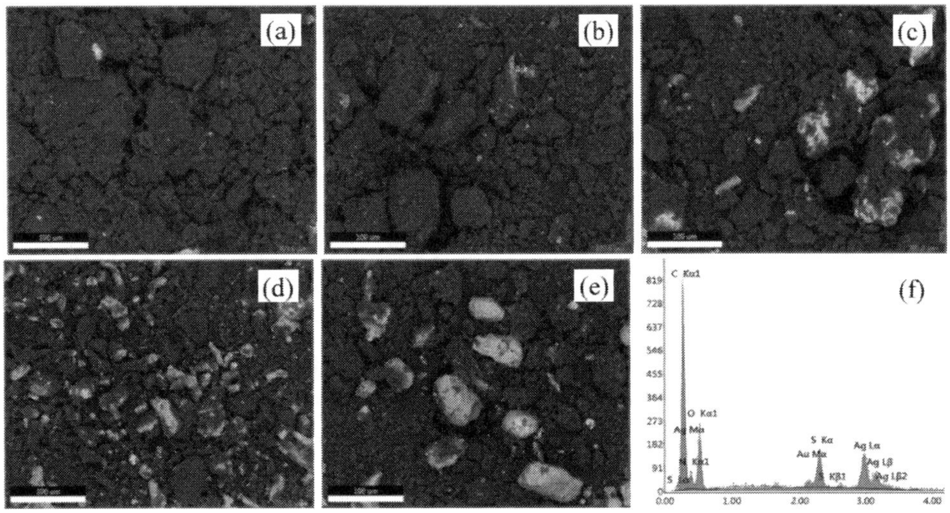


**Figure 4.** EDAX images of the PANI-Ag composites (a) PANI-Ag0.5, (b) PANI-Ag1.0, (c) PANI-Ag1.5, (d) PANI-Ag2.0 and (e) PANI-Ag2.5 and (f) EDAX mapping.

The study of elemental compositions using the EDAX technique are shown in Figure 4. From Figure 4a-e, presented the dispersion of Ag particles in PANI as a matrix. Figure 4d, PANI-Ag2.0 is the most dispersion and distribution and Figure 4e, PANI-Ag2.5 exhibited agglomeration of Ag particles in the PANI matrix because of Ag particles content increasing. The EDAX mapping, Figure 4f, presented %elements of carbon, nitrogen, oxygen and silver.[12]

### *3.4. Cyclic voltammetry*

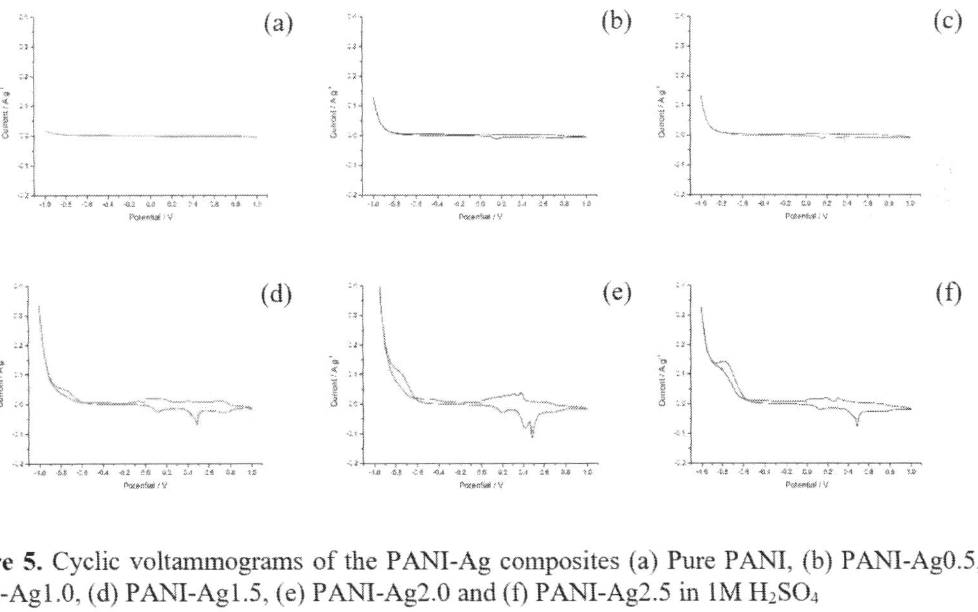


**Figure 5.** Cyclic voltammograms of the PANI-Ag composites (a) Pure PANI, (b) PANI-Ag0.5, (c) PANI-Ag1.0, (d) PANI-Ag1.5, (e) PANI-Ag2.0 and (f) PANI-Ag2.5 in 1M $H_2SO_4$

The results of the study of electrical properties of pure PANI and PANI-Ag composites using the CV technique are shown in Figure 5, at potential range of -1 to +1 V and using 1M $H_2SO_4$ as the electrolyte at scan rate is 50 mV/s. Figure 5a, pure PANI didn't presented redox reaction and low specific capacitance. Figure 5b-c, PANI-Ag0.5 and PANI-Ag1.0 presented only reduction reaction peak and no oxidation reaction peak. In PANI-Ag composites synthesis when Ag particles content increasing, will show oxidation and reduction reaction peak that indicated transition metal (silver) can occur redox reaction which is pseudocapacitors characteristics. The composite materials in different Ag particles content, Ag particles content increasing found the area under the CV curves increasing and specific capacitance also increasing. The specific capacitances of PANI-Ag composites were calculated from equation 1[12], and the results were shown in Table 1:

$$C_{sp} = \frac{\int_{V_i}^{V_f} |I| dV}{2 \times \Delta V \times m \times SR} \tag{1}$$

Where the integral part in the numerator gives the area under the CV curve, $\Delta V$ is the potential range, I is the applied current, m is the weight of the deposited PANI-Ag composites on GCE and SR is scan rate.[12]

**Table 1.** Specific capacitance of the PANI-Ag composites with different mass ratio of Ag.

| Composites | Specific capacitance (F/g) |
|---|---|
| Pure PANI | 18.84 |
| PANI-Ag0.5 | 58.84 |
| PANI-Ag1.0 | 61.34 |
| PANI-Ag1.5 | 185.17 |
| PANI-Ag2.0 | 385.67 |
| PANI-Ag2.5 | 276.94 |

## 4. Conclusion

The preparation of PANI-Ag composites with various silver contents was successful through the interfacial polymerization method. This could be confirmed by the FTIR results, which showed peaks attributed to the functional groups in the polyaniline structure and XRD analysis indicated the crystalline formation of Ag particles, which was confirmed by the presence of sharp peaks. The XRD board peak suggested the amorphous nature polyaniline. In accordance with SEM micrographs, observed PANI formation in Ag particles which act as the growth centers. From studying the dispersion of Ag particles in PANI and compositions of elements with EDAX technique, it was found that PANI-Ag2.0 had the most dispersion and distribution. The results of EDAX were in consistent corresponding with CV technique. The specific capacitance of PANI-Ag2.0 was optimal with the value of 385.67 F/g.

## 5. Acknowledgement

This research was financially supported by Department of Materials Science and Engineering, Faculty of Engineering and Industrial Technology, Silpakorn University.

## References

[1] Rhazi1 EI M, Majid S, Elbasri1 M, Salih1 F E, Oularbi1 L and Lafdi K 2018 *Int. Nano Lett.* **8** 79-99.
[2] Bora C, Kalita A, Das D, Dolui S K and Mukhopadhyay P K 2014 *POLYM INT.* **63**.
[3] Bedre M D, Deshpande R, Salimath B and Abbaraju V 2010 *INT J MIN MET MATER.* 2010 **52** 149-154.

[4] Khawas K, Kumari P, Daripa S, Oraon R and Kuila B K 2017 *ChemistrySelect.* **2** 11783-11789.
[5] Su H, Wang T, Zhang S, Song J, Mao C, Niu H, Jin B, Wu J and Tian Y 2012 *Solid State Sci.* **14** 677–681.
[6] Prasanna B.P., Avadhani D.N., Vinuth Raj, Yogesh kumar K. and Raghu M.S. 2019 *SURRF.ENG.APPL* **55** 463-471.
[7] Ates M, Serin M A, Ekmen and Ertas Y N 2015 *Polym. Bull.* **72**.
[8] Abdolahi A, Hamzah E, Ibrahim Z and Hashim S 2012 *Mater* **5(8)** 1487-1494.

[9] Bedre M D, Basavaraja S, Salwe B D, Shivakumar V, Arunkumar L and *Venkataraman A* 2009 *Polym. Compos.* **30** 1668-1677.
[10] Abdolahi A, Hamzah E, Ibrahim Z and Hashim S 2013 *Adv. Mater. Res.* **686** 86-91.
[11] Singu B S and Yoon K 2018 *J. Alloys Compd.* **742**.
[12] Singu, B. and K. Yoon, *Highly exfoliated GO-PPy-Ag ternary nanocomposite for electrochemical supercapacitor.* Electrochimica Acta, 2018. **268**.

IOP Conf. Series: Materials Science and Engineering **965** (2020) 012016 doi:10.1088/1757-899X/965/1/012016

# Effect of titanium dioxide on the properties of poly(butylene adipate-co-terephthalate) films

**N. Arssanasuwan[1], P. Threepopnatkul[1,*] and A. Sittattrakul[1]**

[1]Department of Materials Science and Engineering, Faculty of Engineering and Industrial Technology, Silpakorn University, Nakhon Pathom, 73000, Thailand

*Corresponding author: poonsubt@yahoo.com

**Abstract.** In this study, the effect of titanium dioxide ($TiO_2$) in poly(butylene adipate-co-terephthalate) was investigated. PBAT/$TiO_2$ biocomposite films containing different weight percentages (0.25, 0.5, 1 and 3 %w/w) of $TiO_2$ were prepared by mixing PBAT and $TiO_2$ in a twin screw extruder and thin films were prepared by cast film extruder. The thickness of each film is in the range of 50-90 micron. PBAT/$TiO_2$ biocomposite films were characterized by Fourier transform infrared spectroscopy, X-Ray diffraction, Tensile testing, Scanning electron microscopy and Differential scanning calorimetry. Incorporating $TiO_2$ (0.25-1 %w/w) in PBAT matrix was found the Young's modulus and Tensile strength were increased while the addition of $TiO_2$ more than 1 %w/w (3 %w/w) exhibited lower Young's modulus and Tensile strength as a result of agglomeration of $TiO_2$ that was shown in SEM images. Thermal properties of biocomposite films showed the percent of crystallinity increased by increasing $TiO_2$ concentration up to 1 %w/w. Moreover, addition of $TiO_2$ concentration up to 3 %w/w the percentage of crystallinity was decreased probably as a result of agglomeration of $TiO_2$ that affected on crystallization of PBAT/$TiO_2$ biocomposite films.

## 1. Introduction

In the present, packagings are widely use in food and postharvest industry. After use, it produces garbage problem because it is not biodegradable. Therefore, in many countries including Thailand have idea to use biodegradable polymer as packaging material. Such as poly(butylene adipate-co-terephthalate) (PBAT) that provides many advantages whether biodegradable, ductile, highly flexible like LDPE and good film-forming ability [1]. In addition, many researchers had investigated the effect of filler in PBAT matrix for improving the properties of polymer for better application in the field of packaging. For example, incorporation of organoclay [2], zinc oxide (ZnO) [3] and titanium dioxide ($TiO_2$) in PBAT was investigated by Raja Venkatesan et al. [4]. Raja Venkatesan et al. studied the effect of adding $TiO_2$ for improving the mechanical properties of PBAT. The results showed that the mechanical properties of PBAT films were improved by increasing $TiO_2$ concentration. However, films were prepared by solvent casting method. In this research, were investigated the PBAT/$TiO_2$ biocomposite films that were prepared by cast film extrusion. Furthermore, $TiO_2$ is a photocatalyst and is useful as ethylene scavenging and antimicrobial agent which is supported by the research of Ubonrat Siripatrawan et al. [5] that find the chitosan film which is incorporated with $TiO_2$ can exhibit ethylene photodegradation and antimicrobial activity against Gram positive, Gram negative bacteria and fungi and also according to the research of Raja Venkatesan et al.[4] that incorporated $TiO_2$ in PBAT matrix which they found the $TiO_2$ could exhibit antimicrobial activity against Gram positive and Gram negative.

The aim of this research is to develop PBAT/$TiO_2$ biodegradable films for food packaging by cast film extrusion and investigate the effect of $TiO_2$ on mechanical, morphological and thermal properties of PBAT matrix.

## 2. Experimental

The PBAT (Ecoflex® F blend C1200) was purchased from BASF SE. Titanium dioxide (anatse, -325 mesh) was purchased from SIGMA-ALDRICH, Co., Canada

PBAT/$TiO_2$ biocomposite films were prepared by mixing $TiO_2$ in different weight percentages (0.25, 0.5, 1 and 3 %w/w) and mixing with PBAT by twin screw extruder which obtained extrudate. Extrudate was cut to pellet and prepared thin films by cast film extruder. The thickness of each film is in range of 50-90 micron.

PBAT/$TiO_2$ biocomposite films were characterized by Fourier transform infrared spectroscopy (FTIR) (NICOLET, Impact). X-ray diffraction (XRD) (SHIMADZU, Lab X, XRD 6100). Mechanical properties of films were performed by Universal Testing Machine according to ASTM D882 (Instron Corp.). Scanning electron microscopy (SEM, Mira3 TESCAN) was performed to observe microstructure of films which was prepared by immersed film in liquid nitrogen and subtract it to crack and sputtered coated with Pd/Au. Thermal properties of films were performed by differential scanning calorimetry (DSC) (TGA/DSC1, Mettler Toledo) under nitrogen atmosphere.

## 3. Results and discussion

### *3.1. Fourier transform infrared spectroscopy*

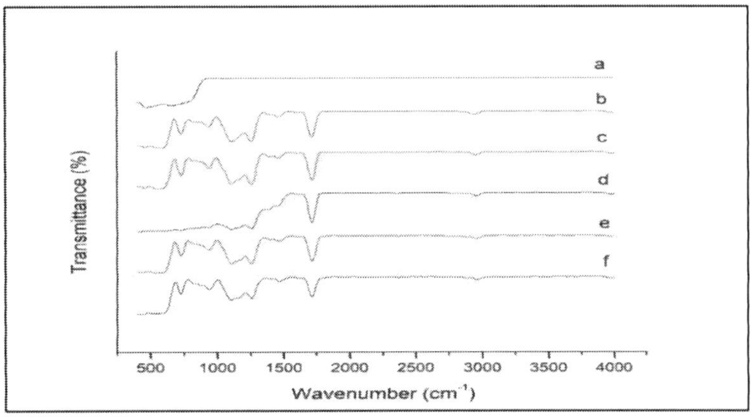


**Figure 1.** FTIR Spectra of a) $TiO_2$ powder and PBAT/$TiO_2$ biocomposite films containing different weight percentage of $TiO_2$ b) 0%, c) 0.25%, d) 0.5%, e) 1% and f) 3%w/w

Fourier transform infrared spectra of $TiO_2$ and PBAT/$TiO_2$ biocomposite films are shown in Figure 1. $TiO_2$ powder in Figure 1a) shows peak in range of 720-530 $cm^{-1}$ correspond to bending vibration of Ti-O-Ti in $TiO_2$ structure [5]. Peak observed at 650 $cm^{-1}$ is attributed to Ti-O vibration [4]. Peak of neat PBAT film is given in Figure 1b). Peak in range of 3200-2700 $cm^{-1}$ is attributed to -OH stretching. The carbonyl groups C=O of ester linkage presented peak around 1730-1715 $cm^{-1}$. Two peaks of benzene substitute were located at 700 and 900 $cm^{-1}$ [4]. FTIR spectra of PBAT/$TiO_2$ biocomposite films containing difference weight percentage of $TiO_2$ were shown in Figure 1c) - f). PBAT/$TiO_2$ biocmposite films show the similar peaks as neat PBAT but the peak shown at 1029 $cm^{-1}$ indicates bending vibration of Ti-O-C which suggests that is the present of $TiO_2$ in PBAT matrix [5]. The peak at 1163 $cm^{-1}$ is attributed to C-O-C stretching might be caused by interaction between -OH groups of PBAT and Titanium [5].

### *3.2. X-ray diffraction*

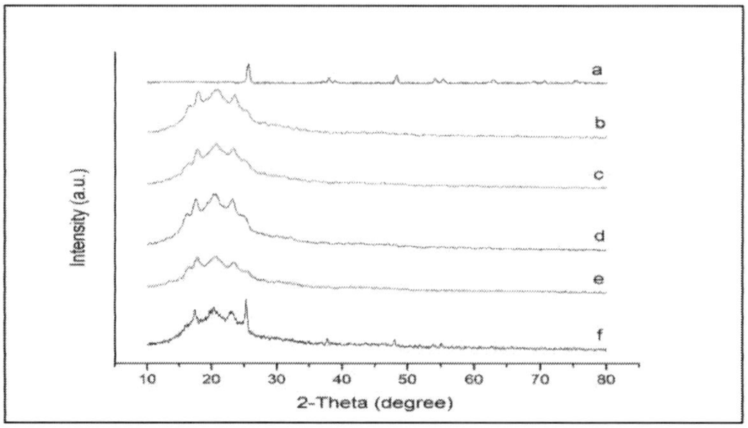


**Figure 2.** XRD pattern of a) $TiO_2$ powder and PBAT/$TiO_2$ biocomposite films containing different weight percentage of $TiO_2$ b) 0% c) 0.25% d) 0.5% e) 1% and f) 3%

The XRD patterns of $TiO_2$ powder and PBAT/$TiO_2$ biocomposite films are presented in Figure 2. The XRD pattern of $TiO_2$ powder shows in Figure 2a) that represented the crystalline peak at 25.3°, 38° and 48° correspond to the crystal plane of (101), (004) and (200) respectively [4]. Peak of neat PBAT film is given in Figure 2b) show a broad peak approximately 18.1° and 20.3° correspond to amorphous structure of PBAT [4]. Figure 2c) - f) show XRD patterns of PBAT/$TiO_2$ biocomposite films containing different weight percentages of $TiO_2$. Neat PBAT film and film containing lower amount of $TiO_2$ (0.25%w/w) represent crystalline peak approximately 17.8° suggestion that is the present of plane (101) which is the plane of orthorhombic of PBAT. The evidenced peaks at 25.3°, 38° and 48° were observed when increasing $TiO_2$ concentration (0.5-3 %w/w) that attributed the plane of (101), (004) and (200) respectively. In addition, the intensity of these peaks was increased when increasing $TiO_2$ concentration [4].

### *3.3. Mechanical properties*

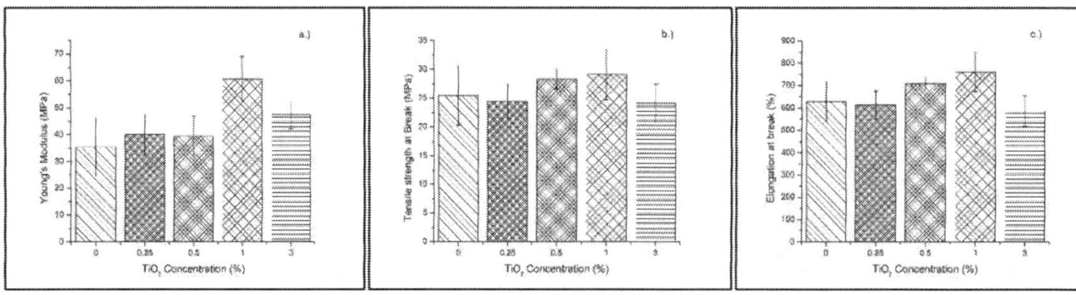


**Figure 3.** Mechanical properties of Neat PBAT and PBAT/$TiO_2$ biocomposite films containing different weight percentage of $TiO_2$ a.) Young's Modulus, b.) Tensile Strength and c.) Elongation at Break

The Young's modulus, tensile strength and elongation at break of neat PBAT and PBAT/$TiO_2$ biocomposite films are shown in Figure 3. Young's modulus and tensile strength of neat PBAT and PBAT/$TiO_2$ biocomposite films have the similar trend. Young's modulus and tensile strength of films increased with increasing $TiO_2$ concentration (0.25-1 %w/w) and then decreased at 3 %w/w of $TiO_2$. Because at 0.25-1 %w/w of $TiO_2$ concentration, $TiO_2$ could uniformly disperse in PBAT matrix and act as reinforcing agent in PBAT films. But the increase of $TiO_2$ concentration up to 3 %w/w, $TiO_2$ tended to agglomerate probably due to decreasing distance between particles [5]. The decrease of Young's modulus and tensile strength was due to the agglomerates of excess $TiO_2$ as presented in cross section of films in SEM images that maybe disrupt the PBAT chain. Elongation at break of films are shown in Figure 3c). Elongation at break of films containing $TiO_2$ at 0.25 %w/w was lower than neat PBAT film. However, elongation at break increased when increasing of $TiO_2$ concentration up to 1 %w/w and decreased when increasing of $TiO_2$ concentration to 3 %w/w.

### *3.4. Scanning electron microscopy*

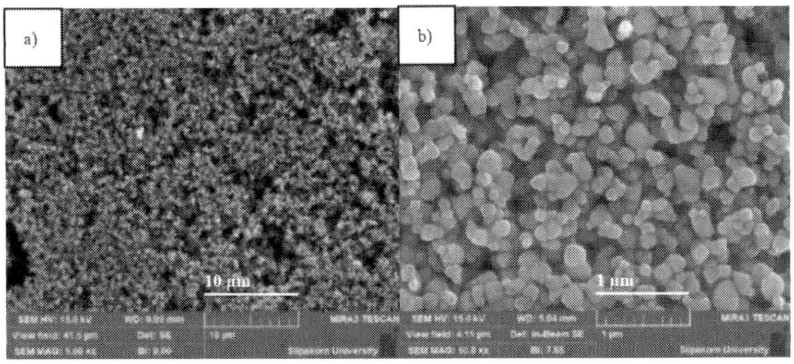


**Figure 4.** Morphology of $TiO_2$ powder at different magnification. a) 5000X and b) 50000X

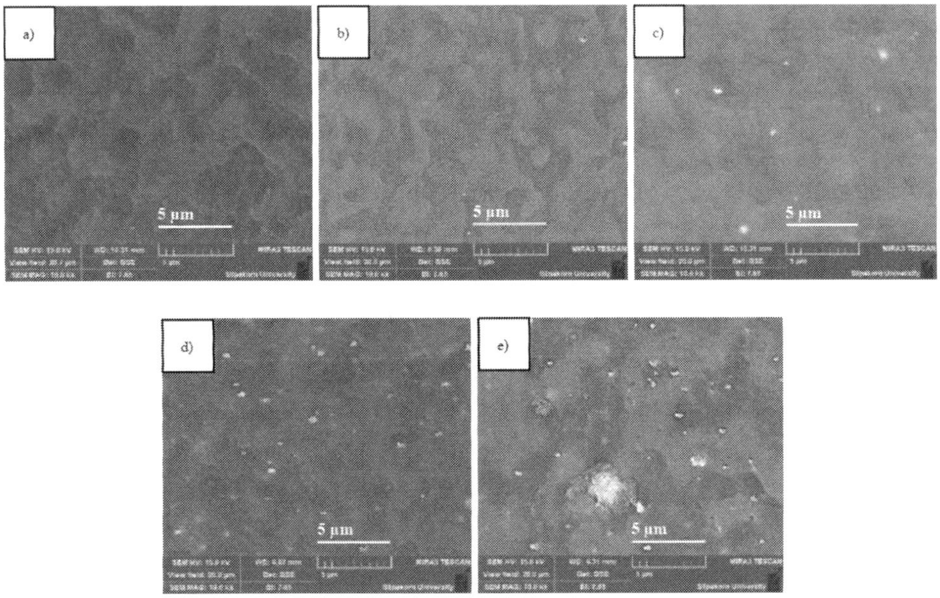


**Figure 5.** Morphology of PBAT/$TiO_2$ biocomposite films containing different weight percentage of $TiO_2$ at 10000X a) 0% b) 0.25% c) 0.5% d) 1% and e) 3%

The morphology of $TiO_2$ powder is shown in Figure 4. $TiO_2$ powder has good dispersion and the size is in the range of 0.1-0.4 micron. The SEM images of PBAT/$TiO_2$ biocomposite films containing different weight percentage of $TiO_2$ are depicted in Figure 5. Neat PBAT film in Figure 5a) shows the smooth surface. When increasing of $TiO_2$ concentration at 0.25-1 %w/w, $TiO_2$ can uniformly disperse in PBAT matrix which are shown in Figure 5b) - d). But the film containing 3 %w/w of $TiO_2$ is shown in the Figure 5e) exhibits the large cluster of agglomerated $TiO_2$ powder [4]. The $TiO_2$ agglomeration can affect the properties of these films.

### *3.5. Thermal properties*

**Table 1.** Thermal properties from DSC of neat PBAT and PBAT/$TiO_2$ biocomposite films

| Sample | $T_m$ (C) | Tc (C) | $\Delta H_m$ | $X_C$ (%) |
|---|---|---|---|---|
| Neat PBAT | 124.72 | 87.20 | 5.86 | 5.14 |
| PBAT/$TiO_2$ 0.25% | 124.64 | 86.62 | 7.77 | 6.83 |
| PBAT/$TiO_2$ 0.5% | 125.45 | 88.04 | 8.03 | 7.08 |
| PBAT/$TiO_2$ 1% | 125.05 | 88.54 | 8.18 | 7.25 |
| PBAT/$TiO_2$ 3% | 125.03 | 86.96 | 7.71 | 6.97 |

Thermal properties of neat PBAT and PBAT/$TiO_2$ biocomposite films are shown in Table 1. The melting temperature and percent of crystallinity of neat PBAT film is 124.72 °C and 5.14 % respectively. An addition of $TiO_2$ concentration at 0.25-3 %w/w found that the melting temperature of the films were 124.64, 125.45, 125.05 and 125.03 ° C. Percent of crystallinity of PBAT/$TiO_2$ biocomposite films containing $TiO_2$ concentration at 0.25-3 %w/w were 6.83, 7.08, 7.25 and 6.97% respectively and confirmed that $TiO_2$ act as nucleating agent in PBAT matrix which supported by the research of Alois Karl Schlarb et al. that $TiO_2$ acts as an efficient nucleating agent for PLA [6]. Melting temperature of neat PBAT and PBAT/$TiO_2$ biocomposite films have the similar values but percent of crystallinity of PBAT/$TiO_2$ biocomposite films was increased when increasing of $TiO_2$ concentration. In addition, percent of crystallinity of the film that contains $TiO_2$ at 3 %w/w is decreased probably due to agglomeration of $TiO_2$ that affects on crystallization of PBAT/$TiO_2$ biocomposite films.

## 4. Conclusion

PBAT/$TiO_2$ biocomposite films were prepared by cast film extrusion. The properties of biocomposite films were characterized by FTIR, XRD, Tensile testing, SEM and DSC. Incorporation of $TiO_2$ in PBAT matrix in the range of 0.25-1 %w/w was found to increase mechanical properties and thermal properties. Because $TiO_2$ were dispersed homogeneously in PBAT matrix that shown in SEM images. Percent of crystallinity is increased when increasing $TiO_2$ concentration. When adding of $TiO_2$ in excessive concentration (3 %w/w) was found to decrease mechanical properties and percent of crystallinity because the agglomeration of $TiO_2$ in the form of cluster made poor dispersion in PBAT matrix and hindered crystallization in biocomposite films.

## 5. Acknowledgement

This research was financially supported by the Department of Materials Science and Engineering, Faculty of Engineering and Industrial Technology, Silpakorn University and National Research and Innovation Information System : NRIIS)

IOP Conf. Series: Materials Science and Engineering **965** (2020) 012016 doi:10.1088/1757-899X/965/1/012016

## References

[1] Ferreira V F, Cividanes S L, Gouveia F R and Lona F M L 2019 *J. Polym Eng Sci.* **59** 7-15.
[2] Falcão G A M, Vitorino M B C, Almeida T G, Bardi M A G, Carvalho L H and Canedo E L 2017 *J. Polym Bull.* **74** 1-14.
[3] Venkatesan R. and Rajeswari N 2016 *J. Polym. Advan. Technol.* **28**
[4] Venkatesan R. and Rajeswari N 2017 *J. Polym. Advan. Technol.*
[5] Siripatrawan U and Kaewklin P 2018 *J. Food Hydrocoll.* **84** 125-134.
[6] Nomai, J, Suksut B, and Schlarb A K 2015 *J. Appl. Sci. Technol.* **8** 251-258.

IOP Conf. Series: Materials Science and Engineering **965** (2020) 012017 doi:10.1088/1757-899X/965/1/012017

# Fabrication of non-woven hybrid natural fiber/poly(lactic acid) composite via prepreg lamination

**J Khieomuang and C Thongpin***

Department of Materials Science and Engineering, Faculty of Engineering and Industrial Technology, Silpakorn University, Nakhon Pathom, 73000, Thailand

*Corresponding author: THONGPIN_C@su.ac.th.

**Abstract.** Hybrid natural fiber-reinforced polylactic acid (PLA) composites were prepared using a prepreg lamination. Two types of the long continuous natural fibers, i.e. sisal fiber (SF) and pineapple leaf fiber (PALF) were used in this work. The SF and PALF were treated using alkalization (NaOH) and then silane coupling agent (APS) to improve the interfacial bonding between fiber and PLA. After the surface treatment, the chemical group on the surface of SF and PALF were analyzed by Fourier transform infrared (FTIR). The FTIR result confirmed that hemicellulose and lignin were removed and also silane was grafted on the fiber surface. The micrographs by scanning electron microscopy (SEM) of the fibers observed morphological change of the fiber surface. The surface treatment improved the thermal stability of fiber as detected by thermogravimetry analyzer (TGA). The tensile strength of single SF and PALF were increased after surface treatment by alkalization and silane, respectively. Moreover, the tensile strength increases as an increase in the fiber length. Thus, the hybrid natural fiber/PLA composites showed high tensile properties compared with SF/PLA, PALF/PLA composites and pure PLA.

## 1. Introduction

Nowadays, the automotive industry is growing and widely used. The automobile parts are generally manufactured from plastics derived from the petroleum industry. When the cycle life is expired or damage occurs from the accident, it will become refuse and affect the environment because it cannot be naturally decomposed. Various agencies have seen problems with the issuance of regulations regarding the production of equipment. For example, end-of-life vehicles or ELVs in Europe declared new rules and regulation for vehicle recycling. [1] In addition, the entrepreneurs are turning to the use of bioplastics instead of plastic from petroleum industrial. The interesting bioplastic is polylactic acid (PLA) because the price is not very high and the physical and mechanical properties of PLA are comp able to other plastics used in the automotive industry. PLA has better young modulus values that can be used to replace these plastics but PLA has brittle. [2] Therefore, reinforcement of PLA by natural fibers to make the material stronger is one of interesting approaches. Natural fiber has high strength and lightweight. [3] Moreover, there are many types of fiber in Thailand such as pineapple leaf fiber and sisal fiber. Both of the fibers have good mechanical properties. The reinforcing natural fiber in PLA matrix can improve mechanical properties of composites and the surface treatment of fiber will improve interfacial properties between fiber and polymer were proposed in the literature. [4-6] Asaithambi et al. studied the effect of fiber surface treatment by benzoyl peroxide (BP) of banana/sisal fiber (BSF) reinforced Polylactic acid (PLA), had on the mechanical properties of the composite. It was found that tensile and flexural modulus of BP treated BSF increase compared with virgin PLA and untreated BSF reinforced composites. [7] The concept of this work is to use the pineapple leaf and sisal fibers in continuous and non-woven fiber to reinforce PLA and study the effect of surface treatments as well as fiber length on mechanical properties of single sisal and pineapple leaf fibers. The mechanical properties of hybrid long natural fiber reinforced PLA composites were also evaluated, which is eco-friendly environment materials for application in automotive application.

## 2. Experimental

### *2.1. Materials*
PLA pellet was provided from Nature Works LLC (Minnetonka, USA). In this work, the sisal fiber (SF) and pineapple leaf fiber (PALF) obtained from Phetchabun and Songkhla Province, Thailand. The length of both fibers was 150±3.0 mm and the diameter was 0.14±0.04 mm (L/D ratio of 1093.5±330.1). Sodium hydroxide pellet supplied by RCI Labscan Limited (Bangkok, Thailand) and 3-aminpropyl triethoxysilane, APS (supplied by Sigma-Aldrich (San Luis, USA)) were used as fiber surface modifiers. Glacial acetic acid was also supplied by RCI Labscan Limited.

### *2.2. Fiber surface treatments*
The SF and PALF, they were treated with 5 wt% NaOH solution for 3 h at room temperature using a fiber/NaOH solution ratio of 1:10 (w/v) to remove hemicellulose and lignin. And then the fibers were washed and dried at 60 °C for 12 h. After that, the alkalized fibers were soaked in silane aqueous solution (5% v/v) at room temperature for 3 h. The pH of solution was adjusted to 3-4 with glacial acetic acid and fiber/silane solution ratio of 1:10 (w/v). Finally, the fibers were washed in reverse osmosis water to neutralize the fiber surface and dried at 60 °C for 12 h.

### *2.3. Preparation of prepreg laminate composites*
Before fabrication of the prepreg laminate composites, The PLA pellet and fibers (SF and PALF) were oven-dried at 60 °C for 12 h. The preparation of prepreg laminate composites contained two stages. In the first stage, PLA sheet was prepared with thickness of 0.8 mm using compression moulding at pressure 800 psi and temperature 170 °C. The fiber were threaded around aluminium frame (15×15×0.15 cm). The PLA sheet was placed on top and bottom of the frame and then pressed inside compression moulding at 170 °C to obtain unidirectional fiber prepreg with PLA as a matrix in thickness 1.5 mm. SF prepreg and PALF prepreg were laminated and pressed as the same condition. Finally, the unidirectional hybrid natural fiber/PLA composite with 15% fiber content (The PALF/SF ratio of 0.5:0.5 (v/v)) was obtained.

### *2.4. Characterization and Testing*
Fourier transform infrared (FTIR) spectroscopy was used to analyse the characteristic functional groups of natural fibers before and after chemical surface treatment. In order to study the thermal stability of untreated and treated natural fiber a TGA (Mettler Toledo) was used and carried out in nitrogen atmosphere between 50 to 600 °C at a heating rate of 10 °C/min. Morphology of fiber surface and fractured surface of prepreg laminate composite were investigated using a Hitachi TM3030 SEM. The tensile test of natural fiber was carried out based on ASTM D3288. Natural single fiber was randomly chosen, mounted and glued on a paper tape before measurement. The tensile testing was performed using Instron 5969 with the sample gage length of 2.5, 5 and 10 cm and crosshead speed of 1 mm/min with a 5kN-load cell. The tensile test of prepreg laminate composites was carried out using Instron 5969 with 5kN-load cell according to standard methods ASTM D638 type IV. The crosshead speed of 50 mm/min and the wide and thickness of tensile specimen are 6 mm and 3 mm, respectively.

## 3. Results and discussion

### *3.1. Fourier transform infrared spectroscopy*
FTIR spectra of untreated sisal fiber (Un_SF) as presented in Figure 1(a) showed the board peak around 3400 $cm^{-1}$ attributed to OH stretching whereas peak at 2900 $cm^{-1}$ is CH stretching of alkyl group in

aliphatic bond of cellulose, hemicellulose and lignin. After Alkaline treatment (Na_SF), peak at 1724 $cm^{-1}$ of C=O stretching vibration and 1222 $cm^{-1}$ of C-O-C symmetrical stretching were decreased. This result confirmed that hemicellulose and lignin were partially removed from the surface of fiber. The fibers after silane treatment (APS_SF) showed peak at 1506 $cm^{-1}$ of $NH_2$ vibration indicating that silane successfully grafted on the fibers. Similar result of untreated pineapple leaf fiber (Un_PALF) and its surface treatment (Na_PALF, APS_PALF) was found in other researchs. [8-9]

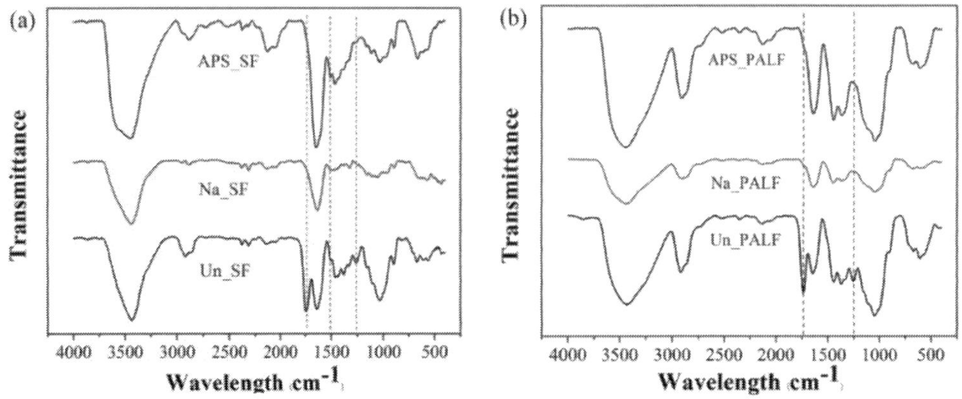


**Figure 1.** FTIR spectra of (a) SF and (b) PALF.

*3.2. Thermogravimetric analysis*

The Figure 2 presents TGA and DTG thermograms of SF and PALF. Both of Un_SF and Un_PALF show three stages of thermal decompositions. The first stage is evaporation of water around 70 °C. The second stage around 295 °C indicating degradation of hemicellulose and the late stage around 358 °C attributed to degradation of cellulose and lignin. After surface treatment fiber with alkaline and silane coupling agent, the thermograms reveals two stages of decomposition which indicates that hemicellulose and lignin were removed, in agreement with FTIR spectra. The thermal decomposition temperation of Na_SF, APS_SF, Na_PALF and APS_PALF were 360.35, 353.27, 363.71 and 359.41 °C, respectively. The thermal stability of treatment fiber was higher than untreatment fiber because remaining cellulose had natural crystalline structure. It can also be seen that the thermal stability of alkali-treated fiber was slightly higher than those of APS-treated and untreated fiber. In addition, the residue of APS-treated fiber has higher char percentage than those of alkali-treated and untreated fiber. [6-9]

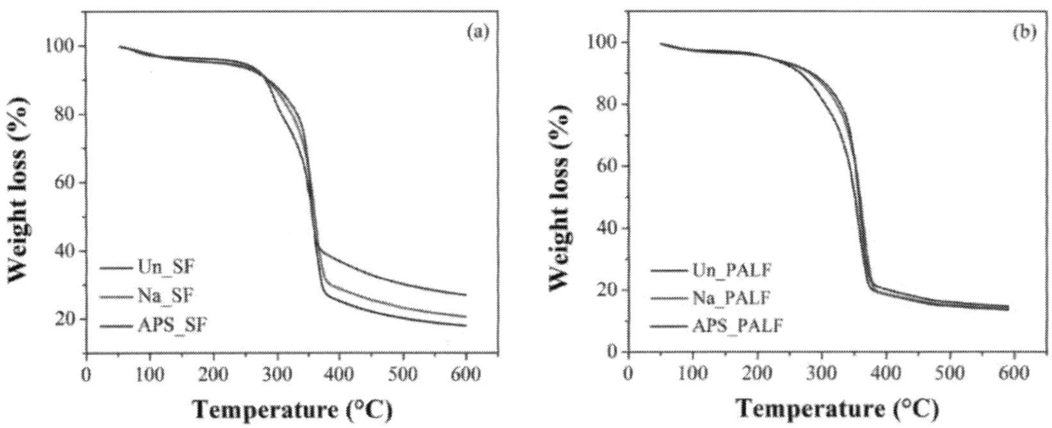

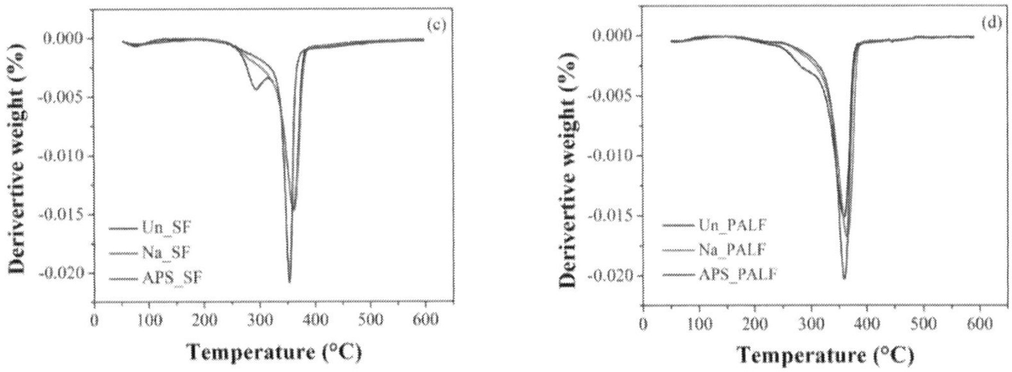


**Figure 2.** (a,b) TGA and (c,d) DTG thermograms of SF and PALF.

### *3.3. Surface morphology of fibers*

The Figure 3 display surface morphology of untreated and treated SF and PALF. Un_SF has rough surface due to a presence of some component such as lignin and hemicellulose covering on the surface. The Na_SF surface in figure 3(b) seems seen become smoother than the untreated fiber due to remove of hemicellulose and lignin on the fiber surface. APS_SF as showed in figure 3(c) has smooth surface and appears siloxane deposited on the surface. The surface morphology of PALF (Figure 3(d)-(f)) like same as sisal fiber. Thus, these morphology results confirm the change of fiber surface after surface treatment. [8-9]

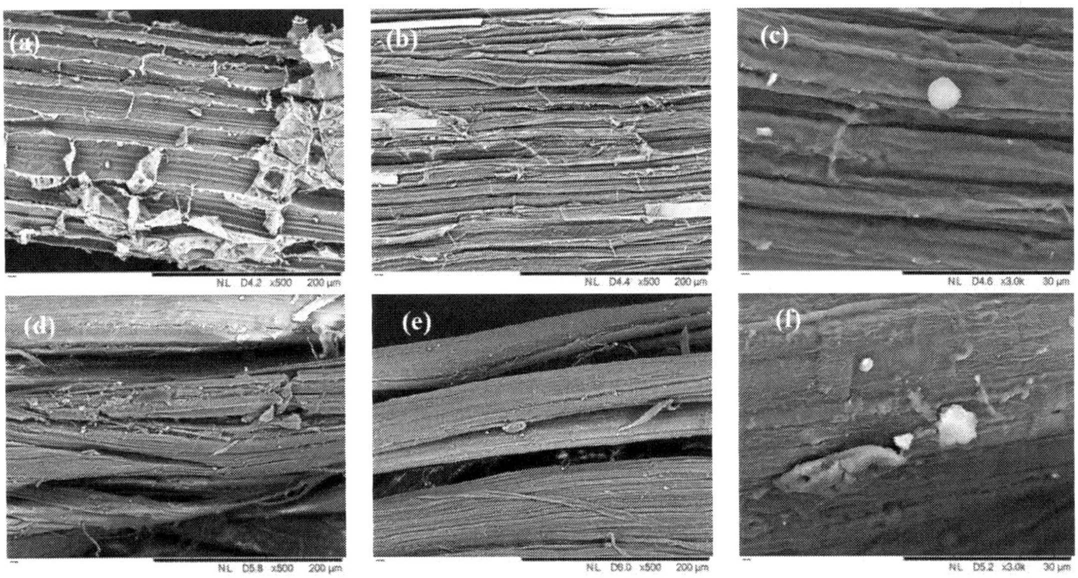


**Figure 3.** SEM micrographs of (a) Un_SF, (b) Na_SF, (c) APS_SF, (d) Un_PALF, (e) Na_ PALF and (f) APS_ PALF.

### *3.4. Tensile properties of fiber*

The effect of surface treatment and fiber length on tensile strength, Young's modulus and elongation at break of SF and PALF is shown in figure 4. The tensile strength and Young's modulus of SF and PALF increased when modified surface because hemicellulose and lignin were removed, resulting in an increase in degree of cellulose. Moreover, the tensile strength and Young's modulus of fiber increased when fiber length increased because long fiber have more surface area than short fiber. Elongation at break of SF and PALF increase after alkaline treatment and decreased after silane treatment. Moreover, the variability of tensile properties was probably related to a microstructure of fiber. [9-10]

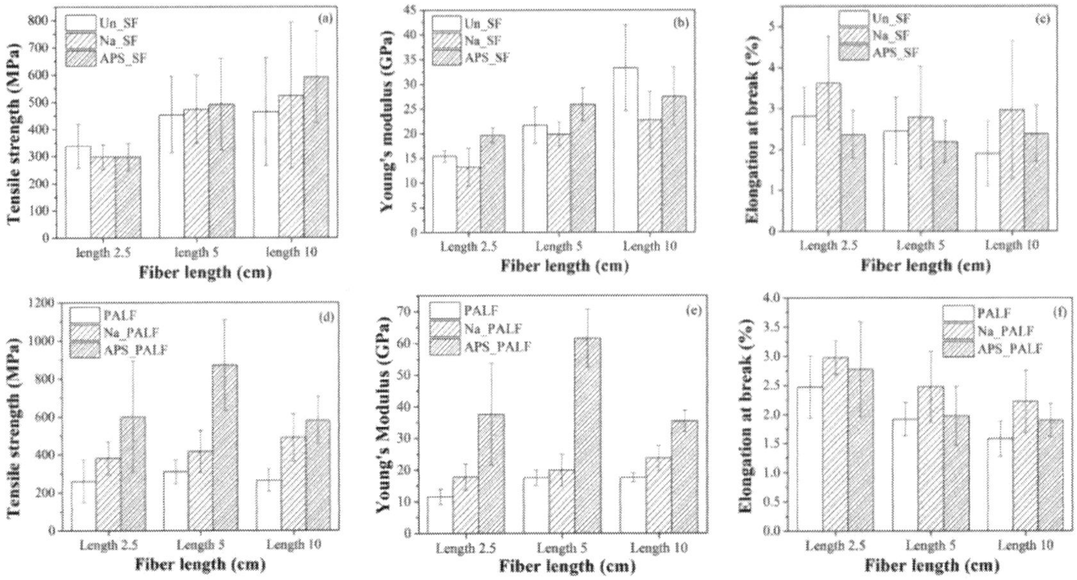


**Figure 4.** The tensile properties of SF and PALF. (a,d) Tensile strength; (b,e) Young's modulus; (c,f) Elongation at break.

### *3.5. Tensile properties of prepreg laminate composites*

The effect of surface treatment on tensile strength, Young's modulus and elongation at break of hybrid fiber reinforcement PLA composite is shown in figure 5. The tensile strengths of composite with untreated pineapple leaf fiber (Un_PALF) and untreated sisal fiber (Un_SF) were higher than that of pure PLA. In the case of composite with untreated hybrid natural fiber (Un_HF) has higher tensile strength and Young's modulus but lower elongation at break compared with pure PLA, Un_SF and Un_PALF. The composite with APS-treated hybrid natural fiber (APS_HF) had lower tensile strength and Young's modulus but higher elongation at break compared with Un_HF composite. This was may be attributed to weak improvement effect of silane on fiber surface. Moreover, acid catalyst condition in silane surface treatment of fiber may be destroyed bond of cellulose structure. [8]

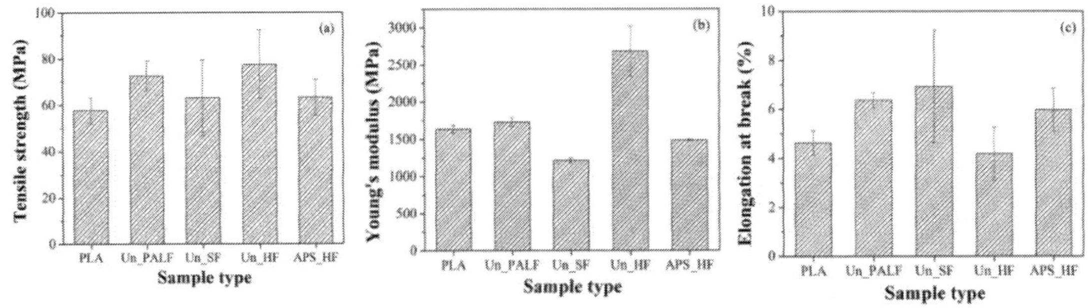


**Figure 5.** The tensile properties of prepreg laminate composites. (a) Tensile strength, (b) Young's modulus and (c) Elongation at break.

### *3.6. Fracture surface morphology of prepreg laminate composites*

The SEM micrographs of fracture surface shown in Figure 7. The Un_PALF, Un_SF and Un_HF laminated composites (Figure 6(a)-(c)) showed a fiber breakage, fiber pull out and matrix debonding, indicating poor interfacial bonding between fiber and matrix. In the case of APS_HF composite shown in Figure 7(d) seen that some fiber well trapped with PLA matrix, which expected that mechanical properties of APS_HF composite should improve. However, the tensile properties of APS_HF composite were lower than Un_HF composite.

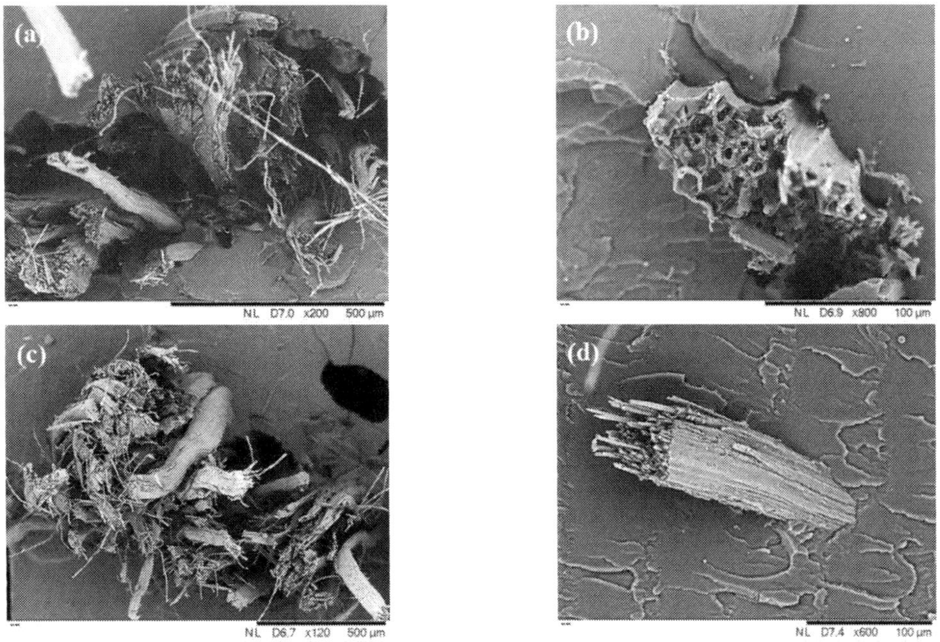


**Figure 6.** SEM micrographs of prepreg laminate composites. (a) Un_PALF, (b) Un_SF, (c) Un_HF and (d) APS_HF.

## 4. Conclusion

In this work, the hybrid long natural fiber/PLA composites were prepared using prepreg laminate method. The natural fiber, i.e. sisal and pineapple leaf fiber were treated with different chemical surface methods. It was found that alkali treatment removed some hemicellulose and lignin as confirmed by FTIR and TGA. Also, the fiber after silane treatment showed amino group that attributed to successfully graft the silane on the fiber. The SEM micrograph revealed smoother surface after the alkaline surface treatment. The tensile properties of the fibers were increased when fiber length increased and after surface treatment. Moreover, the hybrid natural fiber/PLA composites showed higher tensile properties compared with SF/PLA, PALF/PLA composites and pure PLA.

## Acknowledgments

The author would like to thank the financial support from the Department of Materials Science and Engineering, Faculty of Engineering and Industrial Technology, Silpakorn University.

## References

[1] Amornsakchai1 T and Kengkhetkit N 2014 *SWU Sci J.* **30**
[2] Bouzouita A 2016 *Elaboration of polylactide-based materials for automotive application: study of structure-process-properties interactions.* chapter 1 5-6.
[3] Siengchin S and R Srisuk 2017 *The Journal of KMUTN.* **27** 405.
[4] Samouh Z, Molnar K, Boussu F, Cherkaoui O, and Moznine E R 2018 *Polym. Ads. Technol.* 1-9.
[5] Siakeng R, Jawaid M, Ariffin H and Sapuan M S 2018 *Polymer composites* 1-12.
[6] Maya J J and Anandjiwala D R 2008 *Polymer composites* 187-204.
[7] Asaithambi B, Ganesan G and Kumar A S 2014 *Fibers and Polymers.* **15**. 4 847-854.
[8] Orue A, Jauregi A, Pena C, Labidi J, Eceiza A and Arbelaiz A 2015 *Compos. B. Eng.* **73** 132-138.
[9] Huda M S, Drzal L T, Mohanty K A & Misra M 2008 *Compos. Interfaces.* **15**. 2–3 169–191.
[10] Amini-Niaki S R and Zadeh-Kharrat F G 2019 *Environ Prog Sustain.* **38** 5

IOP Conf. Series: Materials Science and Engineering **965** (2020) 012018 doi:10.1088/1757-899X/965/1/012018

# Foamability of Natural rubber (NR)/Poly(butylenes adipate-co-terephthalate) (PBAT) Thermoplastic Vulcanizate with OBSH as a blowing agent

**J Srikham[1,2] and C Thongpin[1,2,*]**

[1]Department of Materials Science and Engineering, Faculty of Engineering and Industrial Technology, Silpakorn University, Nakhon Pathom, 73000, Thailand
[2]Center of Excellence on Petrochemical and Materials Technology, PETROMAT, Chulalongkorn University Research Building, Soi Chula 12, Phayathai Rd., Bangkok 10330, Thailand.

*Corresponding author: THONGPIN_C@su.ac.th.

**Abstract.** Natural rubber, NR, is bio-based material that is generally used for foam preparation as it results in good softness and high resilience. Among many applications, NR foam/sponge is often used for shoe sole either insole or outsole. Likewise, rubber sponge is very important for healing Plantar Fasciitis as rubber sponge is soft, elastic and capable in impact absorption. NR is in fact non-biodegradable resulting in increasing of waste which is a hot issue at present. Huge interest has paid attention in using biodegradable polymer in various products and applications. Polybutylene adipate-co-terephthalate, PBAT, is one of many bioplastics that possesses biodegradability characteristic. It also shows excellent flexibility similar to elastomer. Many attempts have been focus on blending this polymer to improve toughness in brittle plastic. Our research is aimed to incorporate PBAT into NR and producing thermoplastic vulcanizate. NR/PBAT TPV foam is then produced using 4,4-'Oxybis-benzenesulfonyl hydrazide (OBSH) as a blowing agent. The ratios of NR/PBAT are varied as follow: 70/30 65/35 60/40 55/45 50/50 45/55. The polymer blends were mixed on a two-roll mill with the formulation of ZnO 3 phr, luperox 101 1 phr and OBSH 10 phr. Compression molding was used in order to prepare rubber sponge at 160 °C, 1500 psi. Cell size, cell size distribution, density and compression set of NR/PBAT foam were investigated.

## 1. Introduction

Rubber foams, or sponge, have been known for their excellent absorption properties, light weight soft, low density and well supports weight. They often used for products such as pillows, mattresses, seat cushions, insoles, packaging, agriculture, construction, transportation, thermal and acoustic insulation and other fields. Nowadays, plastic foam such as PS foam, PU foam is popular nonetheless they cause huge wastes and pollution. Polybutylene adipate-co-terephthalate, PBAT is semi-crystalline biodegradable polyester but its crystallinity is quite low due to the structural crystallization obstruction. The elastomeric like characteristic of PBAT is interesting that it is able to apply for thermoplastic elastomer. Thongpin and Buaksuntear prepared thermoplastic elastomers from natural rubber and Poly(butylene adipate-co-terephthalate), NR/PBAT blends [1]. They found an excellent elastomeric behavior and they continue their study in preparation of NR/PBAT thermoplastic vulcanizate [2]. It showed also good tension set of the prepared TPV. Previous studies have brought us to this that aims to apply NR/PBAT TPV for foam formation using 4,4-' Oxybis-benzenesulfonylhydrazide, OBSH), as a blowing agent. In recent study, the effect of PBAT phase on foaming ability on foam properties will be reported.

## 2. Experimental

### *2.1. Materials*

Natural rubber (STR5L) was obtained from CHAREON TUT Co. Ltd. Thailand. Polybutylene Adipate-co-Terephthalate (PBAT), Ecoflex F blend with the grade of C1200 was obtained from POLYMAT Co. Ltd. Thailand. 4,4-'Oxybis-benzenesulfonylhydrazides, (OBSH), used as a blowing agent and ZnO were kindly supplied by Chemical Innovation group, Thailand. 2,5-Bis(tert-butylperoxy)-2,5-dimethylhexane) has code name of Luperox 101 was obtained from SIGMA-ALDRICH Co. Ltd. Thailand.

### *2.2. Preparation of PBAT modified*

PBAT and NR were dried in an oven at 80 °C for 6 h before blending to minimize the effect of moisture. PBAT and luperox 101 1 phr were fed into an internal mixer (Brabender Mixer Docking Station W50EHT series) at 160°C for 10 min with a rotor speed of 100 rpm. Confirming the crosslinking reaction between PBAT and luperox 101 by examining its gel content.

### *2.3. Gel fraction measurement*

The gel fraction of PBAT and PBAT modified was determined by equation (1) using dichloromethane as solvent. Reacted PBAT was soaked and kept in a vial. Subsequently PBAT gel was dried in a hot air oven at 80 °C for 6 hrs.

$$\text{Gel fraction} = W_g/W_0 \times 100\% \qquad \text{eq. 1}$$

Where $W_0$ is original weight of sample, $W_g$ is dried insoluble gel.

### *2.4. Preparation rubber compound and rubber foam*

NR was masticated on a two-roll mill (Labtech Co. Ltd. Thailand) for 30 min at 50 °C then PBAT was fed into the masticated on the roll at 130 °C. The ratios of NR/PBAT are as follows: 70/30, 65/35, 60/40, 55/45, 50/50, 45/55. 3 phr of ZnO was then added. After decreasing of roll temperature, 1 phr of Luperox 101 and 10 phr of OBSH were filled and mixed thoroughly. Other compounds were mixed in the same manner except for PBAT in which reacted PBAT was used instead of PBAT pellet. After obtaining NR/PBAT compounds, they were tested using MDR, Gotech2000, Taiwan, at 160 °C to evaluate characteristic parameters. Then foaming process was carried out using compression molding (CHAREON TUT Co. Ltd. series PR1D-W300L350 Thailand) at 160 °C under pressure of 1,500 psi.

### *2.5. Characterization*

#### *2.5.1. Tensile properties*

Tensile properties of NR/PBAT foams were conducted according to ASTM D 412-06a using universal testing machine (Instron Crop, model 5969) with a load cell of 5 kN and crosshead speed of 500 mm/min. Modulus of elasticity, ultimate tensile strength and percentage elongation at break were obtained.

#### *2.5.2. Foam density*

The density of NR/PBAT foams was measured using Mettler Toledo JE 703 C Carat Scale with Internal Calibration, density kits and accessories were used for determined density in water.

#### *2.5.3. Compression set*

Foams were cut into the 2.0 cm × 2.0 cm × 1.0 cm. The compression set was conducted according to ASTM D3574. The specimens were compressed to 50% compressive strain specimens were kept between the plates and put in a hot air oven at 70°C for 22 hrs. After that the samples were taken out from the oven and released from the plates. The specimens were allowed to recover for 30 min at ambient temperature. The compression set value ($C_t$) was then calculated by using equation (2)

$$C_t = (H_0-H_f)/(H_0-H_w) \times 100\% \quad \text{eq. 2}$$

Where $H_0$ is original height of sample, $H_f$ is final height of sample, $H_w$ is height of compressed specimen.

*2.5.4. Morphology Investigation*

Gold sputtered Scalpel cut foams were examined for their morphology in cross-section area using Hitachi TM3030 scanning electron microscopy under an accelerating voltage of 15 kV. The cell size of foam was measured from SEM micrographs using Image J program to determine cell size of foam. The measurement was carried out from the cell foam at least 100-150 cells.

## 3. Results and discussion

### *3.1 Morphology of NR/PBAT TPV foam*

The cell morphology of NR/PBAT TPV foam and NR/modified PBAT TPV foam are displayed in **Figure 1** and **2**. Generally foamability of crosslinked polymer is dependent upon gas produced from blowing agent, distribution of blowing agent in polymer and viscosity of crosslinked polymer [3]. Elasticity of polymer also affect foam characteristic [4]. It is noticeable from **Figure 2** (a) – (c) that cell size of NR/PBAT foam is increased with NR content. This could be due to the solubility of blowing agent in NR is better than in PBAT [5], [6]. Moreover, under processing in compression mold PBAT was in melted state whereas viscosity of NR was increased while crosslinking reaction was increased. Thus, NR phase could hold gas produced from decomposition of blowing agent better than PBAT. As shown, high PBAT content, foam cell skin is thicker than high NR content [7].

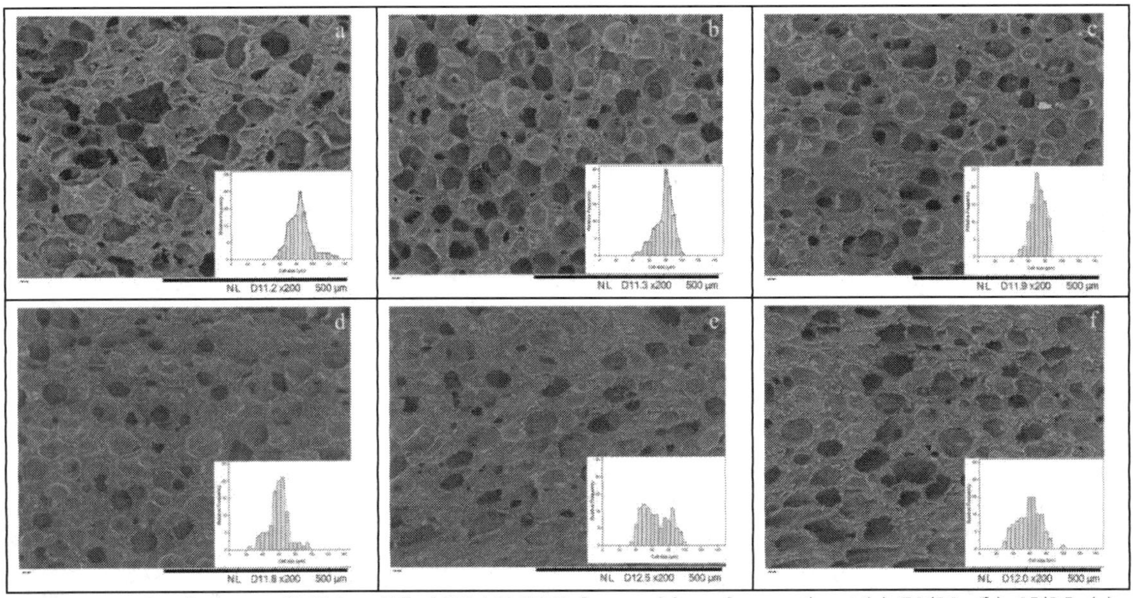


**Figure 1.** Morphology of NR/modified PBAT TPV foam with various ratio at (a) 70/30, (b) 65/35, (c) 60/40, (d) 55/45, (e) 50/50, (f) 45/55

In the case of foam prepared from NR/modified PBAT, cell size distribution seems to show binodal distribution. Structure of foam cell was found to be changed from the previous TPV system. Luperox101 modified PBAT melted mixed with NR compound resulted in the elongated cell foam structure rather than spherical cell foam. In fact, the presence of peroxide in PBAT not only lead to crosslinking of PBAT but also degradation of PBAT [8]. The evidence confirmed by onset temperature under TGA experiment from **Table 1**. The degraded PBAT could act a either plasticization or lubrication during foaming of NR/modified PBAT hence the viscosity of polymers during foaming was lower than NR/PBAT. The pressure gas produced could expand foam skin. Nonetheless foam skin could not expanded in the direction perpendicular to mold surface whereas it could in the transverse direction to the mold. The foam structure obtained could be schematically drawn and depicted in **Figure 4.**

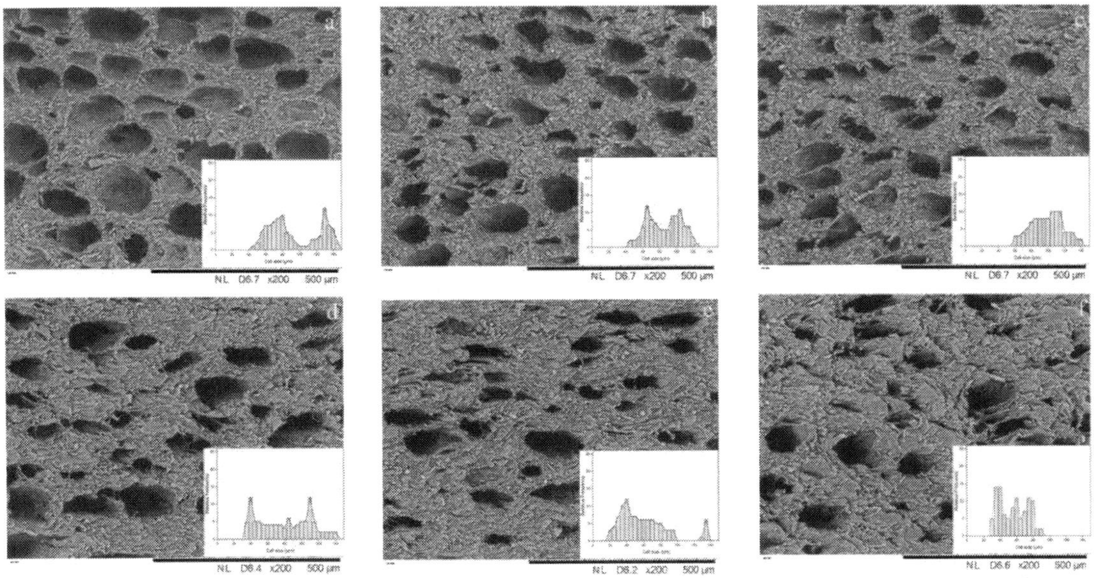

**Figure 2.** Cell morphology of NR/PBAT TPV foam, NR/PBAT TPV ratio follow by (a)70/30, (b)65/35, (c)60/40, (d)55/45, (e)50/50, (f)45/55

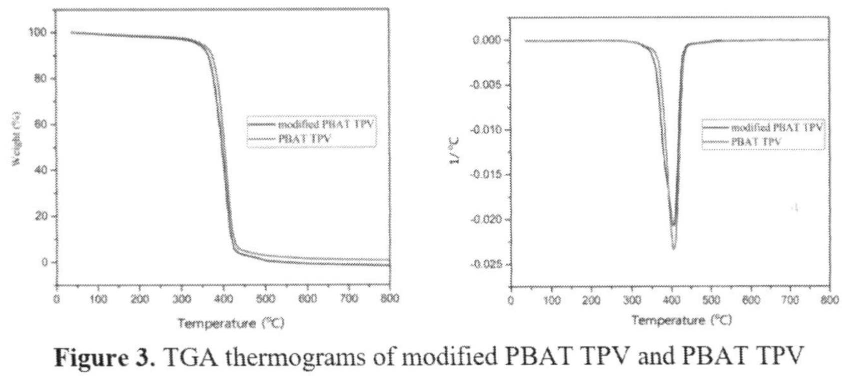


**Figure 3.** TGA thermograms of modified PBAT TPV and PBAT TPV

**Table 1.** TGA data of modified PBAT TPV and PBAT TPV

| Formular | $T_{onset}$ (°C) | $T_d$ (°C) | $T_{endset}$ (°C) |
|---|---|---|---|
| modified PBAT TPV | 360.18 | 404.56 | 427.26 |
| PBAT TPV | 374.17 | 405.85 | 430.96 |

In addition, phase morphology of foam cell skin showed phase separation between NR and modified PBAT. This could be due to the crosslinked PBAT possessed high viscosity intensify incompatibility between NR and PBAT phases.

IOP Conf. Series: Materials Science and Engineering **965** (2020) 012018 doi:10.1088/1757-899X/965/1/012018

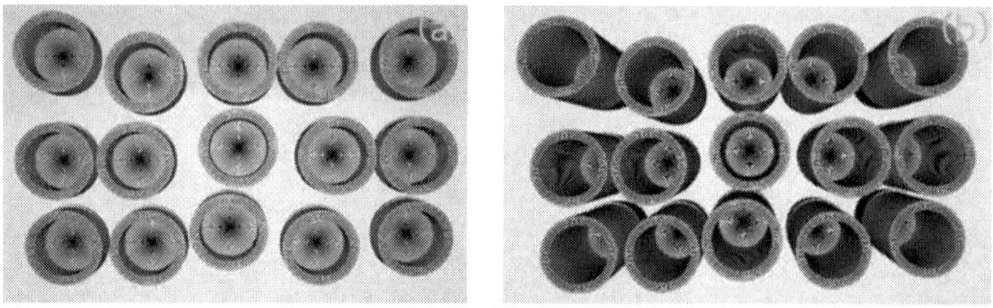

**Figure 4.** Proposed cell structure (a) NR/PBAT TPV (b) NR/modified PBAT TPV

### *3.2 Density of NR/PBAT TPV foam*

Density of NR/PBAT TPV foam compared with NR/modified PBAT TPV foam are presented in **Figure 5**. It should be reminded that cell structure of NR/modified PBAT and its morphology should be elongate and phase separated respectively. These should affect its lowered in bulk density of the foam.

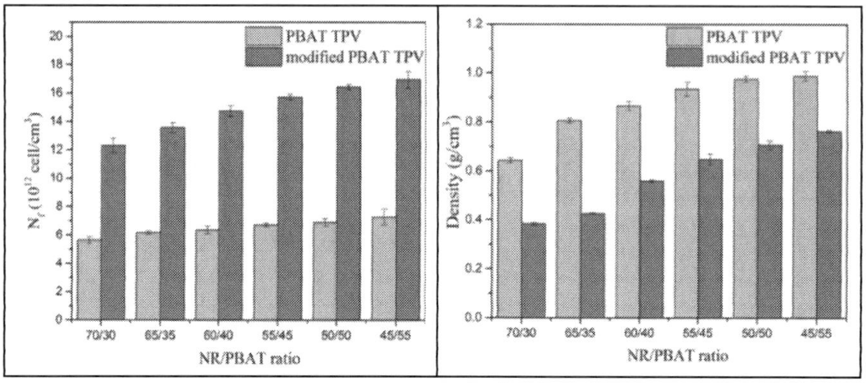


**Figure 5.** Cell density and bulk density of NR/PBAT TPV foam compared with NR/modified PBAT TPV foam.

### *3.3 Tensile properties of NR/PBAT foam*

Tensile properties, i.e. Young's modulus, tensile strength and elongation at break, of NR/PBAT TPV foam compared with NR/modified PBAT TPV foam are present in Figure 6 (a)-(c). Considering Young's modulus of the foams showed in Figure 6 (a), they tend to be increased with PBAT contents due to PBAT having higher modulus than NR [9]. NR/PBAT TPV foams show higher Young's modulus in all NR/PBAT contents that the foam with modified PBAT. As mentioned earlier that PBAT could also be degraded by the addition of peroxide and disadvantaged on both Young's modulus and tensile strength of the foam containing modified PBAT. In term of Elongation at break of the foam, although skin of cell foam thickness was increased with PBAT content, elongation at break was still decreased. This was caused by the phase separation between these two polymers. In the foam system containing modified PBAT, phase separation between crosslinked PBAT and NR are intensified, as seen from SEM micrographs. This consequently resulted in lowering in elongation at break of NR/PBAT TPV foam than NR/modified PBAT TPV foam.

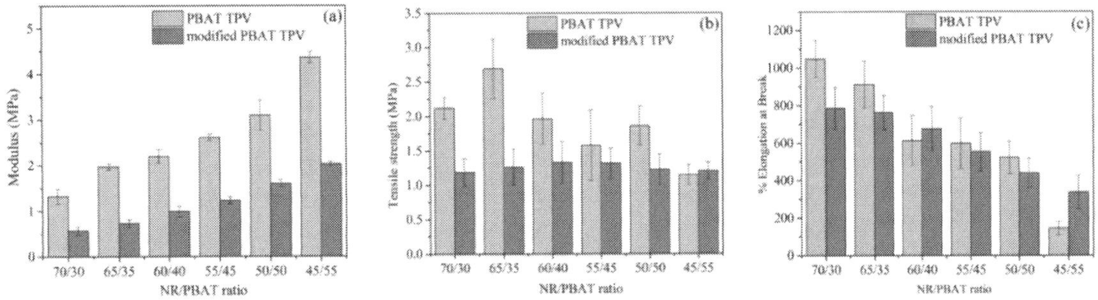


**Figure 6.** The tensile properties of NR/PBAT TPV foam compared with NR/modified PBAT TPV foam (a) tensile modulus, (b) tensile strength and (c) elongation at break.

### *3.4 Compression set of NR/PBAT foam*

It is generally known that factors that affected compression are Low compression set values that indicate excellent elasticity and durability for long-term supporting material [10]. The results from **Figure 7** shows that the higher the NR content the lower compression set. This indicates that NR has excellent elasticity and resilience. After compressive load releasing, foam can be recovered to nearly original dimension. Generally, good TPV should possess compression set less than 50% [11]. In addition, NR/modified PBAT TPV foam seems to showed even low compression set. This could be due to thicker skin of cell foam in NR/PBAT TPV foam than the other foam system. In which, they can support long term compression loading [12].

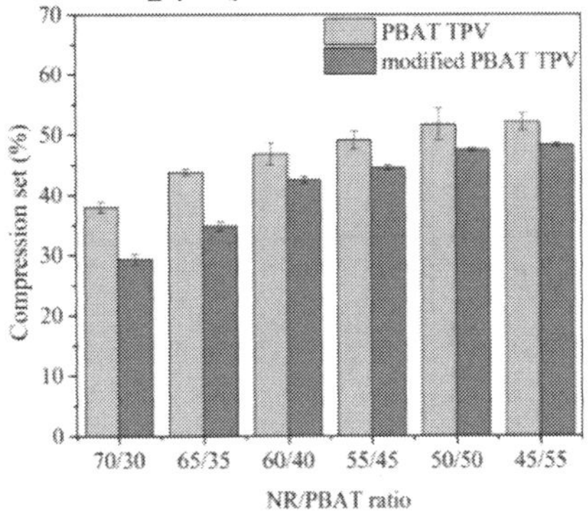


**Figure 7.** Compression set of NR/PBAT TPV foam compared with NR/modified PBAT TPV foam.

## 4. Conclusion

The NR/PBAT TPV and NR/modified PBAT TPV foams were successfully prepared using OBSH as a blowing agent. The foamability of both foams systems was compared. SEM micrographs show foams morphology and cell size of the foams. The results also show that cell morphology and cell size distribution are dependent both on NR/PBAT contents and modification of PBAT. Luperox101 treated PBAT resulted in crosslinking as well as degradation of PBAT and this affected the viscosity of blown polymer during foaming in the mould. The crosslink PBAT domains caused phase separation between NR and PBAT and hence cell foam structure. The cell foam skin thickness also changed from spherical like cell structure to elongated cell foam structure. These also resulted in the lowered density for NR/modified PBAT TPV foam. In term of tensile properties, Young's modulus of NR/PBAT TPV foam is higher than that NR/modified PBAT TPV foam. Due to the elongation at break of both foam systems are decreased thus tensile strength of the foam are also decreased due to stress-induced crystallization of cell foam skin was obstructed by the present of PBAT. Compression set of NR/modified PBAT TPV foam were improved compared to NR/PBAT TPV foam due to the thick foam cell skin. From this primarily study shows that NR/PBAT TPV foam could be able to foam with carefully control condition and our further study will be reported.

**Acknowledgments**
The author would like to thank the Department of Materials Science and Engineering, Faculty of Engineering and Industrial Technology, Silpakorn University and Materials Technology (PETROMAT), The Petroleum and Petrochemical College (PPC), Chulalongkorn University, Thailand, is also gratefully acknowledge.

**References**
[1] Buaksuntear K and Thongpin C 2017 *Pure and Applied Chemistry International Conference (PACCON 2017)* (Bangkok: Thailand)
[2] Buaksuntear K and Thongpin C 2018 *Asian Workshop on Polymer Processing (AWPP 2018)* (Chiang Mai: Thailand)
[3] Tiwary P, Park C.B., and Kontopoulou M 2017 *European Polymer Journal* (Queen's University: Canada) p 283-296
[4] Chen L, Huang M, Li Z, Liu D and Li B 2020 *Journal of Petroleum Science and Engineering* (China University of Petroleum (East China): China) p. 185
[5] Najib N.N, Ariff Z.M, Manan N.A, Bakar A.A., Sipaut C.S. 2009 *Journal of Physical Science Vol. 20* (School of Material and Mineral Resources Engineering: Malaysia) p 13-25
[6] Boonprasertpoh A, Pentrakoon D, and Junkasem J. 2019 *Cellular Polymers* (Chulalongkorn University: Thailand)
[7] Gu R, Konar S and Sain M 2012 *Journal of the American Oil Chemists' Society* (University of Toronto, Canada) p 2103-2111
[8] Song J, Mi J, Zhou H, Wang X and Zhang Y 2018 *Polymer Degradation and Stability* Vol. 157 p. 143-152.
[9] Siyamak S, Lbrahim N and Abdolmohammadi S 2012 *International Journal of Molecular Sciences* (University Putra Malaysia: Malaysia)
[10] Sonnenschein M.F., Prange R. and Schrock A.K. 2007 *Polymer*Vol. 48(2) p. 616-623
[11] Ning N, Li S, Wu H Tian H and Yao P 2018 *Progress in Polymer Science* Vol. 79 p 61–97
[12] Kim D.Y., Bang J.H., Lee C.A., Kim H.Y. and Choi K.Y. 2018 *International Journal of Industrial Ergonomics* Vol. 64 p. 178-187

IOP Conf. Series: Materials Science and Engineering **965** (2020) 012019 doi:10.1088/1757-899X/965/1/012019

# Preparation and characterization of PET/SEBS incorporated with organomontmorillonite

Poonsub Threepopnatkul[1,*], Phakawan Thinsoongnoen[1], Pattaraporn Misut[1], Siriluk Autama[1] and Chanin Kulsetthanchalee[2]

[1]Department of Materials Science and Engineering, Faculty of Engineering and Industrial Technology, Silpakorn University, Nakhon pathom 73000, Thailand
[2]Event and Exhibition Design Department, Suan Dusit University, Bangkok 10300, Thailand

*Corresponding author: poonsubt@yahoo.com

This study is focused on the characterization of polyethylene terephthalate (PET)/Styrene-Ethylene-Butylene-Styrene (SEBS)/Organomontmorillonite (OMMT) blends. SEBS-g-MA is used as compatibilizer and SEBS is the impact modifier at 30 %wt. The compounds were blended by twin-screw extruder and molded for test specimen by injection molding. The effect of OMMT types (Cloisite15A and Cloisite30B) and contents (1 and 3 %w/w) on properties of PET blends with SEBS and SEBS-g-MA were investigated. The PET composites incorporated with Cloisite15A had higher impact strength and %elongation at break whereas lower tensile strength than the one with Cloisite30B. Comparing with PET/SEBS/SEBS-g-MA blends, adding OMMT into PET/SEBS/SEBS-g-MA would increase thermal property. However, it had no significant changes with higher OMMT. %Crystallinity of PET would be increased with OMMT contents. PET/SEBS/SEBS-g-MA had lower %LOI but higher burning rate compared with PET. However, the addition of OMMT could improve flame retardant properties of PET blends. Addition of Cloisite15A and Cloisite30B into PET/SEBS/SEBS-g-MA increased %LOI. From UL 94 tests, polymer composites could be classified in HB class. The burning rate of polymer composites was decreased with increasing OMMT due to decomposition of OMMT act as a barrier to inhibit combustion.

## 1. Introduction

Plastics are commonly used in everyday life because the materials have more distinctive properties and could replace the other materials. Polyethylene terephthalate (PET) is one of engineering material which has excellent processability, rigidity, thermal stability, resistance to oil, recyclability, light weight and relatively low cost. Nevertheless, one major disadvantage of PET is the notch sensitive, which decrease the impact properties of PET and limit the engineering applications of this material [1]. Blending PET with thermoplastic elastomer such as Styrene-Ethylene-Butylene-Styrene (SEBS) is an efficient way to increase its toughness. SEBS is saturated elastomer [2], consisting of a hard end-block polystyrene phase and a soft elastomeric phase. Thus SEBS can act as impact modifier [3]. However, due to the immiscibility between PET and SEBS, SEBS-g-MA was used as the compatibilizer in the system.

The pristine clay minerals are not compatible with most polymeric systems due to hydrophilic nature of clay minerals. However, cation exchange modification of clay minerals with long chain ammonium or phosphoniumquanternary ions increases the inter-layer spaces and also the dipersibiity of layered-silicates by increasing compatibility with polymers. The clay-polymer compatibility as well as the exfoliation of clay-layers strongly depends on type of modifier and the polarity of polymer matrix [4]. In our study, the two different OMMT i.e., Cloisite15A (montmorillonite modified with dimethyl dehydrogenated tallow quanternary ammonium) and Cloisite30B (montmorillonite modified with methyl tallow bis-2-hydroxtethyl quanternary ammonium) are used. The incorporation of this nanoclay will generally improve the Young's modulus and tensile strength of the thermoplastic elastomer nanocomposite while the elongation at break decreased [3]. Additionally, the incorporation of a relatively low quantity of nanoclay in the polymer matrix creates a protective layer during combustion that effectively shields the underlying polymer and limits heat transfer into the material, volatilization of combustible degradation products and diffusion of oxygen into the material. [5] This article aims to study the effect of OMMT types and content on mechanical, morphology and thermal properties and flame retardant of PET blends.

## 2. Experimental

### 2.1 Materials

Polyethylene terephthalate (PET) was obtained from Indorama Ventures Public Co., Ltd. Styrene-Ethylene-Butylene-Styrene (SEBS) was purchased from Toyota Tsusho (Thailand) Co., Ltd. SEBS-g-MA used as a compatibilizer, was obtained from Connell Bros. (Thailand) Co., Ltd. Organically modified montmorillonite is modified with dimethyl dehydrogenated tallow quanternary ammonium called as Cloisite15A and Cloisite 30B is montmorillonite modified with methyl tallow bis-2-hydroxyethyl quanternary ammonium as shown in Table 1.

Table 1. Chemical structure and D spacing of montmorillonite

| Clay type | Chemical structure | D spacing (nm) |
|---|---|---|
| Cloisite 15A | 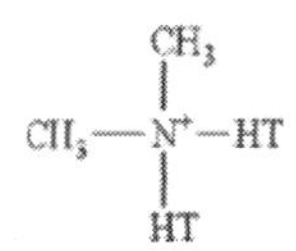 | 3.15 |
| Cloisite 30B | 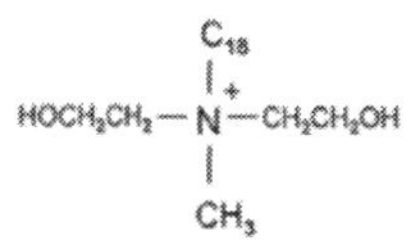 | 1.81 |

### 2.2 Preparation of PET blends

The master-batch was prepared by mixing organoclays with SEBS in chloroform where the concentration of organoclays was 30 phr with respect to neat SEBS. Initially, organoclays and SEBS were dried at 70 °C for 24 h to remove the moisture. The SEBS was dissolved in chloroform until it became homogeneous mixture then added organoclays to the SEBS solution and stirred continuously with a magnetic stirrer for 4 h to disperse the clay-platelets uniformly in SEBS solution. The SEBS/organoclays dispersion was cast on a tray and placed in the oven at 70 °C for 24 h to remove the solvent. After solvent evaporation, a sheet was formed, which was cut to small pieces.

PET was dried at 100 °C for 24 h. Then, PET, master-batch of SEBS/OMMT and SEBS-g-MA were mixed and melt blended using twin screw extruder (EN-MACH, SHJ-25). The temperature setting along the extruder barrel ranged from 155-250 °C at screw speed 30 rpm. Cloisite15A and Cloisite 30B is varied at the content of 1.00 and 3.00 %wt. The pellets were molded for tested specimen by injection molding machine (Sumitomo Heavy industries, SG50M).

### 2.3 Characterization

The mechanical properties were examined by impact tests which performed by Izod testing method according to ASTM D-256 on the notched samples. The work of the pendulum was 1.0 J and the amount of energy required to break the sample were recorded. The tensile tests according to ASTM D-638 were performed on a universal testing machine (Instron 5956) at 50 kN load with 50 mm/min elongation rate. Thermogravimetric analysis (TGA) was performed by using a Thermogravimetric Analyzer (TGA7, Perkin Elmer Co., Ltd) with the temperature scan mode. The samples were heated from 50 °C to 700 °C at heating rate of 10 °C/min in nitrogen atmosphere. Differential scanning calorimeter (DSC) was performed by Perkin Elmer Pyris 1 to obtain the heating thermogram and the percent of crystallinity under nitrogen flow at heating rate of 10 °C/min and temperature range was 50-300 °C. The morphology and fracture surfaces of selected samples after impact and tensile tests were examined via scanning electron microscopy (SEM). The fracture surfaces of the samples were coated with Cu-Pt prior to SEM observation. Flammability of samples was determined according to ASTM D3801 and ASTM D635 for UL-94 vertical and horizontal burning test respectively. The Limiting Oxygen Index (LOI) according to ASTM D2863.

## 3. Results and discussion

### 3.1 Mechanical Properties

Impact strength of PET/SEBS/SEBS-g-MA with different types and content of OMMT (Cloisite15A and Cloisite30B) are shown in Figure 1. The incorporation of OMMT into PET/SEBS/SEBS-g-MA leads to considerably reduce the impact strength because OMMT particles are relatively stiff that could obstruct the mobility of polymer chains.

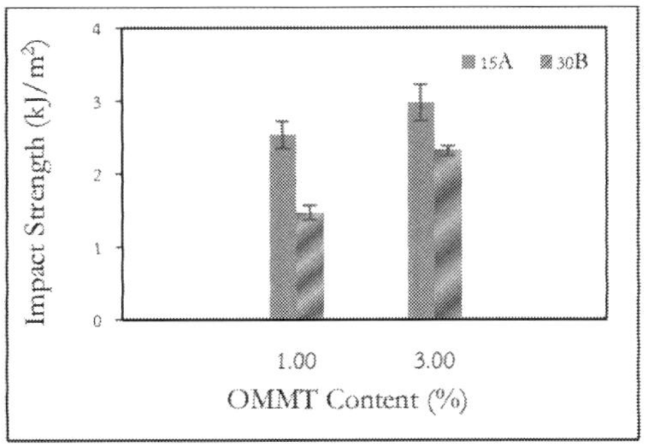


**Figure 1.** Impact strength of PET/SEBS/SEBS-g-MA composites with different types of OMMT (Closite15A and30B).

Considering two types of OMMT, Cloisite 15A is less polarity than Cloisite 30B. In this research, OMMT is dispersed in SEBS phase (Master-batch) before it melt-blended with PET. Cloisite15A is well dispersed in SEBS than Cloisite30B due to polarity, this affect to PET/(SEBS/SEBS-g-MA)/Cloisite15A has higher impact strength than PET/(SEBS/SEBS-g-MA)/Cloisite 30B because Cloisite15A has good compatibility with SEBS so it increases shear force during mixing process that reduce size of SEBS particles. Furthermore, SEBS could well disperse on PET matrix as confirmed by SEM image in Figure 2.

IOP Conf. Series: Materials Science and Engineering **965** (2020) 012019 doi:10.1088/1757-899X/965/1/012019

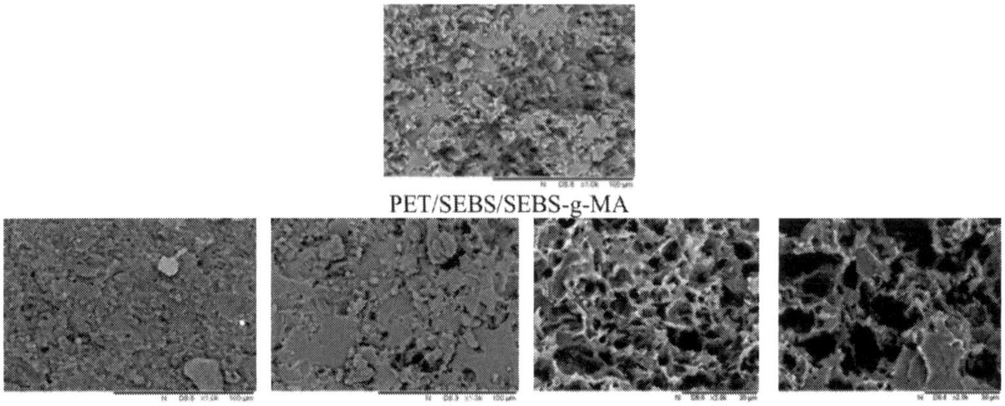


(b) 1% C15A (c) 1% C30B (d) 1% C15A (e) 1% C30B

**Figure 2.** SEM of PET/SEBS/SEBS-g-MA) and PET/(SEBS/SEBS-g-MA) composite with different types of OMMT (Cloisite15A and Cloisite30B) (a-c) before (d-e) after immerse in chloroform

Tensile properties of PET/SEBS/SEBS-g-MA and its composites are presented in Fig. 3. The tensile strength of polymer composites was evidently decreased with the presence of OMMT. However, Young's modulus of PET/SEBS/SEBS-g-MA/OMMT has no significant change compare with the one without OMMT. PET/SEBS/SEBS-g-MA/Cloisite15A possesses lower tensile strength than PET/SEBS/SEBS-g-MA/Cloisite 30B. And tensile strength of PET composites was decreased with increasing the amount of OMMT. Therefore, Cloisite15A has less polarity somewhat similar to SEBS which makes them to be compatible. Therefore, it would increase shear force during melt blending process and enables SEBS to well-disperse in PET. This effect would then reduce tensile strength of polymer composites. Cloisite30B is less compatible with SEBS and cause less shear force during the process. Along with the different viscosity of SEBS and PET, relatively larger size of SEBS would disperse in PET as seen in Fig. 2d-e. This could make it has higher tensile strength and Young's modulus than composites with Cloisite15A especially at 3 %wt of Cloisite. In case of percentage of elongation at break, it decreased for OMMT loading up to 3.00 %wt and increased for higher Cloisite15A loading (3.00 %wt).

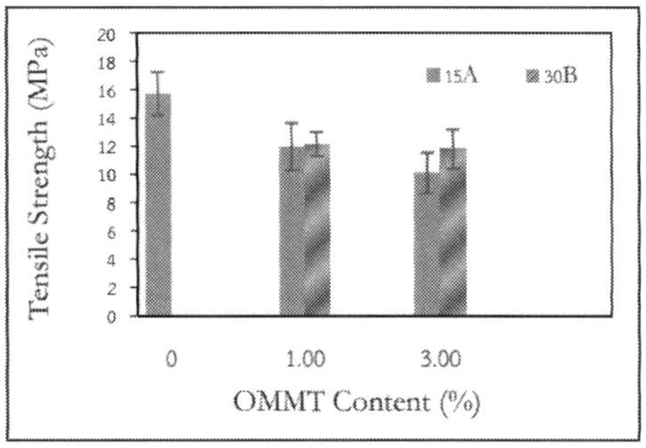


(a)

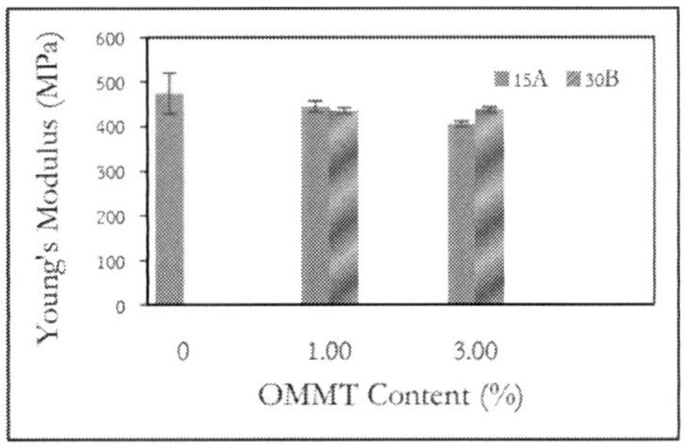


(b)

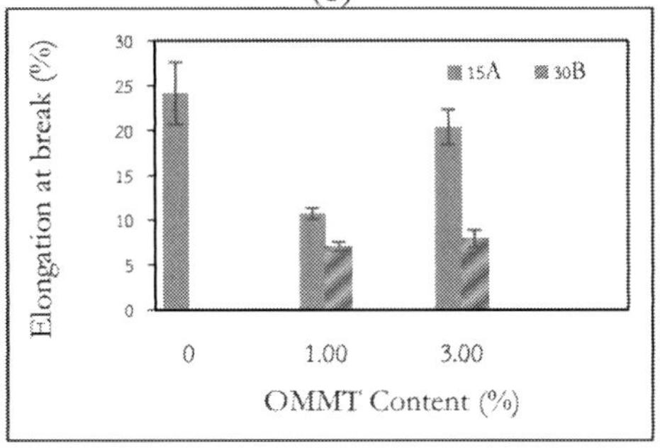


(c)

**Figure 3**. Tensile properties of PET/SEBS/SEBS-g-MA composites with different types of OMMT (Closite 15A and 30B).

### 3.2 Thermal Properties

Table 2 shows thermal properties of PET composites compared with neat PET. Both of onset and degradation temperatures of PET/SEBS/SEBS-g-MA increased with incorporation of both types of OMMT. Nonetheless, it has no significant change with increasing concentration of OMMT.

**Table 2**. TGA data of PET, PET/SEBS(SEBS-g-MA) and PET/SEBS(SEBS-g-MA)/OMMT (Cloisite 15A and Cloisite 30B)

| Samples | $T_{onset}$ (°C) | $T_d$ (°C) |
|---|---|---|
| PET | 408.63 | 429.35 |
| PET/SEBS/SEBS-g-MA | 411.02 | 432.95 |
| PET/SEBS/SEBS-g-MA-1.00-15A | 412.07 | 434.76 |
| PET/SEBS/SEBS-g-MA-3.00-15A | 411.46 | 438.05 |
| PET/SEBS/SEBS-g-MA-1.00-30B | 408.26 | 434.10 |
| PET/SEBS/SEBS-g-MA-3.00-30B | 410.73 | 436.22 |

The addition of SEBS and OMMT (Cloisite15A and Cloisite30B) had nearly no effect on melting temperature of PET. However, it had impact on percentage of crystallinity of composites. SEBS reduces the percentage of crystallinity of composites whereas both of Cloisite15A and Cloisite30B increases percentage of crystallinity compared with PET/SEBS/SEBS-g-MA because OMMT acts as a nucleating agent [6] as shown in Table 3.

**Table 3**. The melting temperature and %crystallinity ($X_c$) of PET, PET/SEBS/SEBS-g-MA and PET/SEBS/ SEBS-g-MA/OMMT (Cloisite 15A and Cloisite 30B)

| Samples | $T_m$ (°C) | $X_C$ (%) |
|---|---|---|
| PET | 250.01 | 26.36 |
| PET/SEBS/SEBS-g-MA | 248.70 | 18.58 |
| PET/SEBS/SEBS-g-MA-1.00-15A | 249.83 | 20.17 |
| PET/SEBS/SEBS-g-MA-3.00-15A | 250.34 | 28.76 |
| PET/SEBS/SEBS-g-MA-1.00-30B | 250.02 | 24.87 |
| PET/SEBS/SEBS-g-MA-3.00-30B | 250.36 | 26.27 |

### 3.3 Flame Retardant

The effect of OMMT on flammability resistance properties is shown in Table 4. PET has the highest limit oxygen index. When adding SEBS, %LOI of composite was dramatically decreased since SEBS itself has lower %LOI and could be burnt quickly. The addition of OMMT increased %LOI of PET/SEBS/SEBS-g-MA/OMMT compared with PET/SEBS/SEBS-g-MA. For UL94 test, all samples were classified as horizontal burning (HB). The OMMT addition reduces the burning rate of the composites when compared with PET/SEBS/SEBS-g-MA. Supposedly, there was organo-modified thermal degradation and strong proton-catalytic sites built-up onto the surface, which can catalyze the formation of a stable char residue. Therefore, the clay accumulation on the surface of the composites acts as a protective barrier that limits heat transfer into the material and diffusion of oxygen into the material. [5]

**Table 4**. %LOI and Burning rate (BR) of PET, PET/SEBS/SEBS-g-MA and PET/SEBS/SEBS-g-MA/OMMT (Cloisite 15A and Cloisite 30B).

| Samples | %LOI | BR (cm/min) |
|---|---|---|
| PET | 25 | 0.50 |
| PET/SEBS/SEBS-g-MA | 13 | 1.80 |
| PET/SEBS/SEBS-g-MA-1.00-15A | 15 | 1.10 |
| PET/SEBS/SEBS-g-MA-3.00-15A | 16 | 0.70 |
| PET/SEBS/SEBS-g-MA-1.00-30B | 17 | 1.80 |
| PET/SEBS/SEBS-g-MA-3.00-30B | 18 | 1.70 |

## 4. Conclusions

In this study, PET/(SEBS/SEBS-g-MA) and PET/(SEBS/SEBS-g-MA)/OMMT (Cloisite 15A and Cloisite 30B) blends were prepared via melt-blending process. The effect of OMMT types and contents on properties of PET blends with impact modifier at weight ratio between PET and SEBS/SEBS-g-MA of 70:30 was investigated. The types and contents of OMMT influenced the mechanical properties. The addition of Cloisite15A into PET/SEBS has raised the impact strength and %elongation at break whereas lower tensile strength. Moreover, Young's modulus had no significant change. For thermal properties, comparing with PET/(SEBS/SEBS-g-MA) blends, adding either Cloisite15A or Cloisite30B into PET/(SEBS/SEBS-g-MA) increased the onset and degradation temperature. PET/SEBS/SEBS-g-MA had lower %LOI but higher burning rate compared with PET. However, the addition of OMMT could improve flame retardant properties of PET blends. Addition of Cloisite15A and Cloisite30B into PET/SEBS/SEBS-g-MA increased %LOI up to 16% and 18%, respectively. From UL 94 testing, polymer composites could be classified in HB class. The burning rate of polymer composites was reduced once increasing OMMT.

**Acknowledgement**
The authors gratefully acknowledge the financial support and materials of following: Department of Materials Science and Engineering, Faculty of Engineering and Industrials Technology, Sipakorn University and Indorama Ventures Public Co., Ltd.

**References**

[1] Tanrattanakul V, Hiltner A and Baer E 1997 *Polym* **38** 2191
[2] Direksilp C and Threepopnatkul P 2014 *Energy procedia* **56** 135
[3] Li X, Yang J, Zhou X, Wei Q, Li J, Qiu B, Wunderlich K and Wang X 2018 *Polym test* **67** 435
[4] Adk B, Butola BS and Joshi M 2018 *Appl Clay Sci* **161** 343
[5] Moolkaew W and Threepopnatkul P 2012 *Proc. Int. 15th European Conf on Composite Materials (Venice)* (Italy) 1
[6] Yang Z, Wu M, Chen G, Li S, Peng P, and Zhang Q 2016 *Polym polym compos* **24** 331

# Low density polyethylene/poly(butylene adipate-co-terephthalate) films: Effect of a compatibilizer on morphology and properties

**S Hemsri[1]*, K Puttiwanit [1], K Saeaung [1] and P Satung[1]**

[1] Department of Materials Science and Engineering, Faculty of Engineering and Industrial Technology, Silpakorn University, Nakhon Pathom 73000, Thailand

*Corresponding author: sudsirihemsri@yahoo.com

**Abstract.** Low density polyethylene (LDPE)/poly(butylene)adipate-co-terephthalate (PBAT) blend films were fabricated using a cast film extruder with different content of PBAT (0,10, 20 and 30 wt%). Polyethylene-graft-maleic anhydride (PE-g-MA) was used as a compatibilizer for the LDPE/PBAT blends. Morphology, mechanical properties, thermal properties, water vapour and oxygen permeability properties of the LDPE/PBAT films with and without adding the compatibilizer were investigated. Scanning electron microscopic (SEM) results revealed that the LDPE/PBAT/PE-g-MA films showed better compatibility and homogeneity between LDPE matrix and PBAT phase. Young's modulus and tensile strength of the LDPE/PBAT/PE-g-MA films were higher than those of the LDPE/PBAT films. The PBAT content and incorporating PE-g-MA did not affect on melting temperature ($T_m$) of the LDPE/PBAT blends. On the other hand, melting enthalpies ($\Delta H_m$) of the polymer blends decrease as increase in PBAT concentration and at same content of the PBAT, the melting enthalpies ($\Delta H_m$) of the LDPE/PBAT/PE-g-MA films were higher than those of LDPE/PBAT films. Water vapor transmission rate (WVTR) of the blend films increased but oxygen permeability (OP) decreased when the PBAT content increased. Compared with uncompatibilized LDPE/PBAT films at the same PBAT loading, the compatibilized LDPE/PBAT films provided a lower water vapor barrier whereas the LDPE/PBAT/PE-g-MA films showed a better oxygen barrier.

## 1. Introduction

Polyethylene (PE) is petroleum-based polyolefin and commonly used in a variety of applications such as packaging, automotive, electrical and agriculture. The PE has many advantages compared with other polyolefins, for examples, high crystallinity, good mechanical properties, water barrier, good oxygen permeability and low price. Nevertheless, the main drawback of PE is non-biodegradable leading to the critical environmental problems from plastic wastes. In the recent year, eco-friendly biodegradable polymer such as poly(lactic acid) (PLA), polyhydroxyalkanoates (PHA), polyhydroxybutyrates (PHB), polybutylene succinate (PBS) and poly(butylene adipate-co-terephthalate) (PBAT) etc. have received attention to replace the conventional polyethylene. However, these biodegradable polymers are more expensive than the PE. Thus, blending the PE and biodegradable polymers is one attractive approach not only to reduce cost for industrial applications, but also decrease an extensive use of the non-biodegradable PE. The poly(butylene adipate-co-terephthalate) (PBAT) is a biodegradable and compostable aromatic-aliphatic copolyester with higher flexibility and elongation than PLA and PBS [1]. Moreover, PBAT has a similar properties like low density polyethylene (LDPE) [2] and PBAT also has good water vapor transmission rate (WVTR), oxygen barrier and fully biodegradable property. Thus, the PBAT are suitable for various packaging and agricultural applications [3]. However, polarity of LDPE and PBAT is significantly different, leading to immiscibility of the LDPE/PBAT blend. Thus, in this research, polyethylene-grafted maleic anhydride (PE-g-MA) was used as a compatibilizer and incorporated into the LDPE/PBAT blends in order to enhance interfacial adhesion between LDPE matrix and PBAT phase. The effect of incorporation of the PE-g-MA compatibilizer and the PBAT content on morphology, mechanical and thermal properties as well as water vapor and oxygen barrier properties of the LDPE/PBAT and LDPE/PBAT/PE-g-MA were studies.

## 2. Experimental

### *2.1 Materials*

Low density polyethylene, LDPE, (InnoPlus LD2426K) with melt flow index (MI) of 3.9 g/10 min was obtained from PTT Global Chemical Company Limited, Thailand. Poly(butylene adipate-co-terephthalate), PBAT, (Ecoflex© F Blend C1200) with MFI of 3.8 g/10 min was supplied by BASF Corporation. Linear low density polyethylene-graft-maleic anhydride, LLDPE-g-MA, (Fusabond® E226) with MFI of 1.5 g/10 min was purchased from Dow Chemical Company as a compatibilizer agent for LDPE/PBAT blends.

### *2.2 Preparation of LDPE/PBAT and LDPE/PBAT/PE-g-MA blend films*

The LDPE, PBAT and PE-g-MA pellets were dried in an oven at 60°C for 8 hours to remove the moisture prior to cast film processing. The LDPE/PBAT and LDPE/PBAT/PE-g-MA blends were prepared at different PBAT contents of 10, 20 and 30 wt% of LDPE with an addition of 10 wt% of PE-g-MA as the compatibilizer in a cast film extruder (Lab Tech Engineering, Thailand). The temperature profile of the cast film extruder was 150, 160, 170, 180 °C with screw speed of 60 rpm and 55°C at chill-roll. The casted film thickness was between 150-200 microns.

### *2.3 Characterization Methods*

*Scanning electron microscopy (SEM) analysis*

Cryo-fractured surface morphologies of uncompatibilized LDPE/PBAT and compatibilized LDPE/PBAT/PE-g-MA blends were investigated by a Scanning Electron microscope (TM3030, Hitachi, Japan). The fractures surfaces of all films were platinum coated to avoid electrostatic charging prior to analysis.

*Tensile properties*

Tensile testing of LDPE, PBAT as well as LDPE/PBAT and LDPE/PBAT/PE-g-MA blends was performed by Universal testing machine (Instron 5969, Instron, USA.) according to ASTM D882: Standard Test Method for Tensile Properties of Thin Plastic Sheeting using a load cell of 5kN with initial distance between the grips of 50 mm and speed rate of 500 mm/min. At least 15 replicates of all blends were tested and Young's modulus, maximum tensile strength and elongation at break were recorded.

*Differential scanning calorimetry analysis*

Thermal properties of LDPE, PBAT, LDPE/PBAT and LDPE/PBAT/PE-g-MA films were performed by using Differential Scanning Calorimeter (DSC; DSC 1 STAR System, Mettler Toledo Co., Ltd.). The heating and cooling cycles of the samples were operated between -80 to 170 °C with scanning rate of 10 °C/min under nitrogen atmosphere.

*Water vapor transmission rate (WVTR)*

The WVTR of the sample films were determined according to ASTM E96: Standard Test Methods for Water Vapor Transmission of Materials with the method A (dry cup method). Glass bottles with a diameter of 3.0 cm containing silica gel (0% RH) were covered with the films (thickness of 0.24 mm), sealed with a parafilm tape and then kept inside a chamber in which relative humidity of 50 ± 3 °C at room temperature. The glass bottles covered with the film were weighted at certain times until 24 days. Weight changes of the glass bottles were plotted vs. times and then slopes of the plot in each sample films were calculated in unit of g/h. The WVTR in unit of g/m$^2$h was defined as the slope divided by test area (m$^2$) of the films. Three replicates of each film formulation were examined to determine the WVTR.

*Oxygen permeability*

Oxygen permeability of all films were measured by Gas permeability tester (GDP-C, Brugger Co., Ltd.) based on an ASTM D1434

## 3. Results and discussion

**Morphology characterization.**

Effect of the PE-g-MA compatibilizer on phase morphology of LDPE/PBAT blends was study by SEM technique. The cryo-fractured surface morphologies of the LDPE/PBAT and LDPE/PBAT/PE-g-MA blends with different PBAT contents of 10, 20 and 30 wt% is presented in Figure 1. For uncompatibilized LDPE/PBAT blends as shown in Fig. 1(a)-(c), phase separation between the LDPE matrix and PBAT phase was clearly observed and the minor PBAT phase formed in spherical particles dispersed in the matrix. Furthermore, the uncompatibilized blends in all PBAT contents showed some cavities from pull-out of PBAT particles indicating poor adhesion between LDPE and PBAT phases. However, in the case of the compatibilized LDPE/PBAT blends, incorporation of the PE-g-MA compatibilizer could dramatically reduce the particle size of the PBAT phase and the PBAT particles became very fine particles embedded in the LDPE matrix, which were not clearly visible observed at magnification of 3000x. Besides, there was no appearance of pull-out of PBAT particles from the matrix. These results confirmed that the addition of PE-g-MA compatibilizer could enhance the interfacial adhesion between LDPE matrix and PBAT phase leading a better compatibility of two polymers in the blend, thereby more homogeneous dispersion of PBAT in the LDPE/PBAT blends.

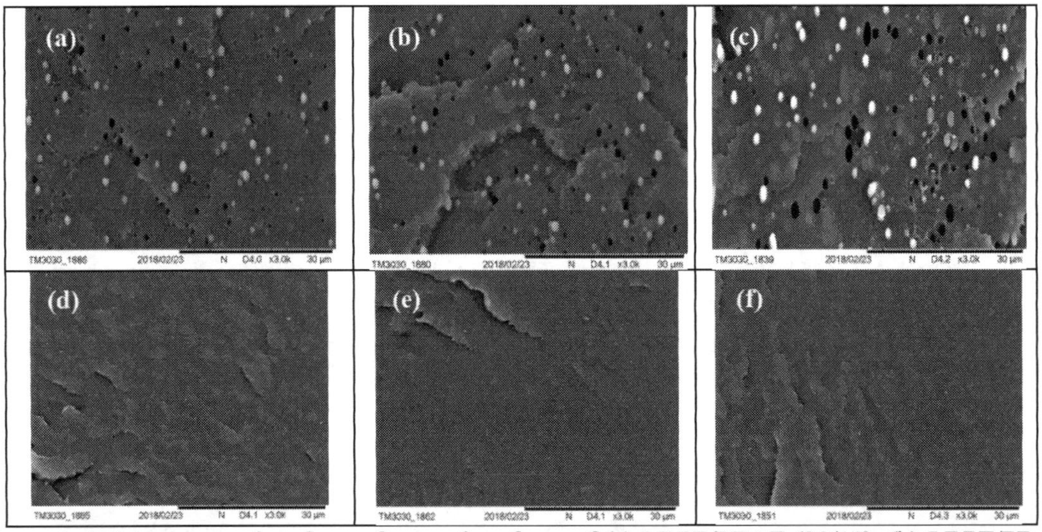


**Figure 1**. SEM micrographs of cryo-fractured surfaces of (a) LDPE/PBAT (90/10), (b) LDPE/PBAT (80/20), (c) LDPE/PBAT (70/30), (d) LDPE/PBAT/PE-g-MA (90/10/10), (e) LDPE/PBAT/PE-g-MA (80/20/10) and (f) LDPE/PBAT/PE-g-MA (70/30/10) films at magnification of 3000x.

**Mechanical properties.**

The Young's modulus, maximum tensile strength and elongation at break of neat LDPE, neat PBAT, uncompatibilized and compatibilized LDPE/PBAT are illustrated in Figure 2 (a)-(c), respectively.

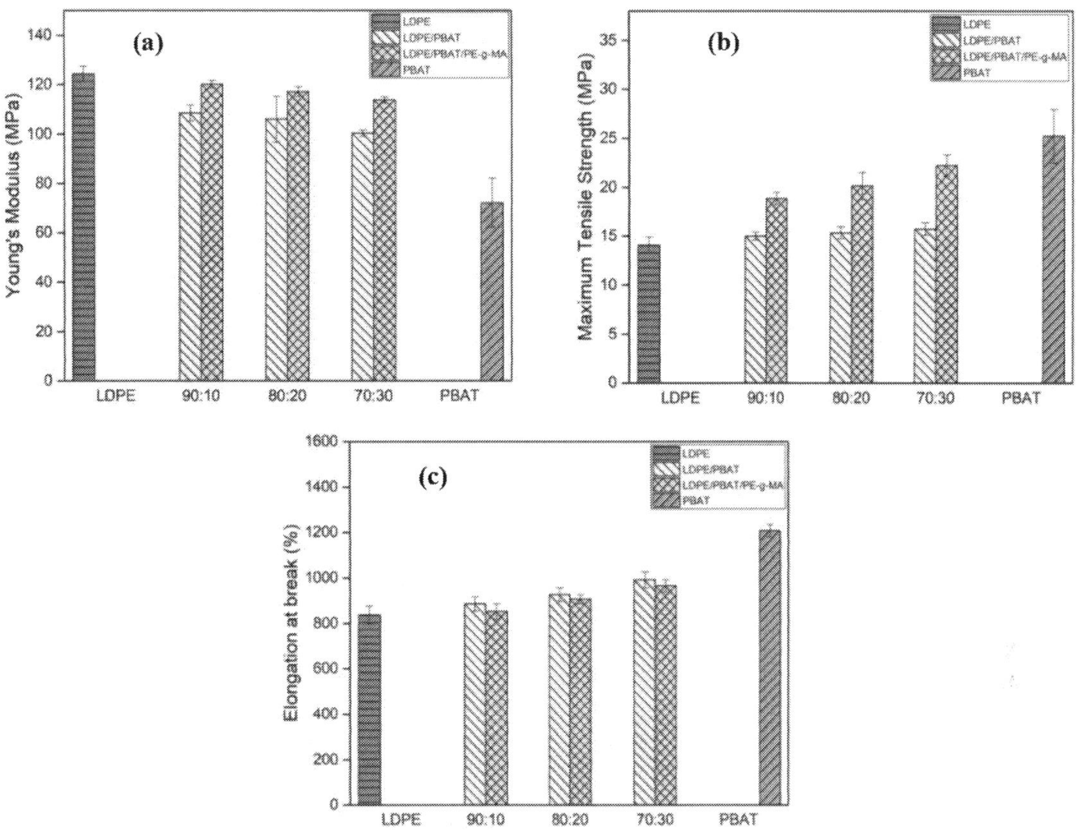


**Figure 2**. Mechanical properties of LDPE, PBAT, uncompatibilized LDPE/PBAT and compatibilized LDPE/PBAT films at various ratios: (a) Young's Modulus, (b)Maximum tensile strength and (c) Elongation at break.

As seen in Fig.2, the PBAT showed a lower modulus (72 MPa) with a higher elongation at break (1206 %) when compared to LDPE with modulus and elongation at break of 124 MPa and 837 %, respectively. This is because the PBAT is a random block copolymer with a small degree of crystallization in the polymer structure, thus leading to a low elastic modulus, high flexibility and toughness with respect to the LDPE. However, maximum tensile strength of the PBAT (25 MPa) was higher than that of LDPE (14 MPa), likely due to a presence of aromatic terephthalate groups in PBAT structure, resulting in high tensile strength [4]. In the LDPE/PBAT films with and without compatibilizer, modulus tended to decrease whereas tensile strength and elongation at break became higher as an increase in PBAT content. Compared with the uncompatibilized blend films at same PBAT concentration, it was clearly seen that the compatibilized LDPE/PBAT films provided greater modulus and tensile strength; in contrast, a lower elongation at break was observed. These results indicated that incorporation of the PE-g-MA compatibilizer enhanced interfacial adhesion between LDPE matrix and PBAT phase, resulting from a formation of physical entanglement between PE chains in PE-g-MA and LDPE matrix chains and chemical reaction of MA groups on PE-g-MA with hydroxyl and carboxyl groups of PBAT. This led to improvement of mechanical properties of the compatibilized LDPE/PBAT films.

**Differential scanning calorimetry analysis**

DSC results including melting temperatures ($T_m$) and melting enthalpies ($\Delta H_m$) of LDPE, PBAT, uncompatibilized and compatibilized LDPE/PBAT films are represented in Table 1. The $T_m$ of neat LDPE and PBAT appeared at 110.62 and 123.41 °C, respectively and the $\Delta H_m$ of LDPE (121.24 J/g) was higher than that of PBAT (15.65 J/g), indicating higher crystallinity in LDPE compared with PBAT. In all polymer blend films, there was a presence of a single peak of Tm around 110-111°C, corresponding to the $T_m$ of LDPE phase in the polymer blends and no observation of $T_m$ peak of PBAT because intensity of the $T_m$ peak of PBAT was quite low and overlapped with the $T_m$ peak of LDPE. Moreover, the DSC results revealed that the $T_m$ of LDPE in all polymer blends did not change when the PBAT content increased. Besides, the $\Delta H_m$ of the polymer blends decreased as the PBAT concentration, which was probably because the PBAT chains containing bulky aromatic aromatic terephthalate groups in the PBAT structure could restrict LDPE crystallization, thus a reduction of $\Delta H_m$ of the polymer blends. Comparing to the uncompatibilized LDPE/PBAT films at the same content of PBAT, the films containing PE-g-MA compatibilizer gave the higher $\Delta H_m$, which inferred more crystallinity content in the polymer blends. This result suggested that the compatibilizer improved interfacial interaction and compatibility between LDPE and PBAT phases, thus promoting crystallization formation in the polymer blends.

**Table 1** Thermal properties of LDPE, PBAT, LDPE/PBAT and LDPE/PBAT/PE-g-MA films

| Sample | Compatibilizer content (%) | Tm (°C) | $\Delta H_m$ (J/g) |
|---|---|---|---|
| LDPE | - | 110.62 | 121.24 |
| 90/10 | 0 | 110.14 | 94.69 |
| 80/20 | 0 | 110.35 | 92.69 |
| 70/30 | 0 | 110.19 | 86.32 |
| 90/10 | 10 | 110.40 | 105.23 |
| 80/20 | 10 | 111.06 | 101.4 |
| 70/30 | 10 | 110.56 | 92.35 |
| PBAT | - | 123.41 | 15.65 |

**Water vapor transmission rate (WVTR)**

The Water vapor transmission rate results of neat LDPE, neat PBAT, LDPE/PBAT films with and without the PE-g-MA compatibilizer at different the PBAT contents are displayed in Figure 3. Among all films, the PBAT film showed the highest WVTR value indicating poor water vapor barrier property of the PBAT films. This was due to more hydrophilic characteristic of polar functional groups in the PBAT chains such as ester, hydroxyl and carboxyl groups forming intermolecular bonding with water vapor molecules via hydrogen bonding and dipole-dipole interaction. In the case of the LDPE/PBAT films with and without incorporating the PE-g-MA compatibilizer, the WVTR significantly increased as an increasing PBAT content in the films. In addition, it was clearly observed that the WVTR of the compatibilized LDPE/PBAT films was higher than that of the uncompatibilized films at the same PBAT concentration. This may be because the addition of the PE-g-MA compatibilizer led to an increase in polarity of the polymer blend system due to possessing maleic anhydride group in the PE-g-MA chains, resulting in a higher degree of WVTR of the LDPE/PBAT films containing the PE-g-MA.

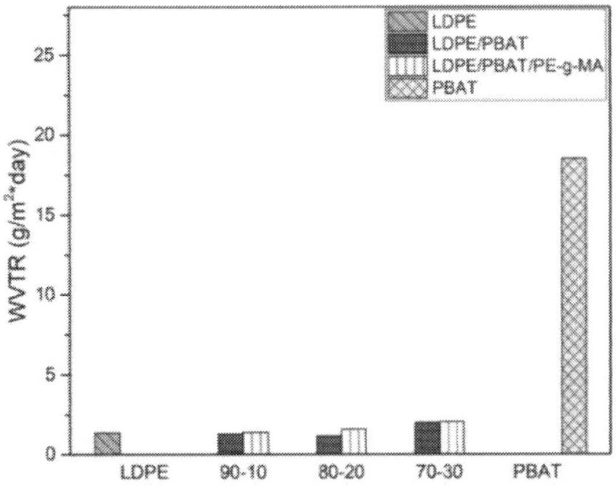


**Figure 3**. The Water vapor transmission rate of neat LDPE, neat PBAT, LDPE/PBAT and LDPE/PBAT/PE-g-MA films at ratios of 90/10, 80/20 and 70/30.

**Oxygen permeability study**

Comparison of the oxygen permeability of neat LDPE, neat PBAT, LDPE/PBAT and LDPE/PBAT/PE-g-MA films was present in Figure 4. The LDPE film exhibited higher oxygen permeability value than that of PBAT film. This result could be explained by considering that solubility parameter ($\delta$) of LDPE ($\delta$ = 3.1 $MPa^{1/2}$) [5] is closed to that of $O_2$ gas ($\delta$ = 6.7 $MPa^{1/2}$) [6] compared with solubility parameter of PBAT ($\delta$ = 18.9 $MPa^{1/2}$) [7], resulting in increment of dissolution of oxygen gas in LDPE film and more penetration of $O_2$ molecules through the PDPE film with respect to PBAT film. Considering the polymer blend films, it was found that the oxygen permeability of the LDPE/PBAT and LDPE/PBAT/PE-g-MA films was lower than the pure LDPE films and decreased proportionally with the increasing order of the PBAT contents. This is possibly because a presence of the PBAT droplets dispersed in the LDPE phase as shown in the SEM images could hinder diffusion path of $O_2$ molecules and prolong the permeation of $O_2$ molecules through the polymer blend films. Besides, it was seen that the compatibilized LDPE/PBAT films provided lower oxygen permeability value, indicating a better barrier property, compared with the uncompatibilized films at the same PBAT loading. These results were be attributed to more homogeneous dispersion of the fine PBAT particles in the LDPE matrix and the better interfacial adhesion between LDPE and PBAT phases which restricted diffusion path of $O_2$ molecules and created longer and more tortuous paths, resulting in a reduction of oxygen permeation through the compatibilized films.

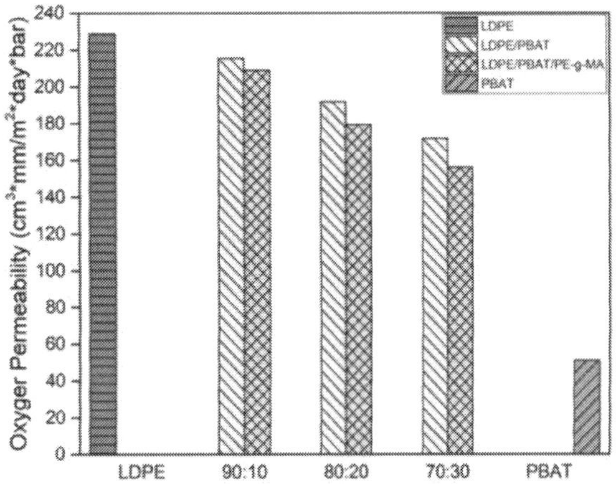


**Figure 4**. The oxygen permeability of neat LDPE, neat PBAT, LDPE/PBAT and LDPE/PBAT/PE-g-MA films at ratios of 90/10, 80/20 and 70/30.

## 4. Conclusions

The SEM results showed more homogeneous in morphologies and better compatibility in the LDPE/PBAT with addition of the PE-g-MA compatibilizer. Moreover, it was found that the compatibilized LDPE/PBAT films exhibited superior mechanical properties to those the LDPE/PBAT films without the PE-g-MA, resulting in a greater interfacial adhesion between LDPE and PBAT polymers. Incorporating PE-g-MA compatibilizer into the LDPE/PBAT blends enhanced crystallization ability in the blends and also promoted the oxygen barrier property as well as slightly increased the WVTR with respect to uncompatibilized LDPE/PBAT blends.

## Acknowledgement

The authors are thankful for the financial support from Silpakorn University.

## References

[1] Ray S S and Okamoto M 2003 *Macromolecules.* **36** 2355.
[2] Fukushima K, Wu M H, Bocchini S, Rasyida A and Yang M C 2012 *Mater. Sci. Eng. C.* **32** 1331.
[3] Averous L 2004 *J. Macromol. Sci. Part C Polym. Rev.* **4** 231.
[4] Jian J, Xiagbin Z and Xianbo H 2020 Adv. *Indus. Eng. Polym. Res.* **3** 19.
[5] Valentina S 2012 *Int. J. Polym. Sci.***2012** 1.
[6] Sato T, Hamada Y, Sumikawa M, Araki S and Yamamoto H 2014 *Indus. Eng. Chem. Res.***14** 19331.
[7] Liu T Y, Lin W C, Yang M C, Chen S Y 2005 *Polymer.***46** 12586.

# Corrosion inhibition effect of 1-thia-3-azaindene on type 304 austenitic stainless steel in 3M $H_2SO_4$ solution

**R T Loto[1*], C A Loto[1] and F Awodein[1]**

[1]Department of Mechanical Engineering, Covenant University, Ogun state, Nigeria
Email: tolu.loto@gmail.com

**Abstract.** Corrosion inhibition of 1-thia-3-azaindene (T3A) on type 304 austenitic stainless steel was studied in 3 M $H_2SO_4$ solution at specific concentrations by potentiodynamic polarization. Results showed T3A performed effectively with optimal inhibition efficiency of 92.58% at 1.5% volumetric concentration with the corresponding corrosion rate of the steel at 0.53 mm/y compared to 6.72 mm/y in the acid solution without T3A compound. Inhibition performance increased with inhibitor concentration. T3A exhibited anodic type chemisorption inhibition according Langmuir, Temkin, Frumkin and Freundlich adsorption isotherms with correlation coefficient generally above 0.9. Polarization plots showed significant passivation behaviour after metastable pitting activity before breakdown at pitting potential signifying extensive surface coverage of the steel. The cathodic portion of the polarization plot showed $H_2$ evolution and $O_2$ reaction mechanisms are under activation control.

## 1. Introduction

Stainless steels generally have higher resistance to corrosion than carbon steels due to the formation of an inert protective oxide on steel surfaces [1, 2]. However, there are industrial condition harmful to stainless steels and significantly decrease their lifespan during operation. The huge financial and economic cost resulting from corrosion of stainless steels is a major problem and is the focus of research into more resistant steels and cost effective corrosion control techniques [3-5]. The huge cost is due to accelerated breakdown of the steel locally leading to unforeseen industrial accidents, high cost of maintenance and constant repairs [6, 7]. Industries such as oil and gas, petrochemical, fertilizer production, chemical processing, marine and shipping, heat exchangers and energy generation experience huge corrosion problems due to the presence of highly reactive corrosive species such as $S_2O_3(2-)$, $Cl^-$, $SO_4^{2-}$, $NO_3$. etc. [8, 9]. Type 304 austenitic stainless steel have extensive application due to its excellent mechanical properties, weldability, formability and corrosion resistance [10, 11] especially for power plant components, chemical containers and heat exchangers. Heterocyclic compounds and compounds of organic origin have been proven to be excellent corrosion inhibitors. [12-22]. However, more research is needed to develop high performance inhibition at very low concentration with chemisorption type inhibition mechanism in industrial environments. In contribution to the research on application of corrosion inhibitor for stainless steels this article focusses on the corrosion inhibition performance of 1-thia-3-azaindene compound on austenitic steel in 3M $H_2SO_4$.

## 2. Experimental methods

Type 304 austenitic stainless steel (AISI 304) rod procured from Lagos, Nigeria was analysed at the Materials Characterization Laboratory, Department of Mechanical Engineering, Covenant University, Ota, Ogun State, Nigeria to determine its composition shown in Table 1. AISI 304 was machined into 7 samples which underwent metallographic preparation before being encased in Versocit acrylic paste. 1-thia-3-azaindene (T3A), a translucent liquid compound was obtained from BOC Chemicals, USA. It has a molecular weight of 135.19 g/cm$^3$. The structural configuration of the compound is shown in Fig. 1. T3A was added to 3 M $H_2SO_4$ (98% analar grade) in volumetric concentrations of 0.25%, 0.5%, 0.75%, 1%, 1.25% and 1.5% T3A concentration per 400 mL of the acid solution.

**Table 1.** Elemental content (wt. %) of AISI 304

| Element | C | Cr | Mn | Si | P | S | Ni | N | Fe |
|---|---|---|---|---|---|---|---|---|---|
| **Composition** | 0.07% | 18.5% | 2% | 1% | 0.045% | 0.0015% | 9% | 0.01% | Balance |

Potentiodynamic polarization was test was utilized with the aid of three electrode system (Pt counter electrode, AISI 304 specimen electrode and Ag/AgCl reference electrode) placed within a visible container with the T3A/$H_2SO_4$ solution and connected to Digi-Ivy potentiostat/computer system for real-time monitoring from -1.5V to +1.5 mV at scan rate of 0.0015V/s. Corrosion current ($C_I$) A, corrosion current density ($C_J$) A/cm$^2$, corrosion potential ($C_P$) V, anodic and cathodic Tafel slope ($B_a$ and $B_c$) V/dec and polarization resistance ($R_p$) Ω were computed from the Tafel plots and the data output. Corrosion rate ($C_R$) was calculated as follows

$$C_R = \frac{0.00327 \times C_J \times E_{WT}}{D} \quad (1)$$

$D$ (g/cm$^3$) indicates density and $E_{WT}$ (g) indicates AISI 304 equivalent weight in grams. Inhibition efficiency ($I_E$) was computer from the equation below;

$$I_E = 1 - \left[\frac{C_2}{C_1}\right] \times 100 \quad (2)$$

$C_1$ and $C_2$ represents corrosion rate with and without T3A addition.

## 3. Results and discussion

### *3.1 Potentiodynamic polarization*

Potentiodynamic polarization plots of AISI 304 in 3M $H_2SO_4$ solution at predetermined concentrations of T3A are shown in Fig. 2. Table 2 shows the data for the electrochemical parameters obtained from the polarization test. Table 2 shows the significant difference between the corrosion rate of AISI 304 at 0% T3A and the corrosion rate values at 0.25% to 1.5% T3A. At 0% T3A the corrosion rate value of AISI 304 (6.72 mm/y) is due to oxidation of AISI 304 surface resulting from the breakdown of the inert protective oxide on the steel surface. $SO_4^{2-}$ anions in the acid solution diffuse through the metal/solution interface displacing the adsorbed $O_2$ atoms responsible for the formation of the protective oxide. This redox electrochemical process exposes the substrate Fe metal to localized corrosion deterioration as the surface deterioration first starts at regions or sites where impurities, inclusions etc. are present. The impurities cause discontinuities in the protective oxide leading to pit initiation and eventual growth of the steel surface area exposed to corrosion. Addition of T3A compound to the acid solution significantly reduced the corrosion rate of AISI 304 with respect to T3A concentration. At 0.25% T3A concentration, the corrosion rate of AISI 304 is 3.59 mm/y corresponding to inhibition efficiency of 49.46% and polarization resistance of 83.3Ω. This value is below the value for effective corrosion inhibition. However, beyond 0.25% T3A concentration, significant decrease in corrosion rate coupled with significant increase in inhibition efficiency was observed. The corrosion rate and inhibition efficiency of AISI 304 and T3A at 0.5% T3A were 2.51 mm/y and 64.76%. At 0.75% T3A, the values are 1.80 mm/y and 74.73% while at optimal T3A concentration of 1.5% the values are 0.53 mm/y and 92.58%. Observation of the polarization resistance values show significant increase in value from 29.26Ω at 0% T3A to 433.0Ω at 1.5% T3A. Variation in corrosion potential from between values at 0.25% to 1.5% T3A and the value at 0% T3A shows significant anodic shift from 0.25% T3A to 1% T3A due to suppression of the oxidation reactions on the steel surface. This assertion is further confirmed from the significant variation of the anodic portion of the polarization plots. Whereas observation of the cathodic portion of the polarization plots shows near similarities signifying that the cathodic reaction mechanisms are under activation control. The maximum variation in corrosion potential is between T3A concentration at 0% and 1% concentration i.e. 135mV in the anodic direction which shows the inhibition mechanism of T3A aligns with anodic type inhibition. This is proven from the passivated region on the anodic portion of the polarization plot from 0.25% T3A concentration. Observation of the plots in Fig. 2 shows the onset and propagation of metastable pitting before passivation, passivated region and pitting corrosion. At 0.25% T3A metastable pitting occurred before passivation. However, immediately after passivation breakdown occurred due to pitting corrosion of the steel surface.

Starting from 0.25% T3A to 1% T3A, it is observed that there is metastable pitting before passivation of the steel, increase in T3A concentration slightly increased the passivated region before breakdown due to pitting. Peak passivation behaviour was observed at 1% T3A. Beyond 1% T3A, significant decrease in the passivated region occurred at 1.25% T3A while at 1.50% T3A the passivated region has almost completely disappeared despite decrease in corrosion rate and increase in inhibition efficiency due to lateral interaction mechanism among inhibitor molecules which influenced their selective precipitation performance.

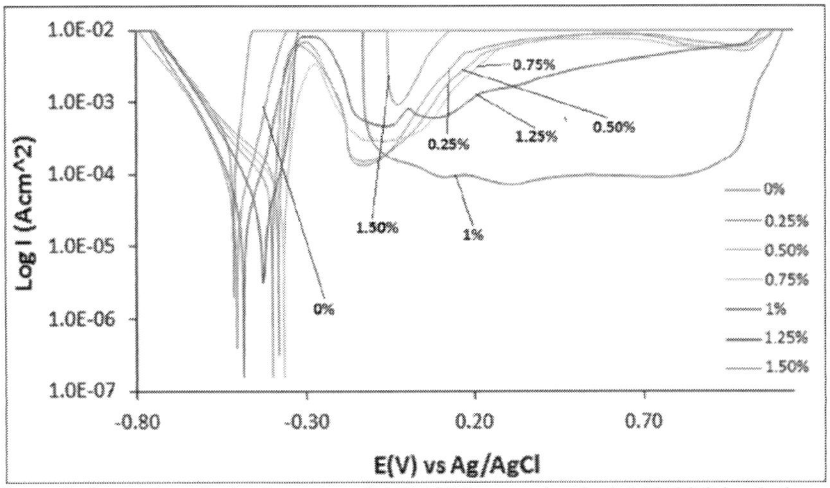


**Figure 2** Potentiodynamic polarization plot of AISI 304 corrosion in 3 M H2SO4 solution at specific T3A concentration

**Table 2.** Data on weight-loss of AISI 304 from 1M $H_2SO_4$/TD (0% - 0.015%) solution

| Sample | T3A Conc. (%) | AISI 304 $C_R$ (mm/y) | T3A *IE* (%) | $C_I$ (A) | $C_J$ (A/cm$^2$) | $C_P$ (V) | $R_p$ (Ω) | $B_c$ (V/dec) | $B_a$ (V/dec) |
|---|---|---|---|---|---|---|---|---|---|
| A | 0 | 6.72 | 0 | 5.15E-04 | 6.54E-04 | -0.335 | 29.76 | 7.803 | 0.035 |
| B | 0.25 | 3.59 | 49.46 | 2.76E-04 | 3.50E-04 | -0.231 | 83.30 | 5.647 | 0.965 |
| C | 0.5 | 2.51 | 64.76 | 1.92E-04 | 2.44E-04 | -0.215 | 102.96 | 4.415 | 0.123 |
| D | 0.75 | 1.80 | 74.73 | 1.38E-04 | 1.75E-04 | -0.200 | 185.80 | 5.178 | 5.193 |
| E | 1 | 1.33 | 81.23 | 1.02E-04 | 1.30E-04 | -0.199 | 262.70 | 4.572 | 0.754 |
| G | 1.25 | 0.85 | 88.03 | 6.53E-05 | 8.29E-05 | -0.327 | 342.30 | 8.780 | 5.150 |
| H | 1.5 | 0.53 | 92.58 | 4.05E-05 | 5.14E-05 | -0.255 | 433.00 | 9.361 | 7.530 |

*3.2 Adsorption Isotherm*

Adsorption and corrosion inhibition effect of T3A on AISI 304 is due to the ability of protonated T3A molecules in $H_2SO_4$ solution to displace $H_2O$ molecules on the steel surface, hinder the electrolytic transport of $SO_4^{2-}$ unto the steel surface and suppress the redox reaction mechanisms responsible for corrosion. Further insight into the inhibition behavior is gotten from mathematical models known as adsorption isotherms which depict show the quantitative variation of protonated T3A ionic species adsorbed unto AISI 304 under equilibrium conditions [23]. The strength of T3A adsorption onto AISI 304 is determined by the strength of the electrostatic attraction between T3A and the steel surface. The presence of polar functional groups with S, and N atoms in the inhibitor molecule of the aromatic heterocyclic compound and pi electrons reacts with the valence electrons on the steel surface resulting in strong covalent bond [24, 25]. AISI 304 in the acid solution causes the accumulation of $SO_4^{2-}$ anions at the metal/solution interface resulting in excess -ve charges on the steel surface. The negative charge on the steel strongly attracts the protonated inhibitor molecule leading to strong impenetrable chemical complex/precipitate on the steel stifling further electrochemical reaction. Results from potentiodynamic polarization were was used to determine the appropriate adsorption isotherm for T3A adsorption. Langmuir, Frumkin, Freundlich and Temkin gave appreciable correlation coefficient values. The values are shown in Table 3.

**Table 3.** Comparison of correlation coefficient values of Langmuir, Frumkin, Freundlich and Temkin adsorption isotherms for T3A adsorption on AISI 304

| **Adsorption Isotherms** | **$H_2SO_4$ Solution** |
|---|---|
| Langmuir | 0.9964 |
| Frumkin | 0.7424 |
| Freundlich | 0.9611 |
| Temkin | 0.9979 |

The Langmuir, Frumkin, Freundlich and Temkin isotherms are shown from Figs. 3(a) to 4(b). Langmuir isotherm states a definite quantity of locations on metallic surfaces exist for adsorption of single layer molecules exhibiting fixed energy value in variable equilibrium according to the equation below;

$$\theta = \left[\frac{K_{ads}C_{T3A}}{1+K_{ads}C_{T3A}}\right] \quad (3)$$

$K_{ads}$ is the equilibrium constant of adsorption during the inhibition process. $C_{T3A}$ is the molar concentration of T3A compound. According to Temkin isotherm the heat of adsorption progressively decreases as molecular increases with equivalent binding energy distribution [26], according to the equation below;

$$qe = B \ln (A+Ce) \quad (4)$$

$$\text{Where } B = RT/b \quad (5)$$

A indicates Temkin isotherm constant (L/g), b indicates Temkin constant related to heat of adsorption, T indicates temperature (K), R represents gas constant (8.314, J/mol K) and Ce represents concentration of adsorbate. B indicates the Temkin constant related to heat of sorption (J/mol). Frumkin isotherm states that molecular coverage on non-homogeneous adsorbent surfaces is similar under high molecular concentrations with respect to the potential of the adsorbent surface. The effect of lateral attraction/repulsion between inhibitor molecules is significant and quantitative according to the equation below;

$$\text{Log}\left[C_{T3A} * \left(\theta/_{1-\theta}\right)\right] = 2.303 \log K_{ads} + 2\alpha\theta \quad (6)$$

$\alpha$ is the lateral interaction parameter. The Freundlich isotherm states that adsorbed molecular interaction exist on adsorbent metal surfaces and the lateral attraction/repulsion between them occurs [27]. The Freundlich equation is stated as follows;

$$\theta=KC^n \quad (7)$$

$$\log \theta = n\log C + \log K_{ads} \quad (8)$$

n represents the constant for the characteristics of adsorbed T3A molecule

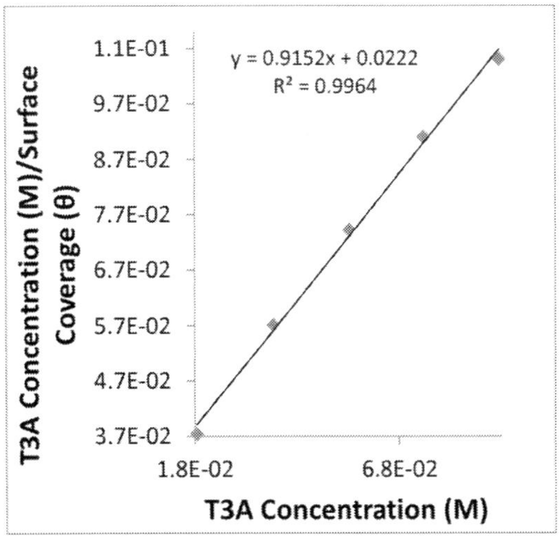


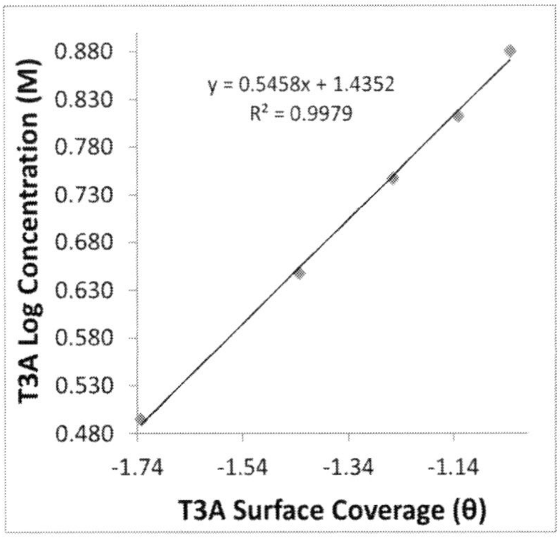


(a) (b)

**Figure 3** Adsorption isotherm plot (a) Langmuir isotherm and (b) Temkin isotherm

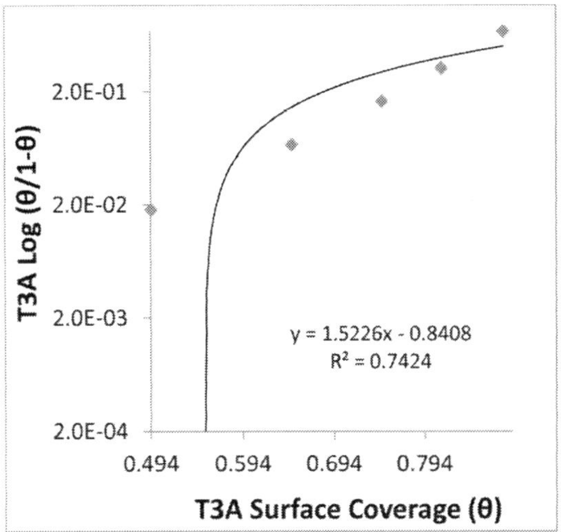


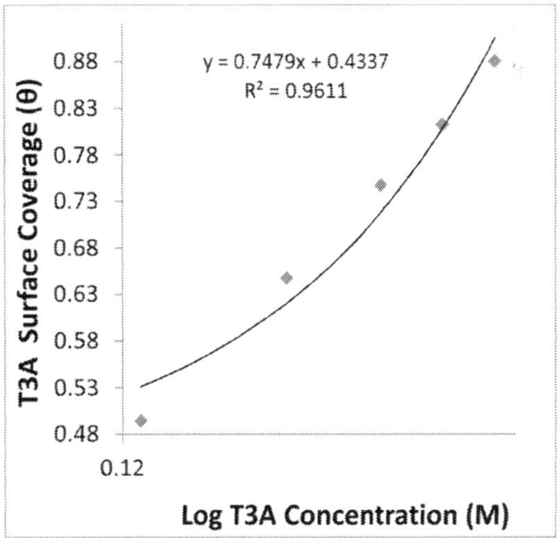


(a) (b)

**Figure 4** Adsorption isotherm plots (a) Frumkin isotherm and (b) Freundlich isotherm

Data for Gibbs free energy ($\Delta G$) representing the adsorption strength of T3A onto AISI 304 surface in $H_2SO_4$ are presented in Table 4. The results were calculated from the formula below. Equilibrium constant of T3A adsorption ($K_{ads}$) was gotten from the Langmuir due to its high correlation coefficient value.

$$\Delta G_{ads} = -2.303RT \log [55.5K_{ads}] \quad (9)$$

55.5 indicates the molar concentration of $H_2O$ in the solution, $R$ is the universal gas constant, $T$ is the absolute temperature. The ΔG values depict chemisorption adsorption of T3A compound unto AISI 304 signifying strong electrostatic attraction and covalent bonding [28, 29]. The values also show T3A effectively displaced $H_2O$ molecules from the aqueous solution and suppression of $SO_4^{2-}$ displacement reactions. The potentiodynamic plots give evidence for competitive adsorption between the inhibitor molecules and corrosive species. The $\Delta G$ values ranges from (lowest to highest) -36.75 KJmol$^{-1}$ to -37.91 KJmol$^{-1}$ signifying chemisorption. However, the values are not proportional to T3A concentration signifying lateral interaction effect occurs irrespective of T3A concentration.

**Table 4.** $\Delta G$ and $K_{ads}$ values for T3A adsorption onto AISI 304

| AISI 304 Samples | T3A Concentration (M) | Surface Coverage (θ) | Equilibrium Constant of adsorption (K) | Gibbs Free Energy, ΔG ($Kjmol^{-1}$) |
|---|---|---|---|---|
| A | 0 | 0 | 0 | 0 |
| B | 1.85E-02 | 0.495 | 52908.9 | -36.90 |
| C | 3.70E-02 | 0.648 | 49692.6 | -36.75 |
| D | 5.55E-02 | 0.747 | 53297.4 | -36.92 |
| E | 7.40E-02 | 0.812 | 58489.4 | -37.15 |
| F | 9.25E-02 | 0.880 | 79520.8 | -37.91 |

## 4. Conclusion

1-thia-3-azaindene compound effectively inhibited the corrosion of type 304 austenitic stainless steel in dilute $H_2SO_4$ solution with inhibition performance progressively increasing with inhibitor concentration. The inhibitor exhibited dominant anodic type inhibition at most concentrations with lateral interaction effective being visible at higher concentration. Passivation of the steel surface varied with inhibitor concentration. The inhibitor adsorbed unto the steel through chemisorption mechanism.

## Acknowledgement

The author is grateful to Covenant University Ota, Ogun State, Nigeria for their support for this project.

## References

[1] Ram C, Sharma C and Singh AK 2015 In-plant corrosion study of steels in distillery effluent treatment plant *J. Mate. Eng. Perform.* **24(5)** 1841-1847.3.
[2] Santamaria M, Di Franco F, Di Quarto F, Pisarek M, Zanna S and Marcus P 2015 Photoelectrochemical and XPS characterisation of oxide layers on 316L stainless steel grown in high-temperature water *J. Solid State Electrochem.* **19(12)** 3511-3519.
[3] Liu G Q, Zhu Z Y, Ke W, Han C I and Zeng, CL 2001 Corrosion behavior of stainless steels and nickel-based alloys in acetic acid solutions containing bromide ions *NACE Corrosion* **57(8)** 730-38.
[4] Collins W D, Weyers R E and Al-Qadi I L 1993 Chemical treatment of corroding steel reinforcement after removal of chloride-contaminated concrete *NACE Corrosion* **49(1)** 74-88.
[5] Ekpe U J, Ibok U J, Ita B I, Offiong O E and Ebenso E E 1995 Inhibitory action of methyl and phenyl thiosemicarbazone derivatives on the corrosion of mild steel in hydrochloric acid *Mater. Chem. Phys.* **40** 87-93.
[6] Ashassi-Sorkhabi H, Masoumi B and Ejbari P E 2009 Corrosion inhibition of mild steel in acidic media by basic yellow 13 dye *J. Appl. Electrochem.* **39(9)** 1497-1501.
[7] Singh D D N, Singh T B and Gaur B 1995 The role of metal cations in improving the inhibitive performance of hexamine on the corrosion of steel in hydrochloric acid solution *Corros. Sci.* **37(6)** 1005-1019.
[8] Ita B I and Offiong O E 1997 The inhibition of mild steel corrosion in hydrochloric acid by 2, 2'-pyridil and α-pyridoin *Mats. Chem. Phys.* **51** 203-210.
[9] Abiola O K, Oforka N C and Ebenso E E 2004 A potential corrosion inhibition for acid corrosion of mild steel *Bulletin of Electrochemistry* **2(9)** 409-413.
[10] M.P. Nascimento, R.C. Souza, W.L. Pigatin, H.J.C. Voorwald, Effects of surface treatments on the fatigue strength of AISI 4340 aeronautical steel, Int J Fatigue, 23 (2001), pp. 607-618
[11] International Stainless Steel Forum Rue Colonel Bourg 120 B-1140 Brussels Belgium. Stainless Steel Applications – Aerospace. 2007/05/23. p. 1–2
[12] Rastogi R B, Singh M M, Singh K and Yadav K M 2005Organolic dithiohydrazodicarbonamides as corrosion inhibitors for mild steel-dimethyl sulphoxide containing HCl *Port. Electrochimica Acta.* **22** 315-332.
[13] James A O, Oforka N C and O.K. Abiola O K 2005 Inhibition of the corrosion of aluminium in hydrochloric acid solutions by pyridoxal hydrochloride *Corros. Eng. Sci. Techn.* **7(21)** 1-10.
[14] Daoud D, Douadia T, Hamania H, Chafaaa S and Al-Noaimi M 2015 Corrosion inhibition of mild steel by two new S-heterocyclic compounds in 1 M HCl: Experimental and computational study *Corros. Sci.* **94** 21-37.
[15] Ahmed S K, Ali W B and Anees A. Khadom A A 2019 Synthesis and investigations of heterocyclic compounds as corrosion inhibitors for mild steel in hydrochloric acid *Int. J. Ind. Chem.* **10** 159–173.
[16] Talati J D and Gandhi D K 1983 N-heterocyclic compounds as corrosion inhibitors for aluminium-copper alloy in hydrochloric acid *Corros. Sci.* **23(12)** 1315-1332.
[17] Singh A, Ansari K R, Quraishi M A, Kayad S and Banerjee P 2019 The effect of an N-heterocyclic compound on corrosion inhibition of J55 steel in sweet corrosive medium *New J. Chem.* **43**, 6303-6313.
[18] Loto R T and Loto C A 2012 Effect of p-phenylediamine on the corrosion of austenitic stainless steel type 304 in hydrochloric acid *Int. J. Elect. Sci.* **7(10)** 9423-9440.

[19] Loto R T 2017 Study of the synergistic effect of 2-methoxy-4-formylphenol and sodium molybdenum oxide on the corrosion inhibition of 3CR12 ferritic steel in dilute sulphuric acid *Results in Phys*. **7** 769-776.
[20] Loto R T, Loto C A and Fedotova T 2012 Inhibition effect of n, n'-dimethylaminoethanol on the corrosion of austenitic stainless steel type 304 in 3M $H_2SO_4$ *Int. J. Elect. Sci.* **7** 10763 – 10778.
[21] Loto C A and Loto R T 2013 Effect of dextrin and thiourea additives on the zinc electroplated mild steel in acid chloride solution *Int. J. Elect. Sci.* **8(12)** 12434-12450.
[22] Loto R T, Leramo R and Oyebade B 2018 Synergistic Combination effect of salvia officinalis and lavandula officinalis on the corrosion inhibition of low-carbon steel in the presence of $SO_4^{2-}$- and $Cl^-$-containing aqueous environment J Fail. Anal. Prev. **18(6)** 1429-1438.
[23] Bockris J O and Swinkels D A J 1964 Adsorption of n-decylamine on solid metal electrodes *J. of Elect. Soc.* **111** 736.
[24] Yaro A S, Khadom A A and Wael R K 2013 Apricot juice as green corrosion inhibitor of mild steel in phosphoric acid *Alexandria Eng. J.* **52(1)** 129-135.
[25] Mahmoud N, El-Haddad M N 2013 Chitosan as a green inhibitor for copper corrosion in acidic medium *Inter. J. Bio. Macromol.* **55** 142-149.
[26] Lowmunkhong P, Ungthararak D and Sutthivaiyakit P 2010 Tryptamine as a corrosion inhibitor of mild steel in hydrochloric acid solution *Corros. Sci.* **52** 30-36.
[27] Zeldowitsch J (1934) Adsorption site energy distribution *Acta Phys. Chim. URSS* **1** (1934) 961–973.
[28] Aharoni C and M. Ungarish M 1977 Kinetics of activated chemisorption. Part 2. Theoretical models *J. Chem. Soc. Faraday Trans.* **73** 456–464.
[29] Loto R T, Loto C A, Joseph O and Olanrewaju G 2016 Adsorption and corrosion inhibition properties of thiocarbanilide on the electrochemical behaviour of high carbon steel in dilute acid solutions *Results in Phys.* **6** 305-314.

# Corrosion inhibition study and statistical data of thymus mastichina and alicium verum oil on low carbon steel in dilute $H_2SO_4$ and HCl solution

**R T Loto[1*] and S Ororo[1]**
[1]Department of Mechanical Engineering, Covenant University, Ogun state, Nigeria
Email: tolu.loto@gmail.com

**Abstract.** Data on the corrosion inhibition of thymus mastichina and alicium verum (TA) on mild steel in $H_2SO_4$ and HCl solution was evaluated by coupon measurement and statistically through analysis of variance. The inhibition performance of TA was observed to decrease significantly with exposure time at 1% TA, beyond which the extent of decrease reduced. In HCl solution TA inhibition efficiency varied significantly with exposure time from 1%-4% TA concentration. Data showed TA effectively inhibited the mild steel after 1% concentration with optimal value of 81.24% and 5% TA concentration. TA performed poorly in HCl with optimal value of 68.33% at 5% TA concentration. Standard deviation showed variation of TA inhibition efficiency from mean inhibition value was significant at 1-5% TA concentration in HCl and at 1% TA concentration in $H_2SO_4$. 88% of TA inhibition data were above 70% inhibition with respect to exposure time and TA concentration at margin of error of $\pm$0.09% compared to 53% in HCl at margin of error of $\pm$0.1%. Analysis of variance showed TA concentration is the dominant source of variation responsible for the inhibition performance of TA with statistical relevance value of 88.83% while TA concentration and exposure time were both influential in HCl solution with values of 67.96% and 26.97%.

## 1. Introduction

Corrosion of carbon steel is a major industrial and economic problem due to its extensive application worldwide. The steel is used as structural material in the manufacture of parts and components. Destruction of the substrate metal leads to industrial accidents, collapse of structures, leakages and product contamination as a result of their high vulnerability to corrosion though they are known for their despite their good mechanical properties [1]. Corrosion is the product of redox electrochemical reactions that occurs on metallic surfaces leading to surface degradation and deterioration of the aesthetic properties of the alloy [2-4]. According to a study in 2000 corrosion damage in addition to maintenance and repairs due to corrosion constitutes about 3.1% of U.S total GDP [5]. Application of corrosion inhibitors is the most economical and effective method of corrosion control [6-8]. Use of non-toxic green compounds for corrosion inhibition are more appropriate alternatives to the conventional toxic derivatives [9-14]. Ongoing investigation on plant extracts application for corrosion inhibition of metallic alloys has yielded promising but mixed results due to a number of factors [15-19]. Essential oil extracts of plants with different phytochemical constituents have yielded results signifying effective performance [20-27]. However, more research is needed to assess the lowest concentration ranges for optimal inhibition performance and the synergistic combination effect of different extracts to combat concentration dependent performance. This manuscript studies the corrosion inhibition effect of thymus mastichina and alicium verum essential oil extracts combined, on mild steel in 0.5 M $H_2SO_4$ and HCl solution.

## 2. Experimental methods

Mild steel (MS) with diameter of 12 mm was sectioned into two sets of 5 samples each for coupon measurement. The prepared samples were subsequently washed with distilled $H_2O$ and $CH_3COCH_3$. Thymus mastichina and alicium verum essential oils (TA) were mixed together (ratio 1:1) and prepared in volumetric concentrations of 0%, 1%, 2%, 3%, 4% and 5% per 200 ml of 0.5 M $H_2SO_4$ and HCl solution prepared from analar grade reagents. MS samples were individually immersed in the acid inhibitor solution and weighed every 24 h for a total of 480 h with Oarhus weighing balance. Corrosion rate, $C_R$ (mm/y) was calculated from the formula below;

$$C_R = \left[\frac{87.6\omega}{DAt}\right] \tag{1}$$

$\omega$ indicates weight loss (mg), $D$ indicates density (g/cm$^3$), $A$ indicates total exposed surface area of MS specimen (cm$^2$), 87.6 is a corrosion rate constant and $t$ indicates is the time (h). Inhibition efficiency ($\eta$) was determined from the following equation;

$$\eta = \left[\frac{\omega_1 - \omega_2}{\omega_1}\right] \times 100 \tag{2}$$

$\omega_1$ indicates weight loss of MS from the acid solution without TA while $\omega_2$ indicates weight loss of MS at specific TA concentrations. Two-factor single level experimental ANOVA test (F - test) was used to evaluate the statistical relevance of TA concentrations and exposure time on their inhibition performance. The evaluation was performed at confidence level of 95% i.e. a significance level of $\alpha = 0.05$ according to the equations below. The sum of squares among columns (exposure time) was obtained from equation 3 while the sum of squares among rows (extract concentration) was obtained from equation 4.

$$SS_c = \frac{\sum T_c^{\,2}}{nr} - \frac{T^2}{N} \tag{3}$$

Sum of Squares among rows (TA concentration)

$$SS_r = \frac{\sum T_r^{\,2}}{nc} - \frac{T^2}{N} \tag{4}$$

Total Sum of Squares

$$SS_{Total} = \sum x^2 - \frac{T^2}{N} \tag{5}$$

## 3. Results and discussion

### *3.1 Coupon measurement*

Corrosion rate data for MS in $H_2SO_4$ and HCl solution without TA compound at specific TA concentration for 480 h are shown in Table 1. Table 2 shows the corresponding inhibition efficiency data for TA on MS in $H_2SO_4$ and HCl solution for 480 h. Fig. 1 shows the comparative plot of TA inhibition efficiency versus exposure time at 1% and 5% TA concentration in $H_2SO_4$ and HCl solution. Observation of Table 1 shows that the corrosion rate data for MS in $H_2SO_4$ and HCl solution at 0% TA is significantly greater than the values obtained at 1% to 5% TA concentration. Secondly, the corrosion rate of MS at 0% TA in $H_2SO_4$ solution decreases progressively with exposure time till 480 h. This is due to weakening of the acid electrolyte overtime as a result of the accumulation of discharged corrosion products from the steel surface. Nevertheless, the final value is 0.0101 mm/y at 480 h which is greater than the values at specific TA concentration. At 1-5% TA in $H_2SO_4$, the corrosion rate value of MS significantly reduced in the presence of TA compound (24 h). However, marginal increase in corrosion rate of MS was observed overtime till 480 h with final value of 0.0071 mm/y. The rate of increase was visibly lower after 1% TA nevertheless, increase in corrosion rate shows MS become more vulnerable to corrosion overtime despite the significant decrease in corrosion rate value. The data for MS corrosion in HCl solution quite contrast the values obtained in $H_2SO_4$ solution. The corrosion rate values at 0% TA decreased with time for the reasons earlier explained while increase in TA concentration results in marginal increase in corrosion rate with respect to time. Nevertheless, it must be noted that increase in TA concentration has limited effect on the corrosion resistance of MS from 1-4% TA as shown in the data in Table 1 where at 0% TA the corrosion rate is 0.0054 mm/y (480 h). At 1-4% TA, the corrosion rate varies from 0.0048 mm/y to 0.0031 mm/y at 480 h. However, at 5% TA the rate of marginal increase in corrosion rate is significantly lower compared to lower values of TA concentration. The final corrosion rate at this TA concentration is 0.0017 mm/y. The corresponding data for inhibition efficiency in Table 2 gives further insight to the inhibition performance of TA compound. The values from $H_2SO_4$ and HCl solution show data indicating effective inhibition performance at 24 h where the data varies between 96.64% and 99.6%. However, the inhibition efficiency data decreases with exposure time. The rate of decrease is minimal from 2% to 5% TA concentration with values between 75.4% and 81.24% signifying effective TA inhibition performance despite decrease in corrosion rate overtime. The inhibition efficiency data at 1% TA in $H_2SO_4$ and at 1% to 5% TA in HCl decreased to values below effective inhibition performance between 11.36% and 68.33%. Observation of Fig. 1 shows inhibition efficiency of TA decreases marginally at 5% TA concentration compared to the plot at 5% TA in HCl, 1% TA in $H_2SO_4$ and 1% TA in HCl solution. The plot shows that inhibitor concentration affects the inhibition performance of TA compound. Secondly it shows the rate of decrease in value over time is influenced by the concentration of TA. It is also observed that TA is generally more effective in $H_2SO_4$ solution than in HCl.

**Table 1.** Corrosion rate data for MS in 0.5 M $H_2SO_4$ and HCl solution at specific TA concentration

| TA Conc. (%) / Exp. Time (h) | $H_2SO_4$ | | | | | | HCl | | | | | |
|---|---|---|---|---|---|---|---|---|---|---|---|---|
| | 0% TA | 1% TA | 2% TA | 3% TA | 4% TA | 5% TA | 0% TA | 1% TA | 2% TA | 3% TA | 4% TA | 5% TA |
| 24 | 0.0788 | 0.0014 | 0.0005 | 0.0003 | 0.0026 | 0.0005 | 0.0679 | 0.0014 | 0.0025 | 0.0008 | 0.0018 | 0.0006 |
| 48 | 0.0524 | 0.0014 | 0.0006 | 0.0002 | 0.0018 | 0.0003 | 0.0378 | 0.0017 | 0.0019 | 0.0006 | 0.0010 | 0.0005 |
| 72 | 0.0370 | 0.0016 | 0.0005 | 0.0003 | 0.0017 | 0.0004 | 0.0256 | 0.0019 | 0.0034 | 0.0006 | 0.0010 | 0.0005 |
| 96 | 0.0280 | 0.0020 | 0.0005 | 0.0004 | 0.0014 | 0.0003 | 0.0193 | 0.0022 | 0.0026 | 0.0009 | 0.0008 | 0.0005 |
| 120 | 0.0225 | 0.0023 | 0.0006 | 0.0005 | 0.0013 | 0.0004 | 0.0155 | 0.0024 | 0.0024 | 0.0011 | 0.0007 | 0.0006 |
| 144 | 0.0211 | 0.0033 | 0.0007 | 0.0004 | 0.0017 | 0.0005 | 0.0129 | 0.0042 | 0.0026 | 0.0011 | 0.0009 | 0.0009 |
| 168 | 0.0187 | 0.0043 | 0.0007 | 0.0006 | 0.0014 | 0.0005 | 0.0119 | 0.0047 | 0.0026 | 0.0019 | 0.0014 | 0.0008 |
| 192 | 0.0164 | 0.0045 | 0.0006 | 0.0005 | 0.0016 | 0.0006 | 0.0105 | 0.0048 | 0.0031 | 0.0019 | 0.0014 | 0.0010 |
| 216 | 0.0146 | 0.0049 | 0.0006 | 0.0004 | 0.0015 | 0.0007 | 0.0094 | 0.0052 | 0.0032 | 0.0018 | 0.0015 | 0.0009 |
| 240 | 0.0134 | 0.0056 | 0.0010 | 0.0008 | 0.0018 | 0.0007 | 0.0086 | 0.0054 | 0.0033 | 0.0021 | 0.0024 | 0.0009 |
| 264 | 0.0153 | 0.0055 | 0.0009 | 0.0011 | 0.0020 | 0.0010 | 0.0078 | 0.0049 | 0.0036 | 0.0021 | 0.0024 | 0.0012 |
| 288 | 0.0147 | 0.0067 | 0.0009 | 0.0010 | 0.0020 | 0.0012 | 0.0072 | 0.0049 | 0.0039 | 0.0029 | 0.0027 | 0.0013 |
| 312 | 0.0137 | 0.0074 | 0.0010 | 0.0010 | 0.0023 | 0.0014 | 0.0067 | 0.0047 | 0.0042 | 0.0033 | 0.0027 | 0.0014 |
| 336 | 0.0130 | 0.0080 | 0.0012 | 0.0010 | 0.0022 | 0.0015 | 0.0063 | 0.0045 | 0.0041 | 0.0036 | 0.0029 | 0.0015 |
| 360 | 0.0122 | 0.0076 | 0.0011 | 0.0014 | 0.0022 | 0.0016 | 0.0063 | 0.0044 | 0.0041 | 0.0042 | 0.0030 | 0.0016 |
| 384 | 0.0116 | 0.0074 | 0.0011 | 0.0014 | 0.0021 | 0.0016 | 0.0060 | 0.0044 | 0.0043 | 0.0043 | 0.0030 | 0.0016 |
| 408 | 0.0110 | 0.0072 | 0.0011 | 0.0013 | 0.0022 | 0.0017 | 0.0057 | 0.0046 | 0.0044 | 0.0043 | 0.0030 | 0.0016 |
| 432 | 0.0104 | 0.0072 | 0.0012 | 0.0013 | 0.0022 | 0.0017 | 0.0059 | 0.0048 | 0.0043 | 0.0042 | 0.0032 | 0.0016 |
| 456 | 0.0099 | 0.0069 | 0.0012 | 0.0014 | 0.0022 | 0.0018 | 0.0056 | 0.0048 | 0.0043 | 0.0042 | 0.0031 | 0.0017 |
| 480 | 0.0101 | 0.0071 | 0.0025 | 0.0025 | 0.0022 | 0.0019 | 0.0054 | 0.0048 | 0.0045 | 0.0040 | 0.0031 | 0.0017 |

**Table 2.** TA Inhibition efficiency data for MS in 0.5 M $H_2SO_4$ and HCl solution at specific TA concentration

| TA Conc. (%) / Exp. Time (h) | 1% TA | 2% TA | 3% TA | 4% TA | 5% TA | 1% TA | 2% TA | 3% TA | 4% TA | 5% TA |
|---|---|---|---|---|---|---|---|---|---|---|
| 24 | 98.19 | 99.32 | 99.60 | 96.64 | 99.40 | 97.95 | 96.29 | 98.81 | 97.40 | 99.04 |
| 48 | 97.26 | 98.82 | 99.67 | 96.61 | 99.46 | 95.58 | 94.91 | 98.36 | 97.49 | 98.71 |
| 72 | 95.59 | 98.69 | 99.27 | 95.33 | 98.96 | 92.75 | 86.76 | 97.54 | 96.21 | 97.99 |
| 96 | 92.76 | 98.21 | 98.42 | 95.01 | 98.82 | 88.75 | 86.37 | 95.12 | 95.91 | 97.20 |
| 120 | 89.97 | 97.25 | 97.97 | 94.11 | 98.13 | 84.79 | 84.31 | 92.72 | 95.62 | 96.41 |
| 144 | 84.58 | 96.85 | 98.17 | 91.91 | 97.69 | 67.62 | 80.11 | 91.44 | 92.78 | 93.21 |
| 168 | 77.03 | 96.21 | 97.03 | 92.46 | 97.18 | 61.02 | 77.95 | 83.70 | 88.06 | 92.99 |
| 192 | 72.49 | 96.17 | 97.03 | 90.04 | 96.37 | 54.15 | 70.97 | 82.15 | 86.53 | 90.60 |
| 216 | 66.54 | 96.14 | 97.03 | 89.63 | 95.40 | 44.26 | 65.56 | 80.78 | 84.17 | 90.30 |
| 240 | 58.08 | 92.80 | 93.80 | 86.44 | 95.06 | 37.18 | 61.28 | 75.47 | 72.00 | 89.88 |
| 264 | 64.04 | 94.14 | 92.96 | 87.18 | 93.45 | 36.85 | 54.24 | 73.10 | 69.00 | 85.06 |
| 288 | 54.31 | 94.23 | 92.92 | 86.35 | 91.71 | 32.59 | 46.03 | 59.55 | 63.33 | 82.70 |
| 312 | 45.90 | 92.43 | 92.93 | 83.43 | 89.54 | 29.59 | 38.09 | 50.55 | 59.23 | 78.90 |
| 336 | 38.60 | 90.89 | 92.65 | 83.27 | 88.15 | 29.12 | 35.21 | 43.50 | 53.53 | 76.70 |
| 360 | 38.17 | 90.88 | 88.48 | 82.30 | 86.51 | 30.40 | 35.11 | 34.28 | 52.35 | 75.24 |
| 384 | 36.59 | 90.88 | 88.18 | 81.63 | 85.94 | 26.78 | 28.55 | 29.17 | 49.55 | 73.99 |

| | | | | | | | | | | |
|---|---|---|---|---|---|---|---|---|---|---|
| 408 | 34.61 | 90.04 | 88.03 | 79.93 | 84.71 | 20.05 | 22.30 | 23.86 | 47.88 | 72.29 |
| 432 | 30.92 | 88.83 | 87.49 | 78.81 | 83.29 | 17.49 | 26.39 | 28.17 | 45.33 | 72.56 |
| 456 | 30.70 | 87.41 | 86.00 | 77.64 | 81.38 | 13.94 | 22.22 | 24.65 | 44.00 | 70.36 |
| 480 | 30.13 | 75.40 | 75.64 | 77.86 | 81.24 | 11.36 | 16.64 | 24.87 | 41.75 | 68.33 |

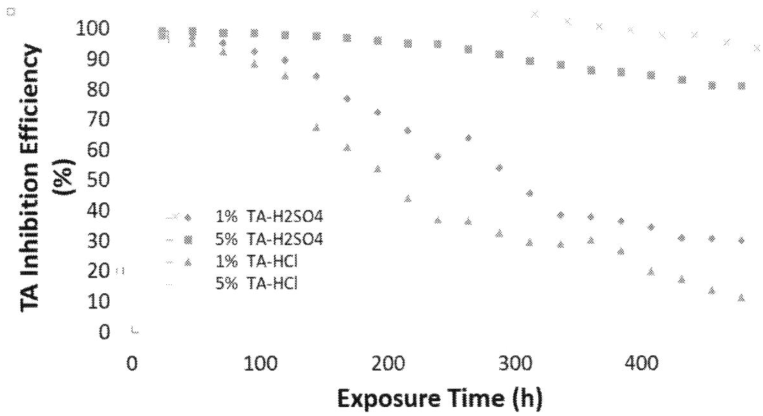


**Figure 1** Comparative Plot of TA inhibition efficiency at 1% and 5% TA concentration in $H_2SO_4$ and HCl solution

*3.2 Statistical evaluation*

Data on the mean, standard deviation and margin of error for the inhibition efficiency data of TA in $H_2SO_4$ and HCl solution are shown in Table 3. The mean inhibition values for TA in $H_2SO_4$ are significantly higher than the values obtained in HCl solution due to greater inhibitor performance in $H_2SO_4$. Observation of the standard deviation values from both acids with respect to inhibitor concentration shows significant variation with the values from $H_2SO_4$ after 1% TA being significantly lower. The high standard deviation value at 1% TA (in $H_2SO_4$) shows the amount of variation of inhibition efficiency varies significantly with exposure time compared to the values at higher TA concentration. The standard deviation values at higher TA concentration in $H_2SO_4$ shows the inhibition efficiency data are generally more stable with respect to exposure time signifying greater stability of the inhibitor molecules over the steel surface. The standard deviation values in HCl from 1% to 4% TA is generally similar to the values obtained at 1% TA in $H_2SO_4$ signifying the instability of the inhibitor molecules due to significant variation with respect to exposure time. However, the value at 5% TA in HCl signifies greater stability though to a lesser extent than the values in $H_2SO_4$. The margin of error data shows 88% of TA dat in $H_2SO_4$ and 53% of TA data in HCl are have inhibition efficiency data greater than 70% at margin of error of $\pm$0.09 and $\pm$0.1.

**Table 3.** Mean, standard deviation and margin of error data for TA inhibition efficiency in $H_2SO_4$ and HCl solution

| **Acid Type** | **$H_2SO_4$** | | | | | **HCl** | | | | |
|---|---|---|---|---|---|---|---|---|---|---|
| **Conc. (%)** | 1 | 2 | 3 | 4 | 5 | 1 | 2 | 3 | 4 | 5 |
| **SD** | 25.22 | 5.52 | 6.10 | 6.50 | 6.45 | 29.49 | 27.10 | 28.99 | 21.18 | 10.72 |
| **Mean** | 61.82 | 93.28 | 93.16 | 87.33 | 92.12 | 48.61 | 56.46 | 64.39 | 71.61 | 85.12 |
| **Margin of Error** | $\pm$0.09 | | | | | $\pm$0.1% | | | | |
| **Data above 70% Inhibition** | 88% | | | | | 53% | | | | |

Statistical evaluation through analysis of variance (ANOVA) was done to assess the statistical influence of inhibitor concentration and exposure time on the inhibition performance of TA. The statistical relevance factor shows the percentage value of the sources of variation while the mean square ratio is the value which must be greater than the theoretical significance factor for the sources of variation to be relevant irrespective of the statistical relevance factor. Table 4 shows TA concentration is the dominant variation responsible for the inhibition performance of TA compound in $H_2SO_4$ solution at 88.83% compared to 7.75% for exposure time. The corresponding data in HCl solution shows both inhibitor concentration and exposure time significantly influenced the performance of TA on MS with values of 67.96% and 26.97%. The mean square ratio for both acids are greater than the theoretical significance factor indicating that the statistical relevance values for both sources of variation in both acids are statistically relevant and influence the performance of TA.

**Table 4**. ANOVA data for TA inhibition performance in $H_2SO_4$ and HCl solution

| **$H_2SO_4$** | | | |
|---|---|---|---|
| **Source of Variation** | **Mean Square Ratio (F)** | **Theoretical Significance Factor** | **Statistical Relevance (%)** |
| TA Conc. | 260.29 | 2.81 | 88.83 |
| Exp. Time | 6.82 | 2.05 | 7.75 |
| Residual | | | |
| Total | | | |
| **HCl** | | | |
| **Source of Variation** | **Mean Square Ratio (F)** | **Theoretical Significance Factor** | **Statistical Relevance (%)** |
| TA Conc. | 134.11 | 2.81 | 67.96 |
| Exp. Time | 15.97 | 2.05 | 26.97 |
| Residual | | | |
| Total | | | |

## 4. Conclusion

Admixture of thymus mastichina and alicium verum effectively inhibited the corrosion of mild steel in $H_2SO_4$ and HCl solution. The performance of the compound was significantly stable with respect to concentration and varied marginally with respect to exposure time in $H_2SO_4$ solution with optimal values generally above 70% inhibition. In HCl the performance of the compound varied significantly with exposure time and concentration attaining optimal values which were generally below values for effective corrosion inhibition. Standard deviation values show the compound varied significantly from mean values in HCl compared to $H_2SO_4$ solution. Statistical data showed exposure time and inhibitor concentration at varying degree influence the inhibition performance of the compound on mild steel.

## Acknowledgement

The authors are grateful to Covenant University Ota, Ogun State, Nigeria for their support for this research project and provision of research facilities.

## References

[1] Sastri V S 2015 Consequences of Corrosion, Chapter 5, in Challenges in Corrosion: Costs, Causes, Consequences, and Control, John Wiley and Sons Inc., https://doi.org/10.1002/9781119069638.ch5.

[2] Barbara A S and Robert G K 2006 What is corrosion? *Electrochem. Soc. Interface* **15(1)** 24-26.

[3] James G S 2014 Oil and Gas Corrosion Prevention, in Surface Facilities to Refineries, Gulf Professional Publishing, Amsterdam, pp. e1-e24. https://doi.org/10.1016/B978-0-12-800346-6.00001-6.

[4] Isecke B, Schütze M and Strehblow H H 2011 Corrosion, in Czichos H, Saito T, Smith L (eds), Springer Handbook of Metrology and Testing, Springer Handbooks, Springer, Berlin, 2011, pp 667-741. https://doi.org/10.1007/978-3-642-16641-9_12.

[5] Vimala J R, Rose A L and Raja S 2011 Cassia auriculata extract as corrosion inhibitor for mild steel in acid medium *Int. J. Chem. Tech. Res.* **3(4)** 1791-1801.

[6] Yurt G B, Balaban A B and Erk E 2005 Effect of schiff bases containing pyridyl group as corrosion inhibitors for low carbon steel in 0.1 M HCl *J. Appl. Electrochem.* **35(10)** 1025-1032.

[7] Loto R T and Olowoyo O 2018 Corrosion inhibition properties of the combined admixture of essential oil extracts on mild steel in the presence of $SO_4^{2-}$ anions *S. Afr. J. Chem. Eng.* **26** 35-41.

[8] Loto R T 2018 Comparative assessment of the synergistic combination of ricinus communis and rosmarinus officinalis on high-carbon and P4 low-carbon mold steel corrosions in dilute acid media *J. Bio Tribo Corros.* **4(47)**. https://doi.org/10.1007/s40735-018-0163-y.

[9] Narayanasamy P, Ramachanderen T, Natesan M and Murugavel S C 2009 Corrosion Inhibition of Mild Steel by Essential Oils in an HCl Environment *Mater. Performance* **48(9)** 52-56.

[10] Hmamou D B, Salghi R, Zarrouk A, Zarrouk H, Errami M, Hammouti B, Afia L and Bazzi L 2013 Adsorption and corrosion inhibition of mild steel in hydrochloric acid solution by verbena essential oil *Res. Chem. Intermed.* **39** 973–989. https://doi.org/10.1007/s11164-012-0609-7.

[11] Loto R T, Olukeye T and Okorie E 2019 Synergistic combination effect of clove essential oil extract with basil and atlas cedar oil on the corrosion inhibition of low carbon steel *S. Afr. J. Chem. Eng.* **30** 28-41. https://doi.org/10.1016/j.sajce.2019.08.001.

[12] Loto R T and Olowoyo O 2018 Corrosion inhibition properties of the combined admixture of essential oil extracts on mild steel in the presence of $SO_4^{2-}$ anions *S. Afr. J. Chem. Eng.* **26** 35-41. https://doi.org/10.1016/j.sajce.2018.09.002.

[13] Loto R T, Leramo R and Oyebade B 2018 Synergistic Combination Effect of Salvia officinalis and Lavandula officinalis on the Corrosion Inhibition of Low-Carbon Steel in the Presence of $SO_4^{2-}$ and $Cl^-$ Containing Aqueous Environment *J. Fail. Anal. and Preven.* **18** 1429–1438. https://doi.org/10.1007/s11668-018-0535-0.

[14] Cutillas A-B, Carrasco A, Martinez-Gutierrez R, Tomas V and Tudela J 2018 *Thymus mastichina* L. essential oils from Murcia (Spain): Composition and antioxidant, antienzymatic and antimicrobial bioactivities, *PLoS One* **13(1)** e0190790. https://doi.org/10.1371/journal.pone.0190790.

[15] Li Y, Zhao P, Liang Q and Hou B 2005 Berberine as a natural source inhibitor for mild steel in 1 M $H_2SO_4$ *Appl. Surf. Sci.* **252** 1245-1253.

[16] Quartarone G, Ronchin L, Vavasori A, Tortato C and Bonaldo L 2012 Inhibitive action of gramine towards corrosion of mild steel in deaerated 1.0 M hydrochloric acid solutions *Corros. Sci.* **64** 82-89.

[17] Ashassi-Sorkhabi H, Majidi M R and Seyyedi 2004 Investigation of inhibition effect of some amino acids against steel corrosion in HCl solution *Appl. Surf. Sci.* **225** 176-185.

[18] ÖZcan M 2008 AC impedance measurements of cysteine adsorption at mild steel/sulphuric acid interface as corrosion inhibitor *J. Solid State Electrochem.* **12** 1653-1661.

[19] Loto RT and Oghenerukewe E 2016 Inhibition studies of rosmarinus officinalis on the pitting corrosion resistance 439LL ferritic stainless steel in dilute sulphuric acid *Orient. J. Chem.* **32(5)** 2813-2832.

[20] Fu J, Li S, Cao L, Wang Y, Yan L and Lu L 2010 L-Tryptophan as green corrosion inhibitor for low carbon steel in hydrochloric acid solution *J. Mater. Sci.* **45** 979-986.

[21] Bouoidina A, Chaouch M, Abdellaoui A, Lahkimi A, Hammouti B, El-Hajjaji F, Taleb M and Nahle A 2017 Essential oil of "Foeniculum vulgare": antioxidant and corrosion inhibitor on mild steel immersed in hydrochloric medium *Anti-Corros. Method M.* **64(5)** 563-572.

[22] Hamdani I, El-Ouariachi E, Mokhtari O, Salhi A, Chahboun N, ElMahi B, Bouyanzer A, Zarrouk A, Hammouti B and Costa J 2015 Chemical constituents and corrosion inhibition of mild steel by the essential oil of Thymus algeriensis in 1.0 M hydrochloric acid solution *Der Pharm. Chem.* **7(8)** 252-264.

[23] El-Ouadi Y, Bouyanzer A, Majidi L, Paolini J, Desjobert J M, Costa J, Chetouani A and Hammouti B 2014 Salvia officinalis essential oil and the extract as green corrosion inhibitor of mild steel in hydrochloric acid *J. Chem. Pharm. Res.* **6(7)** 1401-1416.

[24] El-Ouariachi E, Bouyanzer A, Salghi R, Hammouti B, Desjobert J-M, Costa J Paolini J and Majidi L 2015 Inhibition of corrosion of mild steel in 1 M HCl by the essential oil or solvent extracts of Ptychotis verticillata *Res. Chem. Intermed.* **41** 935-946.

[25] Boumhara K, Tabyaoui M, Jama C and Bentiss F 2015 Artemisia Mesatlantica essential oil as green inhibitor for carbon steel corrosion in 1 M HCl solution: Electrochemical and XPS investigations *J. Ind. Eng. Chem.* **29** 146-155.

[26] Lahhit N, Bouyanzer A, Desjobert J-M, Hammouti B, Salghi R, Costa J, Jama C, Bentiss F and Majidi L 2011 Fennel (Foeniculum Vulgare) essential oil as green corrosion inhibitor of carbon steel in hydrochloric acid solution *Port. Electrochim. Acta.* **29(2)** 127-138.

[27] Loto, RT, Leramo R, Oyebade B 2018 Synergistic combination effect of salvia officinalis and lavandula officinalis on the corrosion inhibition of low-carbon steel in the presence of $SO_4{}^{2-}$- and $Cl^-$- containing aqueous environment *J. Fail. Anal. and Preven.* **18(6)** 1429-1438.

# Influence of Annealing Temperature on Structural Properties and Stability of $CsPbBr_3/TiO_2$ Core/Shell

**Parina Nuket[1], Tetsuya Kida[2], and Paravee Vas-Umnuay[1]***

[1] Center of Excellence in Particle Technology & Material Processing, Department of Chemical Engineering, Faculty of Engineering, Chulalongkorn University, Bangkok 10330, Thailand
[2] Division of Materials Science, Faculty of Advanced Science and Technology, Kumamoto University, Kumamoto 860-8555, Japan

*E-mail: Paravee.V@chula.ac.th

**Abstract**. Although all-inorganic $CsPbBr_3$ perovskites quantum dots exhibit an excellent photophysical properties, which could be applied in various applications, they still suffer from poor structural stability owing to environment factors. Therefore, $TiO_2$ was coated on the $CsPbBr_3$ QDs as a protecting shell to enhance their stability. Moreover, anatase phase of $TiO_2$ shell was obtained by annealing at more than 300°C due to its most effective electron extraction properties among other crystalline forms of $TiO_2$. However, the $CsPbBr_3$ QDs can be degraded under high temperature. Herein, the annealing temperatures ranging from 300-500°C were optimized to obtain the anatase $TiO_2$ shell and good crystallinity of $CsPbBr_3$ QDs core. The morphology and optical properties of $CsPbBr_3$ QDs, $CsPbBr_3/TiO_2$ and $CsPbBr_3/TiO_2$ were studied by using transmission electron microscope (TEM), photoluminescence (PL) measurement and UV-Visible spectroscopy, respectively. Finally, the $CsPbBr_3/TiO_2$ annealed at the optimum temperature was compared with the bare $CsPbBr_3$ QDs to prove the enhancement of their structural stability under ambient air by using X-ray diffraction (XRD).

## 1. Introduction

$CsPbBr_3$ quantum dots (QDs) are one of perovskite materials that have been reported with the excellent optical properties such as adjustable band gap, long diffusion lengths, ambipolar charge transport, and high carrier mobility [1]. $CsPbBr_3$ QDs with a green emission centered at 510-518 nm presented a high photoluminescence quantum yield (PL QY) of 90-95% and high color purity with a narrow full width at half maximum (FWHM) of 16-27 nm [2]. Although their excellent properties led to the success achieved in various fields, they still suffered from the environment factors such as water and heat because of their inherent ionic nature on polar solvents and the direct decomposition under high temperature. For the enhancement of their stability, $CsPbBr_3$ QDs have been reported with an excellent water stability for 12 weeks maintaining 75% of initial PL intensity when they were encapsulated with $TiO_2$ as a protecting shell; furthermore, it was proved that there was charge carrier transfer from $CsPbBr_3$ QDs core to $TiO_2$ shell in photoelectrochemical application [3, 4]. $TiO_2$ is widely used as an electron transport layer (ETL) in photovoltaic devices such as perovskite solar cells because it can absorb light at wavelength of ultraviolet region. It can transport electrons generated from perovskite layer to prevent charge recombination and serve as the block layer to hinder direct contact between the holes and fluorine-doped tin oxide (FTO). Anatase phase of $TiO_2$, in particular, exhibited the most effective electron extraction properties among other crystalline forms of $TiO_2$ due to the closest conduction band minimum (CBM) to the lowest unoccupied molecular orbital (LUMO) of $CsPbBr_3$ QDs, that these anatase $TiO_2$ layers could easily create a quasi-ohmic contact, providing the effective electron transport pathways and enhancing electron extraction [5]. Moreover, $TiO_2$ appeared a high chemical and thermal stability under stressed condition so it has been attracted more attention to stabilize the perovskite materials. In order to obtain the anatase phase of $TiO_2$, the annealing temperature was required in the range of 300 to 500°C before rutile phase of $TiO_2$ formed and grown at higher temperature. Meanwhile, the high temperature can affect the degradation of $CsPbBr_3$ QDs, although thermogravimetric analysis (TGA) has shown the weight loss of the $CsPbBr_3$ QDs only about 5% between 30 and 500°C [3]. In this study, the effects of this heat treatment on the structure of $CsPbBr_3/TiO_2$ core/shell were investigated. The annealing temperatures of 300, 400 and 500°C were optimized in order to obtain the highest purity and crystallinity of anatase phase of $TiO_2$ shell while maintain the good structure of $CsPbBr_3$ core. Morphology, structural and optical properties were investigated using TEM, XRD pattern, PL emission spectra and absorption value, respectively. The stability of $CsPbBr_3/TiO_2$ annealed at the optimum temperature was tested under ambient air to prove the enhancement of their stability.

## 2. Experimental

The experimental includes the preparation of $CsPbBr_3$ QDs, encapsulation of $CsPbBr_3$ QDs with $TiO_2$ at various annealing temperatures, stability testing in ambient air and characterizations of the samples.

### *2.1. Preparation of $CsPbBr_3$ QDs*

16 mg of cesium carbonate ($Cs_2CO_3$), 76 mg of lead acetate trihydrate ($Pb(CH_3COO)_2.3H_2O$), 5.00 mL of octadecene (ODE), 0.45 mL of oleic acid (OA) and 1.00 mL of oleylamine (OLM) were mixed in a three-neck flask, stirred at 250 rpm and heated under vacuum at 130°C for 1.40 h. Then, the solution was heated under argon flow at 170°C for 15 min. 72 μL of benzoyl bromide ($C_6H_5COBr$) was swiftly injected to form yellow colloidal solution of $CsPbBr_3$ QDs. The solution was immediately cooled down in an ice-water bath to room temperature (25°C). The $CsPbBr_3$ QDs solution was collected by using 5 mL of toluene and poured into the centrifuge tube. Afterwards, it was centrifuged at 4,000 rpm for 10 min and the supernatant was discarded to remove the ligands. Then the QDs in precipitates were re-dispersed in 5 mL of toluene. The centrifugation and re-dispersion process was repeated 1 time. The $CsPbBr_3$ QDs were collected in toluene for further use.

### *2.2. Preparation of $CsPbBr_3/TiO_2$*

20 μL of titanium tetraisopropoxide (TTIP) in 1 mL of toluene was dropped into 10 mL of $CsPbBr_3$ QDs in toluene (1 mg/mL) and the solution was stirred for 3 h under 40%RH at room temperature. The solution was centrifuged at 5,000 rpm for 5 min to separate the $CsPbBr_3/TiO_x$, which were dried under vacuum for 1 h. Next, the $CsPbBr_3/TiO_x$ were annealed under nitrogen flow at 300, 400 and 500°C for 1 h with heating rate of 5°C/min to transform amorphous $TiO_x$ shell to anatase $TiO_2$ shell.

### *2.3. Stability testing of $CsPbBr_3/TiO_2$*

The bare $CsPbBr_3$ QDs and $CsPbBr_3/TiO_2$ were stored under ambient air (40%RH, 25°C). The samples were characterized by XRD at day 0, 2, 5 and 11 to investigate their structural stability.

### *2.4. Characterizations of structural and optical properties*

The Fourier Transform Infrared spectrophotometer (FT-IR, FT/IR-4100, JASCO Co.) was employed to investigate the ligand removal of QDs during the process. The morphology of $CsPbBr_3$ QDs was observed by transmission electron microscope (TEM-2000FX, JEOL). The crystallinity of $CsPbBr_3$ QDs core and the anatase $TiO_2$ shell were determined by X-ray diffraction (RINT2100, Rigaku Co., Ltd) with $CuK_\alpha$ radiation source in the range of 10-60°. The optical properties were studied by using a UV-Visible spectrophotometer (V-650, JASCO Co.) at the wavelength range of 300-700 nm. Steady state PL spectra were obtained with a spectrometer (FP-6000, JASCO Co.) with excitation at 330 nm.

## 3. Results and discussion

Li group reported that the structure of $CsPbBr_3$ QDs started to decompose at around 500°C. However, the nucleation and growth of anatase phase was initiated at high temperature. Therefore, $CsPbBr_3/TiO_2$ annealed at 400°C was the most suitable representative of the annealed sample in this study. The sample before annealing, named $CsPbBr_3/TiO_x$, and the sample after annealing at 400°C, named $CsPbBr_3/TiO_2$, were investigated to study the effects of heat treatment on morphology and optical properties. Afterwards, the samples annealed at various temperatures were performed to study the influence of annealing temperature on their structural properties.

The $CsPbBr_3$ QDs were obtained by hot-injection method improved from M. Imran et al. [6] that was based on the dissolution of $Cs^+$ and $Pb^{2+}$ cations in fatty acids and used benzoyl halides as halide precursors. It was possible to precisely tune the composition of the final QDs. Furthermore, the QDs had a high phase purity and a high level of particle-morphology control. In order to control the growth of $CsPbBr_3$ QDs in solution via this method, the long carbon chain organic ligands including of oleic acid (OA) and oleylamine (OLM) have been applied to surround the QDs as a stabilizing shell. However, these ligands have prevented charge transfer as an insulating shell in optoelectronic devices as well as hinder the encapsulation of $TiO_2$ around QDs surface. Therefore, the ligands were inspected by FT-IR measurements on $CsPbBr_3$ QDs, $CsPbBr_3/TiO_x$ before annealing and $CsPbBr_3/TiO_2$ annealed at 400°C.

FT-IR spectra (figure 1) showed that the sample exhibited characteristic modes of oleyl groups corresponding to the OA and OLM as follows: peaks at 2,924 $cm^{-1}$ and 2,854 $cm^{-1}$ correlating with the asymmetric and symmetric CH stretching mode in both OA and OLM, respectively, and peaks at 1,462 $cm^{-1}$ and 1,376 $cm^{-1}$ correlating with the $CH_2$ and $CH_3$ bending in OLM, respectively [7]. After coating the $CsPbBr_3$ with amorphous $TiO_x$, the intensity of the ligand absorption band decreased significantly compared to $CsPbBr_3$ QDs because the ligands surrounding QDs were probably replaced with $TiO_x$ and discarded by centrifugation. However, the ligands removal possibly caused an agglomeration of $CsPbBr_3/TiO_x$ due to the strong ionic nature between the uncapped QDs of neighboring particles [8] as well as between the oxide of titanium. After annealing at high temperature, all ligands were removed due to the decomposition of their structure under an elevated temperature. Moreover, the peaks from 800 to 600 $cm^{-1}$ were occurred due to the presence of Ti-O bond [9] from the anatase $TiO_2$ shell.

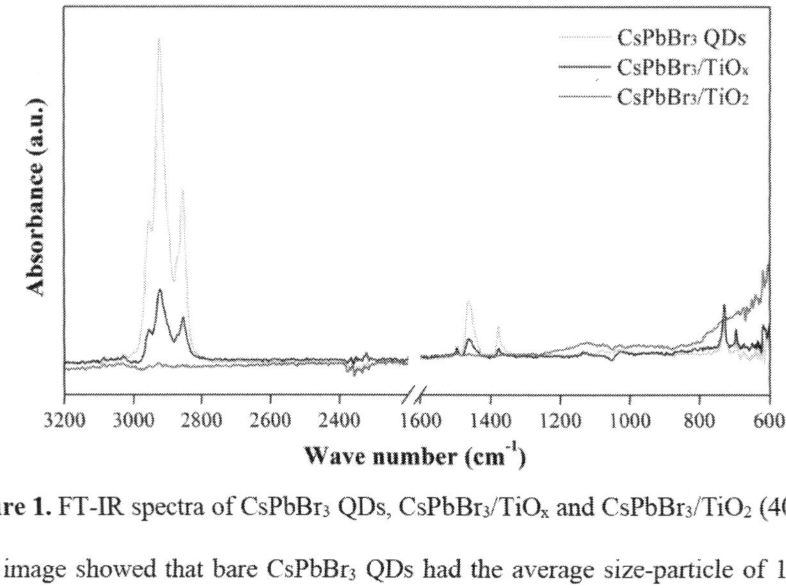


**Figure 1.** FT-IR spectra of $CsPbBr_3$ QDs, $CsPbBr_3/TiO_x$ and $CsPbBr_3/TiO_2$ (400°C).

The TEM image showed that bare $CsPbBr_3$ QDs had the average size-particle of 11±1 nm with a cubic shape (

(a) (b) (c)

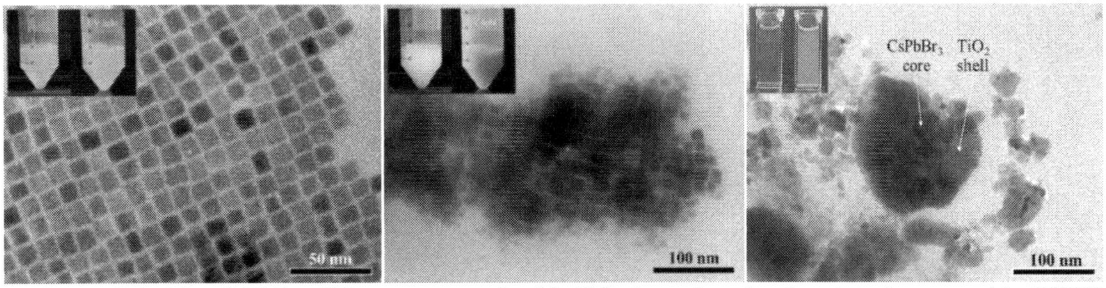


**figure 2**a). The $CsPbBr_3$ QDs were coated with $TiO_2$ by using titanium tetraisopropoxide (TTIP) under 40%RH at ambient temperature. This titanium alkoxide ($Ti(OR)_n$) was hydrolysed when there was a presence of water in an ambient air and subsequently condensed to form an amorphous oxide ($TiO_x.nH_2O$) that covered on the surface of agglomerated $CsPbBr_3$ QDs core as seen as a gray shadow area in TEM (

(a) (b) (c)

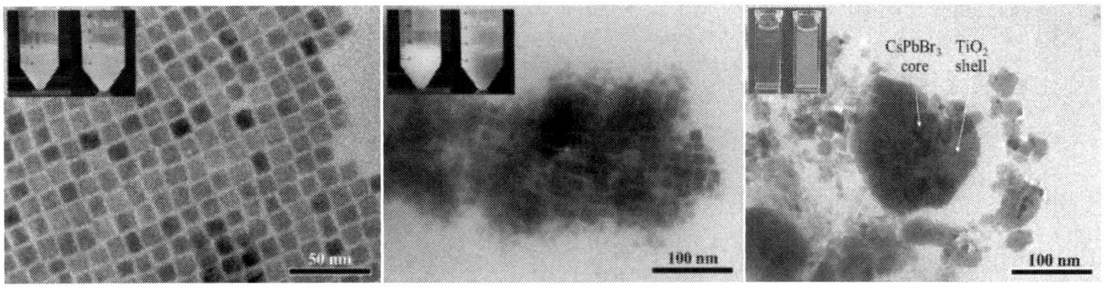


**figure 2**b). After annealing the sample, the particles had a larger size of 160-170 nm with a polycrystalline structure (Figure 2c). Besides the chemical bond interaction between -Ti-O-Ti- network and QDs, the change in morphology of coatings could be attributed to the densification mechanisms activated during the heat treatment leading to grain growth and increase in the particle agglomeration.

(a) (b) (c)

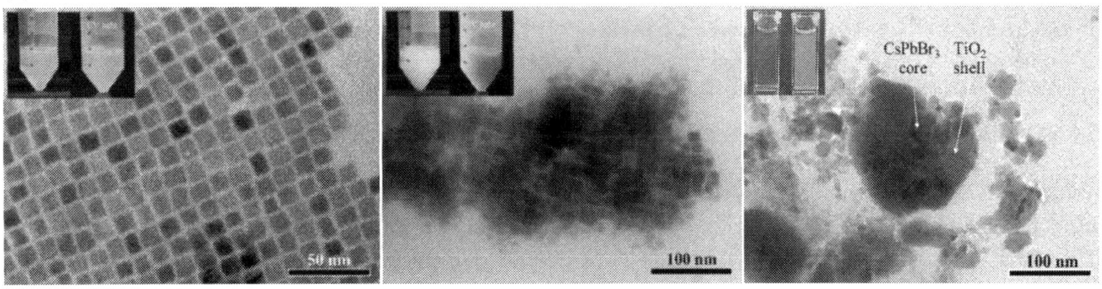


**Figure 2.** TEM images of (*a*) bare $CsPbBr_3$ QDs, (*b*) $CsPbBr_3/TiO_x$ and (*c*) $CsPbBr_3/TiO_2$ (400°C).

PL emission spectra (figure 3a) showed that the bare $CsPbBr_3$ QDs exhibited a sharp green emission peak at wavelength of 515 nm (2.41 eV) with a full width at half maximum (FWHM) of 18.09 nm when excited at 330 nm (3.76 eV). After coating $CsPbBr_3$ QDs with titanium precursor, the color changed from green solution of $CsPbBr_3$ QDs to yellow solution of $CsPbBr_3/TiO_x$ because the emission peak slightly shifts to 518 nm (2.39 eV) with a FWHM of 16.95 nm. The PL intensity of the sample after annealing was very low but the sample still absorbed radiation and has a UV-Vis absorption spectrum. This red-shift of PL emission peak corresponded to the QDs agglomeration. PL quenching of $CsPbBr_3$ QDs after coating $TiO_2$ were possibly attributed to the delocalization of the electron wave function from

$CsPbBr_3$ core into $TiO_2$ shell [4] since the CBM of $TiO_2$ was lower and closer to the LUMO of $CsPbBr_3$ QDs, which provided the effective electron transport pathways. For the absorption spectrum (figure 3b), $CsPbBr_3$ QDs, $CsPbBr_3/TiO_x$ and $CsPbBr_3/TiO_2$ had the excitonic peak at 508 nm, 509 nm and 523 nm, respectively. When compare in term of the intensity, $CsPbBr_3/TiO_x$ and $CsPbBr_3/TiO_2$ possessed a strong absorption into the UV region (< 350 nm) due to the UV absorption of $TiO_2$ shell.

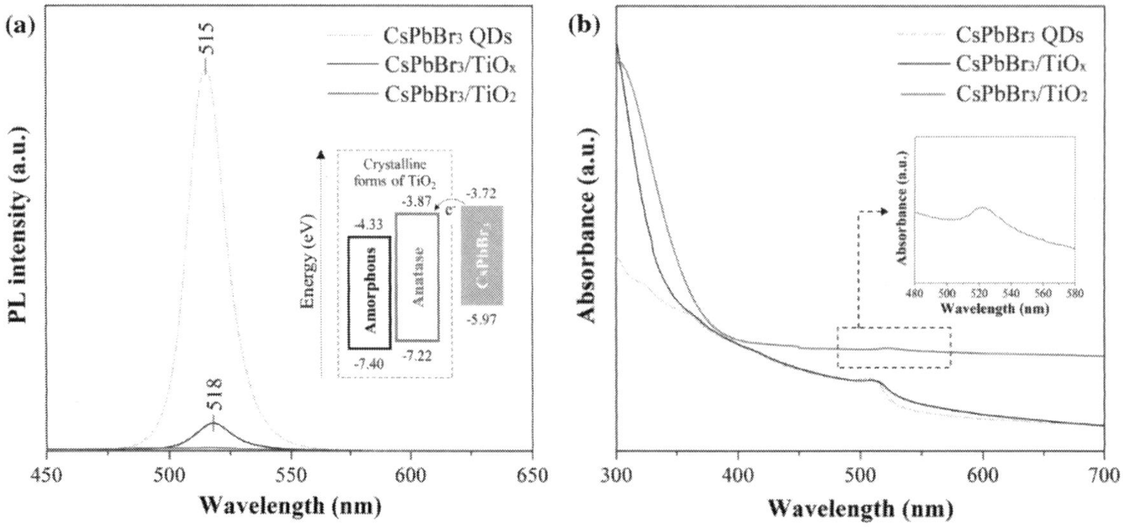


**Figure 3.** (*a*) PL emission spectra and (*b*) UV-Visible absorption of $CsPbBr_3$ QDs, $CsPbBr_3/TiO_x$ and $CsPbBr_3/TiO_2$ (400°C). The excitation wavelength was 330 nm.

The effects of encapsulation and annealing temperature on the structure of $CsPbBr_3$ QDs and $CsPbBr_3/TiO_2$ were studied by XRD (figure 4a-, and 4b). The XRD result confirmed that $CsPbBr_3$ QDs exhibited an orthorhombic phase. For $CsPbBr_3/TiO_x$, the XRD pattern was similar to the bare $CsPbBr_3$ QDs phase according to no phase transition in the QDs structure upon encapsulation process. After coating $TiO_2$ and annealing at various temperatures, the XRD patterns confirmed the existence of $TiO_2$ shell at $2\theta = 25.2°$ assigning to the anatase planes of (101) for $CsPbBr_3/TiO_2$ at 400°C and 500°C. The largest amount of transformed anatase phase was occurred at 500°C. The peak shapes of $CsPbBr_3/TiO_2$ became sharper owing to crystal size growth, in which the crystallinity of $CsPbBr_3$ QDs was improved by 37.2%. In addition, the peak split around $2\theta = 30.5°$ was appeared corresponding to the (040) and (202) inter-reticular planes of *Pnma* orthorhombic phase. As the annealing temperature was increased, these peaks were slightly broader and the (040) peaks shifted to lower angles, which coincided to a thermal cell expansion and there might be a distortion or a preferred orientation during the process.

Although the grain growth of QDs should increase with an increase of annealing temperature, the sample annealed at 400°C showed a higher peak intensity due to its better crystallinity compared to the sample annealed at 500°C because the addition of larger amount of anatase $TiO_2$, which annealed at 500°C, to the coating probably acted to retard the growth of QDs grains during this heat treatment, as predicted by Zener pinning. When counterbalanced between the quality of $CsPbBr_3$ core and anatase $TiO_2$ shell, annealing $CsPbBr_3/TiO_2$ at 400°C improved its crystallinity; meanwhile the amorphous of $TiO_x$ shell was transformed to the excellent charge-transfer crystalline form of anatase phase.

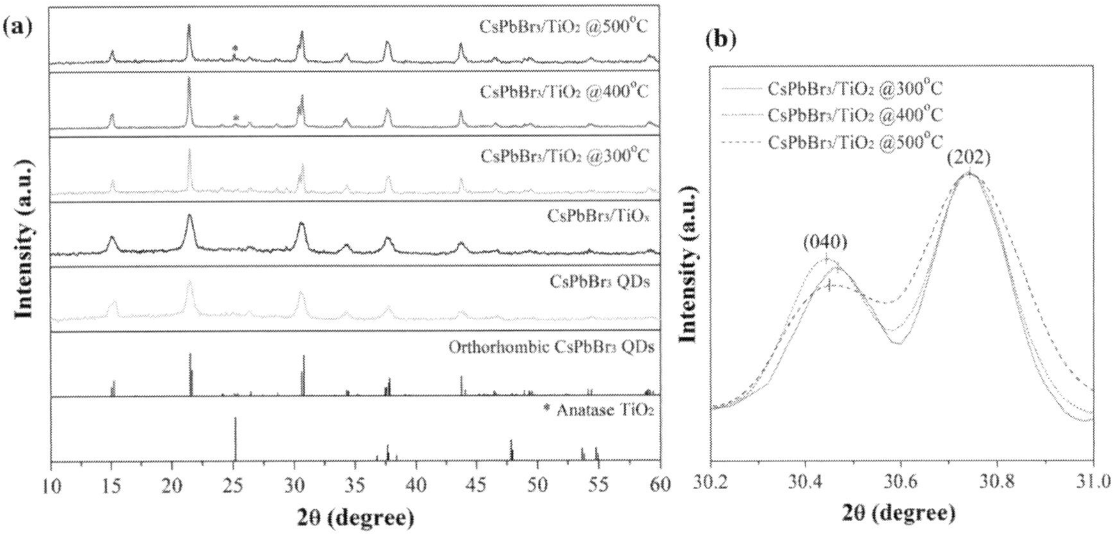


**Figure 4.** (*a*) XRD spectra of $CsPbBr_3$ QDs, $CsPbBr_3/TiO_x$ and $CsPbBr_3/TiO_2$ prepared at various annealing temperatures and (*b*) XRD spectra of (040) and (202) peaks.

To verify the stability enhancement of $CsPbBr_3/TiO_2$ annealed at 400°C with respect to bare $CsPbBr_3$ QDs, XRD measurement was performed to observe the transformation of their structures (figure 5a-, and 5b), in which the samples were stored without an aqueous medium under ambient air with 40%RH. The normalized XRD pattern showed that $CsPbBr_3$ QDs had a poor stability when stored under ambient air for a long time because there was a rapid decrease in crystallinity by 19.3% within 5 days. A decrease of (110) peak intensity at $2\theta = 15°$ and a peak splitting were observed. It was possible that the environment factors such as light, oxygen and humidity have destroyed their structure as studied in previous researches [8]. On the other hand, the $CsPbBr_3/TiO_2$ annealed at 400°C showed the excellent stability for more than a week under ambient atmosphere. Additionally, the change of material composition due to the degradation was not occurred. It can be explained that $TiO_2$ acted as a protecting shell because of its high thermal and water stability under stressed conditions, and prevented the $CsPbBr_3$ QDs core from contacting with degradation factors [3]; at the same time, the structure of $CsPbBr_3$ QDs still maintained a good crystallinity after heat treatment at 400°C.

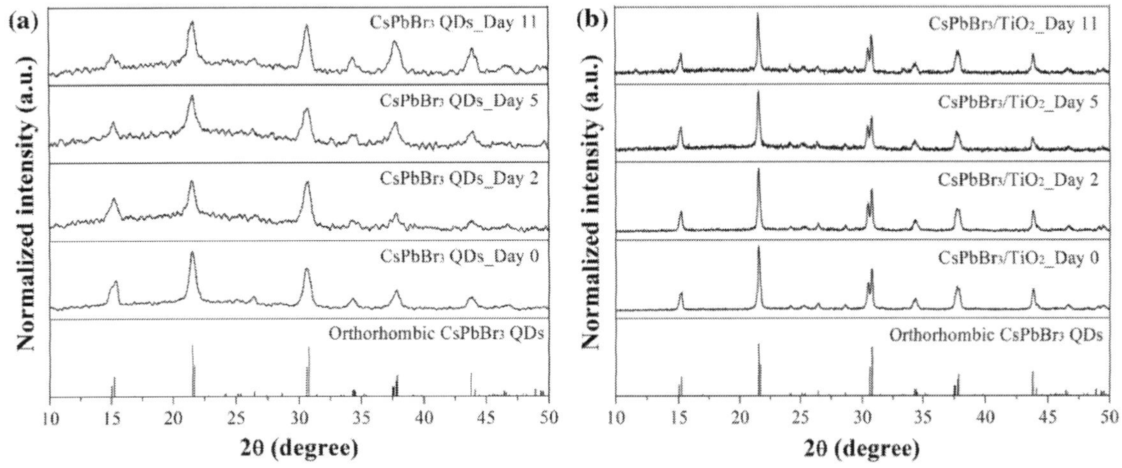

**Figure 5.** (*a*) XRD spectra of bare $CsPbBr_3$ QDs and (*b*) XRD spectra of $CsPbBr_3/TiO_2$ annealed at 400°C, when stored under ambient air for 11 days.

## 4. Conclusion

The encapsulation process of $CsPbBr_3$ QDs with $TiO_2$ caused the ligand removal, which was useful for further optoelectrical applications, as well as the change of morphology and optical properties due to the particle agglomeration. Charge extraction capability can be enhanced owing to the suitable energy level of $TiO_2$ shell with respect to the $CsPbBr_3$ QDs. The structure of orthorhombic $CsPbBr_3$ QDs was maintained after coating with $TiO_2$ and annealing at elevated temperature. Furthermore, the heat treatment enhanced the crystallinity of $CsPbBr_3/TiO_2$ but distorted their structure and caused the polycrystalline particles at the same time. For annealing at 300°C, 400°C and 500°C, the sample annealed at 400°C exhibited the best crystallinity and appeared the desired anatase phase of $TiO_2$ shell. Moreover, $CsPbBr_3/TiO_2$ annealed at 400°C showed the excellent structural stability against the bare $CsPbBr_3$ QDs when stored under ambient atmosphere for 11 days. It has proven that encapsulation with $TiO_2$ shell and annealing at a suitable temperature had a significant influence on both structural property and stability of the $CsPbBr_3$ QDs.

## References

[1] Boyd C C, Cheacharoen R, Leijtens T, McGehee M D 2019 Understanding degradation mechanisms and improving stability of perovskite photovoltaics *Chem. Rev.* **119** (5) 3418-51.

[2] Wei Y, Cheng Z, Lin J 2019 An overview on enhancing the stability of lead halide perovskite quantum dots and their applications in phosphor-converted leds *Chem. Soc. Rev.* **48** (1) 310-50.

[3] Li Z J, Hofman E, Li J, Davis A H, Tung C H, Wu L Z, et al. 2018 Photoelectrochemically active and environmentally stable cspbbr3/tio2 core/shell nanocrystals *Adv. Funct. Mater.* **28** (1) 1704288.

[4] Pramata A D. Photoinduced electron transfer in semiconductor nanocrystal/polyoxometalate systems and its application for electron storage, luminescence mechanism analysis, and photoemission switching. Kumamoto University: Kumamoto University; 2019.

[5] Zhu L, Lu Q, Lv L, Wang Y, Hu Y, Deng Z, et al. 2017 Ligand-free rutile and anatase tio 2 nanocrystals as electron extraction layers for high performance inverted polymer solar cells *RSC Adv.* **7** 20084-92.

[6] Imran M, Caligiuri V, Wang M, Goldoni L, Prato M, Krahne R, et al. 2018 Benzoyl halides as alternative precursors for the colloidal synthesis of lead-based halide perovskite nanocrystals *J. Am. Chem. Soc.* **140** (7) 2656-64.

[7] Moyen E, Kanwat A, Cho S, Jun H, Aad R, Jang J 2018 Ligand removal and photo-activation of cspbbr3 quantum dots for enhanced optoelectronic devices *Nanoscale* **10** (18) 8591-9.

[8] Chen J, Liu D, Al-Marri M J, Nuuttila L, Lehtivuori H, Zheng K 2016 Photo-stability of cspbbr3 perovskite quantum dots for optoelectronic application *Sci. China Mater.* **59** (9) 719-27.

[9] Mahalingam T, Selvakumar C, Ranjith Kumar E, Venkatachalam T 2017 Structural, optical, morphological and thermal properties of tio2–al and tio2–al2o3 composite powders by ball milling *Phys. Lett. A* **381** (21) 1815-9.

# Carbothermic Reduction of the Waste Chromium Oxide Rods from Thermal Spray Processes

**Penpisuth Thongyoug[1,a], Jennarong Tungtrongpairoj[2,b*]**

[1]Department of Industrial Engineering and Management, Faculty of Engineering and Industrial Technology, Silpakorn University, Nakhon Pathom 73000, Thailand
[2]Department of Materials and Production Technology Engineering, Faculty of Engineering, King Mongkut's University of Technology North Bangkok, Bangkok 10800, Thailand
E-mail: [a]penpisuth.ft@gmail.com, [b]jennarong.t@eng.kmutnb.ac.th

**Abstract.** The waste chromium-oxide rods from thermal spray processes contained the valuable components of high chromium. The carbothermic reduction with graphite powders can be used to recover the chromium from the chromium-oxide rod. The waste rods were prepared into small pieces with a half-dimensions of 6.3 mm diameter and 5 mm length, and heated under the reducing atmosphere at 1150 – 1350 °C. Besides, the effect of different types and sizes of reducing agent on the reduction was observed. During the heating and cooling period, argon gas was fed into the atmosphere. Chromium carbide formation was promoted when the temperature and time were increased. Cr-O and Cr-O-Al-Si were observed in the microstructures before and after the reduction. The outermost part of the waste rod can be reduced to chromium because of the interphase zone. Consequently, the carbothermic reduction is one possible recycling method for utilizing ceramic rod wastes from the thermal spray process.

Keywords: Waste chromium oxide rod, Thermal spray, Carbothermic reduction

## 1. Introduction

Ceramic rod containing high chromium oxide is widely used as a raw material in thermal spray coating techniques [1, 2]. There are some remaining parts from the ceramic rod after the process which cannot be reused. However, the waste rods consist of the valuable metals, for example chromium and aluminum. The recycling method is one possible way to recover chromium from the waste rods. Chromium can be extracted by carbothermic reduction is shown in equation (1).

$$M_xO_y(s)+yC(s) = xM(s)+yCO(g) \quad (1)$$

The carbothermic reduction of chromium oxide is a combination of the reactions below.

$$Cr_2O_3(s) = 2Cr(s) + \frac{3}{2}O_2(g) \tag{2}$$

$$3C(s) + \frac{3}{2}O_2(g) = 3CO(g) \tag{3}$$

Combining Equation (2) and Equation (3):

$$Cr_2O_3(s) + 3C(s) = 2Cr(s) + 3CO(g) \tag{4}$$

The high Cr-containing wastes from the metallurgical industries, e. g. the slag in steelmaking processes, had been investigated for recycling by the carbothermic processes at temperatures from 900 °C to 1600 °C [3,4, 6–10] The high chromium vanadium-titanium pellets were investigated in the reduction atmosphere at 900 °C for two hours and showed the reduction degree increased with increasing reduction time [3]. Research of ZHANG et al. [8] reported that chromium percentage from the reduction of chromite ($FeCr_2O_4$) increased from 68.60 to 80.21 wt.% when raising the reduction temperature from 1350 to 1550 °C. Furthermore, the work of MORI et al. [4] showed the slag component from the steelmaking process could be reduced to chromium by the carbothermic reduction at temperature more than 1350 °C [3]. The work of Takano et al. [5] observed metallic phases of Fe-Cr-C in chromium containing slag when it was exposed in carbon at 1500 °C for 10 minutes. In this work, the reduction of the waste rods from thermal spray processes containing high chromium oxide by carbothermic reduction was investigated.

## 2. Materials and Methods

The chromium-oxide rod wastes from the thermal spray processes were used in this study, the composition is shown in table 1. The green waste-rod samples with a half-dimension of 6.3 mm diameter and 5 mm length were prepared and placed in alumina crucibles with the reduction agents such as anthracite, graphite, and carbon powders. The reducing agents showed different particle sizes, as reported in table 2. The samples were exposed in a horizontal furnace at different temperatures; 1150, 1250, and 1350 °C respectively for both time durations of 30 and 60 minutes. Argon gas was flown into the furnace during heating and cooling. The experiment was conducted in two parts, firstly, the reduction temperature and time were changed to investigate their relationship with the reduction products. Secondly, the effect of the reducing agents was investigated when they were used for the reduction of the ceramic rod. During the steps of the reduction testing, the weight of the samples was measured by a 5-digit digital analytical balance, and after the reduction test, the waste ceramic rod was analyzed by a scanning electron microscope with an energy dispersive spectrometer (SEM-EDS).

**Table 1. Chemical composition of the ceramic rod (wt.%) from a technical datasheet of the manufacturer (Siant Gobain Coating Solution).**

| $Cr_2O_3$ | $Al_2O_3$ | $SiO_2$ | $Fe_2O_3$ | MgO |
|---|---|---|---|---|
| 90.33 | 3.67 | 5.62 | 0.27 | 0.11 |

**Table 2. The average particle size of reducing agent.**

| Powder | Particle size (μm) |
|---|---|
| **Anthracite** | 265 ± 90 |
| **Carbon** | 40 ± 15 |
| **Graphite** | 70 ± 20 |

## 3. Results and Discussion

The effect of temperature for the reduction process was revealed when comparing the mass change of the chromium waste rods at temperatures between 1150 and 1350 °C in the graphite powder, as illustrated in figure 1. At the lowest temperature of 1150 °C, the mass of the waste rods was not much changed after the reduction. However, the mass change increased around 0.8 mg when temperature was increased to 1250 °C and held for 30 min. The reason may be related to the transformation of chromium oxide into carbides such as $Cr_{23}C_6$, $Cr_7C_3$, and $Cr_3C_2$, which followed the carbothermic reduction reaction in equation (5) to equation (8). These carbides showed stability of formation with a more negative change in Gibbs free energy (ΔG°) at high temperature (figure 2).

$$Cr_xO_y(s)+2yC(s) = Cr_xC_y(s)+yCO(g) \tag{5}$$

$$23Cr_2O_3(s) + 81C(s) = 2Cr_{23}C_6(s) + 69CO(g) \tag{6}$$

$$7Cr_2O_3(s) + 27C(s) = 2Cr_7C_3(s) + 21CO(g) \tag{7}$$

$$3Cr_2O_3(s) + 13C(s) = 2Cr_3C_2(s) + 9CO(g) \tag{8}$$

After increasing temperature to 1350 °C, the mass change surprisingly increased by 4.75 times compared to the previous temperature (1250 °C). It should be noted that the carbide formation can be promoted at higher temperatures and the results were in good agreement with the previous work which reported that the carbide $Cr_{23}C_6$ could be appeared in pellets containing chromium oxide after reduction at temperature around 1400 °C for 1800 s [4]. Besides, $Cr_2O_3$ still existed because of the high oxide stability ($\Delta G°_{1400\ °C}$ ~ -45,000
J $mol^{-1}$ $O_2$) and a short reduction time.

IOP Conf. Series: Materials Science and Engineering **965** (2020) 012024 doi:10.1088/1757-899X/965/1/012024

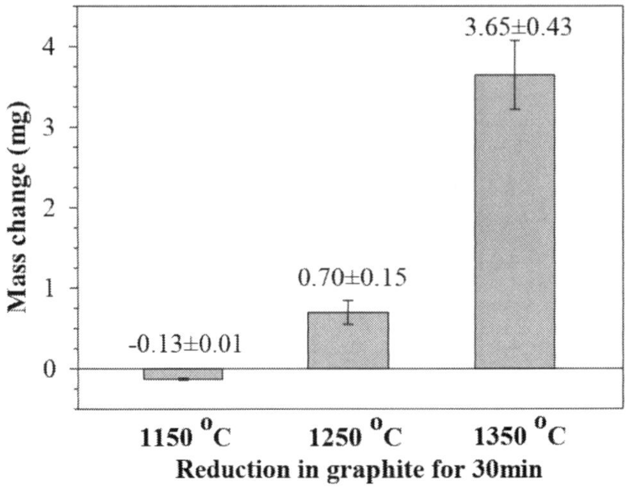


**Figure 1.** Mass change of ceramic rod after reduction in graphite for 30 minutes.

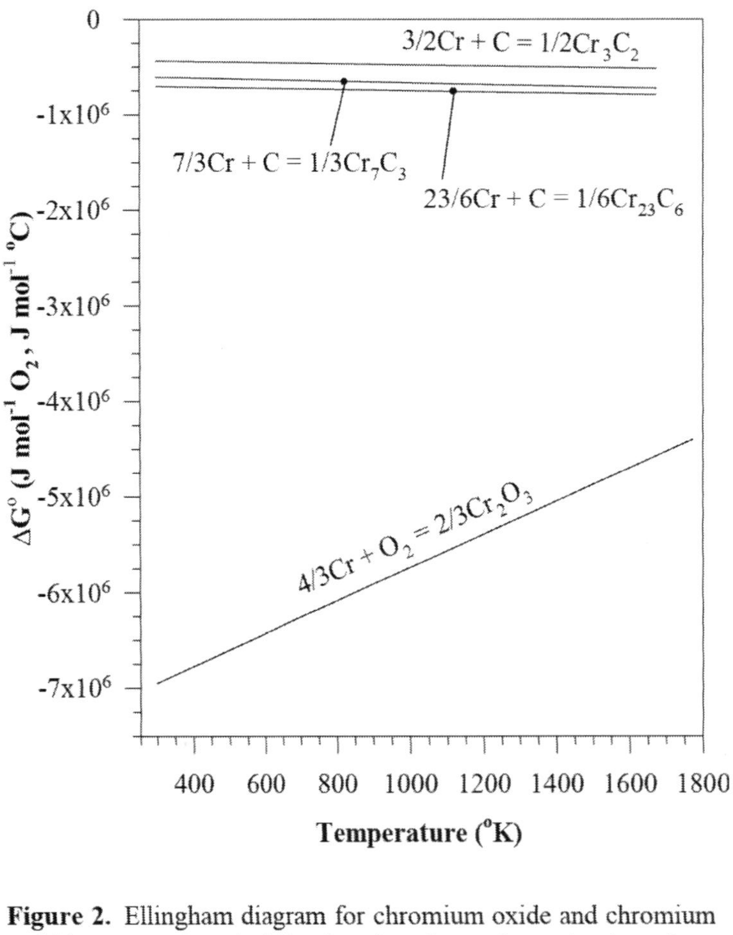


**Figure 2.** Ellingham diagram for chromium oxide and chromium carbides was plotted following by thermodynamic data from Kubaschewski et al. [11]

Figure 3 showed the mass change of reduced chromium rods in different three reducing atmospheres, which compared the effect of reducing agents (carbon, graphite, and anthracite powders). The mass change under anthracite powders showed lowest value at every reduction temperature and time. Concerning the mass loss of chromium rods which heated for 30 min in graphite at 1150 °C and in antracite at 1250 °C for 30 min, it can be concluded that the carburization reaction did not occur and volatile matters in the green chromium rods, such as water, were evaporated following volatilization reaction during heating. Besides, $CrO_2(OH)_2$ and $CrO_3$ gases which are a kind of hexavalent chromium vapors were formed by chromium oxide in dry air and wet air respectively at the tempearature above 400 °C. [12–14]. These reasons can lead to the mass loss of chromium rods for those conditions.
The larger particle of anthracite of around 265 µm may have caused the carbon diffusion in waste rods to be slow in comparison to the other reducing agents. The high value of the mass change was obtained from the reduction in carbon and graphite powders with nearly similar size range. Furthermore, the mass change by carbon powder was small even if temperature changes. However, the mass change by graphite powder at low temperature was lower than that at high temperatures. For this reason, carbon power was more effective for carbothermic reduction over a wide operating temperature range.

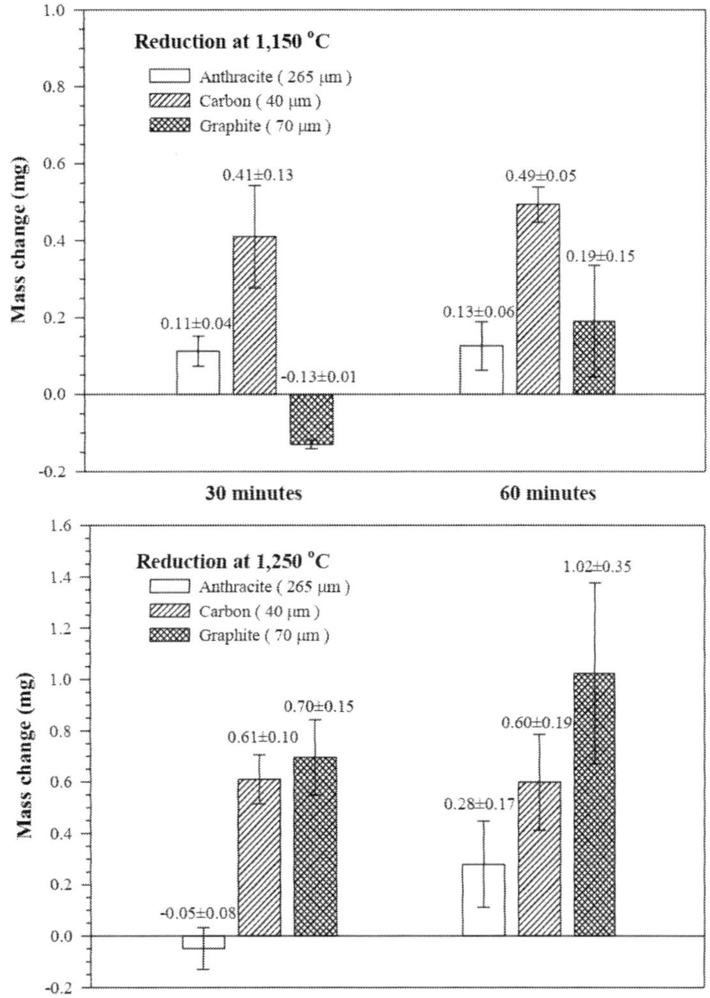


**Figure 3.** Mass change of ceramic rod after reduction test with different reduction agents, time durations, and temperatures.

The microstructures of waste chromium oxide rods before and after the reduction test were compared, as shown in figure 4. Before the reduction, microstructure consisted of two phases; white and black. The grain sizes significantly increased after the reduction test. The SEM-EDS analysis showed the oxide rich in Cr and O in the white phase, and Cr, Si, Al, and O in the black phase. The outermost part of chromium oxide rods, which was next to the reducing agent, was thoroughly observed and revealed high Cr percentage as depicted by a plot of EDS line scans, as shown in figure 5.

Consequently, the white area which is rich in Cr and O was almost reduced to pure chromium at approximately 86.30 and 83.37 wt.% corresponding to the EDS spectrum 1 and 2, respectively, in figure 6. The results were similar to a research work of MORI et al. [4], which presented the reduction degree of 91% for the carbothermic reduction of $Cr_2O_3$ at 1350 °C for 60 minutes. However, they found two compounds i.e. $Cr_7C_3$ and $Cr_{23}C_6$ following the direct reduction. In addition, the reduction of chromite ($FeCr_2O_4$) in the self-reducing agglomerate comprised of petroleum coke, Portland cement and fluxing agents (hydrated lime and silica) [5] showed 58 wt.% Cr at the similar area of this work between the self-reducing agglomerate and solid graphite powder.

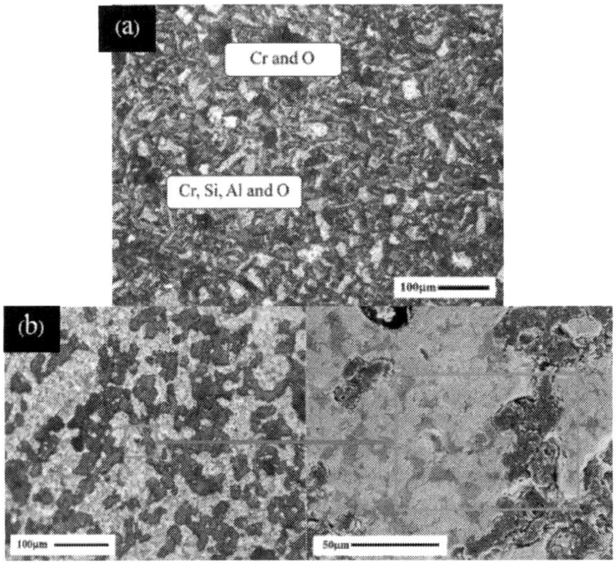


**Figure 4.** Cross section morphology of ceramic rod (a) before reduction, and (b) after reduction in graphite at 1350 °C for 60 minutes.

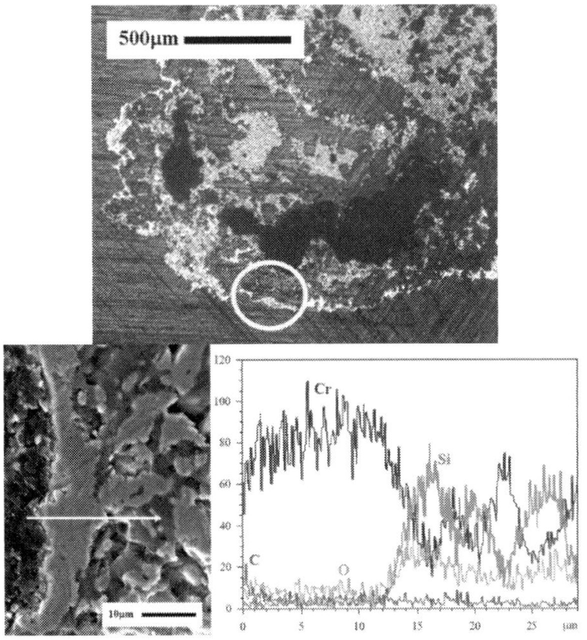


**Figure 5.** SEM cross-section morphology with EDS line scans of the waste ceramic rod after reduction in graphite at 1350 °C for 60 minutes.

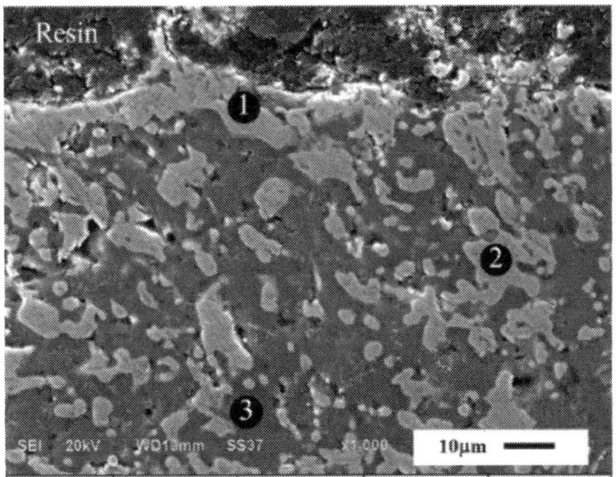


| Position | C | O | Si | Cr |
|---|---|---|---|---|
| 1 | 13.70 | - | - | 86.30 |
| 2 | 15.68 | 0.00 | 0.96 | 83.37 |
| 3 | 8.93 | 48.52 | 35.90 | 6.64 |

**Figure 6.** SEM cross -section morphology and composition in wt.% of the waste ceramic rod after reduction in graphite at 1350 °C for 60 minutes

## 4. Conclusions

- The mass change of waste chromium rods increased when increasing temperature from 1150 to 1350 °C and held for 30 min because of the carbide formation.
- The waste rods had the lowest mass change after reducing with anthracite powder as a result of larger particle size and slow carbon diffusion.
- Carbon powder is suitable agent for carbothermic reduction at temperatures higher than 1150 °C.
- $Cr_{23}C_6$, $Cr_7C_3$, and $Cr_3C_2$ presumably formed after carbothermic reduction at 1150 – 1250 °C, and some Cr formed at 1350 °C when heated for 60 min.
- After the reduction at 1350 °C for 60 minutes, the grain size of the Cr-O and Cr-O-Si-Al rich phases increased because of the effect of high temperature.
- The reduction of the waste rods started to reduce at the outermost part due to the interphase between the entire waste rod and the reducing agent.

## Acknowledgements

This study is funded by the Faculty of Engineering, King Mongkut's University of Technology North Bangkok with the contract no. ENG-NEW-62-23.

## References

[1] Pawlowski L 2008 *The science and engineering of thermal spray coatings* (Chichester: Wiley)
[2] Davis and Joseph R 2004 *Handbook of thermal spray technology* (Materials Park, OH: ASM International)
[3] Cheng G-j, Xue X-x, Gao Z-x, Jiang T, Yang H and Duan P-n 2016 Effect of Cr2O3 on the Reduction and Smelting Mechanism of High-Chromium Vanadium-titanium Magnetite Pellets *ISIJ Int.* **56** 1938
[4] Mori T, Yang J and Kuwabara M 2007 Mechanism of Carbothermic Reduction of Chromium Oxide *ISIJ Int.* **47** 1387
[5] Takano C, Zambrano A P, Nogueira A E A, Mourao M B and Iguchi Y 2007 Chromites Reduction Reaction Mechanisms in Carbon–Chromites Composite Agglomerates at 1773 K *ISIJ Int.* **47**
[6] Shibata E, Egawa S and Nakamura T 2002 Reduction Behavior of Chromium Oxide in Molten Slag Using Aluminum, Ferrosilicon and Graphite *ISIJ Int.* **42** 609
[7] Zhang Y-l, Liu Y and WEI W-j 2014 Products of carbothermic reduction of Fe–Cr–O and Fe–Cr–Ni–O systems *Transactions of Nonferrous Metals Society of China* **24** 1210
[8] Zhang Y-l, Guo W-m, Liu Y and Jia X-l 2016 Reduction mechanism of Fe2O3-Cr2O3-NiO system by carbon J. *Cent. South Univ.* **23** 1318
[9] Hara Y, Ishiwata N, Itaya H and Matsumoto T 2000 Smelting Reduction Process with a Coke Packed Bed for Steelmaking Dust Recycling *ISIJ Int.* **40** 231
[10] Huaiwei Z and Xin H 2011 An overview for the utilization of wastes from stainless steel industries *Resources, Conservation and Recycling* **55** 745
[11] Kubaschewski O and Alcock C B 1979 *Metallurgical thermochemistry (Pergamon international library. international series on materials science and technology* vol. 24*)* 5th ed (Oxford: Pergamon)

# Effect of Sintering Temperature and Cooling Rate on the Microstructure and Hardness of Fe-Cr-Mo-V-W-C Sintered Steel added with $MoS_2$

**V. Songsujaritkul[1], S. Mahathanabodee[1*] and R. Tongsri[2]**

[1] Department of Production Engineering, Faculty of Engineering,
King Mongkut's University of Technology North Bangkok (KMUTNB)
[2] Powder Metallurgy Research and Development Unit,
Thailand National Metal and Materials Technology Center (MTEC)

Corresponding author: sithipong.m@eng.kmutnb.ac.th

**Abstract.** In this work, the Fe-Cr-Mo-V-W-C sintered steels added with $MoS_2$ (1, 3 and 5 wt.%) were produced by powder metallurgy process. The specimens were prepared by mixing, compacting and sintering at temperatures of 1150, 1200 and 1250°C for 45 min in vacuum furnace. After sintering, the specimens were cooled down with $N_2$ at the different cooling rate (0.1 and 5.2°C/s). The density, hardness and microstructure of the sintered Fe-Cr-Mo-V-W-C steels were investigated. The experimental results indicated that the density and hardness of sintered steel could be improved by increase $MoS_2$ addition, cooling rate and sintering temperature. The sintering mechanism was varied from solid state sintering to liquid phase sintering depend on sintering temperatures and $MoS_2$ content. The presence of liquid phases and the increasing of cooling rate resulted in the amount of carbide precipitated decreased in the steel grain but increased precipitation at grain boundaries. The solidified of liquid and eutectic phases were found rising by increased cooling rate. Carbide precipitation in steel grain could be improved by added $MoS_2$ and sintered at low temperature. From the results, it was indicated that the addition of $MoS_2$, increase sintering temperature and cooling rate resulted in the precipitation of carbide, solidified liquid and eutectic phases, and grain growth in the sintered Fe-Cr-Mo-V-W-C steel.

**Keywords:** $MoS_2$, carbide, sintering, liquid phase, cooling rate

## 1. Introduction

The Fe-Cr-Mo-V-W-C powder is used to produce the sintered parts that using in the wear resistance and high temperature applications [1]. This steel powder was developed for using in the conventional compaction process. However, it is lower compressibility than tool steel and high speed steel powder [2]. For the sintered parts, the mechanical properties and wear resistance are influenced by the variation of density and porosity. Therefore, the addition of additive powders for improved mechanical properties and wear resistance were found in many previous works [3-5]. $MoS_2$ powder is used as additive powder in the sintering of high speed steel [4, 6, 7] and stainless steel [8, 9]. $MoS_2$ reacted to the steel matrix and produced the liquid phases at high temperature. The sintered density and porosity can be improved by $MoS_2$ addition. However, the performance of $MoS_2$ on the sintering of Fe-Cr-Mo-V-W-C in vacuum is interesting. Effects of $MoS_2$ addition on sintering temperature and cooling rate and other parameters also need explication. Therefore, in this work, the effect of $MoS_2$ content, sintering temperature and cooling rate on the microstructure and hardness of Fe-Cr-Mo-V-W sintered steel were studied.

## 2. Experimental Procedure

Specimens in this work were prepared by powder metallurgy process. Fe-Cr-Mo-V-W-C powder was added with various amount of $MoS_2$ powder (0, 3 and 5 wt.%) and zinc stearate (1 wt.%). The chemical composition of Fe-Cr-Mo-V-W-C powder is shown in Table. 1. The mixtures were mixed thoroughly in the tumbling mixer for 1 hr. at the speed of 34 rpm. The mixed powders were compacted into the uniaxial die to produce the specimen of Ø 25 mm with 4 mm in thickness. At the compression pressure of 700 MPa, average green density of specimens was about 6.3 g/cm$^3$ by the Archimedes water immersion method. The compacted specimens were sintered in the vacuum furnace at 1150, 1200 and 1250°C for 45 min. Subsequently, specimens were cooled down in different cooling rate of 0.1 and 5.2°C/s with $N_2$. The sintered density was compared with the green density. The sintered specimens were cut, ground, polished and etched with Murakami etchant for microstructure examination. The microstructure of sintered specimens in the different cooling rate were examined under optical microscope and scanning electron microscope with electron dispersive spectroscopy (EDS) for element analysis. Rockwell scale A was used to evaluate the macro-hardness of sintered specimens.

**Table 1** Chemical composition of Fe-Cr-Mo-V-W-C powder

| Element | C | Si | Cr | V | W | Mo | Fe |
|---|---|---|---|---|---|---|---|
| Content (wt.%) | 1.50 | 0.50 | 16.00 | 1.50 | 1.50 | 1.50 | Bal. |

## 3. Results and Discussion

### *3.1 Sintered Density*

Density of Fe-Cr-Mo-V-W-C steel added with $MoS_2$ sintered at 1150, 1200 and 1250°C with different cooling rate of 0.1 and 5.2°C/s is shown in Figure 1. Sintered density was improved by increase sintering temperature and cooling rate. Sintered density of specimens without $MoS_2$ added was in the range 6.33-6.37 g/cm$^3$ at cooling rate of 0.1°C/s and 6.58-7.77 g/cm$^3$ at cooling rate of 5.2°C/s. In addition, sintered density of Fe-Cr-Mo-V-W-C steel was improved by $MoS_2$ added, especially at cooling rate of 0.1°C/s. However, the high sintering temperature resulted in the aggressive sintering and densification on Fe-Cr-Mo-V-W-C steel both with and without $MoS_2$ added. It is indicated that the density of specimen was highly increased after sintered at 1250°C especially at cooling rate of 5.2°C/s. Figure 2 shows shape of sintered specimens at 1150°C and 1250°C. Effect of sintering temperature and cooling rate on porosity of sintered specimens is shown in Figure 3. The great reduction of porosity was found in the specimens at cooling rate of 5.2°C/s. Therefore, the increasing of sinter density resulted in the decreasing of porosity in sintered specimens.

### *3.2 Hardness of sintered specimens*

Hardness of sintered specimens is shown in Figure. 4. The hardness of sintered Fe-Cr-Mo-V-W-C steel added with $MoS_2$ was improved by increased sintering temperature and amount of $MoS_2$. Hardness of specimen was greater improved by sintered at 1250°C. However, the increasing of cooling rate from 0.1 to 5.2°C/s seemed less effect on increasing of hardness of sintered steels.

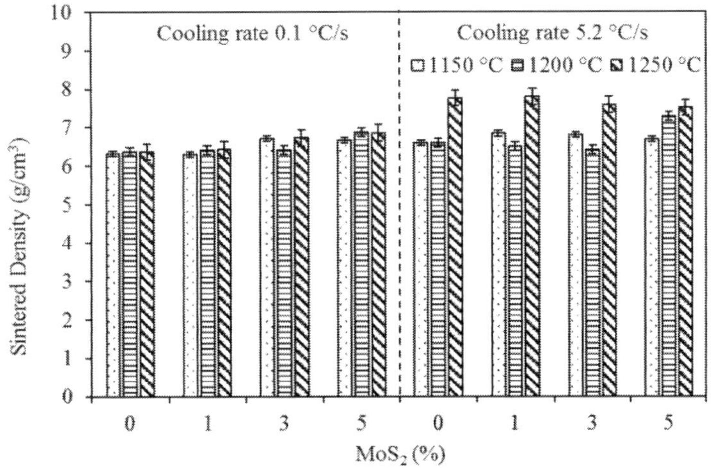


**Figure 1.** Sintered density of Fe-Cr-Mo-V-W-C steel added with $MoS_2$ under different sintering temperature and cooling rate

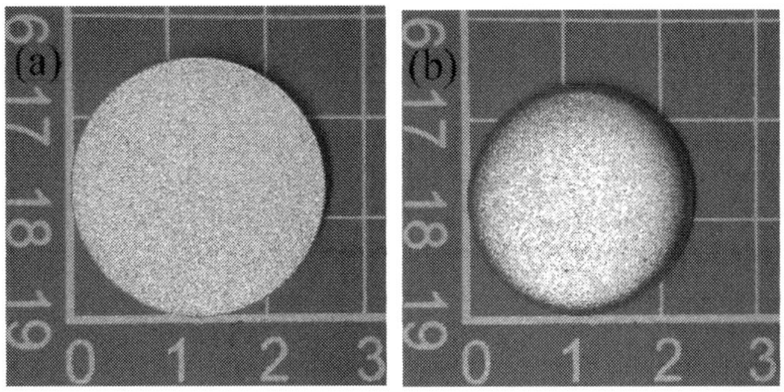


**Figure 2.** Specimens after sintered at (a) 1150°C and (b) 1250°C

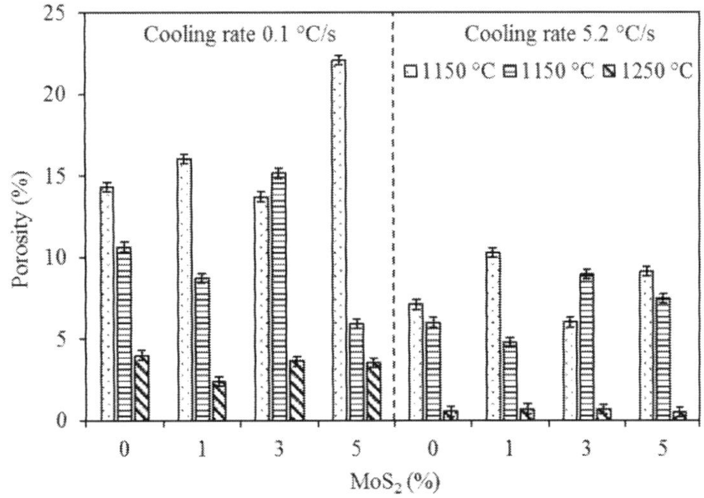


**Figure 3.** Effect of sintering temperature and cooling rate on porosity of sintered Fe-Cr-Mo-V-W-C steel added with $MoS_2$

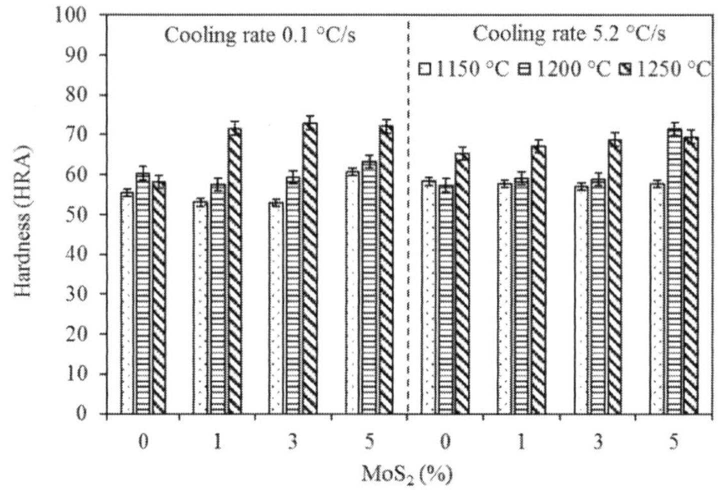


**Figure 4.** Hardness of sintered specimens at different sintering temperature and cooling rate

*3.3 Microstructures*

*3.3.1 Cooling rate at 0.1 ℃/s*

Microstructures of sintered Fe-Cr-Mo-V-W-C steel from cooling rate of 0.1°C/s are shown in Figure 5. Microstructure of Fe-Cr-Mo-V-W-C steel sintered at 1150°C showed distribution of small carbide particles in the microstructure with the good bonding and rounded pores in solid state sintering as shown in Figure 5(a). The bonding of steel matrix was improved and porosity decreased after sintered at 1200°C as shown in Figure 5(b). However, the evidence of liquid phase sintering was found in the microstructure after sintered at 1250°C. The rounded solid grain with solidified liquid at grain boundary was found in the microstructure. It was found that small carbide particles precipitated in the center of grain and at grain boundaries as shown in Figure 5(c). In high chromium-carbon steel system, small liquid fraction can be formed in the microstructure at temperature below 1250°C [10-12]. The liquid phases in the sintered of Fe-Cr-Mo-V-W-C steel at 1250°C resulted in the density and hardness improvement as shown in Figure 1 and Figure 4, respectively. In this system, $M_7C_3$ and $M_{23}C_6$ chromium base carbide were expected to precipitate in the steel matrix [1].

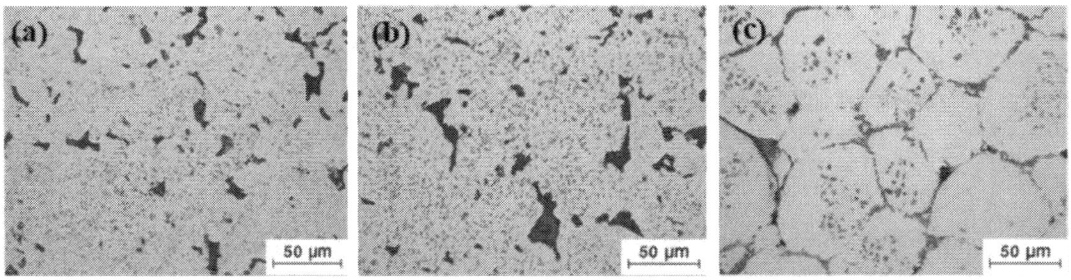


**Figure 5.** Microstructure of Fe-Cr-Mo-V-W-C steel sintered in different temperatures and cooling at 0.1°C/s; (a) 1150°C, (b) 1200°C and (c) 1250°C

Microstructures of Fe-Cr-Mo-V-W-C added with $MoS_2$ are shown in Figure 6. Microstructure of Fe-Cr-Mo-V-W-C steels added with 1 wt.% of $MoS_2$ sintered at 1150°C showed particles bonding and prior particle boundary. Small carbide particles distributed homogeneously in the steel matrix. The dark phases of $MoS_2$ added appeared in the original pores site of sintered steel as shown in Figure 6(a).

Original particle boundaries in microstructure reduced after sintered at 1200°C. It was exhibited that better bonding particles in the microstructure as revealed by rounded pore, prior to particle boundaries and porosity reduction as shown in Figure 6(b). Small carbide particles were found in microstructures. Typical SEM images in Figure 7(a) and (b) showed amount of carbide increased by added $MoS_2$. The addition of $MoS_2$ resulted in improve densification of sintered high speed steel by acting as sinterability enhancer in the sintering process [4]. Therefore, sintered density of $MoS_2$ added specimens in this work was improved as shown in Figure 1. In addition, the addition of $MoS_2$ resulted in increasing of liquid phase during sintering at 1250°C, as indicated by increasing of solidified liquid phase and eutectic phase at grain boundary as shown in Figure 6(c). This is due to the reaction of $MoS_2$ and steel matrix to produce the liquid phase at high sintering temperature [3]. The appearance of liquid phase resulted in density, porosity and hardness of specimens improved as shown in Figure 1, 3 and 4, respectively. The increasing of $MoS_2$ amount and sintered at lower temperature (1150°C) demonstrated that the increasing of porosity and prior particle boundaries as shown in Figure 6(d). However, when increased sintering temperature the densification of sintered steel was improved and carbide particles were found in microstructure. At 1250°C, the eutectic phase and solidified of liquid phases was found at grain boundaries as shown in Figure 6(e). The dark phases and eutectic phases in microstructure of sintered specimens were found similar in the literatures [3, 4]. In this system, $MoS_2$ reacts to the steel matrix and alloy elements. Therefore the mix $M_6C$ carbides were expected to form in the steel matrix [4]. This type of carbide is hard and thermally stability [1].

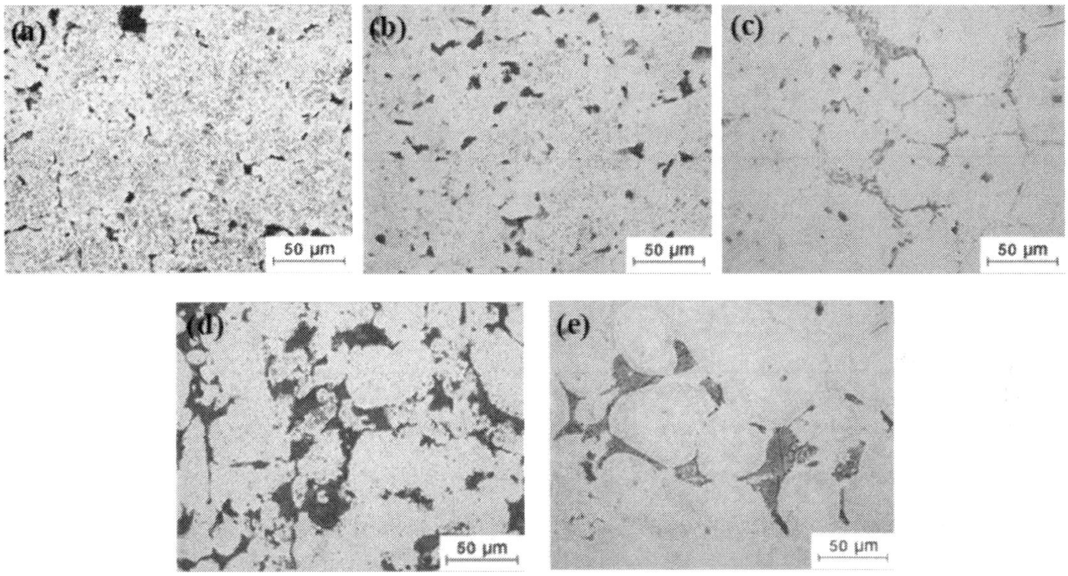


**Figure 6.** Microstructure of Fe-Cr-Mo-V-W-C steel added $MoS_2$ sintered in different temperatures and cooling rate at 0.1°C/s; (a) $MoS_2$ 1 wt.% sintered at 1150°C, (b) $MoS_2$ 1 wt.% sintered at 1200°C, (c) $MoS_2$ 1 wt.% sintered at 1250°C, (d) $MoS_2$ 5 wt.% sintered at 1150°C and (e) $MoS_2$ 5 wt.% sintered at 1250°C

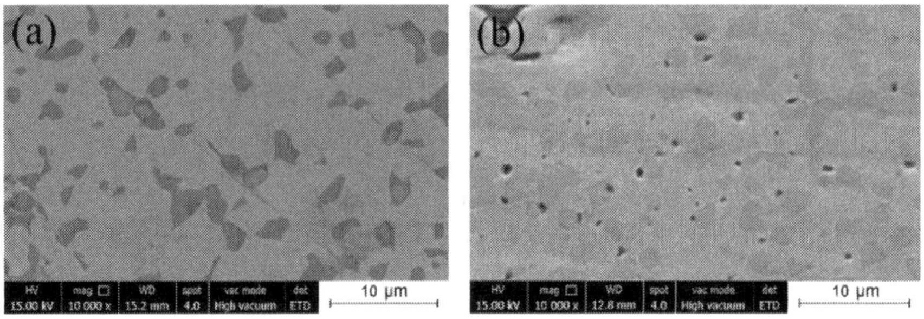


**Figure 7** SEM images of Fe-Cr-Mo-V-W-C added with $MoS_2$ sintered at 1200°C with cooling rates of 0.1°C/s; (a) 0% $MoS_2$ and (b) 3% $MoS_2$

*3.3.2 Cooling rate at 5.2 °C/s*

Figure 8 shows microstructures of sintered Fe-Cr-Mo-V-W-C steel from cooling rate at 5.2°C/s. Small carbide particle precipitated homogeneously in microstructure after sintered at 1150°C. The microstructure showed good bonding and less outline of prior particle boundaries as shown in Figure 8(a). The outline of prior particle and porosity were reduced by increased sintering temperature up to 1200°C as shown in Figure 8(b). The fine carbide particles were found in the microstructures. However, it can be seen that amount of carbide precipitated less than that of lower cooling rate (0.1°C/s). Because of short time for carbide precipitated during cool down at high cooling rate. Effect of cooling rate on carbide precipitation was discussed in previous works [2, 13, 14]. The sintering process changed from solid stage sintering to liquid phase sintering stage when increased sintering temperature to 1250°C. The solidified of liquid phase was found in microstructure as shown in Figure 8(c). The microstructure indicated grain coarsening with rounded grain boundaries and eutectic phases along grain boundaries. It was seen that the eutectic phase fraction was more than that of slow cooling rate. In addition, formation of liquid phases resulted in decreasing of carbide precipitation in the area close to the grain boundaries as shown in Figure 8(c).

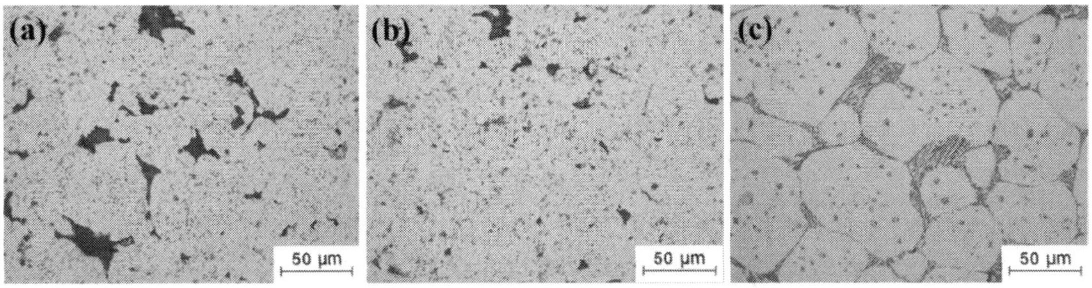


**Figure 8.** Microstructure of Fe-Cr-Mo-V-W-C steel sintered in different temperatures and cooling at 5.2°C/s; (a) 1150°C, (b) 1200°C and (c) 1250°C

Microstructures of Fe-Cr-Mo-V-W-C added with $MoS_2$ sintered in various temperature and subsequent cooled down at 5.2°C/s are shown in Figure 9. In this cooling rate, the bonding of particles and porosity was improved due to increase sintering temperature as shown in Figure 9(a) and (b), respectively. Whereas the microstructure showed the solidified of liquid phases and eutectic phase at grain boundaries after sintered at 1250°C as shown in Figure 9(c). The influence of higher cooling rate (5.2°C/s) showed precipitation of carbide in microstructures was less than that found from cooling rate of 0.1°C/s. Whereas the eutectic phase fraction from high cooling rate was more than that of low cooling rate. The increasing of $MoS_2$ in the sintered steel showed the increasing of carbide precipitation in microstructure as shown in Figure 9(d). When increased sintering temperature, the precipitation of carbide inside the grain was reduced whereas the precipitation of carbide at grain boundaries was increased as shown in

Figure 9(e). Specimens were sintered under liquid phase sintering process after sinter at 1250°C. The eutectic phase, solidified of liquid and coarse grain size with rounded boundaries effected from liquid phase sintering as shown in Figure 9(f). From these results, it is indicated that the variation of $MoS_2$ content, sintering temperature and cooling rate resulted in the precipitation of carbide and the fraction of solidified liquid and eutectic phases in the microstructure of sintered Fe-Cr-Mo-V-W-C steel. However, the characterization of carbide types, crystallography of matrix, effect of tempering time and temperatures, etc. are need to further study.

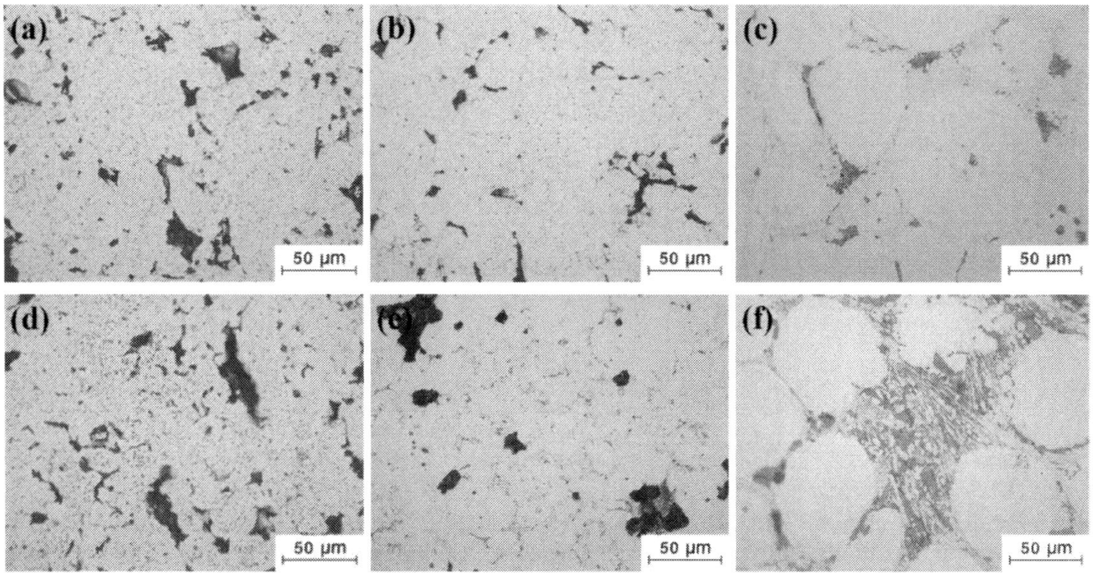


**Figure 9.** Microstructure of Fe-Cr-Mo-V-W-C steel added $MoS_2$ sintered in different temperatures and cooling rate at 5.2°C/s; (a) $MoS_2$ 1 wt.% sintered at 1150°C, (b) $MoS_2$ 1 wt.% sintered at 1200°C, (c) $MoS_2$ 1 wt.% sintered at 1250°C, (d) $MoS_2$ 5 wt.% sintered at 1150°C, (e) $MoS_2$ 5 wt.% sintered at 1200°C and (f) $MoS_2$ 5 wt.% sintered at 1250°C

## 4. Conclusion

In this study, the effect of $MoS_2$ content, sintering temperatures and cooling rate in the sintering of Fe-Cr-Mo-V-W-C steels were investigated. The results indicated that the sintered density and hardness of sintered steels were improved by increase sintering temperature. The sintering mechanism was varied from solid state sintering (1150 and 1200°C) to liquid phase sintering (1250°C). The presence of liquid phase resulted in the decrease of carbide precipitation inside the grains but increase the carbide precipitation at grain boundaries. In addition, the increasing of cool rate showed the decreasing of carbide precipitation in the microstructures. However, the high cooling rate resulted in the increasing of liquid and eutectic phases in the sintered steel. The increasing of $MoS_2$ in the sintered steel showed the increasing of carbide precipitation in microstructure at low temperature whereas $MoS_2$ influenced on the precipitation of carbides at grain boundaries when sintered at higher temperature.

**Acknowledgements**

This research was funded by King Mongkut's University of Technology North Bangkok. Contract no. KMUTNB-60-GOV-036. The authors wish to thank the National Metal and Materials Technology Center (MTEC) Pathum Thani of Thailand for laboratory equipment, characterization facilities and personal help. The Material Manufacturing and Surface Engineering Research Center (MaSE) for laboratory equipment. The courtesy of Acme International (Thailand) Ltd. and Höganäs Great Britain Ltd. for supplied OB1 powder.

IOP Conf. Series: Materials Science and Engineering **965** (2020) 012025 doi:10.1088/1757-899X/965/1/012025

## References

[1] Nurthen P, Bergman O and Hauer I 2008 Modified powders 'point the way to designer alloys' *Metal Powder Report* **63** 24-35

[2] Zheng Y, Wang F, Li C, Lin Y and Cao R 2019 Effect of Martensite Structure and Carbide Precipitates on Mechanical Properties of Cr-Mo Alloy Steel with Different Cooling Rate. In: *High Temperature Materials and Processes,* p 113

[3] Furlan K P, Prates P B, Andrea dos Santos T, Gouvêa Dias M V, Ferreira H T, Rodrigues Neto J B and Klein A N 2015 Influence of alloying elements on the sintering thermodynamics, microstructure and properties of Fe–$MoS_2$ composites *Journal of Alloys and Compounds* **652** 450-8

[4] Sustarsic B, Kosec L, Kosec M, Podgornik B and Dolinsek S 2006 The influence of $MoS_2$ additions on the densification of water-atomized HSS powders *Journal of Materials Processing Technology* **173** 291-300

[5] Selecká M, Salak A and Danninger H 2003 The effect of boron liquid phase sintering on properties of Ni-, Mo- and Cr-alloyed structural steels *Journal of Materials Processing Technology* **141** 379-84

[6] Sustarsic B, Kosec L, Dolinsek S and Podgornik B 2003 The characteristics of vacuum sintered M3/2 type HSSs with $MoS_2$ addition *Journal of Materials Processing Technology* **143-144** 98-104

[7] Sustarsic B, Kosec L, Jenko M and Leskovsek V 2001 Vacuum sintering of water-atomised HSS powders with $MoS_2$ additions *Vacuum* **61** 471-7

[8] Mahathanabodee S, Palathai T, Raadnui S, Tongsri R and Sombatsompop N 2014 Dry sliding wear behavior of SS316L composites containing h-BN and $MoS_2$ solid lubricants *Wear* **316** 37-48

[9] Furlan K P, Binder C, Klein A N and Mello J D B d 2012 Thermal stability of the $MoS_2$ phase in injection moulded 17-4 PH stainless steel *Journal of Materials Research and Technology* **1** 134-40

[10] Balitchev E, Hallstedt B and Neuschütz D 2005 Thermodynamic Criteria for the Selection of Alloys Suitable for Semi-Solid Processing *steel research international* **76** 92-8

[11] Uhlenhaut D I, Kradolfer J, Püttgen W, Löffler J F and Uggowitzer P J 2006 Structure and properties of a hypoeutectic chromium steel processed in the semi-solid state *Acta Materialia* **54** 2727-34

[12] Zelič K, Burja J, McGuiness P J and Godec M 2018 Effect of rare earth elements on the morphology of eutectic carbides in AISI D2 tool steels: experimental and modelling approaches *Scientific reports* **8** 1-8

[13] Sha Q Y, Li G Y, Qiao L F and Yan P Y 2007 Effect of Cooling Rate and Coiling Temperature on Precipitate in Ferrite of a Nb-V-Ti Microalloyed Strip Steel *Journal of Iron and Steel Research, International* **14** 316-9

[14] Luo Y, Guo H and Guo J 2019 Effect of cooling rate on the transformation characteristics and precipitation behaviour of carbides in AISI M42 high-speed steel *Ironmaking & Steelmaking* **46** 698-704

IOP Conf. Series: Materials Science and Engineering **965** (2020) 012026 doi:10.1088/1757-899X/965/1/012026

# Mechanical Properties and Melt Flow Index of Poly (butylene succinate) Blended with a Small Amount of Natural Rubber Compound

**W Singsang[1], P Rumjuan[2], Y Ausungnoen[2], W Charentanom[2], N Srakaew and N Prasoetsopha[2*]**

[1] Department of Aircraft Part Manufacturing Technology, Faculty of Industrial Technology, Rambhai Barni Rajabhat University, Sub-district Ta-Chang, District Muang, Province Chanthaburi 22000, Thailand
[2] Department of Materials Engineering, Faculty of Engineering and Architecture, Rajamangala University of Technology Isan, Sub-district Nai Muang, District Muang, Province Nakhon Ratchasima 30000, Thailand

*Email : natkrita.pr@rmuti.ac.th

**Abstract.** Natural rubber compound (NRC) was blended with biodegradable poly(butylene succinate) (PBS) with PBS/NRC weight ratios of 100/0, 97/3, 94/6 and 90/10. The PBS/NRC blends were melt mixed using an internal mixer at 145°C. Mechanical properties and melt flow index of the blends were investigated. The results showed that the percentage elogation at break and impact strength of the blends impoved when NRC was added. On the other hand, the Young's modulus and tensile strength of the blends decreased with an increase in the amount of NRC. The melt flow index of the blends continuously decreased with an increase in NRC content. This implied that NRC increased the viscosity of the blends. Moreover, the fractured surface morphology of PBS/NRC blends showed a good dispersion of NRC particles in PBS matrix. The average particle size of NRC was 2-5 μm.

## 1. Introduction

Many researchers are concerned with the current ever increasing environmental pollution hence, the development of green or environmental-friendly materials has attracted attention; for example the research on bioplastic materials [1-5]. Bioplastics have attracted a considerable interest to study because they are made from renewable resources such as cellulose, casein, collagen, soy protein, corn. The use of bioplastics is the best way to decrease plastic pollution because they can degrade and decompose in the natural environment. Examples of bioplastics are polycaprolactone [6], poly (lactic acid) (PLA) [2-3], and poly (butylene succinate) (PBS) [4-5]. The most popular material in the bioplastic industry is PLA. However, it is a brittle material and difficult to process. On the other hand, PBS has good mechanical properties and is easy to process. PBS is synthesized from the condensation reaction between 1,4-butandiol and succinic acid. It has a polyester and linear structure. It has thermal stability up to 200°C. After use, PBS degrades into carbon dioxide and water. However, it is still an expensive cost. Hence, there have been many attempts to increase the ability application of PBS i.e. polymer blend [4-5, 7] and/or composites materials [8-9].

Natural rubber (NR) is a bio-based elastomer which has a high elasticity and good tear strength. There have been many attempts to improve the toughness of polymeric matrix by natural rubber [2-5].

Examples of some of the reported studies are in polystyrene [10], poly (lactic acid) [2-3], poly (butylene succinate) [4-5].

This study examined the effect of NRC content on the mechanical properties and melt flow index of biodegradable PBS and natural rubber compound (NRC) blend which were melt blended in an internal mixer.

## 2. Materials and method

### *2.1. Materials*

Natural rubber (NR, STR 5L), stearic acid, zinc oxide (ZnO), sulphur (S), and N-cyclohexyl-2-benzothiazole sulfenamide CBS (CZ) were used to prepare natural rubber compound (NRC). All chemicals used in preparing the rubber compounds were purchased form Chareon Tut Co., Ltd., Thailand. Biodegradable poly (butylene succinate) (PBS), BioPBSTM FZ71PM grade, was supplied by PTT MCC Biochem company limited.

### *2.2. Natural rubber compound preparation*

Natural rubber (NR) and chemicals were mixed on a two-roll mill (Chareon Tut co., ltd., ML-D6L12-INV) with a roller speed ratio between front roll and back roll of 1:1.4 for 35 min in order to make NRC. The order of reagents is shown in Table 1. After being mixed, NRC was kept in room temperature for a night before being blended.

**Table 1.** Natural rubber compound formula.

| | Raw materials | Composition, phr | Mixing time, min |
|---|---|---|---|
| 1. | Natural rubber | 100 | 15 |
| 2. | Stearic acid | 2 | 5 |
| 3. | ZnO | 4 | 5 |
| 4 | Sulphur | 3 | 5 |
| 5 | CZ | 0.7 | 5 |

### *2.3. Poly (butylene succinate)/natural rubber compound blend*

PBS was dried in an oven at 50°C for 24 h before use. NRC and dried PBS were mixed in an internal mixer (Chareon Tut co., ltd., MX500-D75L90) with a rotor speed of 50 rpm at 145°C for 15 min. The ratio of PBS/NRC were 100/0, 97/3, 94/6 and 90/10. The total weight of each batch was 270 g. After completely mixing, the PBS/NRC blend was cooled to room temperature and ground with a grinder.

### *2.4. Hot-compression process*

The ground PBS/NRC blends were dried in an oven at 50°C for 24 h before hot-compression process. In the sample prepared for mechanical test, a hot-compression machine (Chareon Tut co., ltd., PR2D-W300L350 PM-WCL-HMI) was used at a temperature of 145°C and pressure of 1500 psi. The pre-heat time, pressing time and cooling time were 5, 5, and 3 min, respectively. The samples were prepared according to ASTM D638 for tensile test and ASTM D256 for impact test.

### 2.5. Testing and characterization

*2.5.1. Mechanical testing.* For tensile test, universal testing machine (UTM, LLOYD INSTRUMENT, LR10K plus) with a crosshead speed of 50 mm/min was used at room temperature. At least five samples were tested and evaluated. For impact test, the notched Izod impact type was conducted according to ASTM D256 at room temperature by impact tester machine (INSTRON, Ceast 9050). At least ten samples were tested and evaluated.

*2.5.2. Morphological study.* The fractured surface from tested impact sample was investigated by using scanning electron microscope (Phenom-world holding B.V., Phenom pro).

*2.5.3. Melt flow index testing.* Melt flow index (MFI) was investigated according to ASTM D1238. The ground PBS/NRC blends were dried in an oven at 50°C for 24 h before being measured.

## 3. Result and discussions

*3.1. Mechanical testing.*

The impact strength of PBS/NRC blends is shown in Fig 1. The results showed that the impact strength of the PBS/NRC blends in the ratio of 100/0, 97/3, 94/6 and 90/10 were 3.09 ± 0.85, 3.71 ± 0.28, 4.95 ± 0.84 and 5.150 ± 0.57 kJ/m$^2$, respectively. For the blends containing 3, 6 and 10 %wt of NRC, their impact strength were higher than that of the neat PBS. The highest impact strength of the blends showed in the ratio of 90/10 which improved from 3.09 to 5.15 kJ/m$^2$ (around 1.7 times). The reason for this improvement was that NRC which presented as a rubbery particle in the PBS matrix acted as stress absorber in the polymeric matrix. It is a toughened polymer matrix similar with a previous report, which improved toughness of PBS by the blending of the elastomeric acrylonitrile butadiene rubber (NBR) [7].

The tensile properties are illustrated in Fig 2. The elongation at break of PBS/NRC blends showed that NRC contributed to improve this value (Fig 2a). The highest elongation at break of the blends was 20.05 ± 0.03 percentage in the ratio of 90/10. This is due to the elastic property of NRC that is similar to the polymer-rubber system i.e. PBS/NBR blend [7], PBS/NR blend or PBS/epoxidized NR [5]. On the other hand, the incorporation of NRC significantly decreased the tensile strength and Young's modulus of PBS as shown in Fig 2(b) and (c), respectively. As expected, these values decreased with an increase in the amount of NRC due to the rubber particles being a weak point or stress concentration in the system.

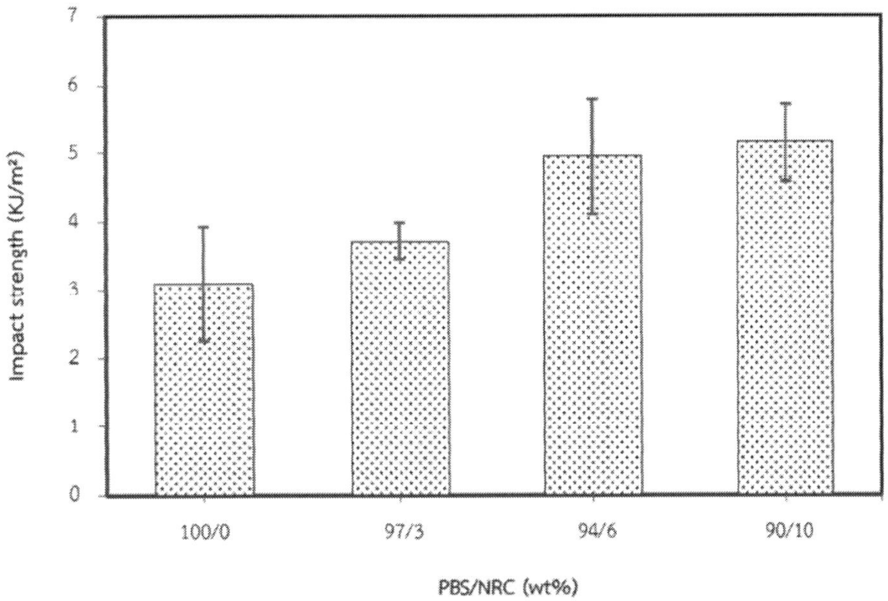


**Figure 1.** Impact strength of poly (butylene succinate)/natural rubber compound (PBS/NRC) blends

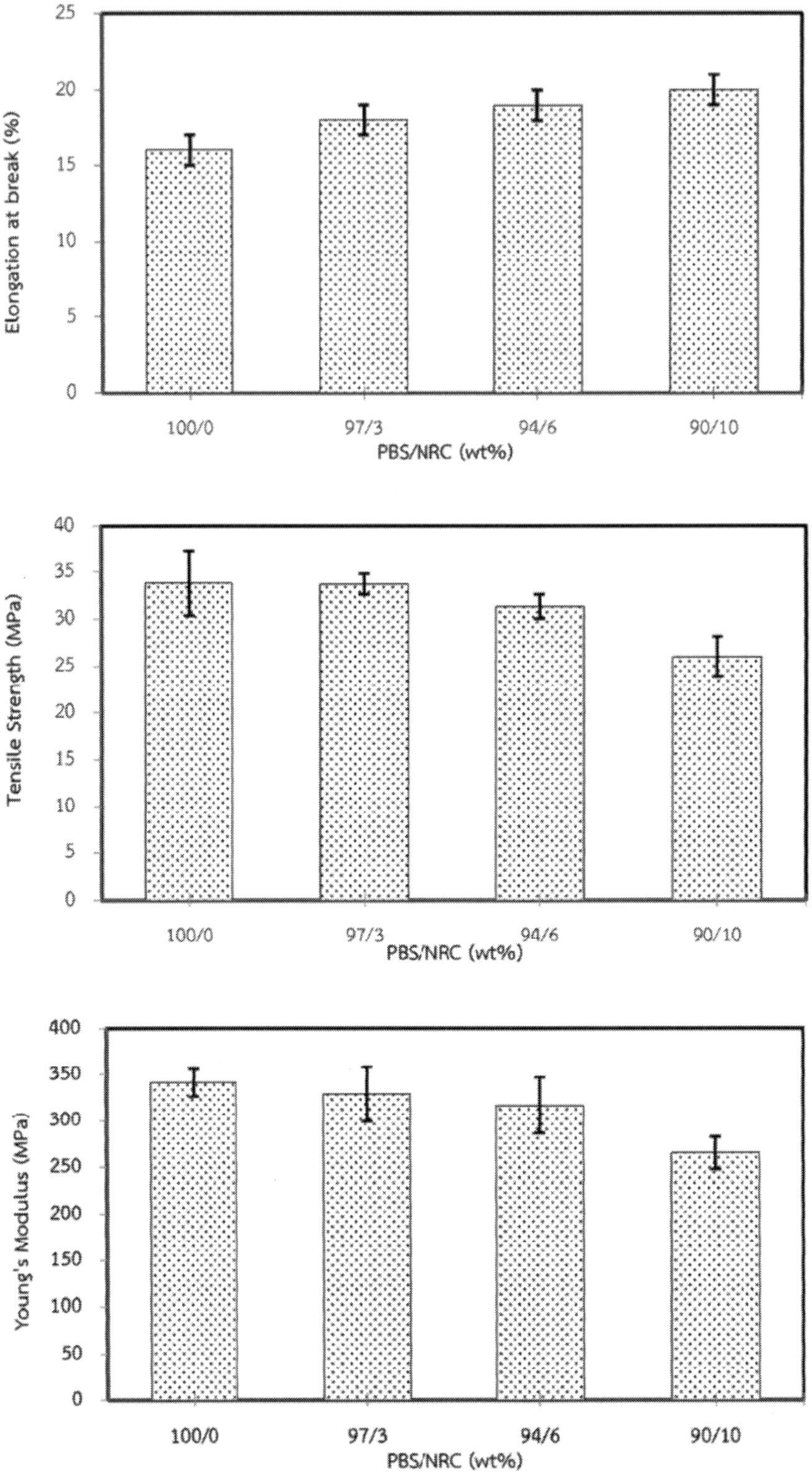


**Figure 2.** Tensile properties of PBS/NRC blends; (a) elongation at break, (b) tensile strength and (c) Young's modulus

*3.2 Morphological property*

Figure 3 illustrates the SEM micrographs of the fractured surface from impact test of PBS/NRC blends with various blend ratios. The result showed that the coarse surface of pure PBS could be observed as shown in Fig. 3(a). All blends illustrated dispersed NRC particles in the PBS matrix (Fig. 3(b)-(d)). The distribution of NRC particle size was approximately 2-5 μm as shown in Fig. 3(b)-(d). These morphologies were involved with the mechanical properties as previously discussed.

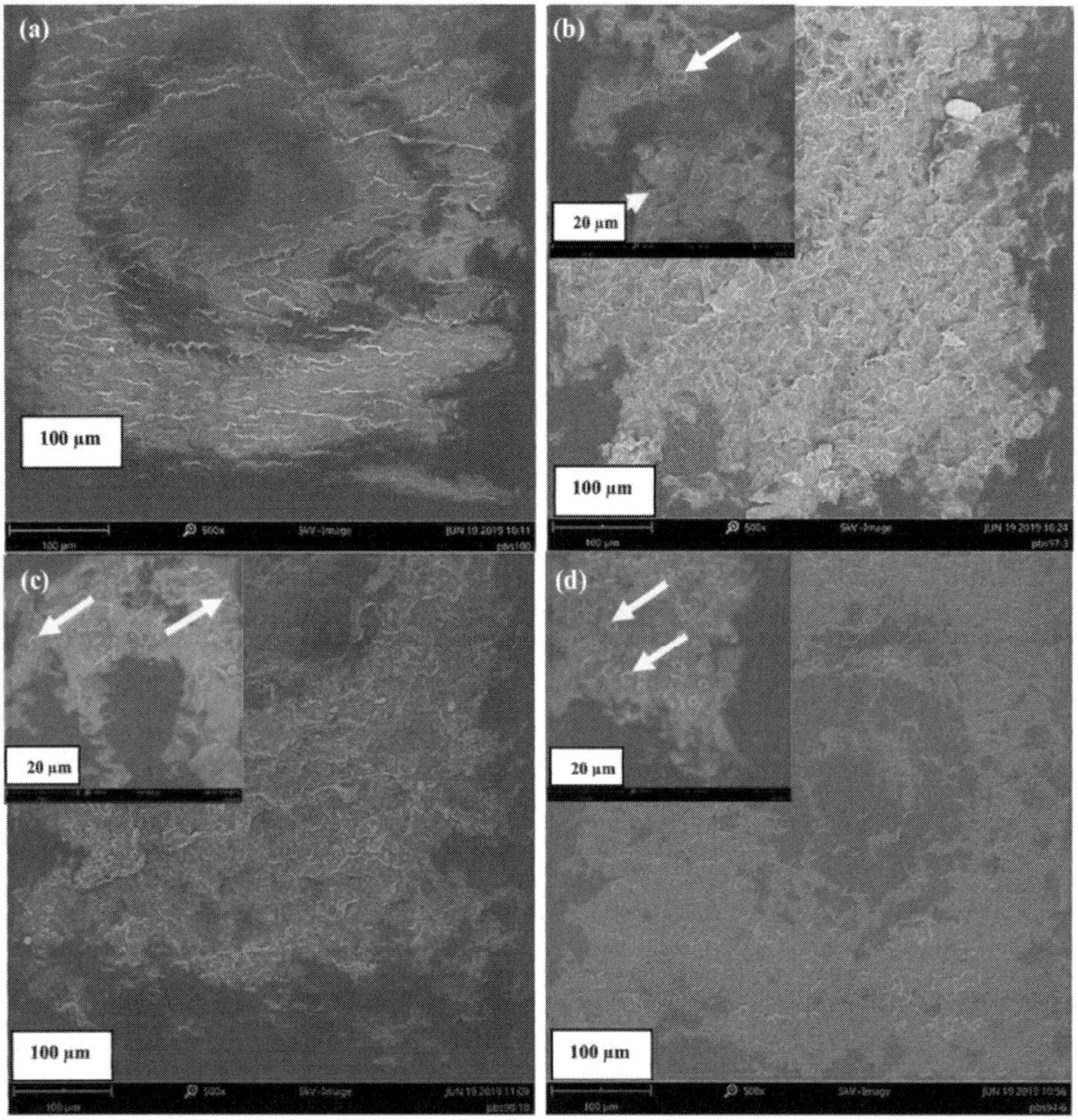


**Figure 3.** Morphology of PBS/NRC blends with a weight ratio of; (a) 100/0, (b) 97/3, (c) 94/6, and (d) 90/10. The arrows mark NRC particles.

### *3.3. Melt flow index*

Figure 4 shows the melt flow index (MFI) as a function of NRC content. The result showed that the MFI was decreased with an increase in the amount of NRC. This implies that the viscosity of the blend increased with an increase of NRC content.

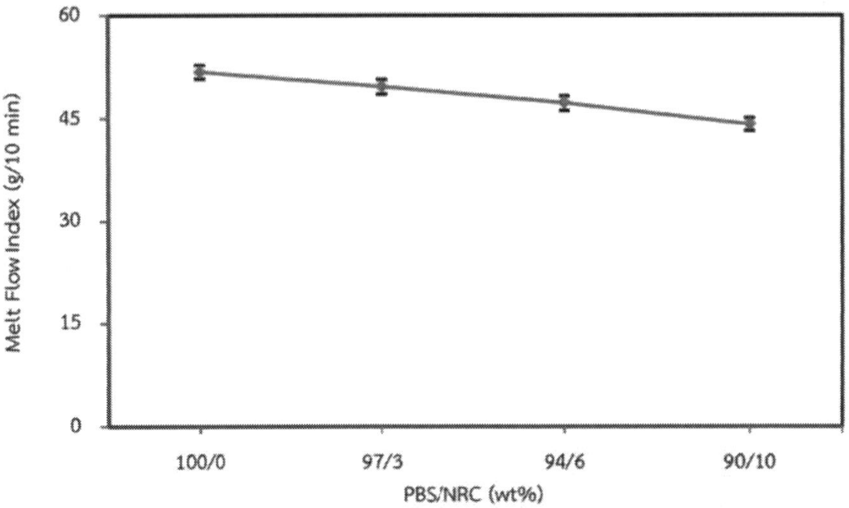


**Figure 4.** Melt Flow Index, MFI of PBS/NRC blends

## 4. Conclusions

PBS/NRC blends with various weight ratios of 100/0, 97/3, 94/6 and 90/10 were prepared via an internal mixer. The amount of NRC had an effect on the mechanical properties, melt flow index and morphology of the blends. The highest impact strength and percentage elongation at break were achieved at 10 wt% of NRC. The tensile strength and Young's modulus monotonically decreased with an increase in NRC content. In addition, the melt flow index of the blends decreased. Finally, NRC might be used to enhance toughness of other brittle polymer matrix or increase viscosity of other polymers.

## 5. References

[1] Maccaferri E, Mazzocchetti L, Benelli T, Brugo T M, Zucchelli A and Giorgini L 2020 *Mater. Design.* **186** 108210
[2] Pongtanayut K, Thongpin C and Santawitee O 2013 *Energy Procedia* **34** 888-897
[3] Dechatiwon Na-Ayutthaya W and Poompradub S 2014 *Macromol. Res.* **22** 686-692
[4] Faibunchan P, Nakaramontri Y, Chueangchayaphan W, Pichaiyut S, Kummerlöwe C, Vennemann N and Nakason C 2018 *J. Poly. Environ.* **26** 2867-2880
[5] Faibunchan P, Pichaiyut S, Chueangchayaphan W, Kummerlöwe C, Vennemann N and Nakason C 2019 *Poly. Adv. Technol.* **30** 1010-1026
[6] Umamaheswara Rao R, Venkatanarayana B, Suman K N S 2019 Mater. Today **18** 85-97
[7] Hemsri S, Thongpin C, Moradokpermpoon N, Niramon P and Suppaso M 2015 *Macromol. Symp.* **345** 145-154
[8] Huang Z, Qian L, Yin Q, Yu N, Liu T and Tian D 2018 *Polym.Test.* **66** 319-326
[9] Calabia B P, Ninoomiya F, Yagi H, Oishi A, Taguchi K, Kuioka M and Funabashi M 2013 *Polymers* **5** 128-141
[10] Asletha R, Kumaran M G and Thomas S 1999 Eur. Poly. J. **35** 253-271

**Acknowledgments**

This research project was supported by Rajamangala University of Technology Isan, Contract No. NKR2561REV033. The author would like to thank the staffs in the Faculty of Engineering and Architecture, and Institute of Research and Development for all kind support.

# Deterioration of Open-cell Aluminum Foam in Strong Sulfuric Acid

**Pat Sooksaen[1, 2]* and Penpisuth Thongyoug[3]**

[1] Department of Materials Science and Engineering, Faculty of Engineering and Industrial Technology, Silpakorn University, Nakhon Pathom 73000, Thailand
[2] Center of Excellence for Petroleum, Petrochemicals and Advanced Materials, Chulalongkorn University, Bangkok, 10330, Thailand
[3] Department of Industrial Engineering and Management, Faculty of Engineering and Industrial Technology, Silpakorn University, Nakhon Pathom 73000, Thailand

*E-mail: sooksaen_p@silpakorn.edu

**Abstract**. Open-cell aluminium foam was fabricated by pressure infiltration and salt leaching method. The processing conditions to obtain uniform interconnected pore structure aluminium foam were melting at 650°C for 1.5 h with the applied pressure of 1.5 bar. Aluminum foams were sectioned into thin pieces to investigate for deterioration behaviour in strong sulfuric acid for 30 days. Significant weight reduction of about 50% was observed when the specimen piece was submersed in 1.0 M $H_2SO_4$ in the first three days after which the weight reduction slightly increased to 60%. Al and Cu were the two main elements that had been leached out from the surface of aluminum foam in contact with the acid resulting in the formation of sulfates.

## 1. Introduction

Aluminum alloys are lightweight and have good mechanical properties as well as good environmental resistance from oxidation corrosion. They can withstand mild acid/base environments therefore are very useful in various industries such as packaging, automotive and heat exchangers. Aluminum foam possesses attractive characteristics when compared to the bulk solid. In the past, when the cast products contained pores, they were considered as defects but in recent years porous metals or metal foams have shown some excellent engineering applications and have been widely investigated. They can be applied as lightweight structures, sound absorbing materials, heat exchanger materials. They also possess high energy dissipation ability. Many fabrication techniques have been developed to produce porous structures and they can be fabricated as open-cell or closed-cell structures [1-4]. Open-cell metal foam has interconnected solid strut network. This structure allows easy fluid or gas flow within it and can be applied in heat exchanger and filter applications [5]. Li *et.al.* [6] investigated heat exchanging behavior in hot fluid in a complex network of metal foams and they found uneven fluid turbulence. The higher open porosity and hence the higher surface area of the thin solid struts resulted in better heat transfer. Hooman and Malayeri [7] mentioned that metal foam can be applied as filter in non-fully combusted engines. However, there must be concerns over the pressure drop, the blockage and also the corrosion from hot gases. Sergej *et al.* [8] investigated corrosion in hybrid composite sandwiched with aluminum foam. They said that it was important to consider for galvanic corrosion and electrolytic environment. In their experiment, the variables consisted of pore size and pore type of aluminum foam and also the tested environment in sprayed salt for 240 h.

The replication method is simple and cheap for fabricating aluminum foams with uniform open-cell structures. Replication method consists of infiltrating a leachable preform which is subsequently removed by dissolution in water or solvent. The effective material to be used as a replication preform is sodium chloride salt since it has a melting point at 800°C which is above the melting point of aluminum. It is non-toxic, chemically inert during melting and infiltration and also can be easily dissolved in water. Pore shape and pore size of open-cell aluminum foams can be controlled by the shape and size of the salt grains. In this current study, the effect of infiltration pressure was investigated in order to find suitable condition for fabricating uniform column aluminum foam. A commercial ADC12 aluminum alloy was investigated in this study since it had lower melting point and better fluidity than pure aluminum. If the aluminum foam was to be applied as a heat exchanger in severe acidic condition it would be interesting to study for the effect on strong acid corrosion. Therefore, deterioration study in sulfuric acid was investigated to see if there were any changes on the physical and chemical characteristics of an aluminum foam.

## 2. Experimental

This study followed the method developed by Goodall *et al.* [1] and the detailed procedures can be found from the work of Lara-Rodriguez *et al.* [3] and Corde *et al.* [4]. The difference from the related literatures was that in this study vacuum was not used during the melting process. It was evident that our technique was able to produce good and uniform column aluminum foams provided that the melting conditions were carefully controlled. Argon gas was used for the infiltration process of molten liquid aluminum in a tightly closed stainless steel mold. The aluminum alloy utilized in this work was commercial ADC12 in which the composition was near to eutectic Al-Si system with low melting point and low viscosity melt. The alloy was supplied by Smooth International Co. Ltd (Thailand) with a chemical composition: 85.84 Al, 10.03 Si, 1.81 Cu, 0.86 Fe, 0.90 Zn, 0.17 Mg, 0.20 Mn and others.

Australian salt was used to make a preform in this process. It had the size range of 10-17.5 mm with angular morphology and the average size was 10 mm. The dimensions of the stainless steel cylindrical mold were 5 cm inner diameter and 15 cm height. Melting temperature at 650°C and melting time of 2 h were sufficient to give good infiltration of the melt through the salt gaps. The only controlling factors in this study were the applied pressure which was varied between 0.5-2.0 bar and the infiltration time between 10-30 seconds. After obtaining suitable conditions for producing uniform aluminum foam sample. The samples were cut into thin pieces with 2 cm thickness and 5 cm diameter for the subsequent deterioration test in strong sulfuric acid of concentrations 0.1, 0.5 and 1.0 M. The specimens were immersed in closed-lid containers, each with 150 ml sulfuric acid in it. One specimen piece was randomly selected for each acid concentration for the test. The weight change was measured every 3 days for total period of 30 days using a 4-digit balance. The physical appearances were observed by a digital camera and microscopic as well as chemical compositions of the important phases were investigated by FE-SEM Tescan Mira3 equipped with an EDX spectroscopy.

## 3. Results and discussion

The ADC12 alloy was a low melting point grade and hence melting at 650°C for 2 h resulted in low viscosity melt. The applied pressure of 2 bar was too high to penetrate the melt at the extent to cause large pores or holes on the top part of the specimen after the salt grains had been leached away (figure 1, in the red circled area). A uniform aluminum foam with reproducibility should have be prepared at pressure lower than 2 bar. This study found that good specimens had to be prepared using pressure of 1.0-1.5 bar. Bulk density and % open porosity of aluminum foams prepared by 1.5 bar pressure were averaged from 10 pieces. The values were 0.82 g.cm$^{-3}$ and 69.6%, respectively.

The pressure lower than 0.5 bar was not have enough to infiltrate the salt preform. Figure 2a showed ADC12 bar and its aluminum foam product with a shiny surface. Figure 2c showed an SEM micrograph on the 1 μm polished and etched surface. Phase morphologies consisted mainly of α-Al and eutectic Si. These were labelled as "A" and "B" respectively. Other minor phases or solid solution may have also existed in the microstructure.

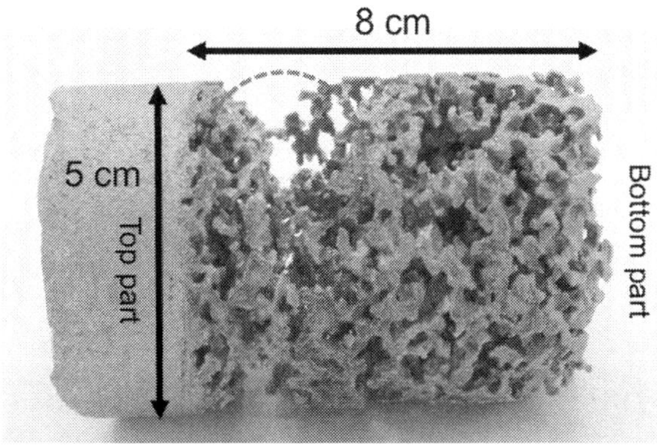


Figure 1 Aluminum foam produced with 2.0 bar argon pressure

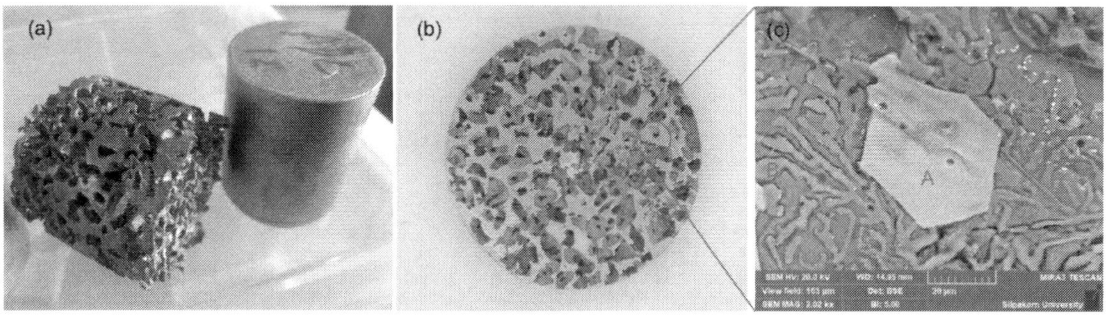


Figure 2 (a) ADC12 bar and its aluminum foam product (b) Cross-section with 1 μm polished surface (c) Phase morphologies by a scanning electron microscope

For deterioration test in sulfuric acid, it can be seen in figure 3 that the % weight reduction was almost not observed when the specimen pieces were immersed in 0.1 M $H_2SO_4$. When the concentration was raised to 0.5 M, the weight reduction of around 10% was seen in the first 3 days. After that, the weight remained quite constant with submerging time. The inset micrograph showed rather rough surface and some surface cracks within the thin solid struts of aluminum alloy. This was different from the starting specimen piece where the surface was smooth. There could be corrosion product/s leached out from the specimen at this concentration. At 1.0 M concentration, it was evident that significant weight reduction occurred. There was fast rate of weight reduction to about 50% at the first 3 days then the weight steadily change in less extent up to about 60% at 30 days. The low corrosion from day 3 to 30 may be related to the formation of oxidized passive film competing with the reaction with strong sulfuric acid.

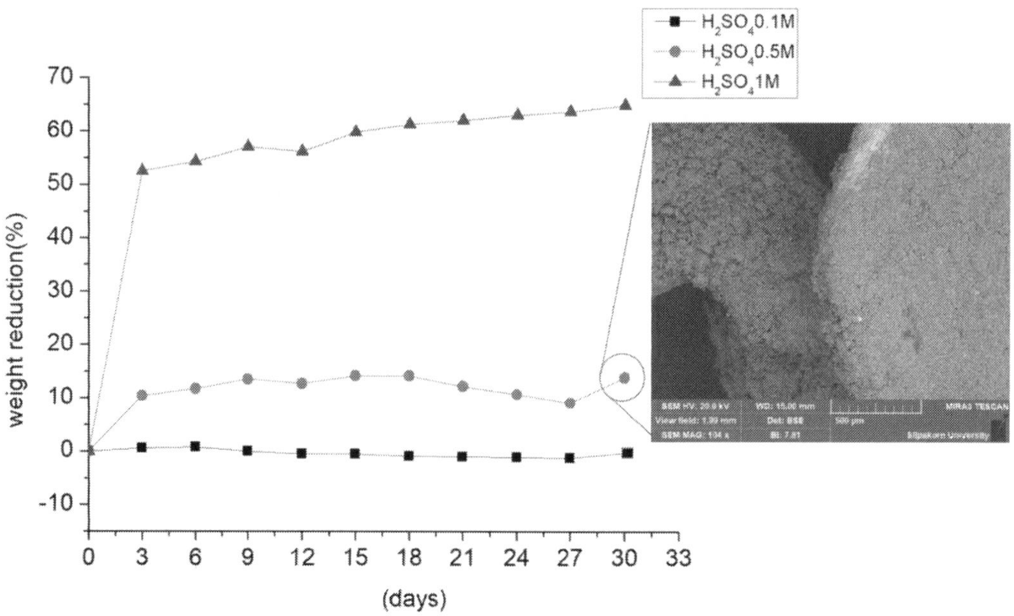


Figure 3 Weight change when immersed the specimen piece in 0.1, 0.5, 1.0 M sulfuric acid

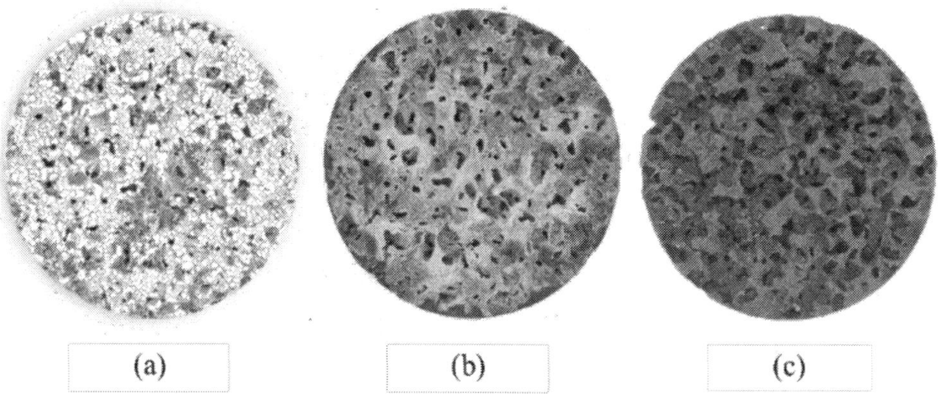


Figure 4 Physical appearance in different $H_2SO_4$ concentrations (a) 0.1 M (b) 0.5 M (c) 1.0 M at 30 days

The visual inspection of the specimens in different $H_2SO_4$ concentrations was show in figure 4. The specimens became dull and darker in colour at higher acid concentration. The loss of shiny surface could be mainly related to aluminum element leached out. The formation of oxidized layer could have also change the reflectivity of the surface. Higher magnification images in figure 5, revealed that the specimen in 1.0 M $H_2SO_4$ had rougher structure and more solid parts were leached out from the thin solid struts. Chemical analysis by small area scan using EDX had been focussed on Si and Cu rich phases in which the two elements were the two most appeared elements in the alloy. Figure 6 showed the difference in the relative X-ray counts in specimens from 0.1 M and 1.0 M $H_2SO_4$ all obtained from 30-day specimens. It was evident that Si content increased at higher acid concentration. This roughly

confirmed that aluminum element and oxide or hydroxide of aluminum were leached out from the specimen. It could be dissolved as ions in the liquid media or precipitated out to form aluminum sulphate, $Al_2(SO_4)_3$. The suggested chemical reaction that had occurred over time period would be:
$2Al\ (s) + 3H_2SO_4\ (aq) \rightarrow Al_2(SO_4)_3(s) + 3H_2\ (g)$

In the Cu-rich area, it was evident that higher acid concentration reacted with Cu significantly leaving less Cu appeared on the specimen surface. This had been confirmed with different analysed areas, although the data in figure 6 were not from the same area. In the case of copper, the corrosion product formed could be solid $Cu(SO)_4$, CuO or copper ions in acid solution. As far as the bright spots (or white particles) in figure 5c was concerned, it could be related to the copper-rich phase which had been leached out from the alloy and still adhered within the surface of solid struts.

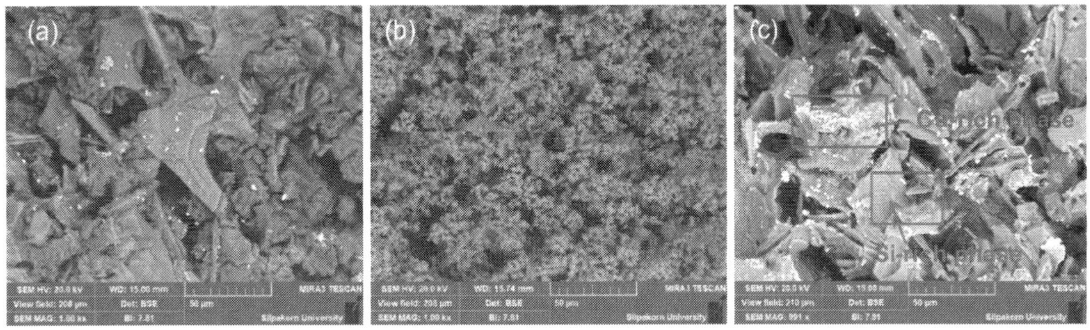


Figure 5 Micrographs of the specimens in different $H_2SO_4$ concentrations (a) 0.1 M (b) 0.5 M (c) 1.0 M at 30 days

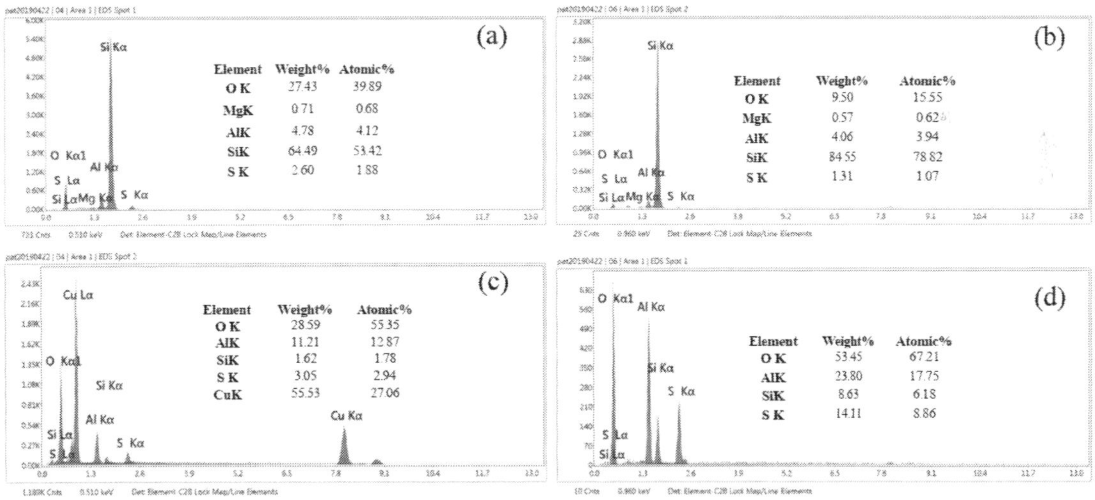


Figure 6 EDX analysis of area with high Si content (a) 0.1 M $H_2SO_4$ (b) 1.0 M $H_2SO_4$ and area with high Cu content (c) (a) 0.1 M $H_2SO_4$ (b) 1.0 M $H_2SO_4$, all from 30-day specimens

After 30 days in 1.0 M $H_2SO_4$, the solvent in the beaker was dried to see if there was any solid precipitated at the bottom. The dried solid was crushed into a fine powder and the morphology under microscope was shown in figure 7. It shows angular morphology which is a characteristic of solid salts which could be related to aluminum sulphate and copper sulphate although the actual phases was not confirmed due to low crystallinity. The associated phase analysis by XRD using Shimadzu D-6100 found unclear diffraction peaks.

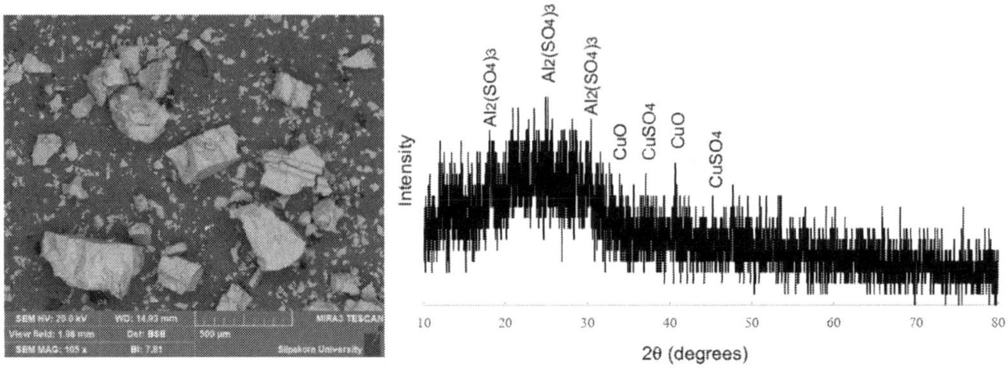


Figure 7 Corrosion products after drying and the XRD for phase analysis

## 4. Conclusion

Open-cell aluminum foams were fabricated in-house design system. The process involved pressure infiltration and salt leaching without the use of vacuum as suggested by other literatures. The uniform interconnected pore structure aluminum foams can be prepared at pressures 1.0-1.5 bar. Deterioration behavior in strong sulfuric acid showed significant weight reduction of about 50% when immersed in 1.0 M concentration. EDX analysis revealed that Al and Cu were the main elements leached out from aluminum foam that formed sulfate salts as the corrosion products.

## 5. Acknowledgements

The authors would like to thank department of Materials Science and Engineering, Silpakorn University, Center of Excellence for Petroleum, Petrochemicals and Advanced Materials, Chulalongkorn University and Smooth International Co., Ltd. for financial support, research equipment and raw materials for this research.

## References

[1] Goodall, R and Mortensen A 2007 Mater. Sci. Eng. A. 465 124
[2] Goodall R and Mortensen A 2014 Elsevier Chapter 24 2399–2595
[3] Lara-Rodriguez G A, Figueroa I A, Suarez M A, Novelo-Peralta O and Alfonso I 2017 Mater. Process. Technol. 243 16
[4] Conde Y, Despois J F, Goodall R, Marmottant A, Salvo L, San Marchi C and Mortensen A 2006 Adv. Eng. Mater. 8 795
[5] Fadhilah S A, Iman A A, Mostafa O and Kamel H 2018 Experimental Thermal and Fluid 99 117
[6] Li Yanzhe, Wang S and Zhao Y 2018 Experimental Thermal and Fluid Science 97 408
[7] Hooman K and Malayeri M 2016 Energy Conversion and Management 1 117 475
[8] Sergej I, Peter R and Kay A W 2018 Composite Structures 202 142

IOP Conf. Series: Materials Science and Engineering **965** (2020) 012028 doi:10.1088/1757-899X/965/1/012028

# The temperature-dependent structural and optical properties of $SrAl_2O_4$-based phosphor

**V Loryuenyong, J Kornsawat, W Nakhowong, W Klinjan, and A Buasri**

Department of Materials Science and Engineering, Faculty of Engineering and Industrial Technology, Silpakorn University, Nakhon Pathom 73000, Thailand

**Abstract**. This research studied the thermal stability of $SrAl_2O_4$ : $Eu^{2+}$, $Dy^{3+}$ phosphor for the fabrication of luminescent glasses. The temperature-dependent structural and optical properties of the phosphor were investigated. The results showed that $SrAl_2O_4$ : $Eu^{2+}$, $Dy^{3+}$ phosphor could emit the green light when excited by ultraviolet (UV) light. It was also found that annealing could induce further growth of $SrAl_2O_4$ crystals, which gave rise to more intense green-light emission. However, at high annealing temperature, the luminescence attained a maximum and started decreasing. An intense luminescence was observed for the $SrAl_2O_4$ : $Eu^{2+}$, $Dy^{3+}$ phosphor annealed at 750°C. The phosphor degradation at temperatures above 750°C could be because of the formation of secondary phases and the oxidation of the rare-earth atoms that were doped in phosphor. The results showed that, with proper fabrication processes, phosphors could be considered as promising materials for various applications.

## 1. Introduction

Phosphors are materials that can emit an intense and long-lasting light when absorbing high external energy. Thus, they are widely used in many applications such as lighting and display devices. In the fabrication of phosphors, rare-earth and transition-metal ions are typically doped in wide-bandgap host compounds such as silicates, phosphates, aluminates, and borates. $SrAl_2O_4$ is one of the promising host materials for long persistent phosphors.[1-2] It is usually co-doped with europium ion ($Eu^{2+}$) and dysprosium ion ($Dy^{3+}$) to exhibit a prolonged afterglow performance, which could be more than ten times compared to other phosphors.[2-3] However, many phosphors including $SrAl_2O_4$-based phosphors tend to lose their luminescence efficiency at high temperatures, which is undesirable for many applications.[4]

Glass-based phosphors are materials containing phosphor powders in glass matrices. These materials have been recently commercialized for their advantages over conventional polymer-based phosphors used in optical devices such as light emitting diodes (LEDs).[5-6] They are prepared from a mixture of glass powders (or frits) and phosphor powders. The typical fabrication processes, however, involve the annealing or sintering at high temperatures to obtain the desired products. As a consequence, various approaches have been attempted to overcome the thermal degradation and to improve the stability of phosphors.

In this study, the temperature-dependent properties of green-emitting $SrAl_2O_4$-based phosphor powders were studied. The samples were characterized using X-ray diffraction (XRD), photoluminescence (PL) spectroscopy, Raman spectroscopy, and scanning electron microscopy (SEM).

## 2. Experimental Procedure

The commercial green-emitting $SrAl_2O_4$ : $Eu^{2+}$, $Dy^{3+}$ phosphor powders (45-55 μm, Fluorescent glow powder, Amhere Intertrade, Ltd.) were used as received. The phosphor powders were first dried in an oven at 120°C for 24 h to evaporate the moisture, and, then, annealed from 700°C to 900°C for 30 min in air to study the temperature stability.

After annealing, the powders were characterized by XRD on a Shimadzu XRD-6100 diffractometer with CuKα radiation (λ=0.154056 nm, 40 kV and 30 mA). The morphology of the samples was examined by scanning electron microscopy (TM3030, Hitachi). Raman measurements were performed at room temperature by The Senterra Raman spectroscopy (Bruker). The optical properties were measured on a photoluminescence spectrometer (AvaSpec-2048TEC, Avantes).

## 3. Results and Discussion

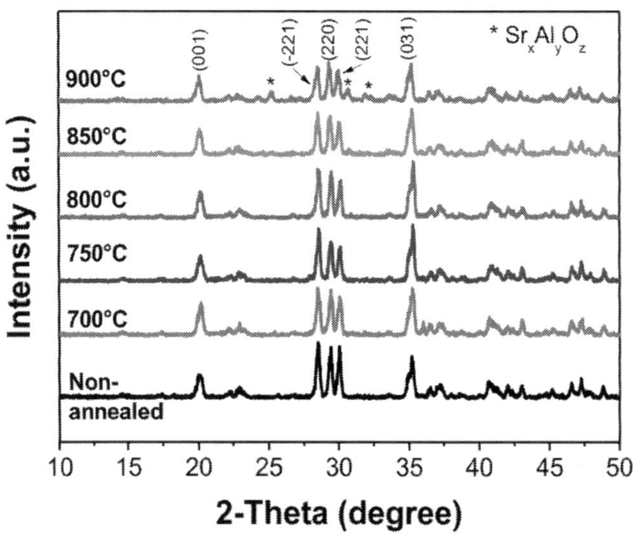


**Figure 1**. XRD patterns of $SrAl_2O_4$-based phosphor powders annealed at various temperatures from 700°C to 900°C for 30 min.

The phase purities and the crystal structures of the phosphor powders annealed at different temperatures (700°C, 750°C, 800°C, 850°C, and 900°C) were characterized by XRD, and the results were shown in figure 1. All of the samples exhibited monoclinic $SrAl_2O_4$ crystal structure as a dominant phase, which could be characterized by five main peaks corresponding to (001), (-221), (220), (221), and (031) planes (JCPDS: 34-0379). The presence of the hexagonal polymorph of $SrAl_2O_4$ was also observed in annealed samples, as indicated by the different intensity ratios of the main peaks. In general, $SrAl_2O_4$ has monoclinic and hexagonal polymorphs with the phase transition occurring at about 650°C.[7] These two polymorphs have the most intense XRD peaks overlapped, and, therefore, it could be difficult to distinguish the two phases. Nevertheless, the result revealed that the presence of dopants had no effect on the crystal structure of $SrAl_2O_4$. This was because the ionic radius of the common $Eu^{2+}$ dopant in $SrAl_2O_4$-based phosphor was similar to that of $Sr^{2+}$ ion, and as a result $Eu^{2+}$ ion tended to be substituted at the $Sr^{2+}$ sites.[8]

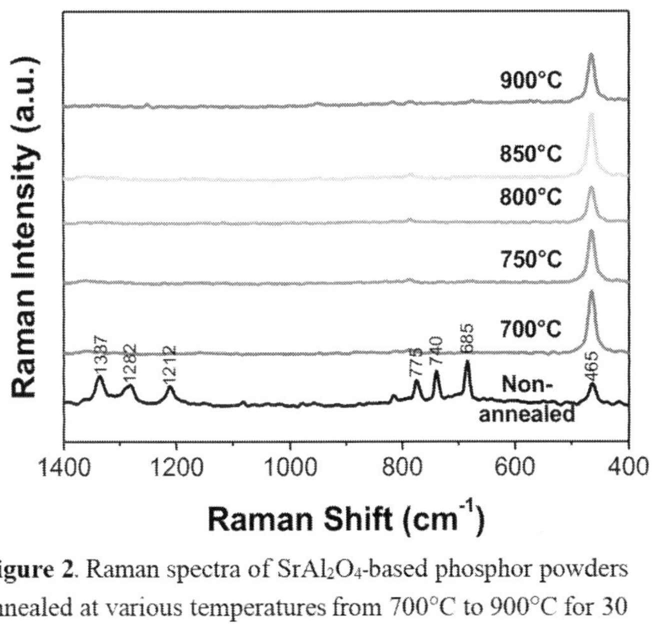


**Figure 2**. Raman spectra of $SrAl_2O_4$-based phosphor powders annealed at various temperatures from 700°C to 900°C for 30 min.

With increasing annealing temperatures, the particle size and the crystallinity of the phosphor were found to increase and reached the maximum at 750°C. When the annealing temperatures was raised to 800°C, it could be seen that the XRD patterns contained additional peaks other than those of $SrAl_2O_4$, indicating the formation of a mixed oxide phase. The XRD peaks of secondary phases ($Sr_xAl_yO_z$) became sharper with an increase in annealing temperatures, indicating the higher crystallinity. Previous studies establish that the luminescent efficiency is a function of the $SrAl_2O_4$ phase, the degree of crystallinity, as well as the existence of secondary phases.[7]

Figure 2 shows Raman spectra of $SrAl_2O_4$-based phosphor powders at room temperature. For non-annealed sample, the spectrum exhibited 7 peaks around 400-1400 $cm^{-1}$. The bands at a frequency higher than 600 $cm^{-1}$ could be attributed to the Al-O stretching vibrations, while the peak at 465 $cm^{-1}$ to the bending vibration of the O-Al-O groups in corner-sharing tetrahedral.[8] However, for the annealed samples, only the peak at 465 $cm^{-1}$ was clearly observed, which was the characteristic peak of the $SrAl_2O_4$ structure. On the contrary, the other peaks depended on the defects or impurities in the crystal lattices.[8-9]

Surface morphology of samples was analyzed using SEM (figure 3). The average particle size of $SrAl_2O_4$-based phosphor powders before annealing was about 128 micrometers. With increasing annealing temperatures, the particle sizes were increased. In addition, the morphologies of the phosphor particles changed from sharp surfaces to irregular and rough surfaces. This could be associated with the particle agglomeration and secondary-phase formation at high annealing temperatures.

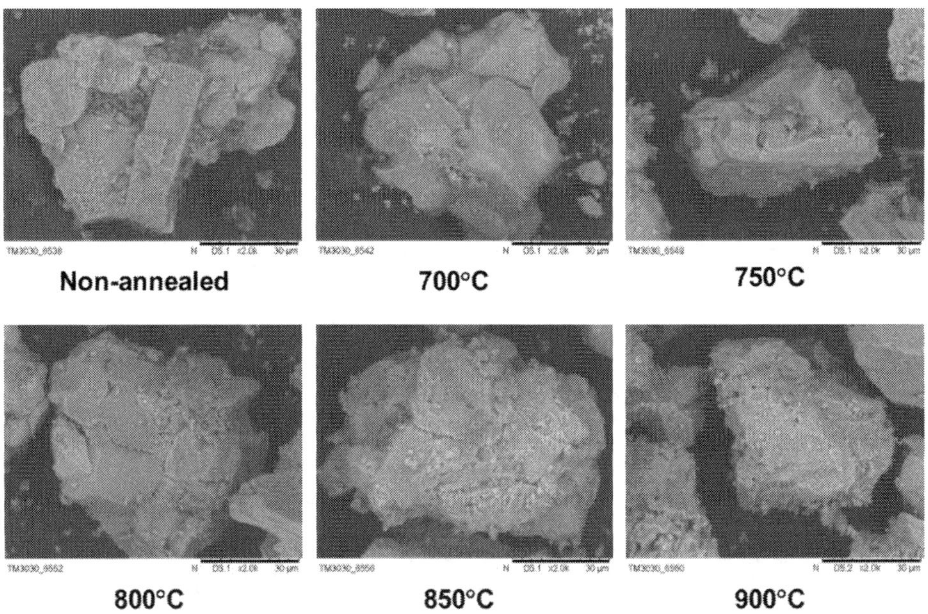


**Figure 3.** SEM micrographs of $SrAl_2O_4$-based phosphor powders annealed at various temperatures from 700°C to 900°C for 30 min.

Figure 4 shows the PL spectra of phosphor powders with the excitation wavelength of 345 nm. The PL spectrum of non-annealed powders showed a maximum at 520 nm, which corresponded to the green-light emission band. This luminescence could be ascribed to the typical $4f^6 5d^1$-$4f^7$ transition of $Eu^{2+}$ ion in the $SrAl_2O_4$. However, the spectrum shape was slightly asymmetric due to the emission of fluorescent pigment, which was contained in the phosphor. For annealed samples, PL intensity obviously increased with increasing annealing temperatures and reached the maximum intensity at around 750°C. Further increase in annealing temperatures led to decreasing luminescent intensity due to the phase transformation and the oxidation of $Eu^{2+}$ ion. In addition, two emission peaks at 500 nm and 524 nm were also observed after annealing. This could be explained in terms of a change in the oxidized $Eu^{2+}$ environment. Previous researches have reported that the allowed 4f-5d transitions are strongly influenced by the chemical environment of $Eu^{2+}$ ion, and, as a result, the emission of $Eu^{2+}$ ion can vary from ultraviolet to red for the aluminates.[9-10] It has been reported that the $Eu^{2+}$ emission shifts to longer wavelength with increasing Sr/Al ratio.[10]

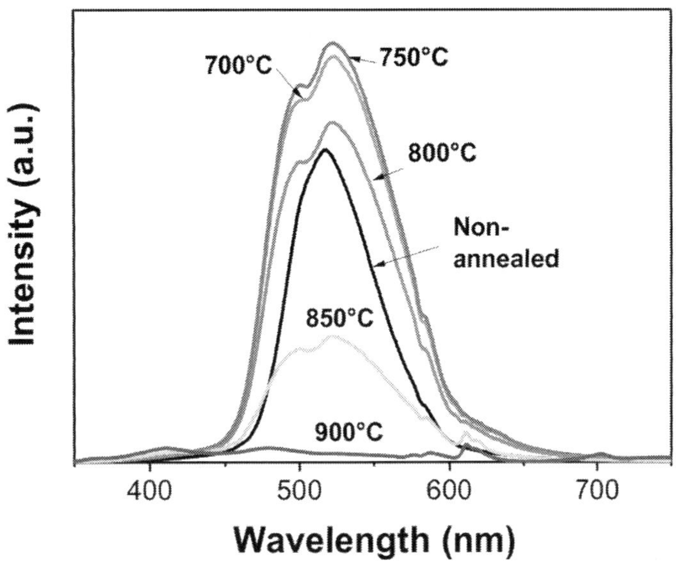


**Figure 4**. PL spectra of $SrAl_2O_4$-based phosphor powders annealed at various temperatures

Figure 5 shows the physical appearance and the glowing effect of $SrAl_2O_4$-based phosphor powders. The samples were exposed to the UV light for 15 min to enable the glowing effect in the dark. It could be clearly seen that the phosphor powders exhibited green-light emission, and the intensity of the emitted light was strongly a function of annealing temperatures. The optimal performance was obtained with the phosphors annealed at 750°C, and the greenish glow was completely lost at 900°C. The main reason for the drastic reduction of luminescent intensity at high annealing temperatures was the thermal oxidation of $Eu^{2+}$.

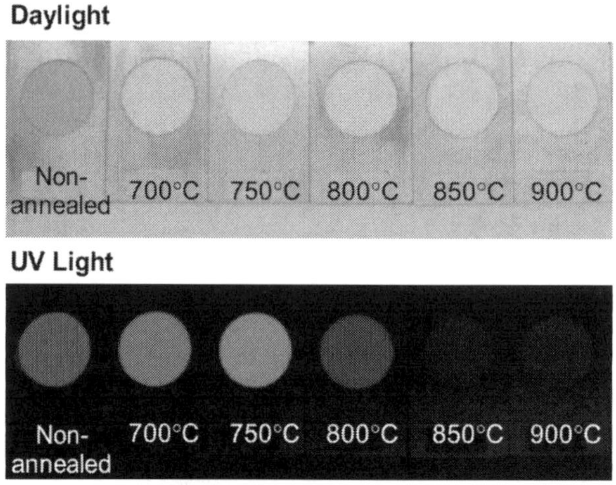


**Figure 5**. $SrAl_2O_4$-based phosphor powders annealed at various temperatures in daylight and under UV light

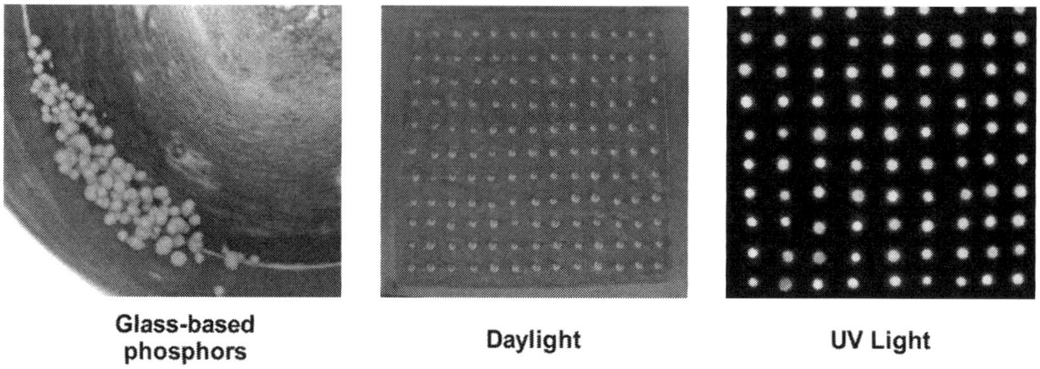


**Figure 6**. (Left) The fabrication of glass-based phosphors using a granular-molding machine, (Middle, Right) Luminescent cements in Daylight and under UV light.

The application of the $SrAl_2O_4$-based phosphor powders as glass-based phosphors and luminescent aggregates in cement is shown in figure 6. Luminescent aggregates were produced by mixing glass frits ($SiO_2$-$CaO$-$Na_2O$-$Al_2O_3$-$MgO$) and $SrAl_2O_4$-based phosphor powders (15% by weight) in granular-molding machine. The resultant spherical-like aggregates were then annealed at 750°C and decorated onto the cement block. The products were found to glow after UV exposure and had good chemical and thermal stabilities.

## 4. Conclusions

In this study, the temperature-dependent properties and the optimum temperature of commercial $SrAl_2O_4$-based phosphor powders were investigated. The temperature-dependent luminescence of the samples was largely determined by their crystallinity, secondary phases, as well as thermal oxidation of doping rare-earth atoms. The powders annealed at 750°C was found to show the most-efficient green luminescence and could be utilized to produce glass-based phosphors, which required high-temperature fabrication processes.

## 5. References

[1] Kim H, Yang J-F, Sekino T and Lee S-W 2009 *Mater. Sci. Forum* **620-622** 473

[2] Hu X, Yang H, Guo T, Shu D, Shan W, Li G and Guo G 2018 *Ceram. Inter.* **44** 7535

[3] Shafia E, Bodaghi M, Esposito S and Aghaei A 2014 *Ceram. Inter.* **40** 4697

[4] Kamiyanagi Y, Kitaura M and Kaneyoshi M 2007 *J. Lumin.* **122-123** 509

[5] Chen D, Yuan S, Li X and Xu W 2017 *RSC Adv.* **7** 36168

[6] Yie H, Kim S, Kim Y and Kim H 2017 *J. Non-Cryst. Solids* **463** 19

[7] Rojas-Hernandez RE, Rubio-Marcos F, Serrano A, Rakhmatullin A, Bessada C, Fernandez JF 2018 *RSC Adv.* **8** 28918-28927

[8] Ayvacıklı M, Ege A, Yerci S and Can N 2011 *J. Lumin.* **131** 2432

[9] Bite I, Krieke G, Zolotarjovs A, Laganovska K, Liepina V, Smits K, Auzins K, Grigorjeva L, Millers DK and Skuja L 2018 *Mater. Des.* **160** 794

[10] Dutczak D, Jüstel T, Rondac C and Meijerink A 2015 *Phys. Chem. Chem. Phys.* **17** 15236

## Acknowledgments

The authors wish to thank Department of Materials Science and Engineering, Faculty of Engineering and Industrial Technology, Silpakorn University for encouraging this investigation.

IOP Conf. Series: Materials Science and Engineering **965** (2020) 012029 doi:10.1088/1757-899X/965/1/012029

# Kinetic study for the co-pyrolysis of water hyacinth biomass with waste polystyrene

**C Phiwchaaum[1], B Ksapabutr[1,2,*], N Chaiyut[1] and M Panapoy[1,2]**

[1]Department of Materials Science and Engineering, Faculty of Engineering and Industrial Technology, Silpakorn University, Nakhon Pathom, 73000, Thailand
[2]Center of Excellence on Petrochemical and Materials Technology, Chulalongkorn University, Bangkok 10330, Thailand

*Corresponding author: kbussarin@gmail.com, ksapabutr_b@su.ac.th

**Abstract**. Co-pyrolysis behavior of water hyacinth (WHC) and waste PS using thermogravimetric analysis to determine kinetic parameters was investigated. TG and DTG results indicated that the pyrolysis of WHC and mixture of WHC and waste PS could be divided into three stages. It was found that co-pyrolysis of WHC and waste PS delayed the pyrolysis process with a wider decomposition temperature range, when compared with the pyrolysis of PS. The results compared with the pyrolysis of individual components revealed a positive synergistic interaction between WHC and waste PS according to the difference in weight loss. Coats-Redfern method was applied in the second stage to evaluate the activation energy ($E_a$) and pre-exponential factor (A) from the data of weight loss using three major reaction mechanisms. Two conversion temperature zones, zone I of 240-400 °C and zone II of 400-530 °C, were identified and a best-fit model describing the co-pyrolysis behavior of mixture was achieved. The addition of waste PS to WHC led to a slight decrease in the activation energy in the first zone and then the activation energy significantly reduced in the second zone. Chemical reaction and diffusion-controlled reaction mechanisms could be kinetically examined and enabled good explanation of the co-pyrolysis behavior of the mixture of WHC and waste PS as well as individual pyrolysis.

## 1. Introduction

The solid waste management and energy crisis are of critical concern and require further attention. Presently, plastics are major contributors in solid waste and are found to be one major pollutant to the environment. Direct combustion of post-consumer plastics is not encouraged because of the formation of toxic pollutants. Therefore, it becomes a great challenge to deal with the large amount of solid waste in an environmentally friendly and pollution-free approach. Waste plastics can be converted into energy to help address issues on both waste disposal and energy needs [1-2]. Among all the renewable energy sources, biomass wastes are considered as one of the attractive sources with the advantages of low greenhouse gas emission and carbon neutral, biodegradable, renewable and huge reserves [3-4]. Biomass conversion to energy is classified into thermochemical and biochemical routes. In the biochemical process, biomass is decomposed into lower molecular-weight compound using enzyme or bacteria as catalysts, requiring a long time for conversion of biomass into value added products. The thermochemical conversion technologies include pyrolysis, combustion, gasification, cofiring, and liquefaction. In the thermochemical process, biomass is decomposed into smaller molecular-weight compounds using heat [5]. Among them, pyrolysis has the potential to decompose biomass very quickly in the absence of oxygen into solid biochar, liquid fuel (known as bio-oil or pyrolysis oil), and gas fuel. However, bio-oil obtained from the pyrolysis of biomass has high oxygen content (~40 wt%) because of high oxygen content contained in the original biomass. Mostly, high oxygen content results in a fuel with low heating value, corrosiveness and thermal instability as well as limits the widespread application of the bio-oil [6]. Plastics with high hydrogen contents (polyethylene, polypropylene and polystyrene) could act as hydrogen donor to the biomass during pyrolysis process which leads to an increase and improvement in yield and quality of the produced oil [1]. Hence, co-pyrolysis of biomass and plastics might be a promising alternative for bio-oil upgrading and sustainable usage of improved bio-fuels. Studies have shown that co-pyrolysis of biomass and plastics compounds could increase bio-oil quantity and quality compared with pyrolysis of separate components [7-9]. Kinetic analysis of biomass and plastic waste is necessary before the conversion of biomass and plastic wastes into fuel, which provides information for the mathematical modeling simplification, process parameter optimization, and design of thermochemical equipment (e.g. pyrolysis reactor). The kinetic parameters include pre-exponential factor (A), activation energy (Ea), and order of the reaction (n). Thermogravimetric (TGA) analysis is the simplest analytical tool to investigate co-pyrolysis characteristics and determine kinetic parameters of organic materials or different types of thermochemical process. For the commercialization of the co-pyrolysis process in bio-oil production, an important further research is needed for better understanding of the reaction kinetics. The objective of current work is to investigate the thermal decomposition behavior, kinetics, pyrolysis of individual wastes, and co-pyrolysis of water hyacinth and waste polystyrene using TGA at different heating rates. Model-fitting methods were applied to evaluate kinetic parameters and find the most appropriate kinetic reaction mechanism using Coats-Redfern model.

## 2. Experimental

### *2.1. Materials and sample preparation*

The biomass sample used in this study is water hyacinth (WHC) and it was collected from a canal located in Kanchanaburi, Thailand. The sample was washed with distilled water to remove impurities and then dried at 80 °C for 72 h. Dried water hyacinth was crushed and sieved to obtain particle size ranging from 75 to 100 μm as well as dried before experiments. Waste polystyrene (oriented PS (OPS)) was collected from several food service locations in Sanam Chandra Palace Campus, Silpakorn University, Thailand. The PS waste was washed and dried, followed by cutting into small pieces with approximately 1 $mm^2$ and using in the thermal pyrolysis reaction.

### *2.2. Experimental technique*

Co-pyrolysis of water hyacinth with waste PS was examined using TGA (Mettler Toledo TGA/DSC1 STARe Thermogravimetric Analysis System) with horizontal balance, which was equipped for simultaneous thermogravimetric analysis. All the tests in this study were run under nitrogen carrier gas flowed at a constant rate of 50 mL/min. The samples were placed in an alumina crucible forming a small fixed bed with an initial mass of approximately 6 mg at mixing ratios (WHC:PS (w/w)) of 100:0, 90:10, 0:100. The samples were heated from 50 °C to 1000 °C and sets of experiments at different heating rates of 10, 15, 25 and 50 °C/min were carried out for every sample. The repeated data obtained from TGA experiment were used for kinetic parameter analysis.

### *2.3. Kinetic study*

The conversion degree ($\alpha$) reflects the thermal decomposition and is defined by the following equation [10]:

$$\alpha = \frac{m_o - m_t}{m_o - m_f} \tag{1}$$

where $m_o$, $m_t$, and $m_f$ are the initial, instantaneous and final masses during thermal decomposition, respectively. The rate of conversion with respect to time ($d\alpha/dt$) for a heterogeneous solid-state thermal decomposition can be expressed as:

$$\frac{d\alpha}{dt} = k(T)f(\alpha) \tag{2}$$

where $f(\alpha)$ is the mechanism function and is only a function of conversion. $T$ and $k$ are the absolute temperature and rate coefficient, respectively. According to the Arrhenius equation, the rate constant $k(T)$ can be established as:

$$k(T) = Ae^{\left(\frac{-E_a}{RT}\right)} \tag{3}$$

where $A$, $E_a$ and $R$ are the pre-exponential factor, apparent activation energy and universal gas constant (8.314 J/mol.K), respectively.
For constant heating rates $\beta = \frac{dT}{dt}$, therefore equations (2) and (3) can be given in the combined form [11]:

$$\frac{d\alpha}{dT} = \frac{A}{\beta}e^{\left(\frac{-E_a}{RT}\right)}f(\alpha) \tag{4}$$

The integrated form for $f(\alpha)$ in equation (4) is generally stated in the following equation:

$$g(\alpha) = \int_0^{\alpha}\frac{d\alpha}{f(\alpha)} = \frac{A}{\beta}\int_{T_0}^{T}e^{\left(\frac{-E_a}{RT}\right)}dT \tag{5}$$

where $g(\alpha)$ is an integral form of reaction model.

A model-fitting method, Coats-Redfern model, is widely used to estimate the pre-exponential factor and activation energy and to predict the order of reaction (n). The basic equation for Coats-Redfern method can be expressed as [11]:

$$ln\left[\frac{g(\alpha)}{T^2}\right] = ln\frac{AR}{\beta E_a} - \frac{E_a}{RT} \tag{6}$$

Activation energy can be achieved by drawing a graph between $1/T$ and $ln[g(\alpha)/T^2]$ and by finding slope from drawn straight line. The pre-exponential factor can be evaluated from the intercept of this graph and $g(\alpha)$ can be varied according to different developed model and reaction mechanisms. Most of the solid-state degradation reactions could fall into three categories which are shown in Table 1. In Criado method, the experimental data were compared with the data calculated by reaction mechanism function. Therefore, the kinetic reaction mechanism of the pyrolysis process could be identified. The equation of this method obtained from equation (4) and (6) can be given below:

$$\frac{z(\alpha)}{z(0.5)} = \frac{f(\alpha)g(\alpha)}{f(0.5)g(0.5)} = \left(\frac{T_\alpha}{T_{0.5}}\right)^2\frac{{d\alpha}/{dt}}{\left({d\alpha}/{dt}\right)_{0.5}} \tag{7}$$

where $T_{0.5}$ and $(d\alpha/dt)_{0.5}$ are the temperature and conversion rate at $\alpha$=0.5. The left-hand side $f(\alpha)g(\alpha)/f(0.5)g(0.5)$ is a theoretical curve referring to the characteristic of each reaction mechanism listed in Table 1. The right-hand side of equation (7) is the reduced rate curve, which was attained from experimental data. By comparing the theoretical curve with the reduced rate curve, it is definitely to determine the most appropriate reaction mechanism [12].

Many methods were developed to determine the pre-exponential factor and reaction order in a direct way due to the variation in reaction models. These methods are explained below: when the model-fitting method is applied for a single heating rate, a large variation in kinetic parameters are acquired when using different reaction models. These values can be correlated by compensation effect since activation energy is partially or completely compensated by changing pre-exponential factor [13]. The relation is shown in the following expression:

$$ln(A) = a + bE_a \tag{8}$$

where a and b are constant known as the compensation effect parameters.

**Table 1.** The common solid-state reaction kinetics models.

| Code | Reaction mechanism | Differential form *f(α)* | Integral form *g(α)* |
|---|---|---|---|
| Fm | Chemical reaction order | | |
| F1 | First order | $1-\alpha$ | $-\ln(1-\alpha)$ |
| F2 | Second order | $(1-\alpha)^2$ | $(1-\alpha)^{-1}-1$ |
| F3 | Third order | $(1-\alpha)^3$ | $[(1-\alpha)^{-1}-1]/2$ |
| Dm | Diffusion | | |
| D1 | 1D diffusion | $1/2\alpha^{-1}$ | $\alpha^2$ |
| D2 | 2D diffusion | $[-\ln(1-\alpha)]^{-1}$ | $(1-\alpha)\ln(1-\alpha)+\alpha$ |
| D3 | 3D Diffusion-Jander | $\frac{3/2\,(1-\alpha)^{2/3}}{\left[1-(1-\alpha)^{1/3}\right]}$ | $\left[1-(1-\alpha)^{1/3}\right]^2$ |
| D4 | 3D, Ginstling-Brounshtein | $\frac{3\left[(1-\alpha)^{1/3}\right]^{-1}}{2}$ | $1-{}^{2\alpha}/_{3}-(1-\alpha)^{2/3}$ |
| D5 | 3D, Zhuravlev, lesokin, Tempelman | $\frac{3/2\,(1-\alpha)^{4/3}}{\left[(1-\alpha)^{-1/3}-1\right]}$ | $\left[(1-\alpha)^{1/3}-1\right]^2$ |
| D6 | 3D, Anti-Jander | $\frac{3/2\,(1+\alpha)^{2/3}}{\left[(1+\alpha)^{1/3}-1\right]}$ | $\left[(1+\alpha)^{1/3}-1\right]^2$ |
| Am | Volume reaction model (Avarami-Erofe'ev) | | |
| A3/2 | m=3/2 | $(\frac{3}{2})(1-\alpha)[-\ln(1-\alpha]^{1/3}$ | $[-\ln(1-\alpha]^{2/3}$ |
| A2 | m=2 | $2(1-\alpha)[-\ln(1-\alpha]^{1/2}$ | $[-\ln(1-\alpha]^{1/2}$ |
| A3 | m=3 | $3(1-\alpha)[-\ln(1-\alpha]^{2/3}$ | $[-\ln(1-\alpha]^{1/3}$ |

## 3. Results and discussion

### 3.1. *Thermal degradation of the mixture of WHC biomass and waste PS*

The conversions of WHC, PS and 90WHC:10PS were shown in Figure 1 at four heating rates (10, 15, 25, and 50 °C/min). Figure 1(a) shows the TGA curves of WHC, PS and 90WHC:10PS at a heating rate of 50 °C/min. The pyrolysis process of WHC could be divided into three stages including dehydration, devolatilization, and decomposition. The major macromolecules (cellulose, hemicellulose and lignin) with the interaction between these moieties in WHC result in a multi-staged decomposition. These findings were in accordance with various reported works [14-15]. The conversion for the thermal decomposition of WHC and 90WHC:10PS demonstrated three stages: stage I from 50 to ~200 °C, stage II from ~200 °C to ~530 °C, and stage III from ~530 °C to 800 °C. These stages indicated various processes during the thermal decomposition of WHC which is the main component in 90WHC:10PS. When the samples WHC and 90WHC:10PS were heated, moisture was first released at stage I, a primitive peak is observed at ~100 °C indicating the removal of moisture and lower volatile matter which accounts for 5.75% for WHC and for 5.29% for 90WHC:10PS of the total weight loss. As the temperature continued to rise, hemicellulose and cellulose in WHC were then decomposed at stage II. This stage mainly involves the generation of volatile compounds with high oxygen content by random fragmentation of glucosidic bonds in hemicellulose and cellulose, leaving a lignin rich carbonaceous matrix (biochar) [16]. This stage was the main reaction stages of WHC and 90WHHC:10PS, with the highest mass loss of 52.46% for WHC and 47.52 % for 90WHC:10PS (~390 °C). The breakdown of hemicellulose occurred in the range of 200-300 °C by dehydrating its methoxy and acetyl bond. Because hemicellulose consists of short-chain heteropolysaccharides with low polymerization and an amorphous and branched structure, it becomes less stable in the thermal decomposition [17]. On the other hand, the dominant peak is observed at 339 °C, which appeared due to breakage of glucosidic bonds in cellulose. Since cellulose is a linear macromolecule polysaccharide containing D-glucosyl groups with high polymerization and hydrolysis resistance, its thermal degradation was more difficult than that of hemicellulose in the thermal conversion [18]. The solid residue was 42.26% for WHC and 46.73% for 90WHC:10PS (~390 °C). In contrast, the thermal decomposition of PS occurred at 360-510 °C. It was found that PS was completely degraded without the formation of a solid residue. At the same time, the weight loss of 90WHC:10PS accounted for 19.83% (390-530 °C), and the solid residual quantity was only 26.90% (36.00% for WHC). These results suggested that the presence of PS had a strong influence on the degradation temperature. The presence of PS inhibited coke formation from cellulose during co-pyrolysis, and the solid residue of the mixture of WHC and PS was also decreased. The lower ash content directed to convenient for burning fuel. Moreover, the ash content has a direct relationship with the heating value of fuel, indicating that lower the ash, higher the heating value of the fuel [19]. Lignin was decomposed throughout the process of pyrolysis. However, the decomposition of lignin became dominant at stage III. Because lignin is a complex organic polymer containing aromatic rings constituents and is present in the cross-linking of cellulose and hemicellulose [18]. Therefore, thermal decomposition of lignin was more difficult than that of hemicellulose and cellulose.

Figure 1 (b)-(d) show the comparison of the conversion at four heating rates. It was found that the temperature decomposition of all the samples increased with the heating rates. The higher heating rate revealed that the samples reach higher temperature with shorter residence time, eventually generating instability. With an increase in heating rate from 10 to 50 °C/min, it was observed that DTG thermograms of all the samples shifted to higher temperature zone. The decomposition rate was slower at the increased heating rate due to heat constraint between WHC biomass and/or PS, unlike the situation at lower heating rates. Furthermore, at the lower heating rate, heat stayed in the particles of biomass and/or PS for a longer time, leading to intense heat transfer between biomass and/or PS particles. Compared with the pyrolysis of individual wastes, the DTG data for co-pyrolysis demonstrated two DTG peaks as shown in Figure 1(d). It was probably because the decomposition of WHC affected the co-pyrolysis process and responded for the first DTG peak at about 200-400 °C (zone I), and the pyrolysis of PS resulted in the second DTG peak at a wider temperature range of 400 to 530 °C (zone II). Additionally, the first DTG peak of the co-pyrolysis was higher than the second DTG peak (Figure 1(a)), which might be attributed to the faster weight loss of WHC at 200-390 °C than the decomposition of PS at 390-530 °C. The maximum temperature from the DTG curves for co-pyrolysis in zone II was higher than that for pyrolysis of PS (Table 2). Similar trend has also been observed by other works [2, 20]. These results described that co-pyrolysis delayed the pyrolysis process in comparison with the pyrolysis of individual PS, indicating the possible interaction between WHC and PS during co-pyrolysis process.

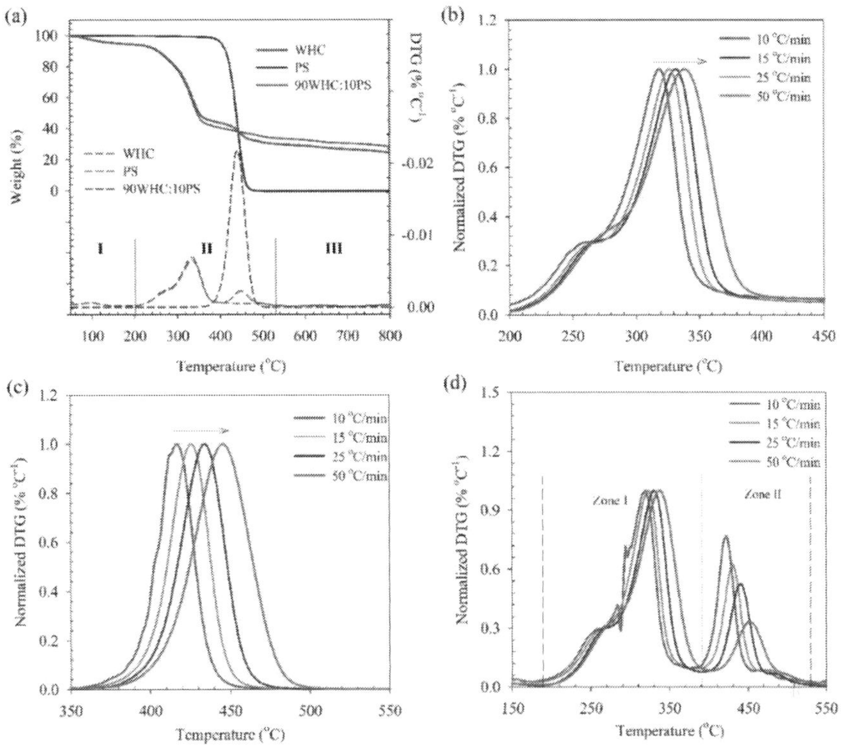


**Figure 1.** (a) TG/DTG curves of individual wastes and their mixture at 50 °C/min and DTG profiles of (b) WHC biomass, (c) waste PS and (d) 90WHC:10PS at various heating rates.

The synergistic effect during the co-pyrolysis of WHC and PS was analyzed using equation (9) by comparing the difference of weight loss from TGA ($\Delta W$, wt%) between the experimental and calculated results:

$$\Delta W = W_{exp} - W_{cal} \tag{9}$$

where $W_{exp}$ is the experimental weight loss from TGA measurement. As shown in equation (9), $W_{cal}$ was calculated from the sum of the weight loss from TGA pyrolysis of individual WHC or PS [21]:

$$W_{cal} = \chi_1 W_{WHC} + \chi_2 W_{PS} \tag{10}$$

where $\chi_1$ and $\chi_2$ are the mass proportion of WHC and PS in corresponding co-pyrolysis sample, and $W_{WHC}$ and $W_{PS}$ are the TGA weight loss of individual WHC and PS, respectively.

To further demonstrate the synergistic effect between WHC biomass and waste PS, the difference of weight loss ($\Delta W$) as the synergistic index parameter for co-pyrolysis of WHC and PS was defined. $\Delta W > 0$ indicates a passive synergistic effect. Meanwhile, $\Delta W < 0$ indicates an accelerated co-pyrolysis process and a positive synergistic effect [20]. Figure 2 represents the variation of $\Delta W$ with temperature at a heating rate of 50 °C/min. It can be seen that for 90WHC:10PS, the weight loss difference is less than ±1% at below ~250 °C. This difference exists because at this temperature stage, the WHC was not decomposed, and no interaction between WHC and PS occurred. It was seen that $\Delta W$ was not equal to zero at this stage, which may have resulted from experimental errors such as different initial weights and thermal conductive conditions. In the temperature range of 280-440 °C, $W_{exp}$ is less than $W_{cal}$ implying that the weight loss was more in the case of the mixture. This indicates a positive synergistic effect between WHC and PS during the co-pyrolysis. The value of $\Delta W$ was started to decrease at 280 °C and increased sharply at 460 °C after peaking at 320 °C. This characteristic behavior of the mixture can be described by the physical state of PS. Firstly, PS was softened and followed by a plastic state which could inhibit the release of volatile compounds from WHC by the formation of a coating layer on the WHC surface. With further increase of temperature, PS began to decompose quickly, leading to the rapid increase of the $\Delta W$. The $\Delta W$ value is less than ±1% at above 460 °C since the devolatilization process was completed.

**Table 2.** Characteristic temperatures of pyrolysis and co-pyrolysis processes.

| Sample | The maximum temperature from the DTG curves at various heating rates | | | | |
|---|---|---|---|---|---|
| | Zone | 10 °C/min | 15 °C/min | 25 °C/min | 50 °C/min |
| WHC | 1 | 319.16 °C | 326.89 °C | 331.7 °C | 338.96 °C |
| PS | 2 | 416.94 °C | 425.72 °C | 434.09 °C | 445.38 °C |
| 90WHC:10PS | 1 | 318.69 °C | 322.68 °C | 330.70 °C | 337.73 °C |
| | 2 | 421.57 °C | 429.94 °C | 439.74 °C | 453.40 °C |

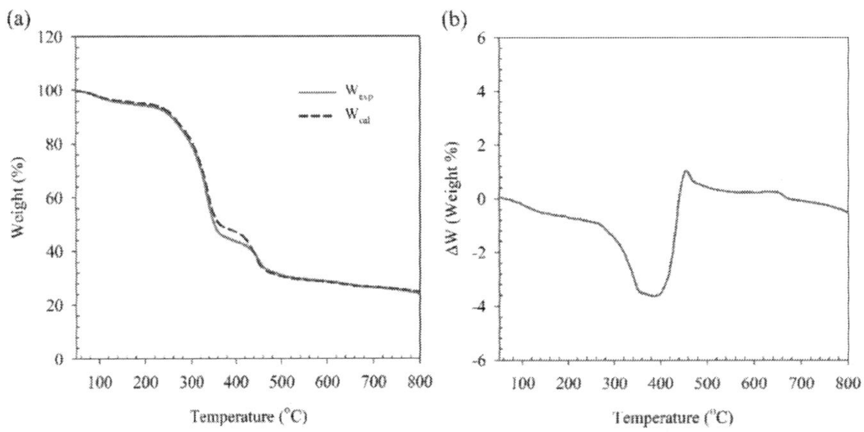


**Figure 2.** Experimental and calculated (from equation (9)) TGA curves for 90WHC:10PS at heating rate of 50 °C/min of (a) weight loss as a function of temperature and (b) $\Delta W$.

IOP Conf. Series: Materials Science and Engineering **965** (2020) 012029 doi:10.1088/1757-899X/965/1/012029

### *3.2. Evaluation of kinetic parameters*

Model-fitting methods provide the activation energy ($E_a$) and pre-exponential factor (A) by fitting different reaction models (Table 1) into the used equation (6). The model that yields the best statistical fit is chosen as the model of choice from which the activation energy and pre-exponential factor can be calculated. This method assumes constant kinetic parameters, as well as uses a single heating rate. The best fit was confirmed by a linear correlation coefficient value of ~ 1 and master plot. The use of different reaction models gives different pairs of $E_a$ and A. However, the compensation effect can be utilized to estimate the actual kinetics parameters. The Coats-Redfern method was used to evaluate the kinetic parameters for single heating value. Figure 3 shows the activation energy for WHC, PS and 90WHC:10PS using the direct differential method. The deviation of the results between different models is because of different assumptions and mathematical derivation of the used reaction models in solid-state kinetics. Table 3 demonstrates the values of the most accurate models with the highest $R^2$ from Figure 3. It can be shown that the models with the highest $R^2$ are the Diffusion models (D1, D2, D3, D4 and D6) and the Reaction-order models (F1) for WHC and 90WHC:10PS (zone I). The highest $R^2$ is approximately 1 and obtained by the Avarami-Erofe'ev and First-order model for PS, whereas the maximum $R^2$ for 90WHC:10PS is obtained by F2 and F3 (zone II). Figure 4 suggests an n-order reaction model for WHC, PS and 90WHC:10PS when comparing the theoretical (z(α)/(z(0.5) vs α) and experimental curves. A reaction model of D3 controls the decomposition steps of WHC (Figure 4(a)) and 90WHC:10PS (zone I) as shown in Figure 4(c), meanwhile F2 controls the decomposition step of 90WHC:10PS in zone II (Figure 4(d)). Model of F1 controls the decomposition step of PS in Figure 4(b). The pre-exponential factor and reaction model were achieved from the compensation effect and master plot, respectively, by using the values of $E_a$ obtained from Coats-Redfern method. The determination coefficients (see Figure 5) of the compensation effect equation were higher than 0.9979 for all the samples. Therefore, the equations in Figure 5 present a linear relationship between the values of A and $E_a$ and the compensation effect parameters can be applied to estimate the values of A. The compensation parameters and $E_a$ value of each pseudo-component were replaced in equation (8). Therefore, Figure 5 illustrates the linear fit obtained by the compensation effect and the master plot curves for each decomposition of WHC, PS and 90WHC:10PS. Compensation effect was used to correlate the obtained values to get the actual values of the kinetic parameters. The pre-exponential factors of WHC, PS, and 90WHC:10PS (zone I and zone II) were 4.21E+10, 9.83E+20, 2.59E+10, and 8.79E+13, respectively. In the pyrolysis zone II, the value of pre-exponential factor (A) was highest for PS and reduced for 90WHC:10PS which may affect changes in Gibbs free energy and in entropy [22].

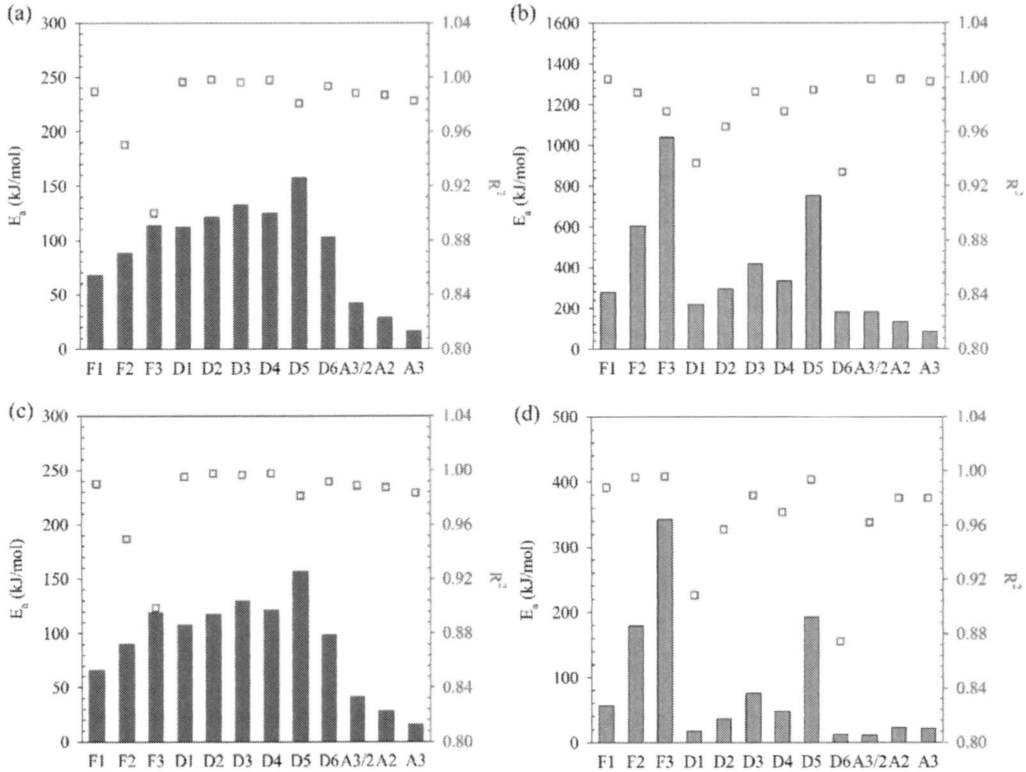


**Figure 3.** Activation energy of (a) WHC, (b) PS, (c) 90WHC:10PS (zone I) and (d) 90WHC:10PS (zone II) obtained by Coats-Redfern method at 50 °C/min.

**Table 3.** The best-fit models of pure glycerol using Coats-Redfern method

| | WHC | | | PS | | | 90WHC:10PS | | | | | |
|---|---|---|---|---|---|---|---|---|---|---|---|---|
| | | | | | | | Zone I | | | Zone II | | |
| | $R^2$ | Ea [kJ/mol] | A [1/min] | $R^2$ | Ea [kJ/mol] | A [1/min] | $R^2$ | Ea [kJ/mol] | A [1/min] | $R^2$ | Ea [kJ/mol] | A [1/min] |
| F1 | 0.990 | 67.623 | 7.47E+05 | 0.999 | 277.230 | 7.88E+20 | 0.991 | 66.455 | 2.90E+06 | – | – | – |
| F2 | – | – | – | – | – | – | – | – | – | 0.996 | 179.103 | 4.04E+14 |
| F3 | – | – | – | – | – | – | – | – | – | 0.996 | 342.815 | 1.50E+27 |
| D1 | 0.997 | 112.778 | 2.19E+09 | – | – | – | 0.995 | 107.972 | 3.81E+09 | – | – | – |
| D2 | 0.998 | 121.849 | 8.82E+09 | – | – | – | 0.998 | 117.534 | 1.72E+10 | – | – | – |
| D3 | 0.996 | 132.882 | 2.45E+10 | – | – | – | 0.997 | 129.95 | 5.87E+10 | – | – | – |
| D4 | 0.998 | 125.491 | 4.51E+09 | – | – | – | 0.998 | 121.467 | 9.38E+09 | – | – | – |
| D5 | - | - | - | 0.991 | 752.271 | 4.67E+54 | | | | 0.994 | 192.364 | 5.40E+14 |
| D6 | 0.994 | 103.599 | 2.83E+07 | | | | 0.992 | 98.776 | 4.90E+07 | – | – | – |
| A3/2 | – | – | – | 0.999 | 180.799 | 7.15E+13 | – | – | – | – | – | – |
| A2 | – | – | – | 0.999 | 132.591 | 2.33E+10 | – | – | – | – | – | – |
| A3 | – | – | – | 0.999 | 84.376 | 7.21E+06 | – | – | – | – | – | – |

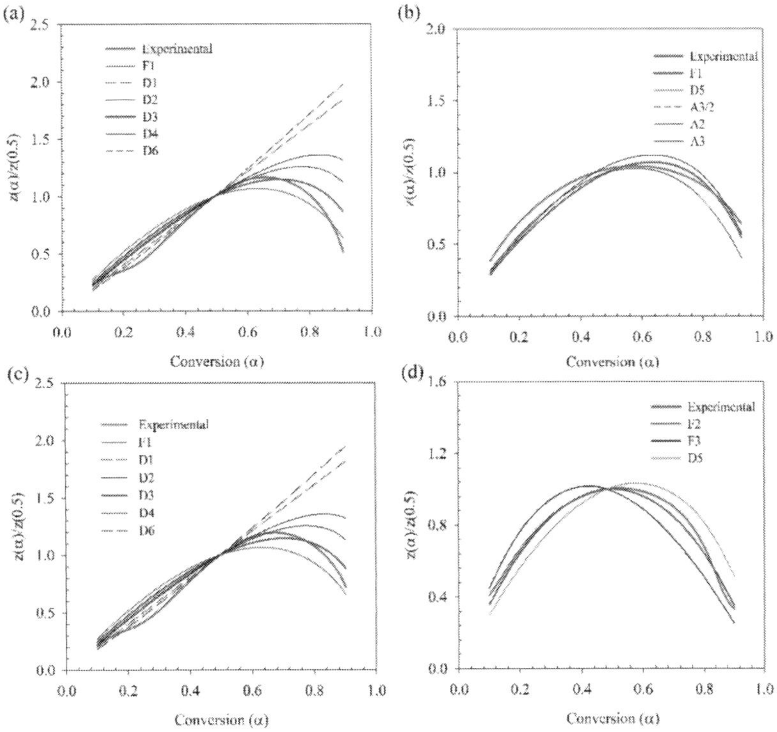


**Figure 4.** The curves obtained from Criado method at a heating rate of 50 °C/min of (a) WHC, (b) PS, (c) 90WHC:10PC (zone I) and 90WHC:10PS (zone II).

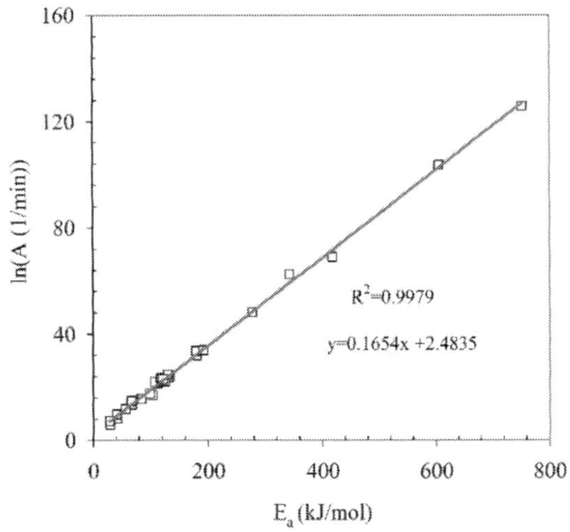


**Figure 5.** Kinetic compensation effect plot showing a linear dependency between $E_a$ and ln (A) for a multi-component model.

## 4. Conclusions

In this study, co-pyrolysis of biomass with waste synthetic polymer was examined to understand the kinetics of the synergistic effects between the waste materials. Oriented polystyrene was used as a model plastic investigated with water hyacinth as a representative of lignocellulosic biomass. TGA/DTG analysis showed that there was a synergistic effect when WHC and PS were pyrolyzed together. During co-pyrolysis, WHC biomass showed an insignificant change in the activation energy compared with individual pyrolysis of WHC, while the activation energy drop of ~21% (48 kJ/mol) with respect to PS was observed for the WHC:PS (90:10) mixture. The apparent activation energy of neat WHC, PS and the mixture of 90WHC:10PS (zone I and zone II) derived from the model-fitting methods was 133, 227, and 130 (in zone I) and 179 (in zone II) kJ/mol, respectively. The thermal decomposition of the mixture 90WHC:10PS followed a diffusion (D3) model in the conversion range of zone I (same as WHC) and second-order chemical reaction kinetics in the conversion range of zone II, whereas the thermal decomposition of PS followed first-order chemical reaction kinetics. The co-pyrolysis of WHC with waste PS enhanced the conversion efficiency of pyrolysis reaction.

**Acknowledgement**

The authors would like to thank the Department of Materials Science and Engineering, Faculty of Engineering and Industrial Technology, Silpakorn University, and the Center of Excellence on Petrochemical and Materials Technology, Chulalongkorn University.

## References

[1] Cao B, Sun Y, Guo J, Wang S, Yuan J, Esakkimuthu S and Wang Q 2019 *Fuel.* **246** 319.
[2] Burra KG and Gupta AK 2018 *Appl. Energ.* **220** 408.
[3] Yuan R, Yu S and Shen Y 2019 *Waste Manage.* **87** 86.
[4] Wang X, Jin Q, Wang L, Bai S, Mikulčić H, Vujanović M and Tan H 2019 *J. Environ. Manage.* **239** 306.
[5] Boateng AA, Weimer PJ, Jung HG and Lamb JFS 2008 *Energ. Fuel.* **22** 2810.
[6] Van Nguyen Q, Choi YS, Choi SK, Jeong YW and Kwon YS 2019 *J. Environ. Manage.* **237** 24.
[7] Lopez G, Artetxe M, Amutio M, Bilbao J and Olazar M 2017 *Renew. Sust. Energ. Rev.* **73** 346.
[8] Uzoejinwa BB, He X, Wang S, El-Fatah Abomohra A, Hu Y and Wang Q 2018. *Energ. Convers. Manage.* **163** 468.
[9] Yang J, Rizkiana J, Widayatno WB, Karnjanakom S, Kaewpanha M, Hao X and Guan G 2016 *Energ. Convers. Manage.* **120** 422.
[10] Vuppaladadiyam AK, Zhao M, Memon M Z and Soomro AF 2019 *Sustain. Energ. Fuels.* **3** 1283
[11] Naqvi SR, Tariq R, Hameed Z, Ali I, Naqvi M, Chen WH and Shahbaz M 2019 *Renew. Energy.* **131** 854.
[12] Xu F, Wang B, Yang D, Hao J, Qiao Y and Tian Y 2018 *Energ. Convers. Manage.* **171** 1106.
[13] Almazrouei M and Janajreh I 2020 *Renewable Energy.* 145 1693.
[14] Huang H, Liu J, Liu H, Evrendilek F and Buyukada M 2020 *Energ. Convers. Manage.* **207** 112552.
[15] Alves JLF, da Silva JCG, da Silva Filho VF, Alves RF, de Araujo Galdino WV and De Sena R F 2019 *Biomass Bioenerg.* **121** 28
[16] Siddiqi H, Kumari U, Biswas S, Mishra A and Meikap BC 2020 *Energy.* **204**. 117933.
[17] Sobek S and Werle S 2020 *Fuel* **261** 116459.
[18] Yu J, Paterson N, Blamey J and Millan M 2017 *Fuel.* **191** 140.
[19] Mishra RK and Mohanty K 2020 *Bioresour Technol.* **311** 123480.
[20] Alam M, Bhavanam A, Jana A, Viroja JkS and Peela NR 2020 *Renew. Energy.* **149** 1133.
[21] Hu Q, Tang Z, Yao D, Yang H, Shao J and Chen H 2020 *J. Clean. Prod.* **260** 121102.
[22] Naqvi SR, Ali I, Nasir S, Ali Ammar Taqvi S, Atabani AE and Chen WH 2020 *Fuel.* **278** 118259.

IOP Conf. Series: Materials Science and Engineering **965** (2020) 012030 doi:10.1088/1757-899X/965/1/012030

# Mechanical properties and foaming behavior of Poly(lactic acid) blend Polybutylene Succinate

**W. Bualuang[1], P. Threepopnatkul[1,*], and A. Sittattrakul[1]**

[1]Department of Materials Science and Engineering, Faculty of Engineering and Industrial Technology, Silpakorn University, Nakhon Pathom, 73000, Thailand

*Corresponding author: poonsubt@yahoo.com

**Abstract.** The extensive use of non-biodegradable plastics has been causing serious environmental pollutions because of the degradability issue from especially consumable food packaging. In this research, biodegradable materials i.e., Poly(lactic acid) (PLA) and Polybutylene succinate (PBS) are chosen for the study. PLA/PBS foam was prepared using Azodicarbonamide (ADC) and Poly(ethylene glycol) (PEG) 20 wt% as a blowing agent and plasticizer, respectively. Foamability of different content PLA/PBS blend were investigated. PLA/PBS foam were characterized by Water absorption, Flexural testing, Scanning Electron Microscope (SEM). Flexural properties of PLA/PBS foam (70/30) showed the highest flexural modulus of 384 MPa due to the smallest cell size while PLA/PBS foam (20:80) showed the highest flexural stress of 11 MPa. PLA/PBS foam showed uniform cell size distribution. PLA/PBS foam had water absorption ability due to its porous structure.

## 1. Introduction

The plastic packaging industry has been continuously developed to suit for the products and consumers needs in the markets and lifestyles. Fresh has been developed and improved to have properties that are more suitable for use. general, the improvement and development of the packaging properties still focus on reducing pollution to the world. In this research, the properties of the packaging will be developed to be more versatile, focusing on the packaging that is used to store fresh food or foods that contain the liquid in the package. Foam packagings are generally porous materials which have a high surface area, high permeability, and lightweight, leading to lower cost of processing. According to Zhou et al.[1], the packaging will be developed in this research that expected to capable of absorbing any excessive liquid to reduce such problem in the fresh produces. However, environment friendly packagings will also be considered as part of this research. In this recently years, the plastic industries slowed down quite considerably due to behavioral changes of environmental concerns and lifestyle. Basic packagings such as plastic bags with handles that are more likely to be used lesser than before due to the global trend including public and private campaigns. Therefore, it is necessary to search for the innovative ideas of creating more versatile yet less polluted for the nuvo packaging. In this study, biodegradable plastics will be used to mold into foam packages. The bioplastics used in this research is biodegradable plastic namely Poly(lactic acid). Currently, Poly (lactic acid) or PLA has been used widely due to its production bases globally. Still, there are some limitations in terms of mechanical properties because of its brittleness if the end product is molded by using only PLA. As a result, the foam packages would be susceptible to crack in the end user's hands. This will definitely cause the hesitation for consumers of not wanting to use PLA for foam packaging. Therefore, it is technically necessary to raise its overall properties mechanically into the packages. Polybutylene Succinate (PBS) is a biodegradable plastic as same as Poly(lactic acid). Only PBS cannot be produced from natural materials. PBS possesses good flexibility and better thermal stability than Poly(lactic acid)[2]. Stability of foaming is another important property related to its elasticity. Therefore, PBS is selected for blending with PLA in this study. In this research, mechanical properties, foaming behavior and water absorption are thoroughly investigated on the PLA/PBS foam.

## 2. Experimental

### Materials

Poly(lactic acid) (PLA) (2003D, $\rho$ = 1.24 g cm$^{-3}$, and MFI = 6 g 10 min$^{-1}$) and Poly butylene succinate (PBS) (FZ91PM, $\rho$ = 1.26 g cm$^{-3}$, and MFI = 5 g 10 min$^{-1}$) was supplied by Nature Works and PTT MCC BioChem, Thailand, respectively. Poly ethylene glycol (PEG) (Mw = 8,000) were supplied by A.C.S.Xenon Limited Partnership was used as a plasticizer. Azodicarbonamide (ADC) (Sunfoam SB00P, Lautan Luas Co., Ltd., Thailand) was used as a chemical foaming agent. PLA, PBS, ADC, and PEG were dried at 70°C for 24 h. before use.

### Compounding and foaming processes

The mixture of various PLA and PBS proportions were prepared as seen in Table 1. The contents of ADC and PEG at 1 phr, 20%wt respectively. PLA/PBS blend was extruded by a co-rotating twin-screw extruder (L/D = 32, LTE20-32, Labtech Engineering Co., Ltd.) with the temperature profile of 110, 120, 130, 140, 145, 150, 160, and 150 °C from hopper to die and the screw speed of 150 rpm. The extrudate was dried at 70°C for 24 h and then sequentially foamed by compression molding (PR1D-W300L350) with the temperature profile of 150 °C preheat 3 min and compression time 5 min.

**Table 1.** The composition of compounding formulations.

| Sample | PLA (wt%) | PBS (wt%) | PEG (wt%) | ADC (phr) |
|---|---|---|---|---|
| PLA100/PBS0 | 100 | 0 | 20 | 1 |
| PLA80/PBS20 | 80 | 20 | 20 | 1 |
| PLA70/PBS30 | 70 | 30 | 20 | 1 |
| PLA60/PBS40 | 60 | 40 | 20 | 1 |
| PLA50/PBS50 | 50 | 50 | 20 | 1 |
| PLA40/PBS60 | 40 | 60 | 20 | 1 |
| PLA30/PBS70 | 30 | 70 | 20 | 1 |
| PLA20/PBS80 | 20 | 80 | 20 | 1 |
| PLA0/PBS100 | 0 | 100 | 20 | 1 |

### Foam morphology analysis

To investigate the foam microstructure, samples were cryo-fractured under liquid nitrogen and then coated using a Gold (Au) sputtering technique. Samples were observed by using a scanning electron microscope (SEM, TM3000, Hitashi). The micrographs were also used to calculate the number of bubbles per cm$^3$ of foam using Eq. (1).

$$N_f = \left( \frac{nM^2}{A} \right)^{\frac{3}{2}} \tag{1}$$

where $N_f$ is the number of bubbles, n is the number of cells, M is the magnification factor, A is the area of the micrographs.

**Water absorption**

Weight of the samples was recorded at different time intervals and compared with the initial dry weight. The weight gain was then calculated. Water absorption ratio was calculated using the following Equation (2).

$$\text{Absorption ratio} = \frac{W_t - W_0}{W_0} \times 100 \tag{2}$$

Where $W_t$ is the weight gain at time t and $W_0$ is the initial dry weight before treatment.

**Mechanical properties**

Flexural tensile strength and Flexural modulus of the PLA/PBS foam were tested according to ASTM D790 using a universal testing machine (Instron model 55R4502) at the crosshead speed of 10 mm $min^{-1}$ with 5 kN.

## 3. Results and discussion

### *3.1. Mechanical properties*

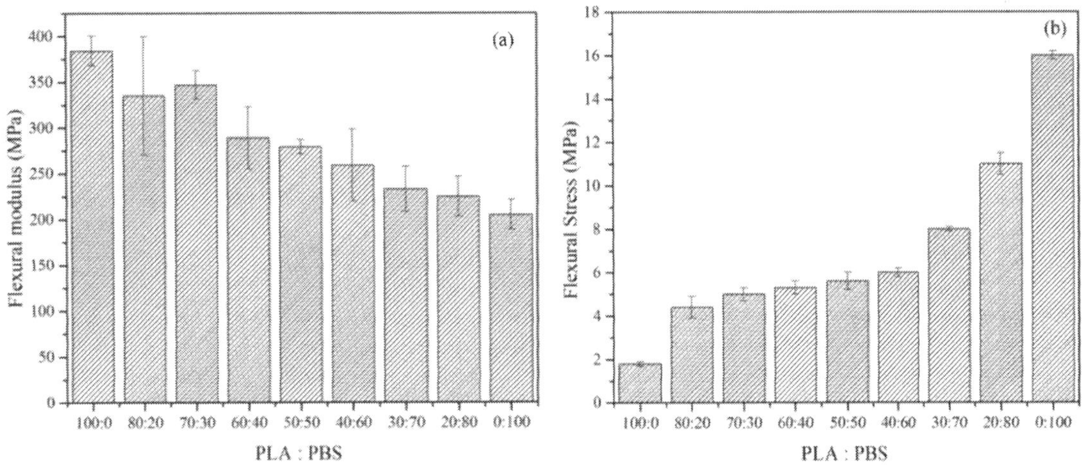


**Figure 1.** (a) Flexural modulus, (b) Flexural stress of PLA-PBS blend foam samples

The mechanical properties of foam samples are shown in Figure 1. The results showed that flexural tensile strength and flexural modulus of PLA/PBS foam. Flexural tensile strength increased with increasing PBS fraction because the structure of PLA/PBS foam at PBS-rich has porous less than PLA/PBS foam at PLA-rich due to large pore size may be stress concentration[3]. Moreover, the Flexural modulus of PLA/PBS blend foam follows approximately the rule of mixtures over the whole composition range. Flexural modulus of PLA/PBS blend foam decreased with increasing PBS fraction due to pore size. The small pore size has a larger cell wall than the large pore size which PLA-rich has pore size larger than PBS-rich therefore PLA-rich has higher flexural modulus than PBS-rich[4].

IOP Conf. Series: Materials Science and Engineering **965** (2020) 012030 doi:10.1088/1757-899X/965/1/012030

### *3.2. Scanning Electron Microscope*

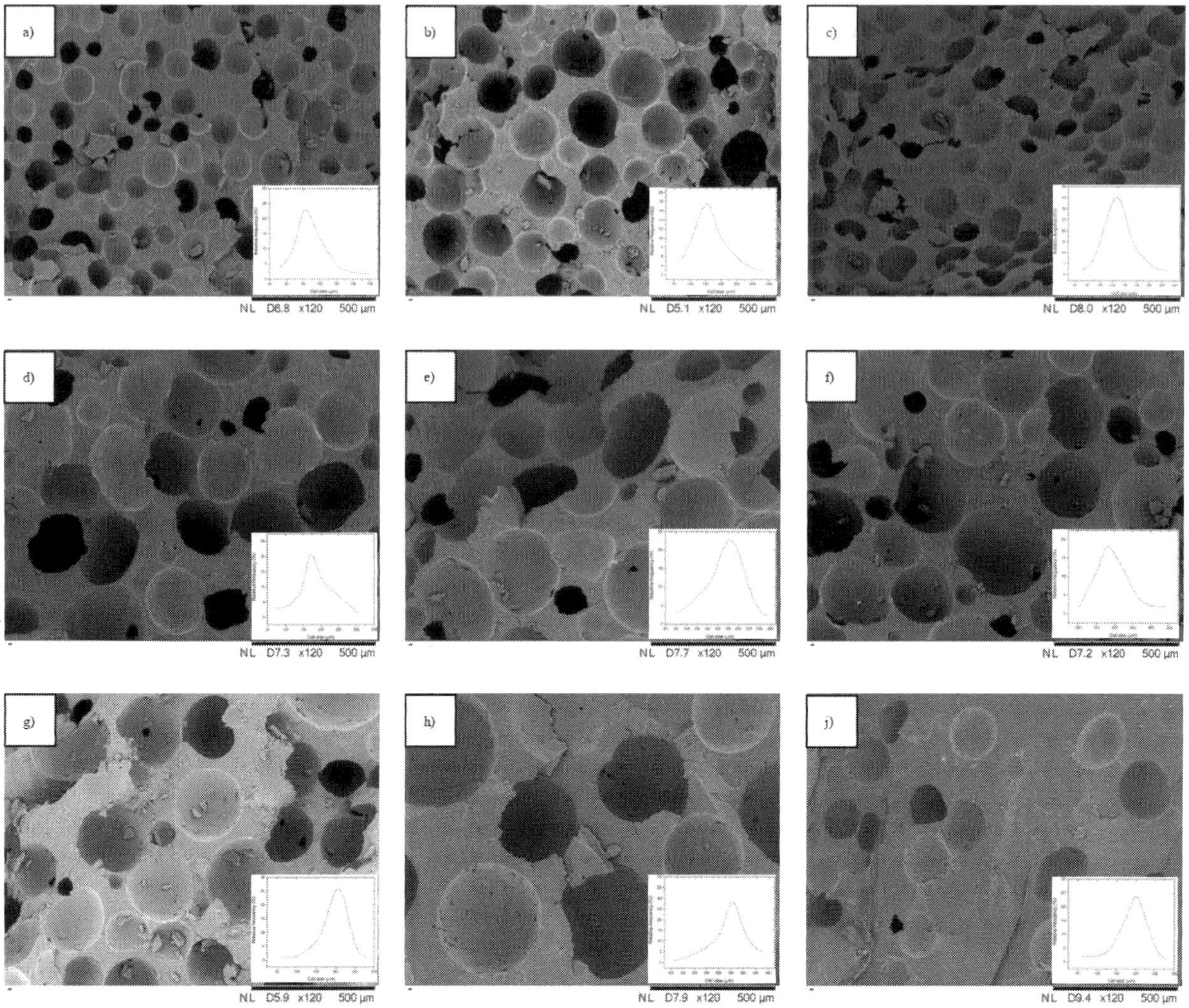


**Figure 2.** SEM micrographs different weight percent of PLA-PBS blend at 120x magnification a) 100:0, b) 80:20, c)70:30, d) 60:40 e) 50:50, f) 40:60, g) 30:70, h) 20:80, j) 0:100

PLA-PBS blend foam morphology was investigated as shown in Figure 2. All samples showed uniform cell size distribution. PLA foam has a cell size of about 125 microns and cell density about 240 x$10^4$ cell/cm$^3$ while PBS foam has cell size about 150 microns and cell density about 180 x$10^4$ cell/cm$^3$ as shown in Figure 3.The cell size of PLA-PBS foam increases with PBS fraction increase while the cell density of PLA-PBS foam decreased with PBS fraction increase as shown in Figure 3 as a result of the difference between viscosity of PLA and PBS phase and temperature processing of PLA/PBS. PLA has a melting temperature at 160°C and PBS have a melting temperature at 115°C when forming the specimen, using temperature processing at 150°C leading to PBS-rich couldn't hold and agglomerated into larger cell size when compared to high PLA fraction results in higher cell size and lower cell density in PLA-PBS foam with higher PBS fraction but cell size of PLA-PBS at 80:20 have the larger more than pure PBS because ADC is compatible with PLA more than PBS thus number of porous in PLA-PBS at 80:20 more than pure PBS and agglomerated. One more reason PBS showed lower viscosity than PLA[5], PBS-rich have larger pore size more than PLA-rich as a result of the viscosity PLA-PBS foam decreased with PBS fraction increase[6].

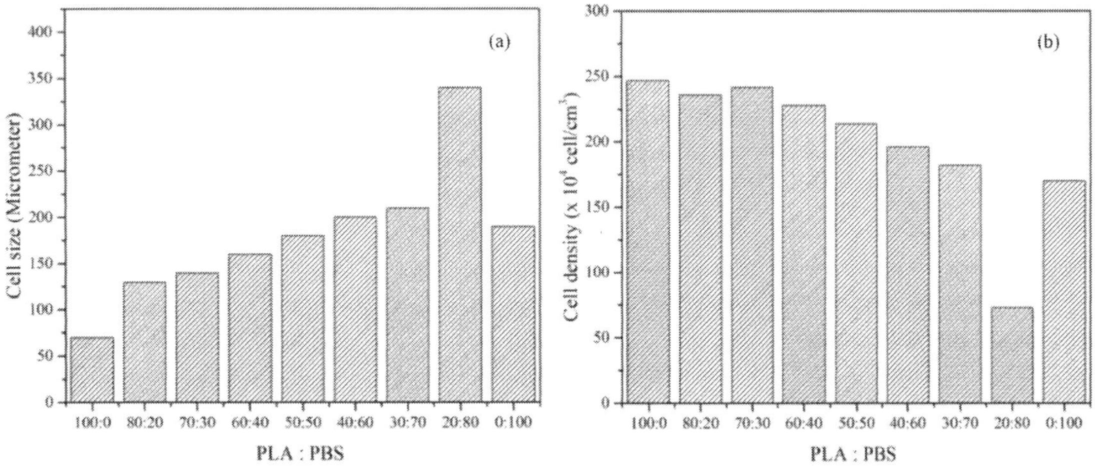


**Figure 3.** (a) cell size and (b) cell density of PLA-PBS blend foam sample

### *3.3. Water absorption*

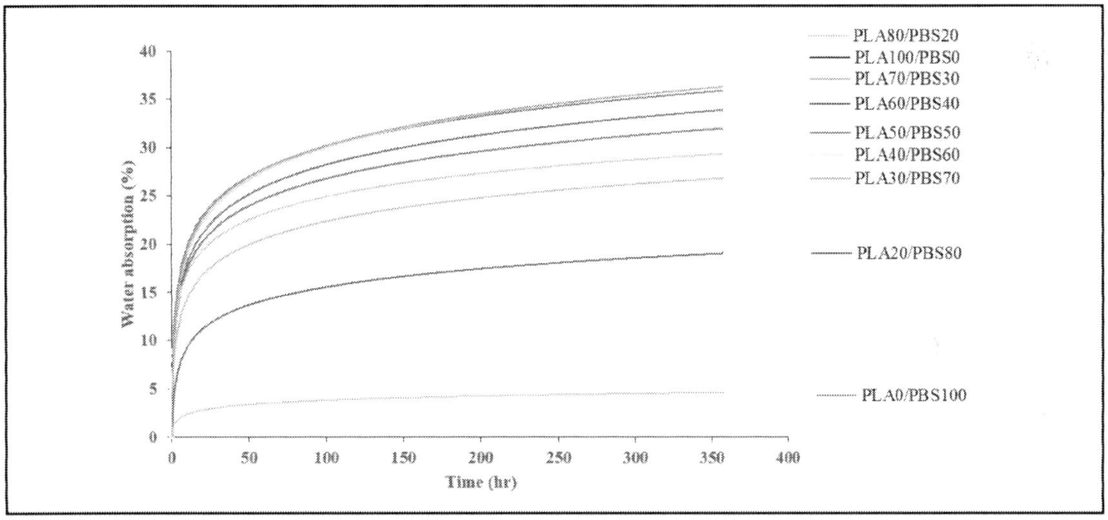


**Figure 4.** The water absorption of PLA/PBS blend foam samples

The water absorption of PLA/PBS blend foam is shown in Figure 4. All samples had water absorption ability due to its porous structure. Water absorption of PLA/PBS foam with PLA-rich and PBS-rich has been investigated and it was found to decrease with increasing PBS content due to the polarity of PBS less than the polarity of PLA[7], leading to water can absorb into PLA-rich more than PBS-rich. PLA100/PBS0 increased in weight by 35%, it's the highest absorption ability while PLA0/PBS100 increased in weight by 4.98%, it's the lowest absorption ability.

## 4. Conclusion

PLA/PBS foams at different PLA and PBS ratios were prepared by compression molding. Foam samples showed more uniform cell morphologies and higher flexural modulus of PLA-rich samples, whereas flexural stress of PLA/PBS foam increased with increasing PBS fraction. The cell size of PLA/PBS foam increases with PBS fraction increase while the cell density of PLA-PBS foam decrease with PBS fraction increase Water absorption of PLA/PBS foam was found to increase with increasing PBS content.

## 5. Acknowledgment

This research was financially supported by the Department of Materials Science and Engineering, Faculty of Engineering and Industrial Technology, Silpakorn University.

## References

[1] J. Zhou, Z. Yao, C. Zhou, D. Wei, S. Li, Mechanical properties of PLA/PBS foamed composites reinforced by organophilic montmorillonite, Journal of Applied Polymer Science. 131 (2014).

[2] Joël R., Richard G. and Michel F. Champagne, Extrusion Foaming of Poly(Lactic acid) Blown with CO2: Toward 100% Green Material, pp. 83-116 (2005)

[3] J. Charoenta., C. Tongpin, Thermoplastic Vulcanizate Foam Prepared from NR/PBS Blends, The 2019 Pure and Applied Chemistry International Conference, pp. 77-82 (2019)

[4] K. Ketkul, P. Threepopnatkul., D. Aussawasathien, K. Hrimchum, Poly(Lactic Acid) (PLA) - Polybutylene Succinate (PBS) - Activated Carbon Composite Foams, Key Engineering Materials Vol. 751, pp. 344-349 (2017)

[5] V. Philsaard, S. Naimlaung, S. Pavasupree, S. Philsaard, Preparation of Polymer Blends between Polylactic acid and Poly(butylenes succinate) for Injection Molding Application, Thailand, Rajamangala University of Technology Thanyaburi (2014)

[6] B. Panyad, N. Hongsripun, Modification of mechanical properties and flame retardancy of Polybutylene succinate for using in electrical housing, Thailand, Silpakorn University (2011)

[7] D. Preechawong, M. Peesan, P. Supaphol, R. Rujiravanit, Preparation and characterization of starch/poly(l-lactic acid) hybrid foams. Carbohydrate Polymers, 59(3), 329–337. (2005)

IOP Conf. Series: Materials Science and Engineering **965** (2020) 012031 doi:10.1088/1757-899X/965/1/012031

# Effect of poly(ethylene oxide) on the properties of poly(lactic acid)-based blends

**R. Khanteesa[1], P. Threepopnatkul[1,*] and A. Sittattrakul[1]**

[1]Department of Materials Science and Engineering, Faculty of Engineering and Industrial Technology, Silpakorn University, Nakhon Pathom, 73000, Thailand

*Corresponding author: poonsubt@yahoo.com

**Abstract.** Poly(lactic acid) (PLA) was blended with poly(ethylene oxide) (PEO) with four different molecular weights of $1\times10^5$, $2\times10^5$, $3\times10^5$, and $4\times10^5$ g/mol to improve the flexibility of PLA. The blends were compounded in an internal mixer at 190 °C with 50 rpm for 15 minutes and using a hot press to mold rectangular sample sheets. The mechanical properties of PLA blends were analyzed to prove the ductility. Additionally, phase morphologies of blends were investigated by using tensile fracture surface. The addition of PEO into PLA matrix leads to slight increase in tensile strength, while the elongation at break was decreased by increasing the molecular weight of PEO. However, the elongation at break of PLA/PEO blends was higher than the one of neat PLA. In other words, PEO enhanced the flexibility of PLA blends, according to scanning electron microscope images. In brief, PLA-based blends greatly increased in ductility.

## 1. Introduction

Currently, biopolymers have been considered as a substitute for traditional petroleum-based polymers. Poly(lactic acid) (PLA) is biodegradable aliphatic polyester that is derived from renewable resources such as corn, cassava, potato and sugarcane. PLA has a wide range of applications such as in properties such as high mechanical strength, high modulus, biodegradability, biocompatibility, bioabsorbability, transparency, low toxicity and easy process ability [1]. However, PLA have been limited for applications due to poor thermal stability and low impact strength [2].

The toughness of PLA can be improved by blending with other polymers, such as poly(butylene succinate) (PBS) [3], poly(ethylene glycal) (PEG) [4], and poly(ethylene oxide) (PEO) [5]. PEG and PEO have a similar chemical structure and they are biodegradability, biocompatibility, and nontoxicity. It will be called PEO when the molecular weight was over 100,000 g/mol. Meanwhile, the incorporation of PEG can still maintain biodegradability, biocompatibility and low toxicity of PLA [6].

Recently, many studies on properties of PLA-based blends. Jacobsen et al. [7] investigated the effect of different plasticizers (i.e., glucose monoesters, PEG [$M_w$ = 1500 g/mol], and partial fatty acid esters) on the mechanical properties of PLA. They found that the best amount of the three plasticizers was 10 %wt. The incorporation of PEG to PLA caused the greatest reductions in glass transition temperature ($T_g$), tensile modulus and strength, whereas the largest increments in elongation at break (up to 180%) and impact strength. Feng-Jiao Li et al. [8] plasticized PLA using PEG with different molecular weights (6,000, 10,000, and 20,000 g/mol) and different contents (0-20 %wt). They reported that the tensile modulus and strength increased, while tensile stress at break and strain at break decreased when increasing the molecular weight of PEG. Zhang et al. [6] blended PLA and PEG with five different molecular weights (200–20,000 g/mol) and various content of PEG-10,000 ranged from 0 to 20 %wt and reported the increases in both V-notched Izod and Charpy impact strengths of PLA/PEG-10,000 blends were 206.10% and 137.25%, respectively. Meanwhile, the crystallinity of PLA/PEG-10,000 blends increased from 3.95% to 43.42%. For 10 %wt PEG content, the crystallization and impact properties of PLA/PEG blends mainly depended upon PEG molecular weight. The introduction of PEG reduced the intermolecular force and enhanced the mobility of PLA chains, thus improving the crystallization capacity and flexibility of PLA. Yoojun et al. [9] have investigated the mechanical and thermal properties of PLA films blended with PEO ($M_w$ = 100,000 g/mol). The blend films exhibited a elongation at break was increased by 4 times at 5 %wt PEO loading, while the tensile strength was decreased by 17%. In addition, the glass transition temperature ($T_g$), melting temperature ($T_m$), and cold crystallization temperature ($T_{cc}$) decreased with adding 5 %wt PEO.

Furthermore, the high molecular weight of PEO is an important factor that affects on properties of polymer blends. However, there have been very few researches. Therefore, the major objective of this research is to study the effects of high molecular weight of PEO on the mechanical properties and morphologies of PLA/ PEO blends.

## 2. Experimental

### Materials

PLA (4043D, $M_w$ = 190,000 g/mol) was purchased from NatureWorks. Density, melt flow index, glass transition temperature ($T_g$), and melting temperature ($T_m$) were 1.24 g/cm$^3$, 6 g/10 min, 60 °C, and 160 °C, respectively. PEOs were purchased from Sigma Aldrich. Molecular weight ($M_w$) and melting temperature ($T_m$) were $10^5$ - $4\times10^5$ g/mol and 65 °C, respectively.

### Melt blends and sample preparation

The blends were carried out in internal mixer (Brabender Mixer Docking Station W50EHT series) at 190 °C with a capacity of 50 g and rotor speed of 50 rpm for 15 minutes. The sample sheets were obtained using a hot press at 190 °C into rectangular sample sheets. The thickness of each sheet was in range of 300-350 microns. The specimens were then kept in plastic bags before testing. The composition of PLA blends were shown in Table 1.

**Table 1** Compositions of PLA-based blends with PEO

| Sample | Molecular weight ($M_w$) of PEO (g/mol) | Composition (%wt) | |
|---|---|---|---|
| | | PLA | PEO |
| Neat PLA | - | 100 | 0 |
| PLA_1PEO | 100,000 | 90 | 10 |
| PLA_2PEO | 200,000 | 90 | 10 |
| PLA_3PEO | 300,000 | 90 | 10 |
| PLA_4PEO | 400,000 | 90 | 10 |

### Characterizations

Mechanical properties of the blends were performed by Universal Testing Machine according to ASTM D882 (Instron 5969, USA). A 5 kN load cell was used. The crosshead speed was 12.5 mm/min. The morphology of the tensile fractured surface of specimens was investigated using a Scanning Electron Microscope (SEM) (Hitachi TM3030, Japan). The samples were sputtered coated with Pd/Au.

IOP Conf. Series: Materials Science and Engineering **965** (2020) 012031 doi:10.1088/1757-899X/965/1/012031

## 3. Results and discussion

### *3.1. Mechanical properties*

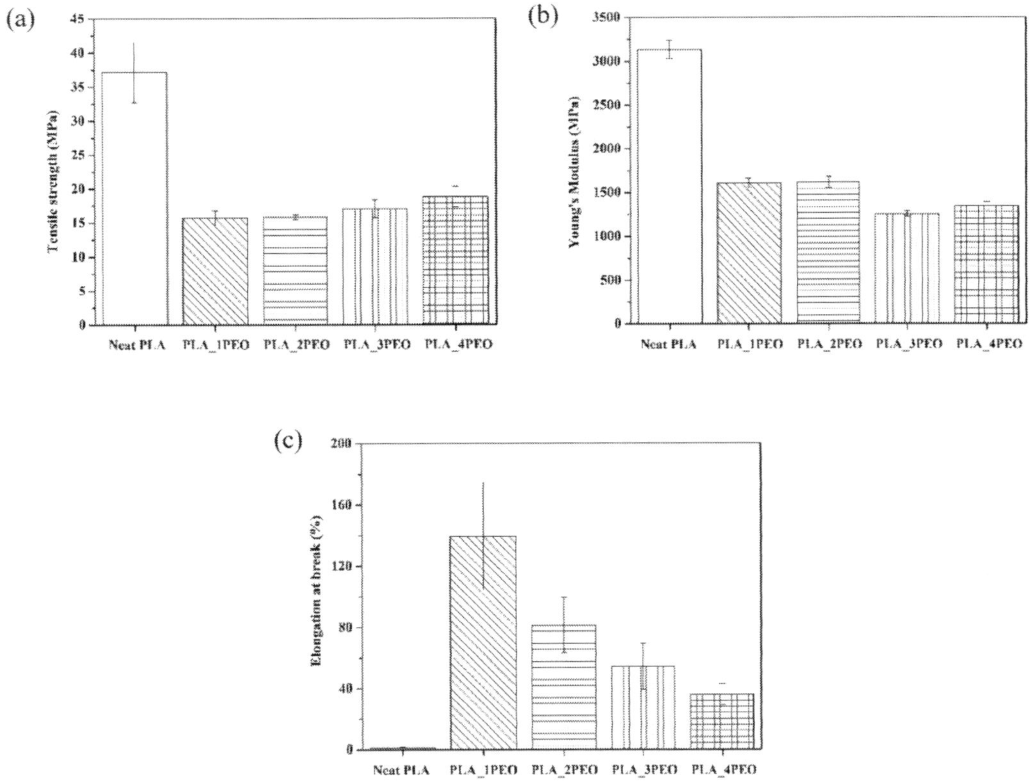


**Figure 1.** Effect of PEO molecular weight on tensile properties (a) Tensile strength, (b) Young's Modulus and (c) Elongation at break of PLA/PEO (90/10) blends

Figure 1 presents the mechanical properties of neat PLA compared with PLA blended with different molecular weight of PEO. Tensile strength is an importance parameter to characterize the largest stress limit of materials under tensile load. Tensile strength of neat PLA was 37.16 MPa. Similar mechanical values of neat PLA have been reported previously, i.e., tensile strength of neat PLA was from 36 to 52 MPa [5, 10]. The tensile strength of PLA/PEO blends gradually increases with an increase in molecular weight of PEO. The increase of intermolecular forces, surface energy and molecular chain entanglement were caused by the incremen of the molecular weight of PEO during this study. This could reduce the slippage of PLA molecular chains under tensile loading which resulted in an increase of tensile strength. The addition of PEO into the PLA matrix reduced the tensile modulus of the blends in comparison with neat PLA. This is due to the interactions between plasticizer and polymers which weaken the polymeric intermolecular interactions and therefore improved the rearrangement between polymer chains under external force, and flexibility to the PLA. The elongation at break of PLA/PEO blends was improved compared with the one of neat PLA. However, the elongation at break of PLA blended with lower molecular weight PEO was higher than the one with higher molecular weight PEO. This illustrated that lower molecular weight PEO can more efficiently improve the ductility of PLA. Generally, the miscibility of PEG with other polymers deteriorated with an increase in PEG molecular weight [6].

*3.2. Fracture surface morphology analysis*

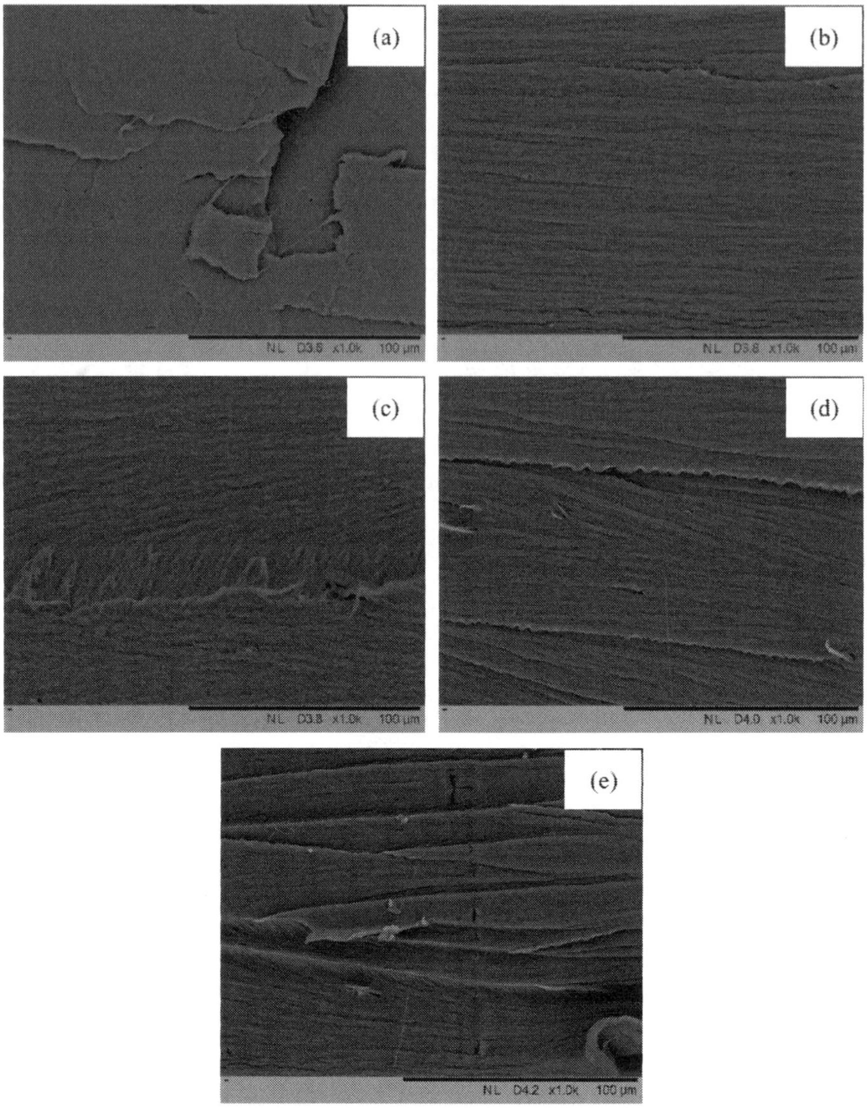


**Figure 2.** Scanning electron micrographs of tensile fracture surfaces of (a) Neat PLA, (b) PLA/1PEO, (c) PLA/2PEO, (d) PLA/3PEO, and (e) PLA/4PEO blends containing 10 %wt PEO content.

SEM was utilized to investigate the morphologies of the tensile fractured surface of PLA blended with four different molecular weight PEO. From Figure 2, neat PLA (Fig. 2a) shows a smoother surface than PLA/PEO blends (Fig. 2b-2e). When the molecular weight of PEO was increased, the fracture surface of PLA/ PEG blend becomes rougher because of the increment of entanglement between higher molecular weight PEG and PLA molecular chains [6].

## 4. Conclusion

To improve the ductility of PLA and reduce brittleness, PLA was blended with four different molecular weight PEO at a constant concentration of 10 %wt. The PLA/PEO blends were prepared using an internal mixer and then hot press into specimens. The properties of the blends were characterized by

Tensile testing and SEM. The ductility of PLA was significantly increased when PEO was added into PLA matrix. The tensile modulus of PLA blends film declined but the tensile strength increased by addition of higher molecular weight of PEO. In this study, the elongation at break of PLA/PEO blends increased when compared with neat PLA, but decreased with increasing molecular weight of PEO. SEM images reveal that when the molecular weight of PEO was increased, the fracture surface of the PLA/PEO blends becomes rougher. Besides, PEO enhanced the flexibility of PLA blends.

## 5. Acknowledgement

This research was financially supported by the Department of Materials Science and Engineering, Faculty of Engineering and Industrial Technology, Silpakorn University.

## References

[1] Öz A, Süfer Ö and Celebi sezer Y 2017 *Food Sci Technol Int.* **2** 000131.
[2] Auras R, Harte B and Selke S 2004 *Macromol Biosci.* **4** 835-864.
[3] Pivsa-Art W, Fujii K, Nomura K, Aso Y, Ohara H and Yamane H 2016 *J. Appl. Polym. Sci.* **133** 43044.
[4] Bijarimi M, Ahmad S, Rasid R, Khushairi M A and Zakir M 2016 *AIP Conference Proceedings.* **1727** 020002.
[5] Saha D, Samal S K, Biswal M, Mohanty S and Nayak S K 2019 *Polym. Int.* **68** 164-172.
[6] Li F, Zhang S, Liang J and Wang J 2015 *Polym Advan Technol.* **26** 465-475.
[7] Jacobsen S and Fritz H G 1999 *Polym. Eng. Sci.* **39** 1303-1310.
[8] Li F, Liang J, Zhang S and Zhu B 2015 *J. Polym Environ.* **23**
[9] Eom Y, Choi B and Park S 2018 *J. Polym Environ.* **27**
[10] Park B, Song J C, Park D H and Yoon K B 2012 *J. Appl. Polym.* **123** 2360-2367.

IOP Conf. Series: Materials Science and Engineering **965** (2020) 012032 doi:10.1088/1757-899X/965/1/012032

# The effect of rotor speed on EVA/PBAT blend using Twin screw extruder and foamability of EVA/PBAT blend

**Chanchai Thongpin[1,*], Kanyapak Theanchai[1], Pattama Riangpha[1] and Weena Anantawut[1]**

Department of Materials Science and Engineering, Faculty of Engineering and Industrial Technology, Silpakorn University, Sanamchandra Palace Campus, Nakhon Pathom, 73000, Thailand.

*Corresponding author: THONGPIN_C@su.ac.th

**Abstract**. This research is aimed to study EVA/PBAT blends with various EVA and PBAT composition and screw speeds. At PBAT contents 40% EVA and PBAT present as co-continuous phase morphology. The morphology did not change due to screw speed. At PBAT 50% by weight, co-continuous phase occurred and the morphology becomes dispersed as screw speed was increased to 80 and 100 rpm. Extrudates obtained from twin screw extruder were tensioned under tensile force. The results show elastomeric behavior of the blends. Elongation at break was higher than 800 % strain and exhibited strain hardening characteristic. Secant modulus of EVA/PBAT blends was increased with PBAT contents due to the high modulus of semi-crystalline PBAT. Content of EVA controls elongation ability of the blends. EVA/PBAT blends with the content of 40/60, 50/50, 60/40, 70/30 and 80/20 were form for foam and found that all composition could be able to be foam. This could be further studied and carried on for shoes sole application.

## 1. Introduction

Dispersion of ethylene-vinyl acetate copolymer (EVA) are randomly disperse in main chain by ethylene copolymer, polar vinyl acetate (VAc) units. Different VAc content in copolymers changes EVA from plastics to elastomers in several applications. A good example of this is solar cell encapsulant[1], foot ware midsole and toy industry. Each application chosen are dependent upon the grades of EVA. Each type of the EVA is determined according to the type of curing agent, which can be of three types: standard cure (SC), fast cure (FC) and ultra-fast cure (UFC). In order to increase the yield of the rolling process, the fast or ultra-fast cure formulations are more common as fast cure materials can achieve higher yield due to a reduction of almost 40% of the lamination process. The EVA cure is needed to change the physical properties of an encapsulant by chemical reaction through polymerization, in the presence of heat and catalyst, alone or in combination [2]. In addition, the blends of EVA/PBAT possess high thermo shrinkage and high stability when subjected to aging, weathering, and aggressive media. Therefore, the EVA/PBAT blends are used in many applications including foam material, multilayer packaging films and sheets, automobile parts, agricultural films, medical tubes, and so on [3]. Nevertheless, an important shortcoming in blending different polymers is that the polymers often form immiscible and incompatible blends, displaying a coarse morphology with particles of dispersed phase badly adherent to the matrix and badly distributed. Consequently, a compatibilization process is necessary in order to modify the interfacial properties in immiscible polymer blend, thus leading to a reduction of the interfacial tension coefficient and the formation and stabilization of the desired morphology [4].

Poly (butylene adipate-Co-terephthalate) (PBAT), also known under the trade name of Ecoflex, has attracted great interest as potential substitution of low density polyethylene (LDPE), especially for packaging materials such as lawn waste bags, food containers or film wraps, among others. Apart from its biodegradability, PBAT offers attractive physical properties, including high flexibility, high strength and good tear resistance, and it is approved for food contact applications [5]. In addition, it has short moulding cycles and excellent sealing performance, while it is easy to process at high extrusion speeds [6]. As a wildly used PBAT blend which can blend with other polymers such as PLA/PBAT blends [7], PBAT/EVOH blends [8], ATC/PBAT blends [9] and PP/PBAT blends [10].

In this part of our research EVA/PBAT blends will be prepared and their morphology, mechanical properties will be evaluated. The chain extender such as joncryl4300 will be added in order to investigate its effect of EVA/PBAT blends. As EVA has been mainly applied for foam using as shoes sole application, this early work will preliminarily study the foamability of these blends.

## 2. Experiment

### *2.1. Materials*

Polyethylene vinyl acetate, EVA, N8038 was purchased from TPI POLENE Co. Ltd., Thailand, of which the Tm, density and melt flow index are 86 °C, 0.941 g/cm$^3$ and 2.8 g/10 min respectively. Polybutylene adipate-co-terephthalate, PBAT, Ecoflex F blend C12000 was obtained from POLYMAT Co. Ltd., Thailand, of which its Tm, density and melt flow index are 190°C, 1.260 g/cm$^3$ and 3.8 g/10 min respectively.

### *2.2. Melt Blending of EVA/PBAT*

Before melt blending, EVA and PBAT were dried at 50°C under oven for 6 hours. EVA/PBAT with various composition ratios was fed into a twin screw extruder (SHJ-25 ENMACH, Thailand) with the temperature profile from zone 1-7 and die as follows: 120/120/120/130/130/140/150/120 and various screw speed at 60, 80 and 100 rpm. EVA/PBAT extrudate was collected and cooled to room temperature before being kept in a plastic bag until used. In the case of blend with joncryl4300, the blend was mixed with joncryl4300 before being fed into the extruder.

### *2.3 Characterization*

*Morphology*: The morphology of the EVA/PBAT blend was investigated using scanning electron microscope, SEM (TM3030 Hitachi, Japan). The cryogenic fractured surface of the extrudates were platinum coated using sputtering before investigation under SEM using 15.0 kV accelerated of electron beam produced from tungsten filament.

*Tensile test*: Tensile properties were performed by using Instron universal testing machine (Instron 5900, USA) at room temperature, crosshead speed at 500 mm/min and under load 5 kN follow ASTM D 412 type C.

## 3. Results and Discussion

### *3.1 Morphology of EVA/PBAT*

SEM micrographs of EVA/PBAT with various ratios and different rotor speed of mixing are shown in Figure 1-3. Figure 1 show EVA/PBAT at the composition ratio 10-40 % EVA of which EVA was found dispersed in PBAT phase seen as dark spherical particles except for as high as 40 % EVA where EVA particles appeared as sea-island morphology where EVA presents as island. It could be noticed that the sea-island morphology of EVA in PBAT could be overcome when the screw speed was increased to 100 rpm due to high shearing during melting. For blend systems containing low contents of PBAT shown in Figure2, phase inversion occurred. PBAT was found dispersed in EVA being as major phase and PBAT particles size was increased with PBAT content as shown in Figure2(b). Co-continuous and sea-island morphology were present in the blend with 60% PBAT content [11]. This is caused by the higher MFI of PBAT (3.8 g/10 min) than that of EVA (2.8 g/10 min) resulting larger dispersed phase comparing between 60/40 and 40/60 of EVA/PBAT. In addition, morphology of the blend was changed to co-continuous phase with small PBAT phase size at higher screw speed. At EVA content 70-90 % by weight, PBAT become disperse in EVA. Also its particles were reduced with the high shear rate applied by high screw speed. In fact co-continuous phase morphology has also occurred at screw speed 60 rpm with the content of 50/50 as seen in Figure 3. The higher screw speed led to high shearing and hence sheared PBAT into smaller particles.

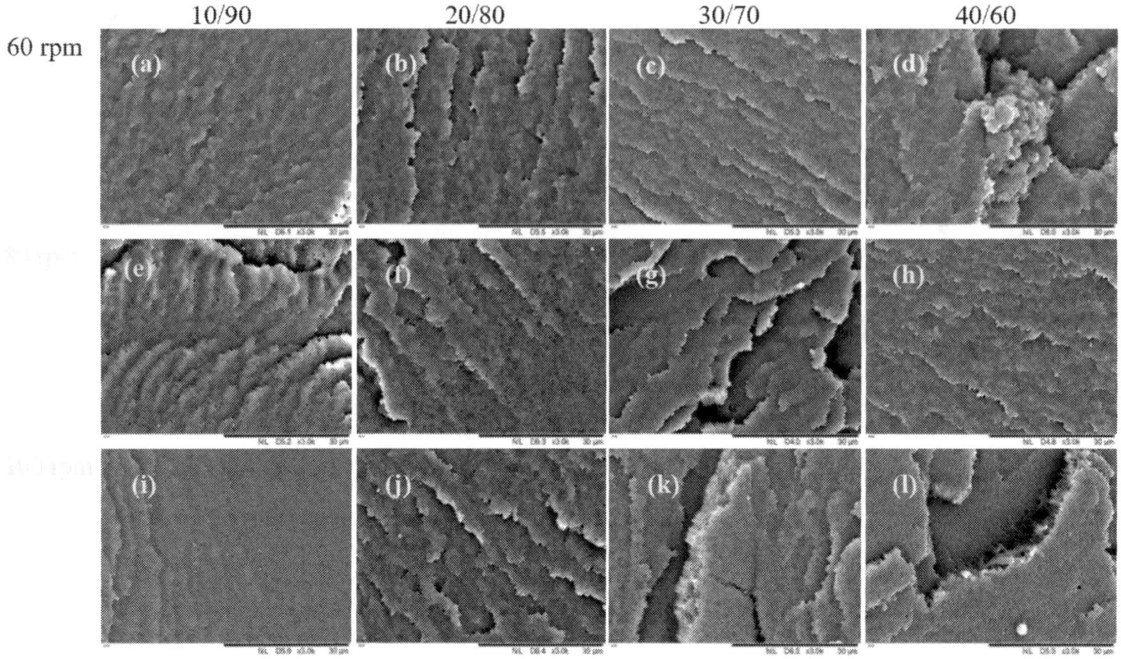


**Figure 1** SEM micrographs of high PBAT composition blends speed of mixing at 60, 80 and 100 rpm.

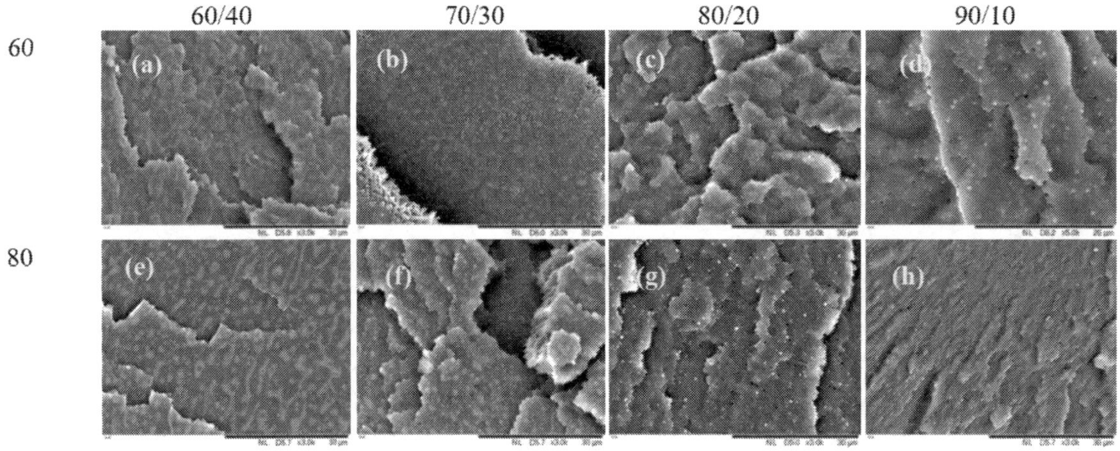

100

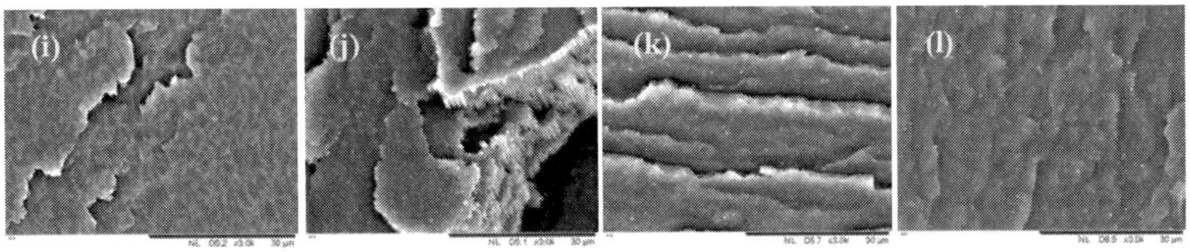


**Figure 2** SEM micrographs of low PBAT composition blends speed of mixing at 60, 80 and 100 rpm.

Joncryl4300 is in fact epoxidized soybean oil (ESO) with molecular weight of 5500 [12], its structure is shown in Figure 6 (c). As shown in Figure 6 (b) that addition of joncryl4300 leads to larger EVA particles (dark domain) compared with unfilled joncryl4300 blends. This could be due to interaction between EVA molecule was increased by the interaction between ESO and EVA molecules as found by the MFI of EVA/ESO [13]. In our research, MFI of PBAT filled with joncryl4300 was decreased from 18.564 g/10min to 11.888 g/10 min where that of EVA was increased from 3.244 to 3.490 g/10 min. These results indicated that addition of joncryl4300 results in high viscosity of PBAT and resulting in shearing applied on EVA phase. As a results EVA phase become changed as shown for in Figure 3.

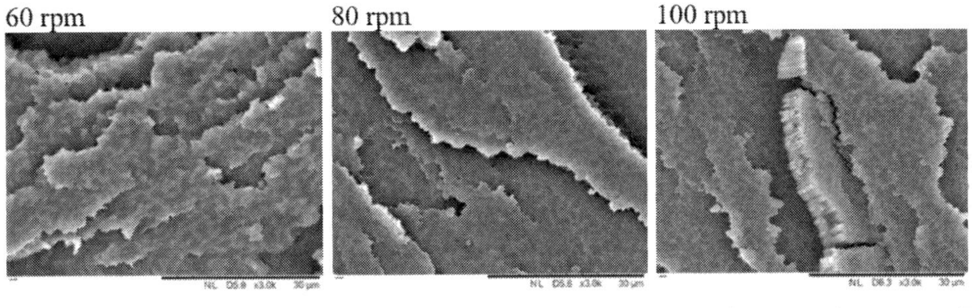


**Figure 3** SEM micrographs of EVA/PBAT blend with composition of EVA and PBAT 50/50.

### 3.2 Tensile Properties

***Effect of rotor speed of mixing***

Tensioning of EVA/PBAT blends were performed at the crossed head speed of 500 mm/min using extrudate extruded from twin screw extruder. Their stress-strain of the representative blend composition of 30/70, 50/50 and 70/30 mixed at different screw speeds at 60, 80 and 100 rpm are exhibited in Figure 4. Tensile properties obtained from stress-strain curves obtained from the blends are also depicted in Figure 5. It could be noticed that stress-strain behavior of the blends shows strain hardening behavior due to their elastomeric character. As the screw speed was increased, dispersed particles become smaller and hence tensile strength was increased especially for the blends containing high EVA contents.

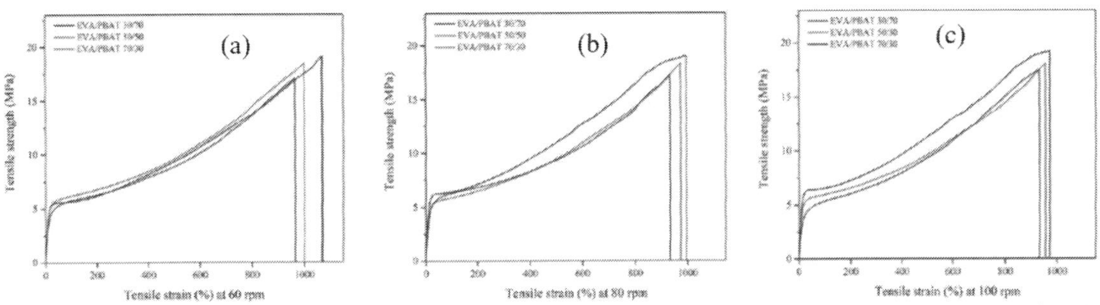


**Figure 4** Stress strain curves of EVA/PBAT blends represented by the ratios 30/70, 50/50 and 70/30 rotor speeds of mixing at 60 (a), 80 (b) and 100 rpm (c).

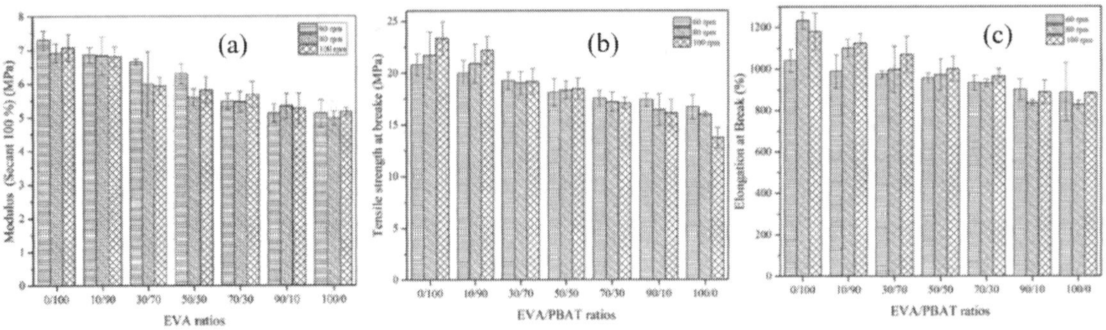


Figure 5 Secant modulus (a), tensile strength at break (b) and elongation at break (c) of EVA/PBAT blend with various blend compositions and different screw speeds.

***Effect of joncryl4300 on EVA/PBAT blend***

As shown in mechanical behavior, it was found that the properties were controlled by EVA domain in the blend. Addition of joncryl43000 was added in order to improve compatibility of the blends. As explained earlier that addition of joncryl4300 led to decreased MFI of PBAT and slightly increased MFI of EVA this results in phase changed of EVA/PBAT as show in Figure 6. The more joncryl4300 added the lower secant modulus of EVA/PBAT blend was obtained, as elucidated in Figure 7. However, tensile strength and elongation at break were not significantly changed by the presence of joncryl4300. This should be due to the elastomeric behavior of EVA/PBAT blends, where strain hardening control strength and elongation at break of the blends.

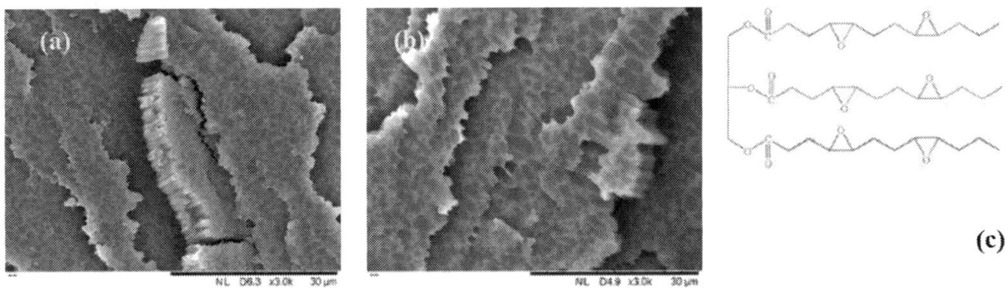


**Figure 6** SEM micrographs of EVA/PBAT blends without processed at 100 rpm. (a) and without (b) Joncryl4300) depicted in (c)

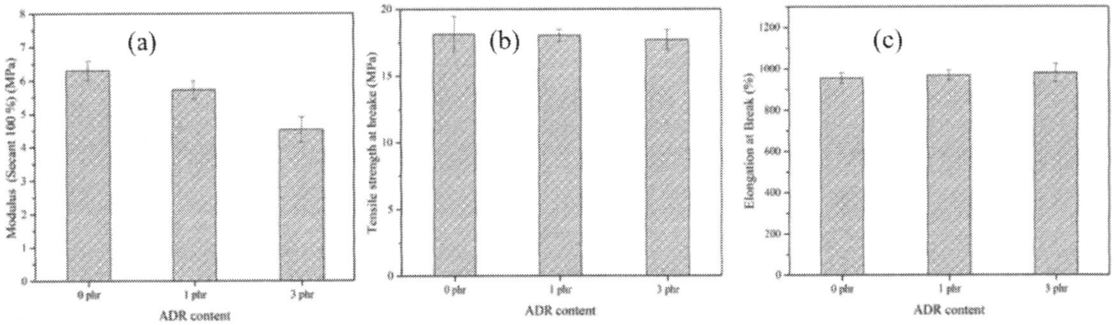


**Figure 7** Secant modulus (a), tensile strength at break (b) and elongation at break (c) of EVA/PBAT blend, composition 50/50 with joncryl4300 content of 0, 1 and 3 phr.

***Foam ability of EVA/PBAT blends***

As mentioned earlier that these blends will be applied to foam formation thus some blends were preliminarily selected and kneaded on a two roll mill with dicumyl peroxide and blowing agent and consecutively form foam under compression molding at 160 °C. Foam morphology od EVA/PBAT blends is shown in Figure 8. Morphology of foam formed show that blends of containing PBAT from 20-60 % could be processed. More information about the foam is going on and will be studied in detail and be reported in our further work.

## Conclusion

EVA/PBAT blends with various EVA and PBAT composition were prepared using twin screw extruder and screw speeds were varied as 60, 80 and 100 rpm. At PBAT contents 40% EVA, PBAT present as co-continuous phase morphology. The morphology did not change due to screw speed. At PBAT as high as 50% by weight, co-continuous phase occurred and the morphology becomes dispersed as screw speed was increased to 80 and 100 rpm. Extrudates obtained from twin screw extruder were tensioned under tensile force. The results show elastomeric behavior of the blends. Elongation at break was higher than 800 % strain and exhibited strain hardening characteristic. Secant modulus of EVA/PBAT blends was increased with PBAT contents due to the high modulus of semi-crystalline PBAT. Content of EVA controls elongation ability of the blends. EVA/PBAT blends with the content of 40/60, 50/50, 60/40, 70/30 and 80/20 were form for foam and found that all composition could be able to be foam. This could be further studied and carried on for shoes sole application.

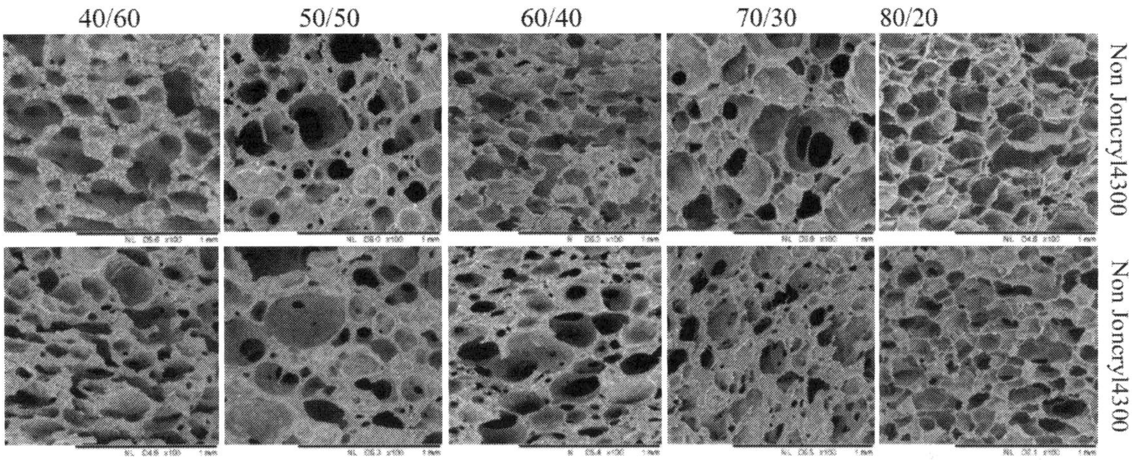


**Figure 8** Foam prepared from EVA/PBAT blends at the content 40/60 50/50 60/40 70/30 and 80/20 without joncryl4300 and with joncryl4300.

## References

[1] Jin, J., Chen, S. and Zhang, J. 2010 *Polymer Degradation and Stability* **95** p 725-732.
[2] Cândida, M. Oliveira, D.C., Sonia, A., Diniz, A., Marcelo, C., Viana, A., Freitas, D.V. and Lins, C. 2018 *Renewable and Sustainable Energy Reviews* **81** p 2299-2317.
[3] Alothman, Y.O. 2012 *Advances in Materials Science and Engineering*. p 1-10.
[4] Mantia, F.P. Morreale, M., Botta, L., Mistretta, C. M., Ceraulo, M. and Scaffaro, R. 2017 *Polymer Degradation. and Stability* **145**(1) p 79-92.
[5] Ferreira, V.F., Cividanes, S.L., Gouveia, F.R. and Lona, L. 2019 *Polymer Engineering & Science* **59** p 7-15.
[6] Liu, W., Liu, S., Wang, Z., Liu, J., Dai, B., Chen, Y. and Zeng, G. 2020 *European Polymer Journal* **122** p 1-8.
[7] Nofar, M., Salehiyan, R., Ciftci, U., Jalali, A. and Durmuş, A. 2020 *Composites Part B: Engineering* **182**(1) p 1-35.
[8] Elhamnia, M., Motlagh, H.G. and Jafari, H.S. 2020 *Polymer Testing* **82** p 1-31.
[9] Sellami, F., Kebiche, S.O., Marais, S., Couvrat, N. and Fatyeyeva, K. 2019 *Reactive and Functional Polymers* **139** p 120-132.

[10] Oliveira, A.T., Oliveira, R.R., Barbosa, R., Azevedo, B.J. and Alves, S.T. 2017 *Carbohydrate Polymers* **168** p52-60.
[11] Dil, J.E., Carreau, J.P. and Favis, D.B. 2015 *Polymer* **68** p 202-212.
[12] Reddy, M. 2014 joncryl ADR (epoxy-functional polymer or oligomers) (online) Available: https://www.slideshare.net/MuniswamyReddy [2020, February 23].
[13] Khankrua, R., Pongpanit, T. Paneetjit, P. Boonmark, R. Seadan, M. and Suttiruengwong, S. 2019 *Polymer MDPI* **11** p 1-17.

# Effect of Glycerol on Gluten filled in Natural Rubber/WG Blend

**C Thongpin[1,*], C Wongprat[1], P Srithavorn[1] and P Muenchee[1]**

[1]Department of Materials Science and Engineering, Faculty of Engineering and Industrial Technology, Silpakorn University, Nakhon Pathom, 73000, Thailand

*Corresponding author: THONGPIN_C@su.ac.th.

**Abstract.** In recently, a large number of utilization of NR has been found in several applications such as foam, insole but they do not provide good properties. In order to improve mechanical properties, in this work, wheat gluten (WG) which is a complex protein derived from wheat was used as a reinforcement. Furthermore, NR with WG (plasticized WG) can reduce costs of materials and easily degrade. NR-WG blend was prepared by mixed NR and WG, plasticized by glycerol, then mixed each other by using two rolls mill and compressed to get specimens. SEM, MDR, tensile testing and tear testing were used to characterize compounds. The effect of glycerol contents added and acting as plasticizer were also studied. The results show that the presence of glycerol also affected vulcanization behavior of rubber matrix. The increasing of glycerol content can also improve dispersion of WG in rubber matrix and indeed improve mechanical properties of NR.

## 1. Introduction

Natural rubber (NR) is widely used in several because its available is chosen for its resilience, resistance to fatigue, low heat built up, to be able to apply over a wide range of operating temperatures. Nevertheless, its vulcanizate is not good enough properties any applications [1-2]. WG is low cost, has good film forming ability and oxygen barrier properties, and has been explored as a biodegradable packaging material. WG can be compressed in a compression mold, which encourages hydrophobic interactions and can induce aggregation and crosslinking, thereby increasing tensile strength while lowering elongation to break [3]. Commercial applications require the stability and reliability for their raw materials sources. Wheat is desirable feedstock for commercial applications because its supply is stable. Wheat gluten is among the largest and most complex protein networks in nature with molecular weight of more than 10 million and is consisted of spherical gluten particles. Wheat gluten is also a low cost raw material and is rarely studied as reinforcements in rubbers. Dry wheat gluten is rigid with a storage modulus of 1 GPa and therefore should provide reinforcement for soft polymers such as rubbers. Reinforcement fillers from natural sources are sustainable, light weight, carbon neutral, biodegradable, and renewable [4-5].Wheat gluten containing plasticizer is able to form network upon thermosetting. Glycerol is usually used as a plasticizer to reduce intermolecular forces and the glass transition so as to increase chain mobility of proteins [6].

As noted above, some studies have reported about blending WG with various thermoplastic polymers. However, based on our knowledge, there have been no reports on blending WG into NR. Thus, the objective of this research is to fabricate a new NR/WG blend which is expected to present good mechanical properties with respect to NR pure. In this work, mechanical properties, including tensile mechanical properties and tear mechanical properties of the blends were investigated. This part of our work will concentrate on the effect of glycerol content on NR/WG blends in term of mechanical properties and morphology including cure characteristic of NR/WG vulcanization.

## 2. Experimental

### *2.1 Materials*

Wheat gluten (WG) was supplied by Chemipan Co. Ltd., Thailand. Natural rubber was produced by Muang Mai Guthrie Public Co. Ltd., Thailand. In this work, to improve mechanical properties of blends, conventional vulcanization (CV) was used to cure NR phases in blends. All chemicals for vulcanization in the rubber phase were used as received including activators, Zinc oxide (ZnO) and stearic acid, accelerators, tetramethyl thiuram disulfide (TMTD) and 2-Mercaptobenzothiazole (MBT) and sulfur as a vulcanizing agent.

### *2.2 Preparations*

Firstly, wheat gluten (WG) was mixed with glycerol with the content of glycerol of 25, 30, 35, 40, 45 and 50 phr. with respect to WG. WG and glycerol were hand-mixed in a bowl to form glycerol plasticized WG. The samples were rested overnight at room temperature before mixing with NR. The 30 percent by weight of WG (after plasticization) was mixed with NR and masticated on a two roll mill (Lab Tech, Thailand) at 60 °C for 40 min. After that the temperature of rolls was set to 40 °C before the incorporation of the following vulcanizing agents which are ZnO (3 phr.), stearic acid (2 phr.), MBT (1.5 phr.), TMTD (0.2 phr.) and sulfur (2 phr.) were added consecutively into the masticated NR/WG blends. The vulcanization behavior of all blends were investigated using MDR (Moving Die Rheometer, Go Tech M2000), at 150 °C, to estimate scorch time ($t_{s2}$) and cure time ($t_{c90}$) which the latter characteristic time will be used as vulcanization time in compression molding machine. All NR/WG with various glycerol contents were compressed at 150 °C under 1500 psi using compression molding machine (Lab Tech. Thailand). The vulcanized NR/WG slabs were obtained and used to prepare test specimens for tensile testing, tear testing.

### *2.3 Characterization*

*2.3.1 Tensile testing*

Vulcanized NR/WG was shaped into dog bone specimens (follow ASTM D412) using pneumatic cutting machine (Chareon Tut, Thailand). Tensile test was conducted according to ASTM D-412 using a universal machine (Instron Crop, model 5969) at room temperature with a 5 kN load cell and test speed of 500 mm/min.

*2.3.2 Tear testing*

Tear specimens were also obtained by cutting vulcanized NR/WG blends using pneumatic cutting machine. Tear test was conducted according to ASTM D-624 using a universal machine at room temperature with a 5 kN load cell and test speed of 500 mm/min.

*2.3.3 Morphological testing*

A scanning electron microscope (SEM, Hitachi TM3030) was used to investigate the surface morphology of cryogenic fractured vulcanized NR/WG blends. The fractured surfaces of all specimens were gold sputtered in order to induce surface conductivity before investigation under SEM.

## 3. Results and discussion

### *3.1. Cure Characteristic parameters of NR/WG blends*

In term of rubber goods, rubber compounds are needed to be examined for their vulcanization characteristic such as, scorch time ($t_{s2}$), cure time ($t_{c90}$). These parameters are related to reaction induction time and curing time used for vulcanization of rubber compounds. Moreover rheological properties of rubber compounds such as minimum torques, ML, and maximum torque, MH, of rubber compound and vulcanized rubber under disc rotating are also investigated. NR/WG blend with various glycerol examined under MDR obtain $t_{s2}$, $t_{c90}$, ML, MH and cure rate index, CRI. The results are showed Figure 1 and Figure 2. The difference between MH and ML denoted as different torque, DT, which is related to modulus of vulcanized rubber, is also obtained ad showed in Figure 2.

It could be seen from the results in Figure 1 (a) that the presence of WG results in slightly decreased of scorch time. Kongkaew and coworkers [7] has reported that the presence of protein in rubber could shorten in scorch time due to mallard effect of protein in which protein could undergo crosslinking. While the glycerol content is increased, the decreased in scorch time was resulted. This could be due to the plasticization of WG and could be able to be sheared during mixing to small particles. The small particles led to being increased surface area of WG particles. This results in the increase in the interaction between WG and curing agents. In the case of NR/glycerol 30 phr., scorch time appeared to be very short. It is important to note that glycerol could also plasticize NR matrix and facilitate the dispersion of curing additive to localize in rubber matrix. Hence the vulcanization induction time is shortened. It was not the case for cure time, Figure 1 (b) where additives undergo chemical reaction with rubber molecules. Glycerol could interact with fatty acid as well as with accelerator thus vulcanization reaction was retarded led to prolong of $t_{c90}$.

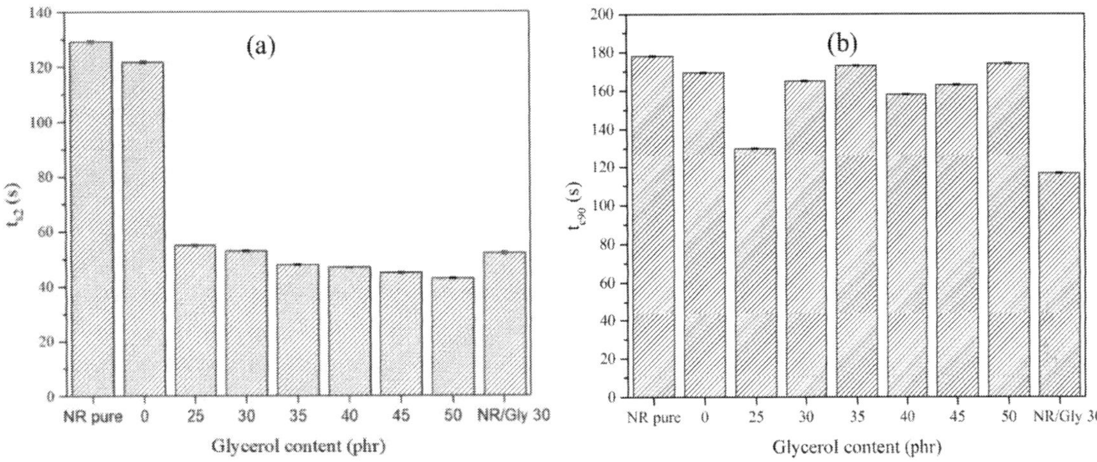


**Figure 1** Cure characteristics of NR/WG compound with various glycerol contents in WG (a) Scorch time, $t_{s2}$ (a) and (b) cure time for 90% vulcanization, $t_{c90}$.

In term of rheological parameter, as shown in Figure 2 (a), i.e. minimum torques, ML, which indicated the viscosity of rubber compound including filler distribution during disk oscillation in MDR. NR compound shows ML around 0.125 N-m whereas the presence of WG leads to increasing of ML. This is due to the obstruction of large WG particle, as can be seen from SEM micrograph in the next section, and hence results in slightly increased ML. The well distribution of WG would also results in difficulty of rubber flow during oscillating shear and results in increased ML. The more dispersion the more rubber flow obstruction occurred. Hence high glycerol content lead to better dispersion and distribution of WG occurred and consequently increased ML. However, ML of NR/WG with glycerol contents seem to show relatively unchanged at glycerol content from 30 phr. In the case of NR/Glycerol, ML is very low compared to other NR/WG compounds. This should be due to the plasticization effect of glycerol on NR.

In term of maximum torque, MH, it is important to note that maximum torque is related to stiffness of vulcanized rubber. The addition of reinforcement into vulcanized rubber should result in high MH. In addition, degree of crosslink should also lead to high MH. As shown in Figure 2 (b), addition of WG leads to increased in MH. This is owing to the high strength of WG. The improved dispersion and distribution of glycerol plasticized WG in NR result in better adhesion and hence promoted reinforcing effect of WG in NR. The increased of maximum torques, MH then obtained. Generally gluten could be crosslinked during kneading [8] that also result in increasing of MH. In fact delta torque, DL, relates to crosslink density of the vulcanized rubber. Thus it should show the same trend as MH. Excess glycerol could also act as plasticizer for NR [9] and this could result in slightly decreased in MH for NR containing WG plasticized with glycerol 50 phr.

In the case of cure rate index which obtain from the following equation; $100/(t_{c90}-t_{s2})$, relate to curing reaction rate of rubber compound. The result is presented in Figure 2 (d) and indicated that vulcanization reaction might be retarded by glycerol localized in WG phase.

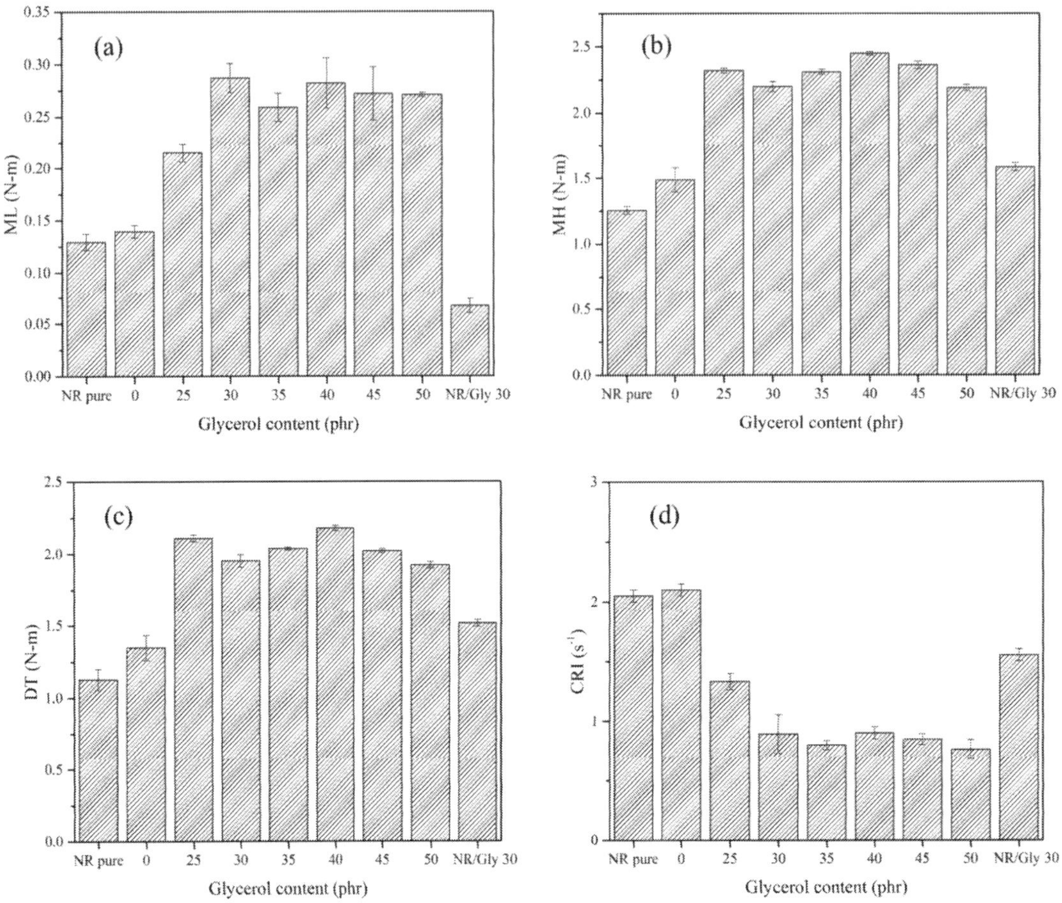


**Figure 2** (a) Minimum torque (ML), (b) maximum torque, (c) delta torque (DT) and (c) Cure rate index (CRI) of NR, NR/30Glycerol, NR/WG with various glycerol contents.

### *3.2. Morphology of NR-WG*

SEM images of cryogenic fractured surface of NR/WG with various contents of glycerol are shown in Figure 3. The results show that NR and WG are immiscible blend due to the high polarity of WG therefore they are not compatible. Glycerol has been one of many plasticizer used for WG. The addition of glycerol into WG, it could be soften and consequently broken to smaller particles and enhance distribution of WG in NR as could be elucidated from Figure 3 (a)-(f). The particles size of WG was decreased with the contents of glycerol. As could be seen from WG particle presents in Figure 3 (a) that it was reduced from 60 μm to about 3 μm in Figure 1 (e). With high plasticizer content, such as at 50 phr., WG started to swell and could be sheared into smaller particles as seen in Figure 3 (f) where the particle size are less than 3 μm. In addition, at high glycerol content could lead to elongated WG particles, as seen from the Figure. Excess of glycerol could also plasticizes NR matrix as shown by the tensile properties, especially elongation at break, see section 3.3.

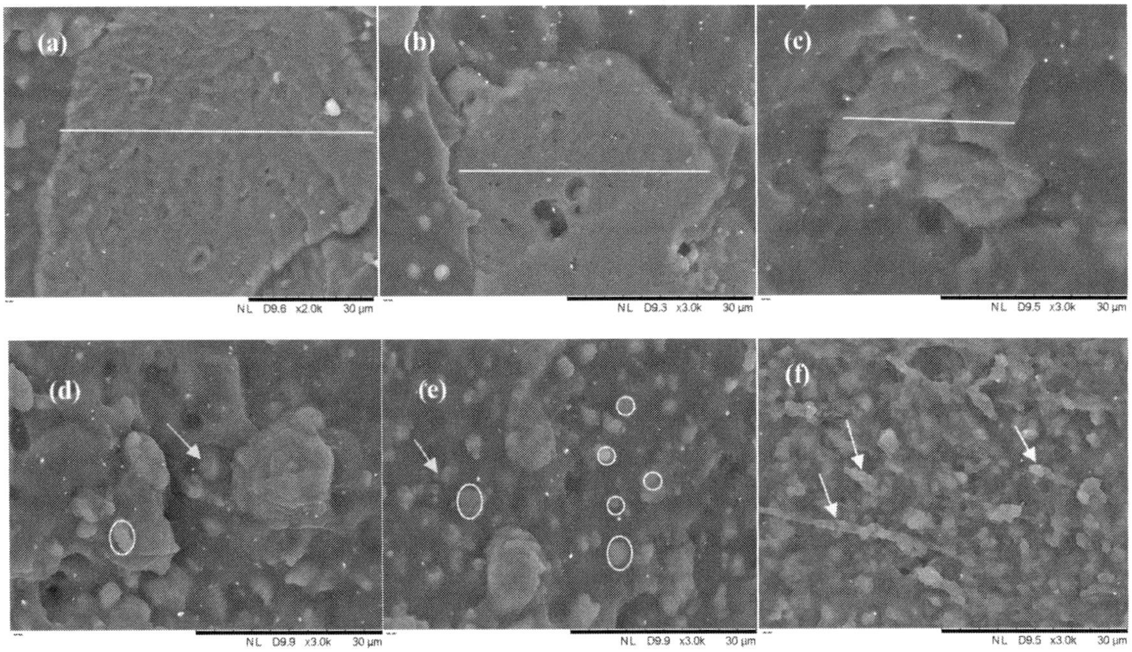


**Figure 3** SEM micrographs of cryogenic fractured surface of vulcanized NR/WG with glycerol contents varied from 25 (a), 30 (b), 35 (c), 40 (d), 45 (e) and 50 (f) phr. with respect to WG.

### *3.3. Tensile Properties*

The results from tensile stress and tensile strain curves of NR/WG blends with various contents of glycerol depicted in Figure 4 (a). The results show that the increased in glycerol content self reinforcement of vulcanized NR/WG increased. This could be due to the plasticization effect of WG by the addition of glycerol. As shown in SEM that the higher the glycerol added the WG particles will be sheared to small particles and well distributed in NR matrix. This results in increasing reinforcing effect of WG on NR matrix, as shown in the increasing of secant modulus at 100% tensile strain. Furthermore, increased contents of glycerol enhance tensile stress due to the stress induced crystallization of NR matrix which was promoted also by the well distribution of WG. The ability to elongate, inducing rubber chain to be aligned parallel with stress direction, should be facilitated by the plasticization effect of glycerol. The plasticization effect including well distribution and dispersion of WG particles results in increased elongation at break of NR/WG blend. However, at 50 phr. of added glycerol, with respect to WG, gave rise to largest elongation at break due to excess plasticization effect of glycerol on NR matrix, i.e. facilitating rubber molecules to be slippage under tension. At this glycerol content may slightly affect secant modulus of the vulcanized rubber. This is general for polymer that the addition of plasticizer could result in low modulus [10-11]. For stress induced crystallization of vulcanized rubber, it is not the same case as ductile plastic, as vulcanized rubber is an elastomer which it could be stretched to more than 1000 % elongation. This is because the polymer chains, i.e. rubber chains, are in rubbery state. Hence the crosslinked chains could be conformed in to a large number of conformations and the stretched specimen size is reduced during being stretched. Crosslink points in the specimen act as linkage to resist stretching of the chain. This is the reason why rubber chains that being stretched to the higher strain will exhibit high resistance to elongate. Schematic picture explains how crosslink in vulcanized rubber affect the high stress at break, tensile strength, is showed in Figure 5. After understanding in stress (or strain) induced crystallization of vulcanized rubber under tension (stretching), one should consider effect of plasticizer that it facilitates the mobility of chain segments [12]. Hence polymer molecules should be able to be stretched to higher strain and consequently enhanced stress induced crystallization. It should also comprehend that stress (or strain) induced crystallization could also know as stress (or strain hardening).

IOP Conf. Series: Materials Science and Engineering **965** (2020) 012033 doi:10.1088/1757-899X/965/1/012033

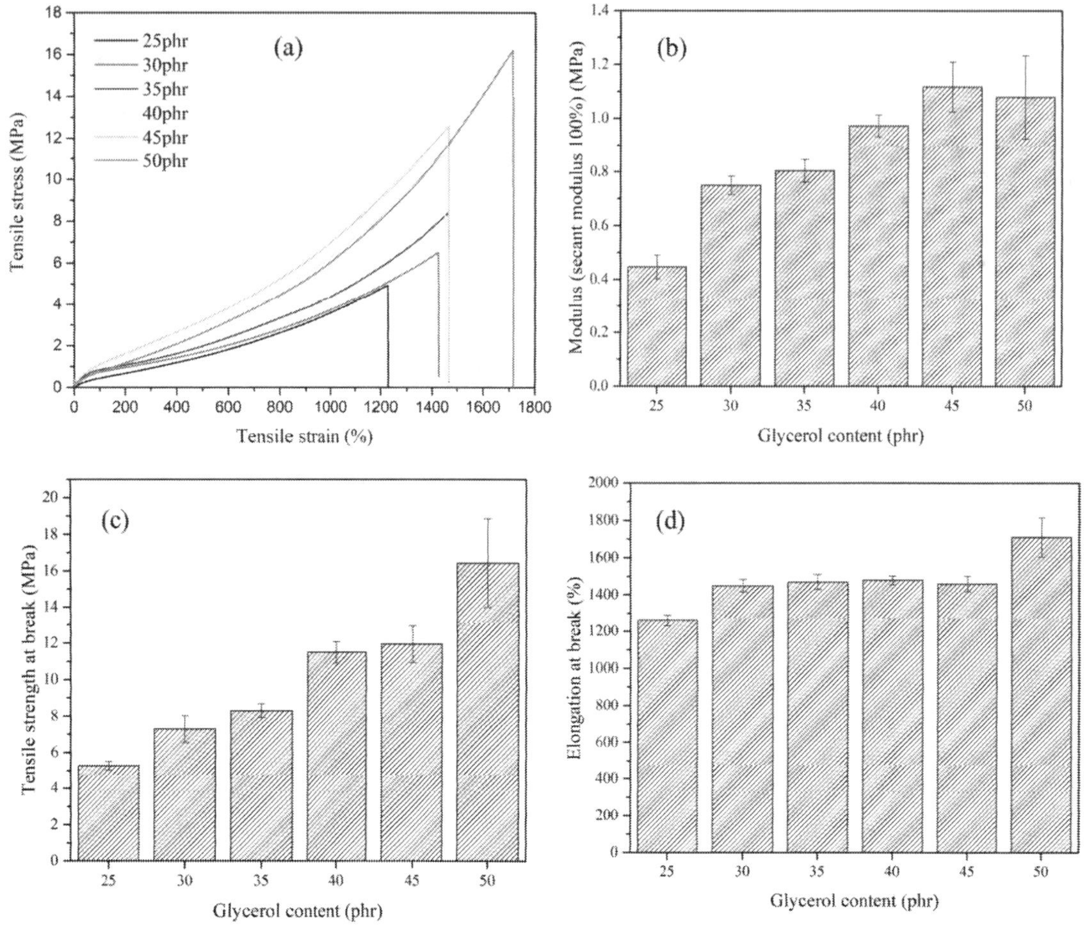


**Figure 4** (a) Stress-strain curves, (b) secant modulus, (c) tensile strength and (d) elongation at break of NR/WG with glycerol content at 25-50 phr. with respect to WG.

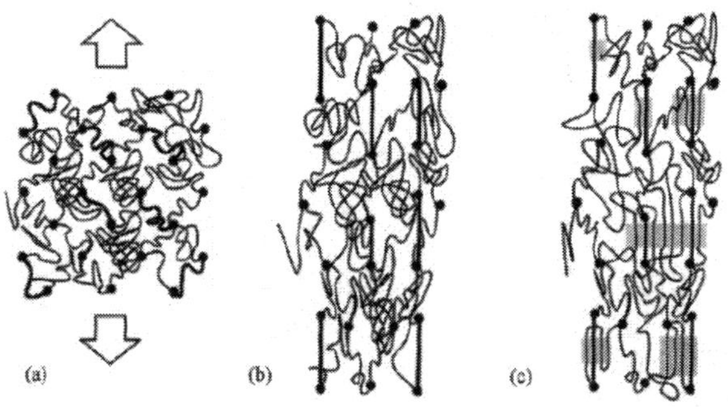


**Figure 5** Relatively short chains are drawn as thick lines. Filled circles represent crosslinks. (a) Before deformation (b) after deformation just before the onset stress induced crystallization and (c) the fully stretched chains have acted as chains alignment (shaded parts). [12]

### *3.4. Tear Testing*

Tear resistance, presented by tear strength, of NR/WG with various plasticization of WG by adding various content of glycerol is depicted in Figure 6. It can be seen that tear strength was increased with the increased contents of glycerol added into WG. This could be due to two reasons. Firstly, as glycerol content was increased WG would be well distributed and dispersed in NR matrix resulting in the promoted adhesion between WG and NR. Also the mallard reaction in WG could lead to the increased in tear resistance of NR/WG blend.

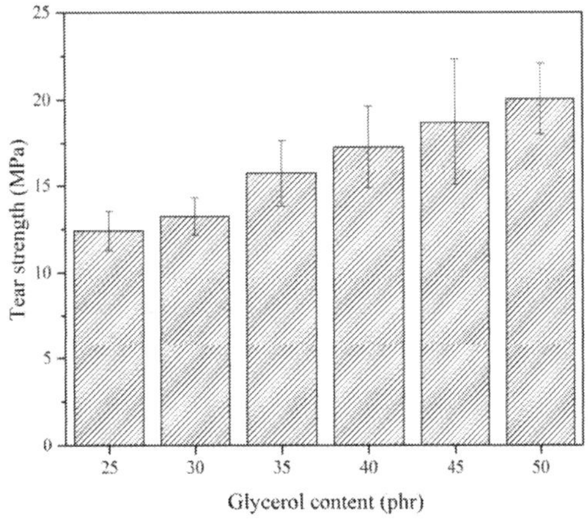


**Figure 6** Tear strength of NR/WG blend with various contents of glycerol plasticization in WG.

## 4. Conclusions

It should be bear in mind that this part of the work, we studied effect of glycerol contents on NR/WG blend vulcanizate where WG content is fixed at 30 % by weight. The fabrication of NR/WG blends with various contents glycerol in WG was varied as 25, 30, 35, 40, 45 and 50 phr. with respect to WG were carried out. The higher contents glycerol added to WG lead to more plasticization of WG and softening of WG. This gives rise to well dispersion and distribution of WG in NR as evidence by SEM micrographs. The distribution and dispersion of WG in NR will result in adhesion and compatibility between WG and NR results. The adhesion between NR and WG led to enhancement in tensile properties and tear properties.

## Acknowledgments

This research is supported by the department of Materials Science and Engineering, Faculty of Engineering and Industrial Technology, Silpakorn University, Thailand.

## References

1. Davies B 1986 Materials & Design **7** 68.
2. Hota A, Adhikari B and Maiti M 2003 Die Angewandte Makromolekulare Chemie **153** 113.
3. DeButts B, Spivey C R and Barone J R 2018 ACS Sustainable Chemistry & Engineering **6** 2422.
4. Tomasik P and Schilling C H 2004 Advances in Carbohydrate Chemistry and Biochemistry **59** 175.
5. Peterson S C and Jong L 2008 Composites Part A: Applied Science and Manufacturing **39** 1909.
6. Song Y and Zheng Q 2008 Food Hydrocolloids **22** 674.
7. Kongkaew C, Intiya W, Loykulnant S and Sae-oui P 2017 *KGK* Kautschuk Gummi Kunststoffe rubber point **70** 37-41.
8. Lucas I, Becker T and Jekle M 2018 Polymers **10** 617.
9. Hemsri S, Thongpin C, Suppatti N, Manomai P and Socharoentham A 2016 Energy Procedia **89** 264.
10. Petrovic ZS, Milic J, Ionescu M and Halladay JR 2017 Rubber Chemistry and Technology, **90** 667.
11. Thongpin C, Sangnil C, Suerkong P, Pongpilaiprertti A and Sombatsompop N 2009 Advanced Materials Research **79** 2171.
12. Huneau B 2011 Rubber Chemistry and Technology **84** 425.

IOP Conf. Series: Materials Science and Engineering **965** (2020) 012034 doi:10.1088/1757-899X/965/1/012034

# Magnetic force microscopy characterization of write field profiles of magnetic recording heads having a pole-tip protrusion

**A Phuagprapun[1], K Saengkaew[1], I Cheowanish[2] and B Damrongsak[1,*]**

[1]Department of Physics, Faculty of Science, Silpakorn University, Nakhon Pathom, 73000, Thailand
[2]Western Digital (Thailand) Co. Ltd., Bang Pa-In, Ayutthaya, 13160

*Email: damrongsak_b@su.ac.th

**Abstract**. Almost all previous studies on a thermal-induced pole-tip protrusion (PTP) were mainly focused on the extension of the main pole which in turn affects the flying height of magnetic recording heads and the hard-disk drive reliability. None was reported the PTP effect on a variation of write filed profiles that is directly related to the track width of the magnetic data. In this work, a method to characterize the magnetic field distribution of perpendicular recording heads with a protrusion in a pole-tip region was proposed and implemented. A magnetic head under test was biased with two current sources, one for the write coil and the other for the heater. The write field was then observed by magnetic force microscopy (MFM). MFM images were post-processed in order to investigate the field profiles. Two models of magnetic write heads having different heater designs were employed as test samples. We observed the variations in the area of the write field for the write heads having the protrusion of the pole tip, while the magnitude of the field intensity was almost constant.

## 1. Introduction

As there is a demand to increase the data storage density in hard disk drives (HDDs) in order to make them competitive to the rapid growth of solid state drives, the physical spacing, or the so-called flying height (FH), between the main pole and the surface of the recording media tends to be decreased [1],[2]. A FH adjustment is typically controlled by thermal heating by applying the DC current to the heater element or by Joule heating due to the biased write current during writing of data on recording media [3],[4]. The heat generated by the FH adjustment causes thermal expansion of the main write pole, referring to as the pole-tip protrusion (PTP) that in turn reduces the spacing gap between the recording head and the magnetic disk. Both thermal-induced and write current-induced PTPs have extensively been studied by many researchers, by either finite element modeling (FEMs) [5] or experimental analysis [6],[7]. Almost all works are mainly focused on thermal expansion of the main pole and the effects of the PTPs on the disk stability. There is no previous work reporting the effects of the PTP on the writing performance.

In our previous work, we employed magnetic force microscopy (MFM) to observe the write field [8] as well as the remanent field on the write element [9]. In this study, we developed additional function for imaging the magnetic write field generated from a thermally protruded main pole of magnetic recording heads. This can be used to study the effects of the pole-tip protrusion on the intensity and width of the write field. Moreover, the image of the write field profiles can be further utilized to analysis the failure in the magnetic writer fabrication. In the next section, we will discuss the experiment setup and results of two different models of magnetic write heads.

## 2. Experimental Setup

The MFM setup used for the characterization of the write field profile was illustrated in Figure 1. Magnetic write heads in a form of sliders or rowbars were equipped on the xyz piezoelectric positioner. A 4-pin probe fixture was used as the interconnection between the write heads and two current sources (Keithley SMU2400), one is for the write coil and the other is for the heater. During MFM imaging, the write coil was biased by a DC current in order to generate the magnetic write field from the tip of the main pole. The DC current can be applied to the heater to activate the pole-tip protrusion. A soft magnetic coated MFM probe, scanning at 10 nm height across the slider surface, was employed to detect the magnetic stray field. The scan size is 800 nm x 800 nm with a resolution of 256 pixels x 256 pixels.

A MFM image of the write field generated by magnetic head samples as shown in Figure 2(a) was constructed from a phase shift between the piezoelectric driving force and the oscillation of the MFM probe. The area shown in red represents large phase shift corresponding to the high intensity of the magnetic write field; on the other hand, the area shown in blue represents the low magnitude in the phase shift, implying that there is no magnetic field in the area. Figure 2(b) shows the post-processed contour plot of the MFM image for a comparison of the write fields between the protruded and unprotruded main pole.

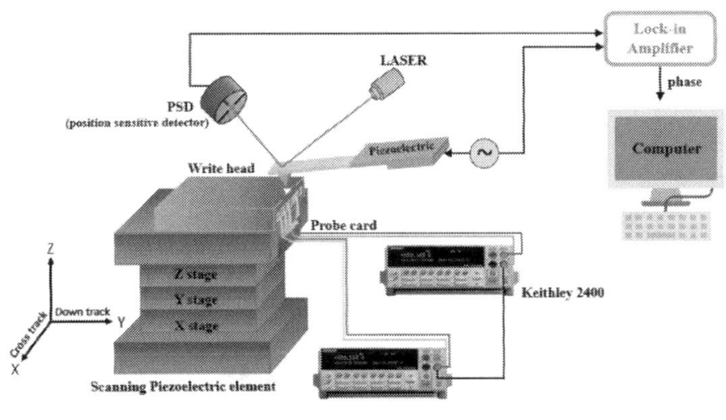


**Figure 1.** Schematic diagram of MFM system for the characterization of the magnetic write field.

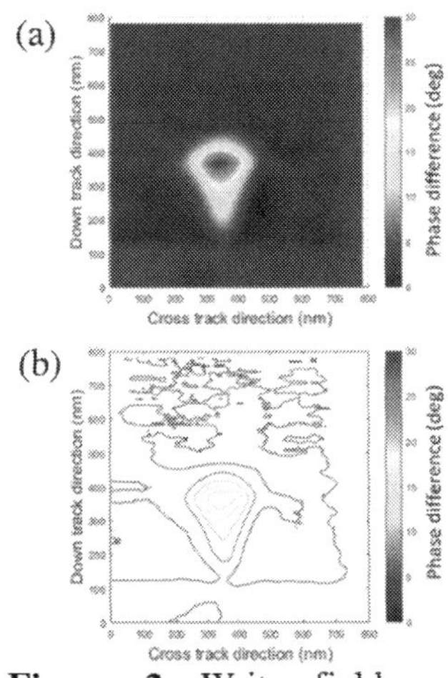


**Figure 2.** Write field profile observed at a main pole tip of magnetic write heads when the write coil is biased with a DC current. (a) MFM phase contrast image and (b) post-processed contour plot of the MFM image.

## 3. Results and discussions

Perpendicular magnetic write heads with a fully wrap around shield were used as test samples for the experiment. Two head designs, i.e. model A and B, were investigated. These samples have different pole and heater designs. For the model A, the write heads have shorter main pole and larger gap spacing between the pole and the side shields. The pole width of the model A is in a range of 40 – 70 nm. On the other hand, the main pole of the model B is longer but the gap is significantly narrower. The pole width of the model B is relatively larger, ranging from 60 – 80 nm, approximately. The view of the write heads on the air-bearing surface (ABS) are illustrated in Figure 3. In addition, the heater design of the model A is a dual heater in order to make the protrusion in both the writer and the reader regions simultaneously. For the write head in model B, the heater can cause the protrusion only on the main writer pole. Those model A and B are required different magnitudes of the biased DC current to make the pole fully saturated, i.e. 40 mA for model A and 30 mA for model B. The MFM images of the write field profile for the writer having model A and B are shown in Figure 4. No DC current was applied to the thermal heater yet. As can be seen, the shape of the write field profiles was similar to the physical geometry of the main pole.

For the sake of demonstration, the write coil was biased with the required DC current and the pole-tip protrusion (PTP) was activated by applying the DC currents to the heater coil. DC currents ranging from 10 – 40 mA, which are equivalent to the heater power of 60 mW to 100 mW, were applied to the heater.

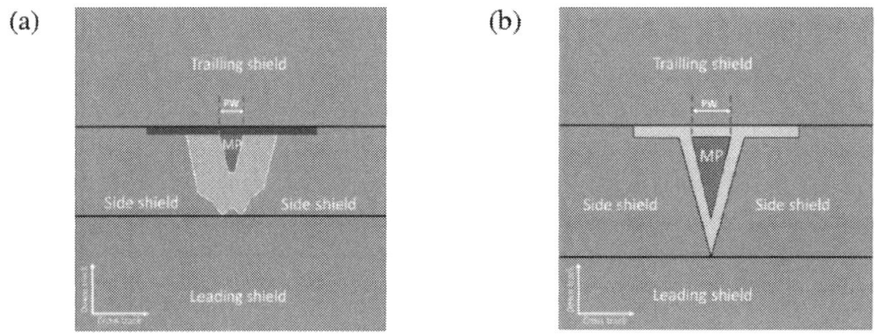


**Figure 3.** Drawing showing the ABS view of magnetic write heads (a) Model A and (b) Model B. MP is stand for a main pole and PW is for a pole width.

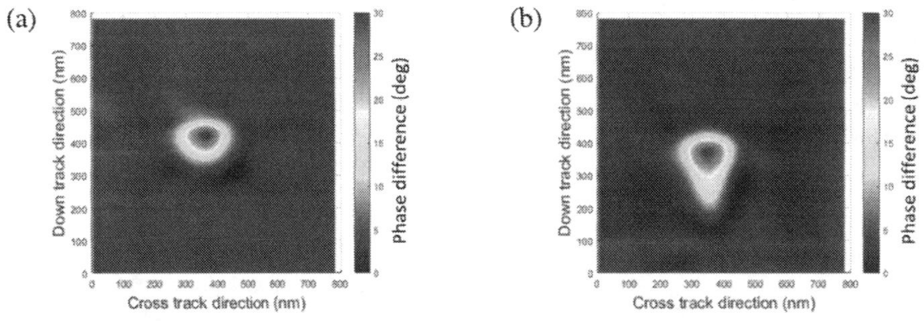


**Figure 4.** MFM images of the magnetic write field for the writer with (a) model A and (b) model B.

Figure 5 shows MFM images and contour plots of the write field generated from the model A write heads having 3 different values of magnetic write width (MWW), i.e. 58.8, 65.5 and 68.7 nm.

For the sake of comparison, the area of the write field profiles was calculated at 5 and 15 degree phase shifts. We then compared the area for the writer with different heater currents from 0 to 40 mA. It was found that the area of the write field for the model A write heads was narrower when the heater element was biased and the area was independent to the amount of applied heater currents. For the model B, on the other hand, the area of the write field profile of the protruded writers was relatively wider than ones without the PTP. We also observed the changes of the write profile area when the heater current increases. Results for the model B are shown in Figure 6. This finding can lead to the further investigation on the effects of the PTP on the writing performance as well as the effects of heater designs on the device stability and reliability.

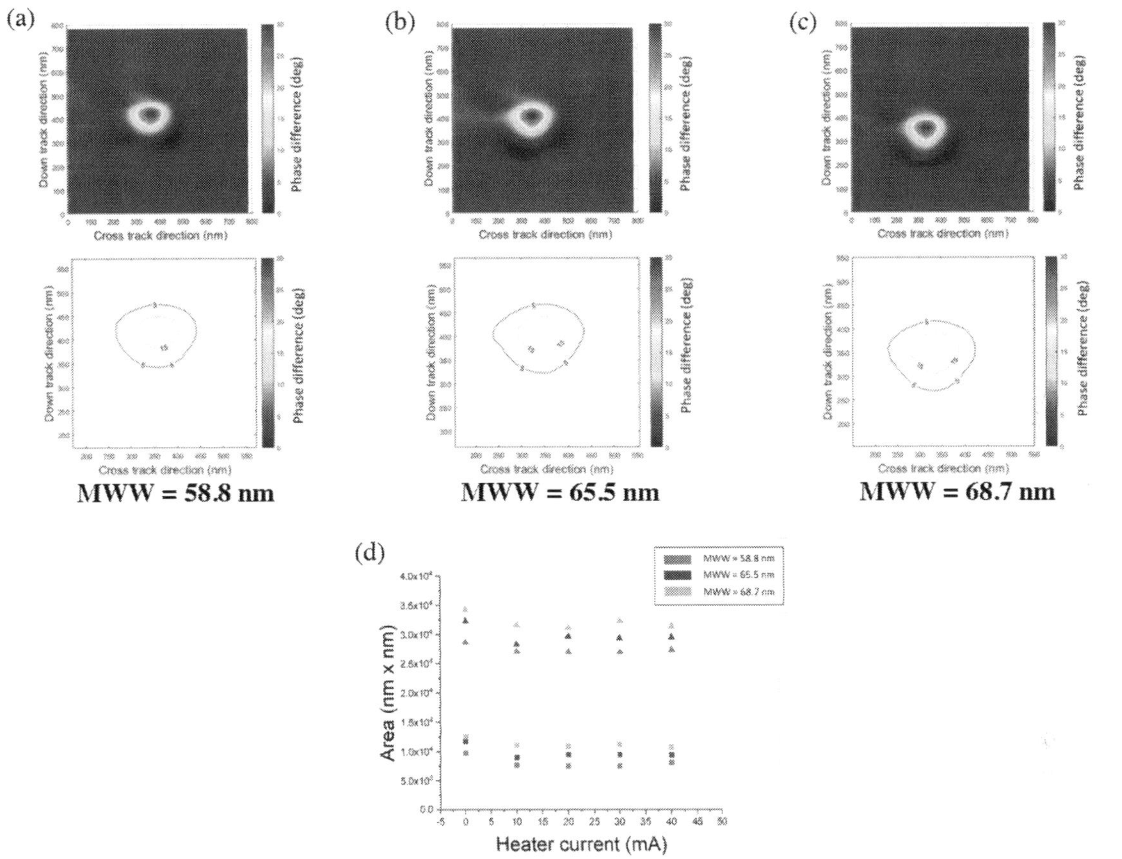


**Figure 5.** MFM images contour plots of the write field of model A having 3 different values of MWW. (a) MWW = 58.8 nm (b) MWW = 65.5 nm and (c) MWW = 68.7 nm. (d) Calculated area of the write field profiles corresponding to different heater currents at 5 degree (▲) and 15 degree (■) phase shifts.

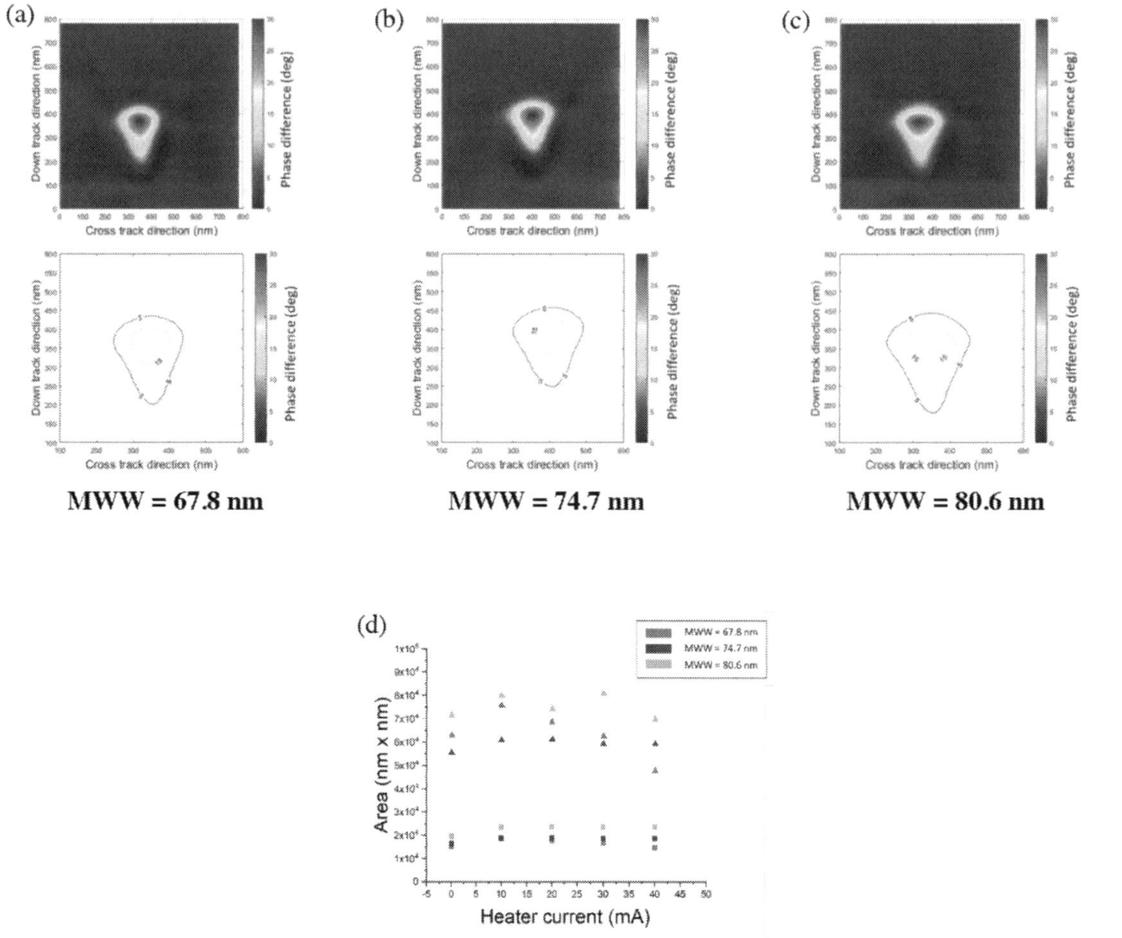


**Figure 6.** MFM images contour plots of the write field of model B having 3 different values of MWW (a) MWW = 67.8 nm (b) MWW = 74.7 nm and (c) MWW = 80.6 nm. (d) Calculated area of the write field profiles corresponding to different heater currents at 5 degree (▲) and 15 degree (■) phase shifts.

## 4. Conclusions

MFM images of the write field for magnetic heads having the PTP were demonstrated in this study. A given DC current as high as 30 – 40 mA was applied to the write coil to generate the saturated magnetic field on the main pole. The pole protrusion was activated by applying the DC current to the heater element. Results of the demonstration revealed that the profiles of the write field were altered when the protrusion was present. Moreover, we observed the variation of the write field profile between the model A and model B writers. This can lead to furture investigation of the PTP effects on the write field distribution in the future development of magnetic write heads.

## 5. References

[1] Boettcher U et al. 2011 *IEEE Trans. Magn.* vol 47 no. 7 pp 1823-1829
[2] Zhang C et al. 2014 *IEEE Trans. Magn.* vol. 50 no. 11 pp 3303104
[3] Aoki K et al. 2005 *IEEE Trans. Magn.* vol 41 no. 10 pp 3043-3045
[4] Ong C, Ang S and Yuan Z 2012 *IEEE Trans. Magn.* vol 48 no. 11 pp 4598-4601
[5] Juang J, Forrest J, and Huang F 2011 *IEEE Trans. Magn.* vol 47 no. 10 pp 3437-3440
[6] Chen Y, Peng J, David B 2014 *IEEE Trans. Magn* vol 50 no. 11 p 3302805
[7] Nikitin V et al. 2004 *IEEE Trans. Magn.* vol 40 no. 1 pp 326–331
[8] Sealim B et al. 2017 *J. Phys.: Conf. Ser.* **901** 012093
[9] Dilekrojanavuti P et al. 2017 *J. Phys.: Conf. Ser.* **901** 012091

**Acknowledgments**

Authors would like to thank Western Digital (Thailand) Co., Ltd. and Department of Physics, Faculty of Science, Silpakorn University for financially supports of this work. Special thanks to WD engineers and technical staffs in test engineering department for all technical contributions and helpful discussions.

IOP Conf. Series: Materials Science and Engineering **965** (2020) 012035 doi:10.1088/1757-899X/965/1/012035

# A Dual-Band Printed Dipole Antenna for Unidirectional Radiation

**Sompongpinit Suwanachart[1,a], Duangarthit Srimoon[2,b], and Jhirat Mearnchu[3,c]**

[1] Student in Master of Electrical and Computer Engineering Program, College of Engineering, Rangsit University, 52/347 MuangAke, Phaholyothin Road, Lakhok, Pathumthani, 12000, Thailand.

[2] Department of Computer Engineering, College of Engineering, Rangsit University, 52/347 MuangAke, Phaholyothin Road, Lakhok, Pathumthani, 12000, Thailand.

[3] Department of Electrical Engineering, Faculty of Engineering and Industrial Technology, Silpakorn University, 6, Rajamankha Nai Road, Muang, Nakornpathom, 73000, Thailand.

[a] sompongpinit.s@rsu.ac.th, [b] duang-arthit@rsu.ac.th, [c] jhirat@su.ac.th

**Abstract**. This paper presents a dual-band printed dipole antenna for unidirectional radiation. The antenna is designed for the Wireless Local Area Network (WLAN) system that operating on both the frequency ranges of 2.400 – 2.483 GHz and 5.15 – 5.85 GHz. Radiated elements of the antenna are flat conductors on the surface of FR4 printed circuit board. All elements are on the same side of FR4. Coaxial line is chosen for feeding. It makes the antenna good matching and easy to assembly. The antenna is simulated by computer program before fabricating a prototype. It provides omni-directional radiation but has a direction with the gain about 5 dBi at the central of both operating frequencies. It is appropriate for the system that desired the signal to radiate surround but has some direction together. The size of the antenna is 44.0 x 60.0 $mm^2$. Advantages of the antenna are low profile structure, light weight, and can be fabricated with cheap material. So, it is a low-cost antenna.

## 1. Introduction

There are many wireless communication systems such as the Digital Video Broadcasting - Second Generation Terrestrial (DVB-T2) operate with the frequency range of 470 - 862 MHz, the Digital Cellular System (DCS) operate with the frequency range of 1.71 - 1.88 GHz, the Personal Communications System (PCS) operate with the frequency range of 1.85 - 1.99 GHz, the Universal Mobile Telecommunications System (UMTS) with the frequency range of 1.92 - 2.17 GHz, the Worldwide Interoperability for Microwave Access (WiMAX) operate with the frequency range of 2.496–2.690 GHz, and the Wireless Local Area Network (WLAN) operate with the frequency ranges of 2.400 - 2.483 GHz and 5.15 – 5.85 GHz. However, all systems must use the antenna for transmitting and receiving signals. Thus, many antennas are developed to support each system above. The antenna that operated cover more than one system is called broadband antenna [1] while the antenna that can be operated cover only 2 frequency bands together is called dual-band antenna [2].

Most of broadband antennas are wideband antennas. Wideband antennas had been proposed in many research works. For example, the planar monopole [3] that radiated in omni-direction, it provides a very wide bandwidth but had distort beam at the high frequency of its band. For the bi-directional radiation such as the wide slot antenna [4], although it has a wide bandwidth but squeeze beams occurs at the high frequency of the band as well. For uni-directional radiation antennas, there are many types of the antenna that provide the uni-directional radiation but their bandwidths are not wide enough. Form the problem of distort-beam at the high frequency of wideband antenna, the dual-band antenna is proposed.

Recently, various antenna designs such as those of micro-strip line, coplanar waveguide, and printed dipole antennas have been proposed in the literature. A printed dipole antenna with U-slotted arms was demonstrated in [5]. From the aid of an embedded U-shaped slot, it can generate a new resonant mode at 5.2 GHz. In [6], printed dipole antenna with parasitic elements was presented. The measured results show that parasitic strips can be coupled with dipole antennas for providing additional resonance modes. In [7], a broadband printed dipole antenna with a step shaped feed gap was presented. The wide operating band is controlled by the different lengths of the radiation arms. Therefore, it was shown that a dipole antenna with parasitic elements has the ability to give multiband operation.

This paper describes the dual-band antenna that can be used to operate with WLAN system in the frequency ranges of 2.400–2.483 GHz and 5.15–5.85 GHz. A prototype that operates in the both frequency bands is developed. By adding 3 parasitic strips with a dipole printed on the same side of the FR4, a WLAN antenna that actives dual-band operation from 2.400 – 2.483 and 5.15 - 5.85 GHz is realised. The dipole and parasitic elements are able to produce three resonant modes. Additionally, the reflection coefficient, far-field radiation pattern and antenna gain of the proposed antenna are presented.

## 2. Antenna description

The analysis and design was conducted using the simulation software at first. After the optimization by software, we received the antenna dimensions through the parameters which depend on the reflection coefficient such as the length of the driven dipole. Few parameters are approximated such as the size of the substrate. To confirm the accuracy of the simulation, the prototype of the antenna is fabricated with the same dimensions and materials as receive from the simulation except the feeding.

A structure of the proposed antenna is shown in Figure 1. (a) while the prototype of the antenna is shown in Figure 1. (b). The antenna was contributed on an FR4 substrate with a thickness of 1.6 mm and a permittivity of 4.8. The overall dimensions of the structure were 44.0 x 60.0 ($a$ x $b$) mm$^2$. The antenna was fed at the centre by a 50 Ω coaxial line (for the prototype while discrete port for the simulation) that connects to two symmetrical rectangular dipole arms (driven element). The length of the dipole, $l_1$, is used to determine the low frequency (2.400 – 2.483 GHz) band, and the width $s_1$ is adjusted to vary the impedance for the desire of impedance matching. To increase the resonant modes of the high frequency (5.15 – 5.85 GHz), 2 parasitic elements were added to the rectangular dipole. The long parasitic element (director), $l_3$, determines the high frequency band of the low-resonant mode (at 5.15 GHz). And the width $s_3$ is adjusted to vary the impedance for the desire of impedance matching. The short parasitic element, $l_4$, determines the high-resonant mode (at 5.85 GHz), while the width $s_4$ is adjusted for the desired impedance as well. The resonant band is obtained through coupling between the dipole and the 2 parasitic strips. In addition, the band of low-resonant mode (5.15 GHz) was able to combine with the band of high-resonant mode (5.85 GHz), resulting in operation from 5.15 to 5.85 GHz.

For the high frequency, we increased the gain of the antenna into the z-direction by extending the width of the director parasitic strip, $s_3$, in the z-direction. For the low frequency, we increased the gain of the antenna into the z-direction by adding the reflector parasitic strip width of $s_2$ and length of $l_2$ as in Figure 1. (a). The reflector is far from driven dipole $d$. In this way, a dual-band dipole antenna for a unidirectional radiation for WLAN applications was designed. The optimal antenna parameters for WLAN are given as in Table 1.

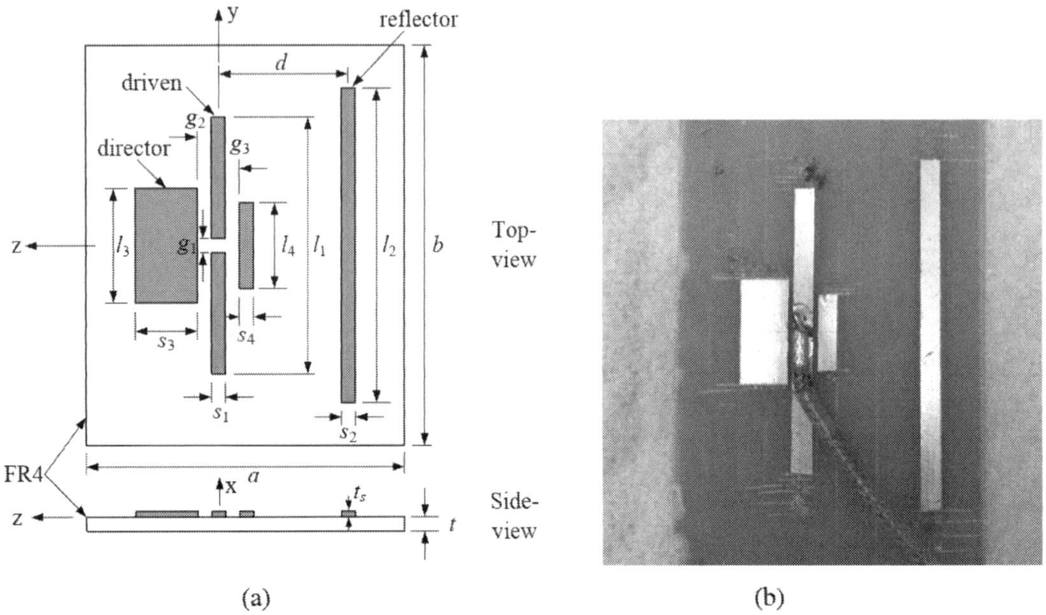


**Figure 1.** (a) Structure of the proposed dual-band printed dipole antenna (b) Prototype of the proposed dual-band printed dipole antenna

**Table 1.** Parameters and their dimensions of the proposed dual-band printed dipole antenna

| Parameter | Dimension (mm) | Parameter | Dimension (mm) |
|---|---|---|---|
| $s_1$ | 3.0 | $g_1$ | 1.5 |
| $l_1$ | 43.0 | $g_2$ | 0.5 |
| $s_2$ | 3.0 | $g_3$ | 0.5 |
| $l_2$ | 54.0 | $a$ | 44.0 |
| $s_3$ | 8.0 | $b$ | 60.0 |
| $l_3$ | 15.0 | $d$ | 18.0 |
| $s_4$ | 3.0 | $t$ | 1.6 |
| $l_4$ | 10.0 | $t_s$ | 0.05 |

## 3. Simulated and measured results

### *3.1. Simulated and measured reflection coefficients*

The simulated and measured reflection coefficients are shown in Figure 2. The simulated result of the reflection coefficients is agrees with the experimental. The measured reflection coefficients of the antenna provides the bandwidths of 6.3% and 12.4% (VSWR<1.5:1) between the frequency range of 2.3 - 2.5 GHz and 5.0 - 5.9 GHz, respectively. However, the range of measured bandwidth is slightly wider than the simulated result. The discrepancy is mainly due to the simulation of the feeding that using the discrete port while the fabricated antenna is fed by 50 Ω coaxial line.

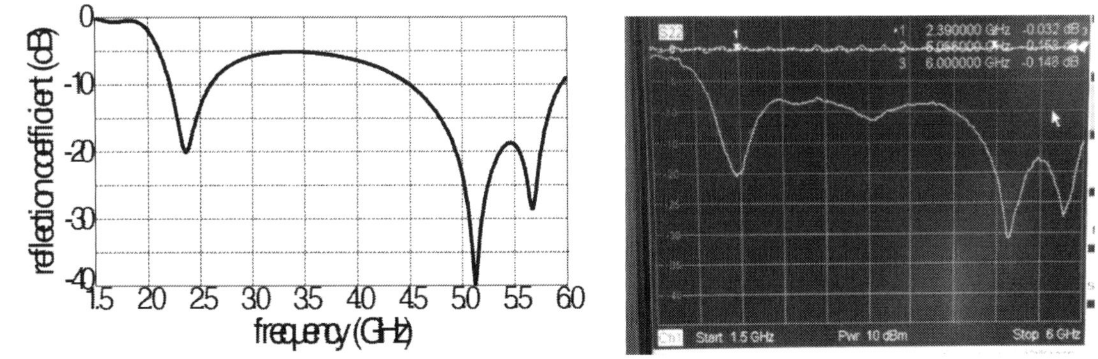


(a) (b)

Figure 2. Simulated (a) and measured (b) reflection coefficient of the proposed antenna

Almost parameters are varied to show the characteristics of the proposed antenna. The first parameter is $l_1$ or the length of dipole. The results of reflection coefficient with different $l_1$ are showed in Figure 3. It is found that $l_1$ is influence to the low frequency (2.400 – 2.483 GHz) band. When $l_1$ is increase from 40.0 mm to 43.0 mm and to 26 mm, the low frequency band is shelved to low.

The second parameter is $l_2$ or the length of reflector. The results of vary $l_2$ is showed in Figure 4. It is found that $l_2$ is little influence to the low frequency band. When $l_2$ is increase from 50.0 mm to 54.0 mm and to 58 mm, the impedance of the low frequency band is increase.

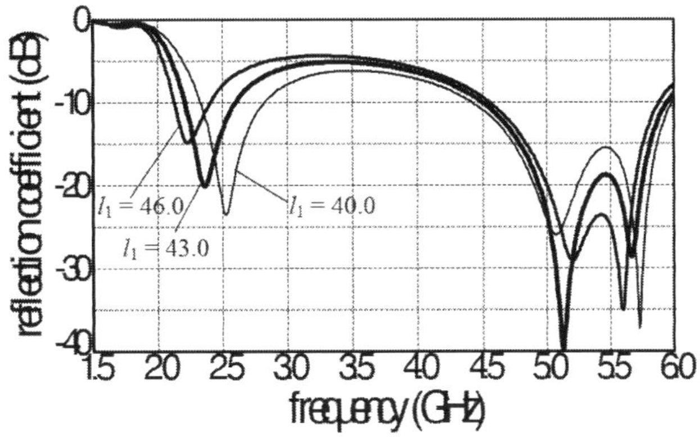


**Figure 3.** Simulated reflection coefficients when changed parameter $l_1$

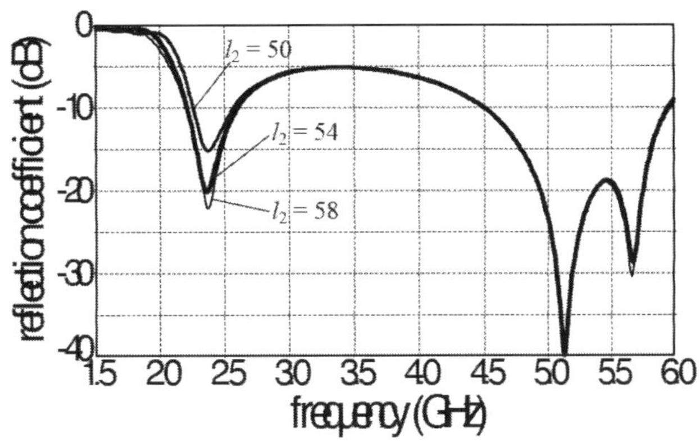


**Figure 4.** Simulated reflection coefficients when changed parameter $l_2$

The third parameter is $l_3$ or the length of the director. The results of reflection coefficient with different $l_3$ are showed in Figure 5. It is found that $l_3$ is influence to the high frequency (5.15 – 5.85 GHz) band, especially for low-resonant mode. When $l_3$ is increase from 14.0 to 15.0 and to 16 mm, the low-resonant mode is shelf to high make the bandwidth decrease.

The next parameter is $s_3$ or the width of the director. The results of vary $s_3$ is showed in Figure 6. It is found that $s_3$ is little influence to both the low and high frequency bands but it is used to determine the gain of the antenna to direct to the z-direction at the high operating frequency. While $g_2$ is explicitly effect to the high frequency band as showed in Figure 7.

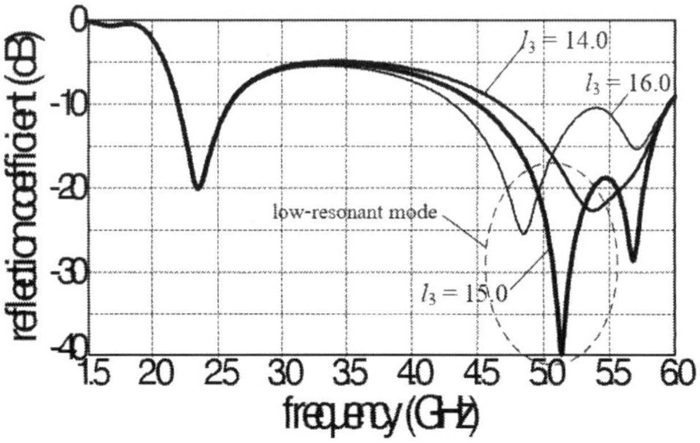


**Figure 5.** Simulated reflection coefficients when changed parameter $l_3$

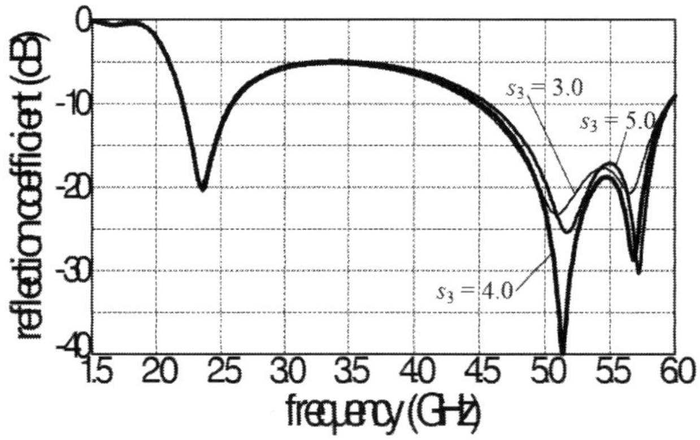


**Figure 6.** Simulated reflection coefficients when changed parameter $s_3$

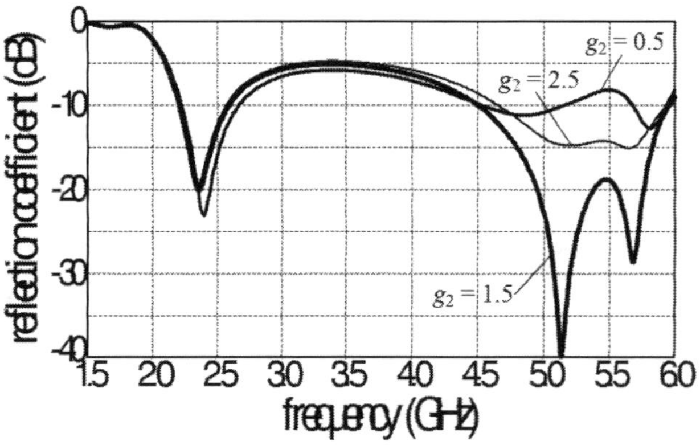


**Figure 7.** Simulated reflection coefficients when changed parameter $g_2$

The next parameter is $l_4$ or the length of the shortest parasitic. The results of reflection coefficient with different $l_4$ are showed in Figure 8. It is found that $l_4$ is used to determine the high frequency (5.15 – 5.85 GHz) band, especially for high-resonant mode. When $l_4$ is increase from 8.0 to 10.0 and to 12 mm, the high-resonant mode is shelf to low make the bandwidth decrease.

The next parameter is $s_4$ or the width of the shortest parasitic. The results of vary $s_4$ is showed in Figure 9. It is found that $s_4$ is little influence to both operating frequency bands but it is used to adjust the radiation pattern direct to the z-direction at the high operating frequency. While $g_3$ is explicitly effect to the high frequency band as showed in Figure 10.

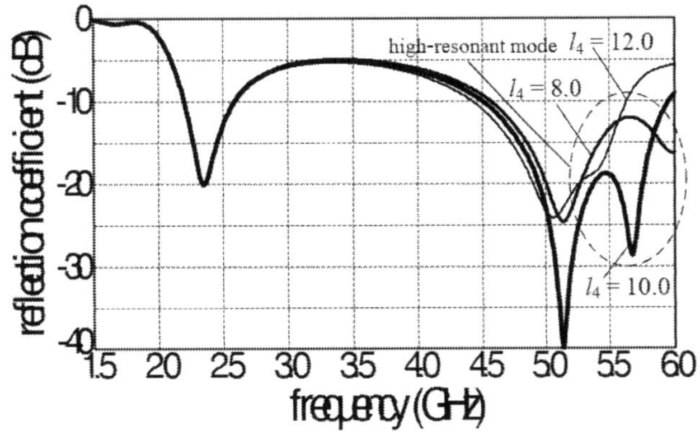


**Figure 8.** Simulated reflection coefficients when changed parameter $l$.

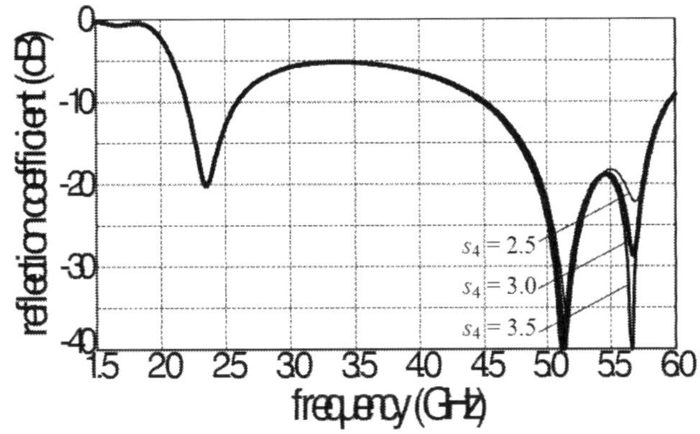


**Figure 9.** Simulated reflection coefficients when changed parameter $s$.

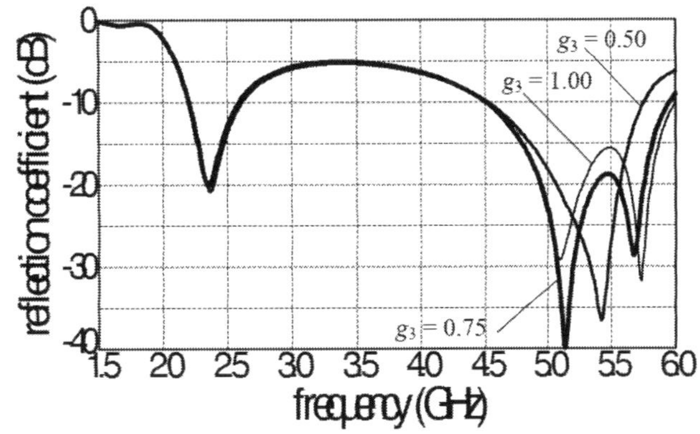


**Figure 10.** Simulated reflection coefficients when changed parameter $g_3$

IOP Conf. Series: Materials Science and Engineering **965** (2020) 012035 doi:10.1088/1757-899X/965/1/012035

### *3.2. Simulated and measured radiation patterns*

The experimental was operated in open area to decrease the effect of the reflection. The observed operating frequencies are 2.44, 5.15, and 5.85 GHz as shown in figure 11, 12, and 13, respectively

Figure 11 shows the normalized of simulated H plane and E plane (or *yz*-plane and *xz*-plane, respectively, as indicated the coordinate and direction in Figure 1) co-polarization patterns compare with the measured results at 2.44 GHz. It was found that the front-to-back ratio is 8 dB. The half power beam-widths (HPBW) are 176° for yz-plane and 77° for xz-plane. The patterns in both plane are symmetry.

Figure 12 shows the simulated and measured co-polarized radiation patterns at 5.15 GHz. It was found that the front-to-back ratio is 12 dB. The HPBW are 177° for yz-plane and 68° for xz-plane. Figure 13 shows the simulated and measured co-polarized radiation patterns at 5.85 GHz. It was found that the front-to-back ratio is 14 dB. The HPBW are 160° for yz-plane and 68° for xz-plane.

However, gains of the antenna at the three points are 4.8 dBi, 5.5 dBi, and 8.0 dBi, respectively.

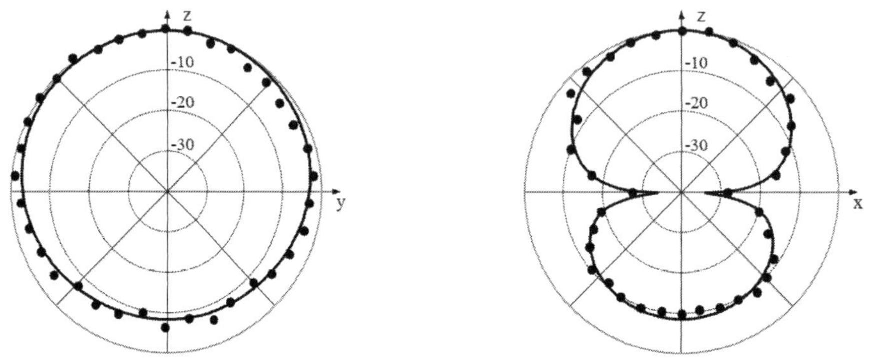


(a) (b)

**Figure 11.** Simulated (solid line) and measured (doted) co-polarized radiation patterns at 2.44 GHz (a) yz plane (b) xz plane

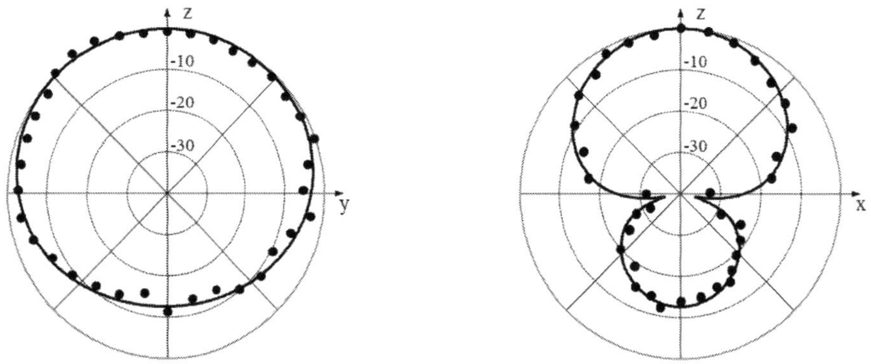


(a) (b)

**Figure 12.** Simulated (solid line) and measured (doted) co-polarized radiation patterns at 5.15 GHz (a) yz plane (b) xz plane

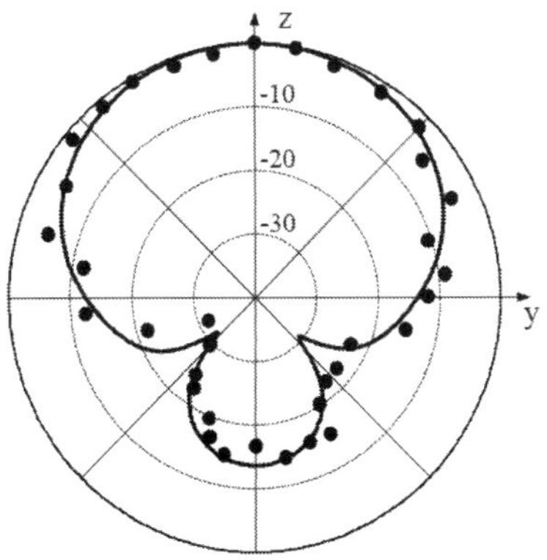


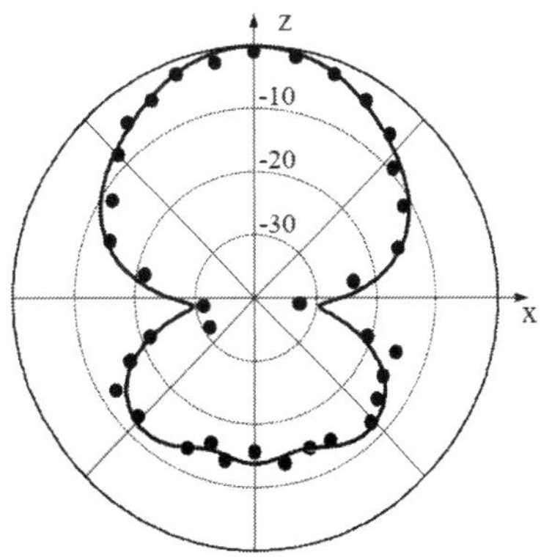


**Figure 13.** Simulated (solid line) and measured (doted) co-polarized radiation patterns at 5.85 GHz (a) yz plane (b) xz plane

## 4. Conclusions

This paper presented the dual-band printed dipole antenna for unidirectional radiation. The antenna is fabricated from conductor on the surface of FR4 printed circuit board in one-side and fed by a coaxial line. The elements of the antenna composed of dipole, 2 parasitic, and reflector. From the analysis, we found that the antenna has the measured bandwidth cover both operating frequencies of 2.400 – 2.483 GHz and 5.15 – 5.85 GHz. Antenna gains are 4.8 dBi, 5.5 dBi, and 8.0 dBi at the operating frequencies of 2.44 GHz, 5.15 GHZ, and 5.85 GHz, respectively. The advantages of the antenna are low profile structure, easy to fabricate, light weight, and can be fabricated with the cheap material. It is the low-cost antenna that suitable for using both base station and receiving antenna with a linear polarized system.

## References

[1] R. Jordan and C. T. Abdallah, "Communications and networking: an overview," *IEEE Trans. Antennas Propagat. Mag.*, vol. 44, no. 1, pp. 185-193, Feb. 2002.
[2] N. P. Agrawall, G. Kumar, and K. P. Ray, "Wide-band planar monopole antenna," *IEEE Trans. Antennas Propagat.*, vol. 46, no. 2, pp. 294-295, Feb. 1998.
[3] H. D. Chen, "Broadband CPW-fed square slot antennas with a widened tuning stub," *IEEE Trans. Antennas Propagat.*, vol. 51, no. 8, pp. 1982-1986, Aug. 2003.
[4] D. Srimoon, "Broadband Dipole on a U-Shaped Cross Sectional Ground Plane Antenna," APMC2008 with the IEEE Catalogue Number of CFP08APM-USB, Hong Kong, CHINA, 2008.
[5] C. M. Su, H. T. Chen, and K. L. Wong, "Printed dual-band dipole antenna with U-slotted for 2.4/5.2 GHz WLAN operation", *Electronic Letters*, vol. 38, no. 22, pp. 1308–1309, 2002.
[6] K. Chang, H. Kim, and Y. J. Yoon, "A triple-band printed dipole antenna using parasitic elements", *Microwave Optical Technology Letters*, vol. 47, no. 3, pp. 221–223, 2005.
[7] Y. W. Chi, K. L. Wong, and S. W. Su, "Broadband printed dipole antenna with a step-shaped feed gap for DTV signal reception", *IEEE Trans. Antennas Propagat.*, vol. 55, no. 11, pp. 3353–3356, 2007.

IOP Conf. Series: Materials Science and Engineering **965** (2020) 012036 doi:10.1088/1757-899X/965/1/012036

# A Wideband Printed Monopole Antenna with a Co-Planar Waveguide Feeding

**Chanapon Surin[1,a], Duangarthit Srimoon[2,b], and Jhirat Mearnchu[3,c]**

[1]Student in Master of Electrical and Computer Engineering Program, College of Engineering, Rangsit University, 52/347 MuangAke, Phaholyothin Road, Lakhok, Pathumthani, 12000, Thailand.

[2]Faculty of Computer Engineering, College of Engineering, Rangsit University, 52/347 MuangAke, Phaholyothin Road, Lakhok, Pathumthani, 12000, Thailand.

[3] Department of Electrical Engineering, Faculty of Engineering and Industrial Technology, Silpakorn University, 6, Rajamankha Nai Road, Muang, Nakornpathom, 73000, Thailand.

[a]bor_n_benz@hotmail.com, [b]duang-arthit@rsu.ac.th, [c] jhirat@su.ac.th

**Abstract**. A wideband printed monopole antenna with a co-planar waveguide feeding is presented. This antenna provides a frequency range of 1.7 – 6.0 GHz. It is a flat conductor on FR4 printed circuit board. The shape of ground and pole are rectangle that the 2 corners near the feeding point are 25 degree chamfered, symmetrically. The ground is etched to form a co-planar waveguide for feeding. For this work, the antenna is simulated by using computer program before fabricating a prototype. The antenna provides an omni-directional radiation same as donut shape at the low frequency of the operating band. The radiation pattern is change when the frequency is shift to high but still omni-direction. The bandwidth of the antenna is 112% between 1.7 GHz and 6.0 GHz with the reflection coefficient less than -10 dB across the band. Gains of the antenna are 2.3, 4.9, and 3.4 dBi at the frequencies of 1.8, 3.8, and 5.8 GHz, respectively. They are the 3 points of low-, central- and high-frequency of the operating band. The size of the antenna is 70.0 x 40.0 mm$^2$. Advantages of the antenna are wideband, low profile structure, light weight, and can be fabricated by FR4 make it low-cost.

## 1. Introduction

Wireless communication systems such as DCS (Digital Cellular System) has the frequency range of 1.71 - 1.88 GHz, PSC (Personal Communications System) has the frequency range of 1.85 - 1.99 GHz, UMTS (Universal Mobile Telecommunications System) has the frequency range of 1.92 - 2.17 GHz, WiMAX (Worldwide Interoperability for Microwave Access) has the frequency range of 2.496 – 2.690 GHz, and WLAN (Wireless Local Area Network) has the frequency ranges of 2.400 - 2.483 GHz and 5.15 – 5.85 GHz. All systems as indicated must use the antenna for transmitting and receiving signals. Therefore, many antennas are developed for supporting systems above. However, it is noted that each system has operating frequency close to each other. If we have the antenna that can operate covers all system bands, the antenna is interested. Thus, the combination of the operating band covers 1.71 – 5.85 GHz. The antennas which operate cover more than one system is called broadband antenna [1] while the antenna that operated cover 2 frequency bands together is called dual-band antenna.

Broadband antenna is wideband antenna. Wideband antenna had been proposed in many research works. For example, wideband planar plate monopole antenna [1] that radiated in omni-directional, they provide a very wide bandwidth but had distort beam at the high frequency of the operating band. Same as bi-directional antenna in [2], although it has a wide bandwidth but squeeze beams occurs at the high frequency of the band. Various antenna designs such as those of micro-strip line, co-planar waveguide (CPW), printed monopole antenna, and printed dipole antennas have been proposed in the literature [3]. Therefore, it was shown that a monopole antenna has the ability to give multiband operation.

This paper discussed the wideband printed monopole antenna that can be used to operate with various systems between the frequencies of 1.71 GHz and 5.85 GHz. A prototype that operates in the frequency bands is developed. By etching to form shapes of monopole, ground, and CPW printed on the same side of the FR4, a printed monopole antenna that actives wideband operation from 1.71 - 5.85 GHz is realised. The monopole is able to produce enough operating frequency band. Additionally, the reflection coefficient, far-field radiation pattern and antenna gain of the proposed antenna are demonstrated.

## 2. Antenna description

Structure of the proposed antenna is shown in Figure 1. (a). The antenna was contributed on an FR4 substrate with a thickness of 1.6 mm and a permittivity of 4.8. The overall dimensions of the structure were 44.0 x 70.0 ($a$ x $b$) mm$^2$. The antenna was fed by a CPW by etching copper ground plane that coated on the surface of FR4 to form double slots and a strip-line which appropriate to the impedance of 50 Ω. The width of the CPW strip-line is $s$, while the width of the 2 slot-lines is g. Dimensions of the ground plane were $w$ x $l$ where $w$ is the width and $l$ is the length. The 2 upper corners of the rectangle ground are 25° chamfered by the depth of $c$ for adjusting radiated beam as shown in Figure 1. (a). Over the rectangle ground is the flat printed monopole. The width of the flat printed monopole is equal to the width of the ground. Gap between ground and monopole is $g$. The length of the monopole, $l$, is used to determine the low frequency start at 1.7 GHz, and the width $w$ is adjusted to vary the impedance for the desire of impedance matching. The 2 corners of monopole near ground are 25° chamfered by the depth of $c$ as well. The antenna is designed by using a ground plane to form a CPW as proposed which has three functions: (1) a ground plane for the monopole and CPW, (2) radiating element and (3) component to form the distributed matching network with the monopole [3].

For comparing and to confirm the accuracy of the simulation, the prototype of the proposed antenna is fabricated with the dimensions and materials as indicated in Table 1. The prototype of the antenna is shown in Figure 1. (b).

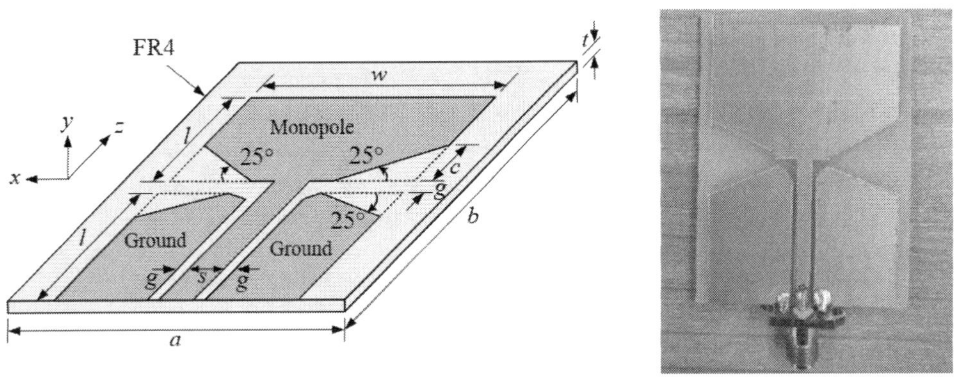


(a) (b)

**Figure 1.** (a) Structure of the wideband printed monopole antenna (b) Prototype of the antenna

**Table 1.** Parameters and their dimensions of the proposed wideband printed monopole antenna

| Parameter | Dimension (mm) |
|---|---|
| *a* | 44.0 |
| *b* | 70.0 |
| *w* | 40.0 |
| *l* | 29.0 |
| *c* | 7.0 |
| *g* | 0.5 |
| *s* | 3.0 |
| *t* | 1.6 |

## 3. Simulated and measured results

### *3.1. Simulated and measured reflection coefficients*

The simulated and measured reflection coefficients are shown in Figure 2. The simulated result of the reflection coefficient agrees with the measured result. The measured reflection coefficients of the antenna provides the bandwidths >112% (<-10 dB) between the frequency range of 1.7 GHz and 6.0 GHz. However, the simulated result is better than the measured one at the high frequency. The discrepancy is mainly due to the error of dimensions between the simulation and the fabricated antenna.

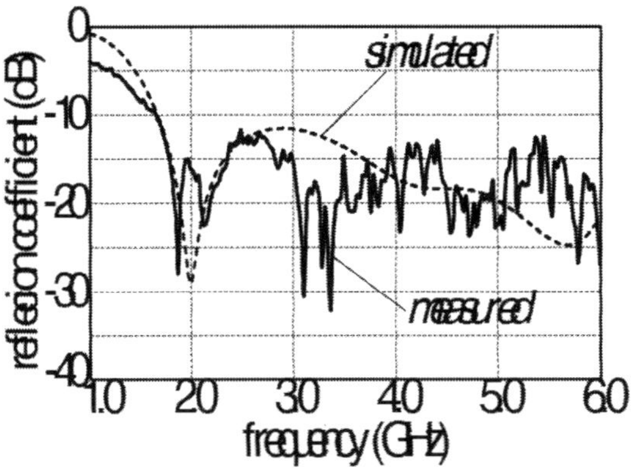


Figure 2. Simulated and measured reflection coefficients of the proposed antenna

Many parameters are varied to show the characteristics of the proposed antenna. The first parameter is *w* or the width of the flat monopole and ground. The results of reflection coefficient with different *w* are showed in Figure 3. It is found that *w* is influence to the impedance matching. When *w* is increase from 36.0 to 40.0 and to 44.0 mm, the reflection coefficient between 2.0 and 3.5 GHz is increase. However, if *w* is greater than 44.0 mm, the operating band is shift to high, especially at the start of the band, it greater than 1.7 GHz.

The second parameter is *l* or the length of monopole and ground. The results of vary *l* is showed in Figure 4. It is found that *l* is influence to impedance matching and operating band. When *l* is increase

IOP Conf. Series: Materials Science and Engineering **965** (2020) 012036 doi:10.1088/1757-899X/965/1/012036

from 27.0 to 29.0 and to 31.0 mm, the reflection coefficient between 2.0 and 5.0 GHz is decrease and the bandwidth is wide to the low frequency.

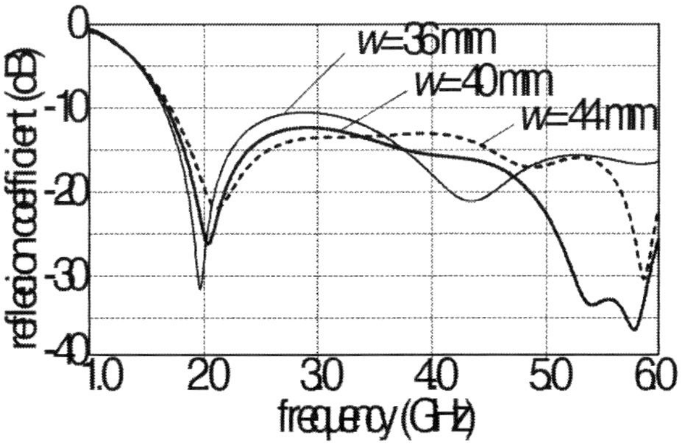


**Figure 3.** Simulated reflection coefficients when changed parameter *w*

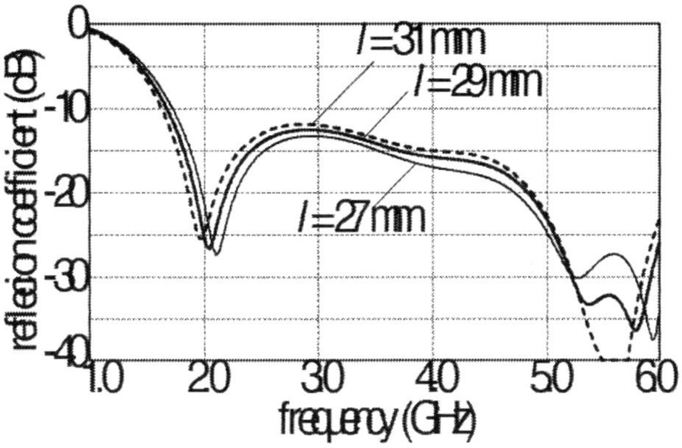


**Figure 4.** Simulated reflection coefficients when changed parameter *l*

The third parameter is *c* or the 25° chamfer of monopole and ground. The results of reflection coefficient with different *c* are showed in Figure 5. It is found that *c* is influence to the high frequency (3.0 – 6.0 GHz) band. When *c* is increase from 5.0 to 7.0 and to 9 mm, the best matching is occur.

The next parameter is *g* or the gap of slot-etching. The results of vary *g* is showed in Figure 6. It is found that an appropriate of *g* is about 0.5 mm.

The last parameter is *s* or the width of the strip-line feeding. The results are showed in Figure 7. It is found that an appropriate of *s* is about 3.0 mm that is corresponding to the impedance of 50 Ω.

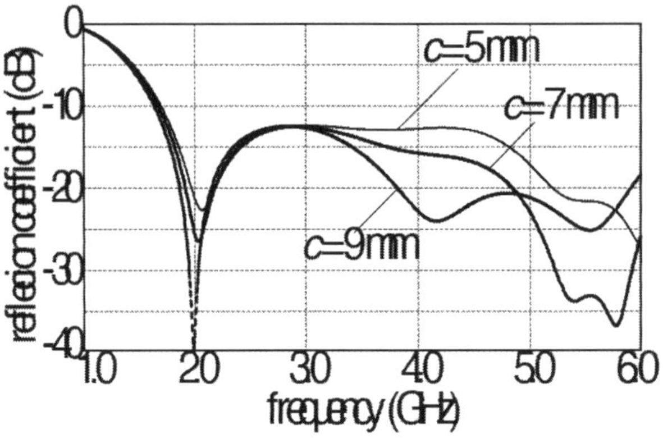


**Figure 5.** Simulated reflection coefficients when changed parameter *c*

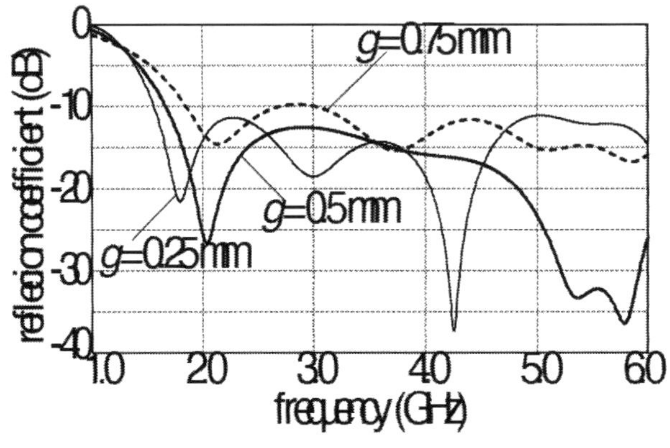


**Figure 6.** Simulated reflection coefficients when changed parameter *g*

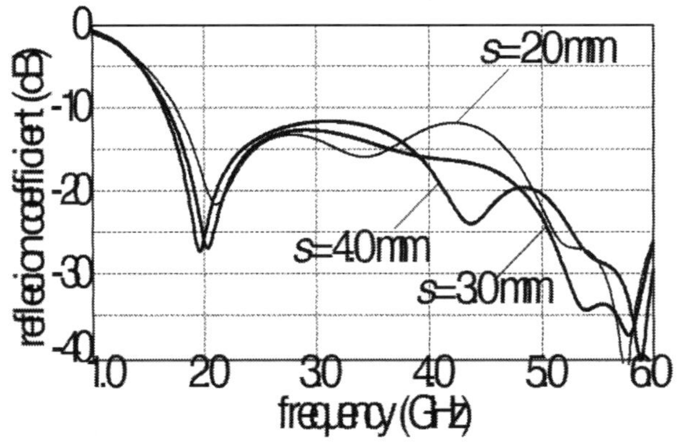


**Figure 7.** Simulated reflection coefficients when changed parameter *s*

### *3.2. Simulated and measured radiation patterns*

The experimental was operated in open area to decrease the effect of the reflection. The observed operating frequencies are 1.8, 3.8, and 5.8 GHz as shown in figure 8, 9, and 10, respectively. The 3 observed frequency points are low, mid, and high of the band.

Figure 8 shows the normalized co-polarization patterns of simulated in 3 planes as indicated the coordinate and direction in Figure 1 compare with the measured results at 1.8 GHz. It was found that the half power beam-widths (HPBW) are 79° for both yz- and xz-plane. The patterns in both planes are symmetry. Gain for this frequency is 2.3 dBi.

Figure 9 shows the simulated and measured co-polarized radiation patterns at 3.8 GHz. It was found that the half power beam-widths (HPBW) are 98° for yz-plane and 2 x 40° for xz-plane. The patterns in both planes are not symmetry. Gain for this frequency is 4.9 dBi in y-direction.

Figure 10 shows the simulated and measured co-polarized radiation patterns at 5.8 GHz. It was found that the half power beam-widths (HPBW) are 68° for yz-plane and 40°, 30° for xz-plane. The patterns in both planes are not symmetry. Gain for this frequency is 3.4 dBi in y-direction.

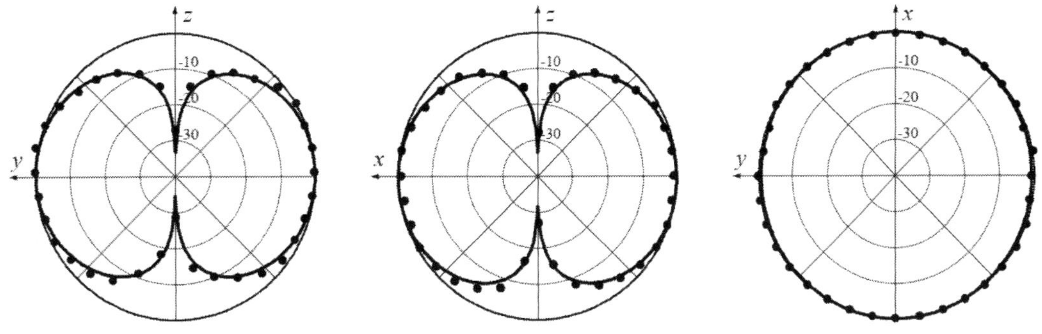


(a) (b) (c)

**Figure 8.** Simulated (solid line) and measured (doted) co-polarized radiation patterns at 1.8 GHz (a) yz plane (b) xz plane and (c) xy plane

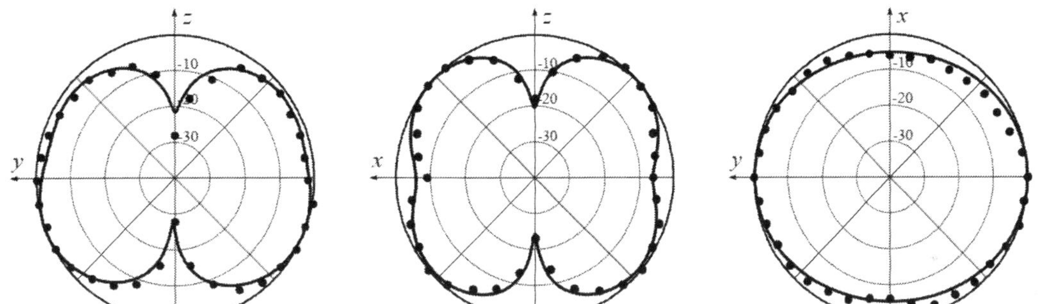


(a) (b) (c)

**Figure 9.** Simulated (solid line) and measured (doted) co-polarized radiation patterns at 3.8 GHz (a) yz plane (b) xz plane

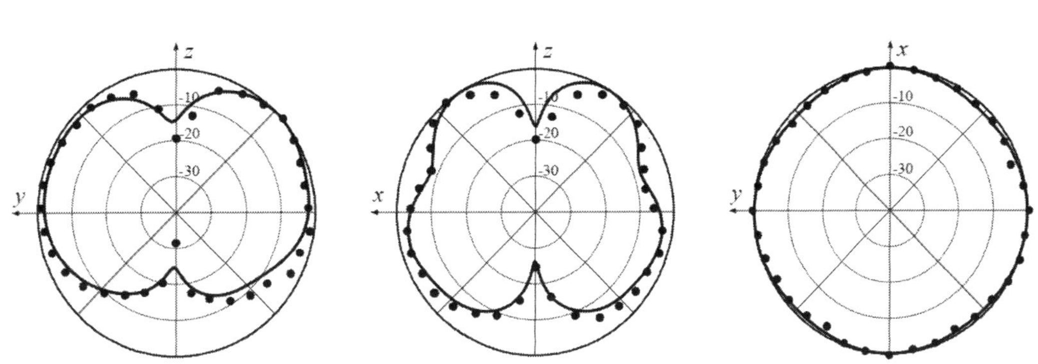

(a) (b) (c)

**Figure 10.** Simulated (solid line) and measured (doted) co-polarized radiation patterns at 5.8 GHz (a) yz plane (b) xz plane

## 4. Conclusions

This paper presented the wideband printed monopole antenna with a co-planer waveguide feeding. The antenna is fabricated on an FR4 substrate in one-side. We were found that the antenna has the measured bandwidth cover various systems between the operating band of 1.7 – 6.0 GHz. Antenna gains are 2.3, 4.9, and 3.4 dBi at the operating frequencies of 1.8, 3.8, and 5.8 GHz, respectively. The advantages of the antenna are low profile structure, light weight, good matching, and can be fabricated with the cheap material. It is the low-cost antenna that suitable for using in various systems.

## References

[1] H. R. Hassani and S. M. Mazinani, "Wideband Planar Plate Monopole Antenna, Passive Microwave Components and Antennas," Vitaliy Zhurbenko (Ed.), ISBN: 978-953-307-083-4, InTech, Available from: http://www.intechopen.com/books/passive-microwave-components-and-antennas/wideband-planar-platemonopole- antenna, 2010.

[2] H. D. Chen, "Broadband CPW-fed square slot antennas with a widened tuning stub," *IEEE Trans. Antennas Propagat.*, vol. 51, no. 8, pp. 1982-1986, Aug. 2003.

[3] Cheolbok Kim, "Ultra-Wideband Antenna, Microwave and Millimeter Wave Technologies Modern UWB antennas and equipment," Igor Mini (Ed.), ISBN: 978-953-7619-67-1, InTech, Available from: http://www.intechopen.com/books/microwave-and-millimeter-wave-technologies-modern-uwb-antennas-andequipment/ultra-wideband-antenna, 2010.

IOP Conf. Series: Materials Science and Engineering **965** (2020) 012037 doi:10.1088/1757-899X/965/1/012037

# Simulating Elliptic Low-pass Filter based on MOCCCII

**Nanan Chomnak, Siradanai Srisamranrungrueang and Natapong Wongprommoon**

Department of Electrical Engineering Faculty of Engineering and Industrial Technology Silpakorn University Nakhon Pathom, Thailand

wongprommoon_n@su.ac.th

**Abstract**. Multiple output current-control current conveyer (MOCCCII) based Elliptic low-pass filter by using a simulating of RLC ladder prototype is presented in this work. The good performances of RLC ladder as well as low-sensitivity are inherited. Signal flow graph (SFG) method is applied to convert the RLC Elliptic ladder low-pass filter prototype to be the different function blocks. These function blocks can be realized by integrator and differentiator. Lossy integrator, lossless integrator and lossless differentiator realized from MOCCCII are straightforward replaced in SFG. Due to the tuning ability of MOCCCII, the proposed Elliptic low-pass filter can be tuned by adjusting the bias current ($I_B$). In addition, the frequency responses can be varied from 50kHz to 1MHz by tuning $I_B$ from 1μA to 100 μA to prove the electronic tuning performance of the proposed filter. The simulation results are carried out by PSpice that the performances are totally agreed with the theory.

Keywords-MOCCCII, Elliptic, Low-pass, Electronic tuning

## 1. Introduction

Continuous time signal processing is very important process in telecommunication works. High performance filter plays important role for selecting the desired signal and removing unwanted signal. Many types of filter are developed for specific applications as well as noise reduction in electronic circuit and digital to analog conversion. However, the characteristic of the filters should be selected which suitable for using in applications. Some usage required only simple and low-performance filter such as first-order passive filter [1-2] to complete the requirement. When the requirements are more complex, higher performance filters are required to achieve the requirement. Second-order filters were introduced with many active building blocks, such as OPAMP, OTA or CCII [3-5] with limitations, for example, required addition resistance and consist of floating capacitor. These limitations lead to large die area, complex structure and high sensitivity.

In high performance applications, low-order of filters and the mentioned first and second-order filters provided insufficient performances. High-order passive ladder filters were considered due to their performance and additional reliability as well as low sensitivity [6-8]. Chebyshev and Elliptic approximation functions are interested in high-order ladder filters realization. The RLC ladder Chebyshev type is preferred approximation type and it also called all poles filter because the transfer function has only poles without zeros. The Elliptic approximation type has both of poles and zeros in the transfer function. The cutoff frequency of RLC Elliptic ladder filter type is sharper than Chebyshev counterpart. Unfortunately, the passive RLC circuit has no electronic tuning and integration abilities. Few years ago, the filter circuits using active building blocks [9-11] were reported by using passive ladder filter prototype. These circuits were suffered from the complex structures and used of large number of passive elements. Recently, Chebyshev [12] and Elliptic [13] ladder filters were proposed with electronic tuning feature. The performances were good for the Chebyshev, but the Elliptic structure was rather complex. Due to the previous Elliptic ladder filter required the extremely low current gain which the current gain was very difficult to design and optimize. The poor performance of current gain was inevitable that meant the realized Elliptic filter had limitation for integration.

This paper focuses on a new synthesis of Elliptic ladder filter that has a simple structure by using lossless differentiator directly replace in SFG. Due to the simple SFG, its able to realized higher order filter. The proposed circuit uses only grounded capacitors which suits to further integrated circuit fabrication. The selected active building block is a MOCCCII which are versatile devices with electronic tuning ability that leads to the frequency tunable feature in the proposed Elliptic filter.

## 2. Principle and Theory

### 2.1. MOCCCII

MOCCCII [14] is versatile active building block which is developed from CCCII [15], its electronic symbol and structure are shown in Figure1. and Figure 2., respectively. Current and voltage relationship can be written as

$$\begin{bmatrix} i_y \\ v_x \\ i_z \end{bmatrix} = \begin{bmatrix} 0 & 0 & 0 \\ 1 & R_x & 0 \\ 0 & \pm 1 & 0 \end{bmatrix} \cdot \begin{bmatrix} v_y \\ i_x \\ v_z \end{bmatrix} \tag{1}$$

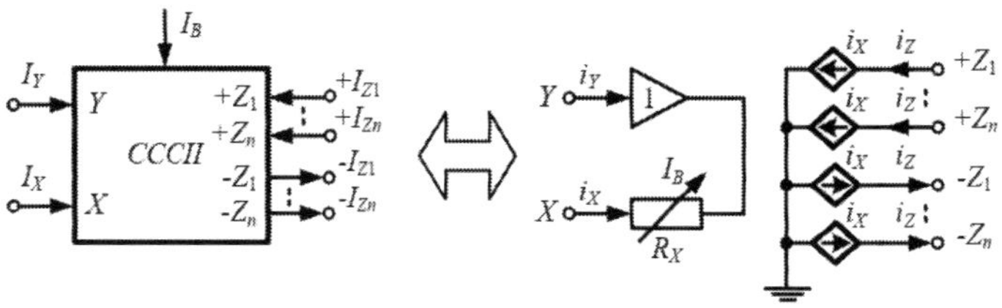


**Figure 1.** Electronic symbol of CMOS-based MOCCCII

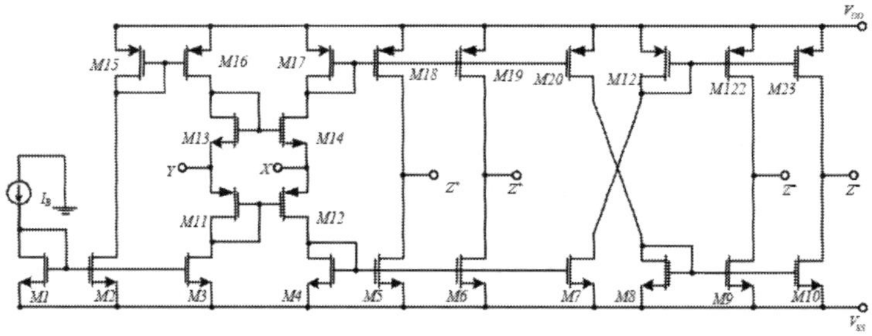

**Figure 2.** Structure of CMOS-based MOCCCII

From (1) positive and negative signs at port $Z$ illustrate the positive and negative signal. $R_x$ is intrinsic resistance at port $X$ which can be described by the transconductance ($g_m$) of $M_{12}$ and $M_{14}$ in (2)

$$R_X \approx \frac{1}{g_{m12} + g_{m14}} \tag{2}$$

$g_{mi}$ is a transconductance of MOS number $i$. Assume that transistors $M_{12}$ and $M_{14}$ are matched, then $g_{m12}$=$g_{m14}$ the equation can be rewritten to

$$R_X \approx \frac{1}{2g_m} = \frac{1}{2\sqrt{8\mu C_{OX}(W/L)I_B}} \tag{3}$$

Where $\mu$, $C_{ox}$, $W$ and $L$ are surface mobility, oxide capacitance, channel width and length of MOS transistor, respectively. It can be seen that $R_x$ can be electronically tuned by bias current ($I_B$).

### 2.2. Lossless and Lossy Intergrators

MOCCCII based lossless integrator can be easily realized by connecting a grounded capacitor to port $Y$, and connecting port $X$ to ground as shown in Figure3. The current transfer function can be expressed as

$$\frac{I_{O1}}{I_{IN}} = \frac{1}{sCR_x}, \frac{I_{O2}}{I_{IN}} = \frac{-1}{sCR_x} \tag{4}$$

MOCCCII based lossy integrator can also be easily realized by connecting grounded capacitor to port X, and connecting port Y to ground as shown in Figure 4. The current transfer function can be expressed as

$$\frac{I_{O1}}{I_{IN}} = \frac{1}{sCR_x+1}, \frac{I_{O2}}{I_{IN}} = \frac{-1}{sCR_x+1} \tag{5}$$

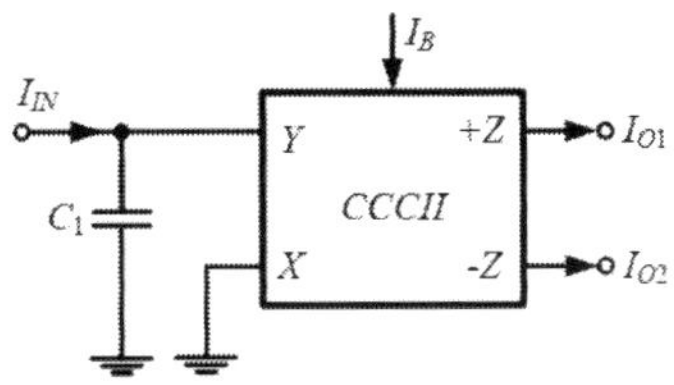


**Figure 3.** MOCCCII based lossless integrator

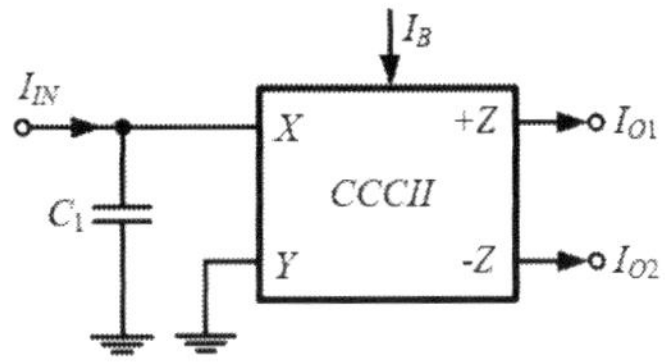


**Figure 4.** MOCCCII based lossy integrator

### 2.3. Lossless Differentiator

To realized lossless differentiator, we use lossy differentiator or high-pass filter [16] with positive feedback to the input as shown in Figure 5.

By using concept in Figure 5., we can realize MOCCCII based lossless differentiator as shown in Figure 6. and its current transfer function can be expressed as

$$\frac{I_{O1}}{I_{IN}} = sCR_x, \frac{I_{O2}}{I_{IN}} = -sCR_x \tag{6}$$

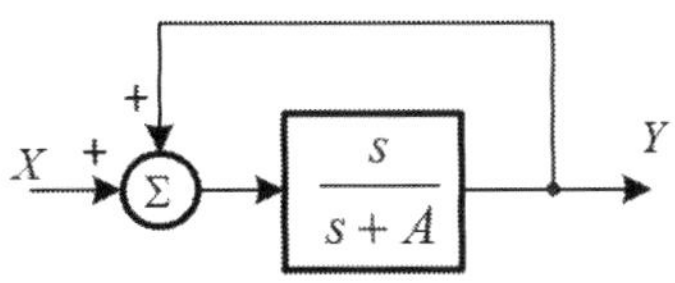


**Figure 5.** Realization of lossless differentiator form lossy differentiator

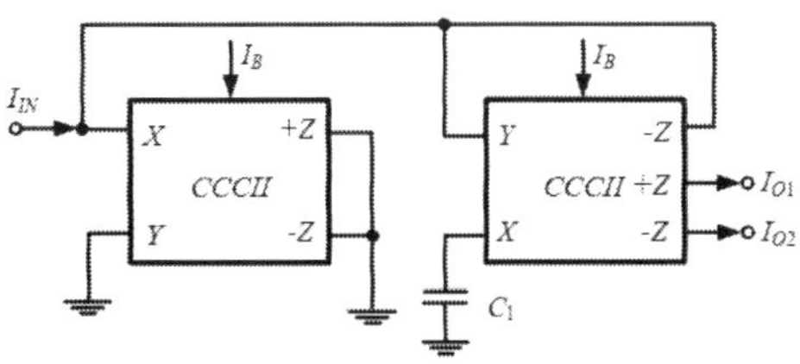


**Figure 6.** MOCCCII based lossless differentiator

IOP Conf. Series: Materials Science and Engineering **965** (2020) 012037 doi:10.1088/1757-899X/965/1/012037

## 3. MOCCCII based Elliptic Ladder Low-pass Filter

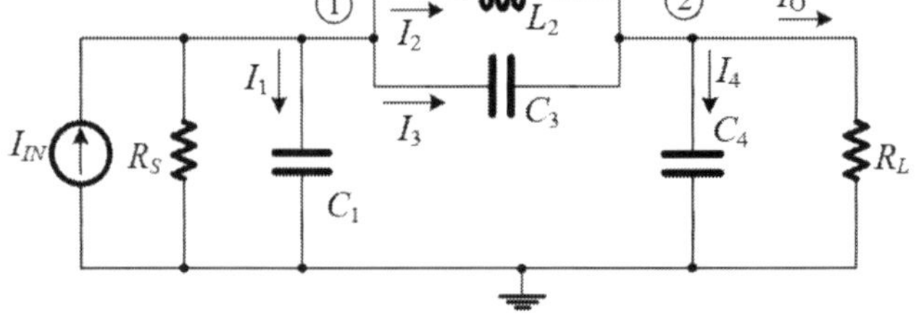


Figure 7. RLC Elliptic ladder low-pass filter prototype

Using the RLC Elliptic prototype [17], applying Kirchhoff's Current Law, (KCL) to the RLC Elliptic ladder low-pass filter prototype in Figure 7., the current and voltage relationships can be written as

$$I_1 = I_{IN} - V_1/R_S - I_2 - I_3 \tag{7}$$

$$I_2 = \frac{(V_1 - V_2)}{sL_2} \tag{8}$$

$$I_3 = (V_1 - V_2)sC_3 \tag{9}$$

$$V_1 = \frac{I_1}{sC_1} \tag{10}$$

$$V_2 = \frac{I_4}{sC_4} \tag{11}$$

$$I_4 = I_2 + I_3 - I_O \tag{12}$$

$$I_O = \frac{V_2}{R_L} \tag{13}$$

From equations (7) to (13), The SFG can be written as Figure 8.

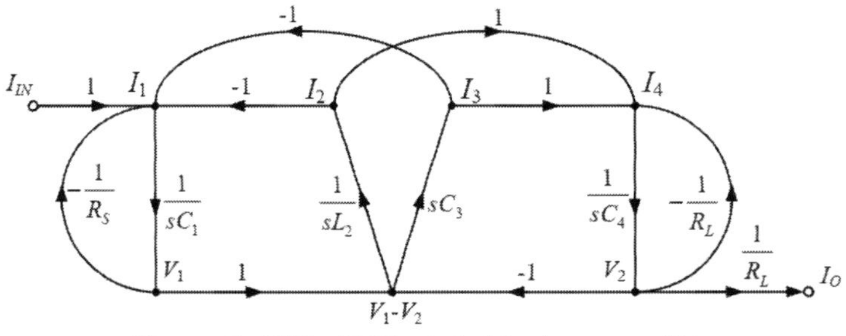


**Figure 8.** SFG of Elliptic ladder low-pass filter

From Figure 8., it consists of various variables especially in the junctions of voltage and current variables. In the realization all junctions should be transformed into the same type. Transforming all junctions of SFG in Figure8 into current variables by using resistance ($R_x$). Considering at the beginning and the end paths of SFG, the feedback paths $-1/R_S$ and $-1/R_L$ if we set $R_S$=$R_L$=$R_x$ the functions perform as lossy integrators. Then we can simplify the SFG to Figure9.

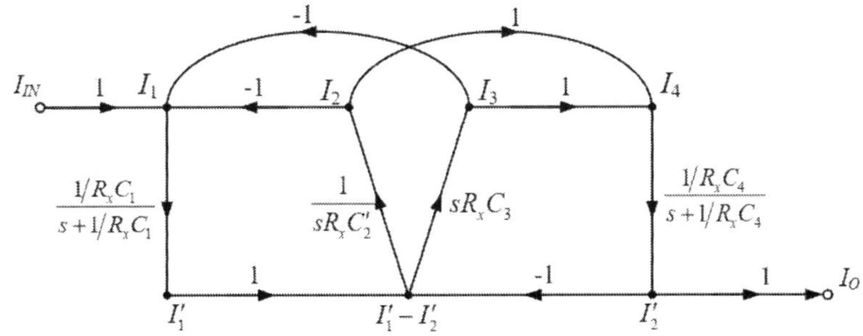


**Figure 9.** Simplified SFG of Elliptic ladder low-pass filter

Considering SFG in Figure 9., the different junctions are incorporated as the transfer functions of lossy integrator, lossless integrator and lossless differentiator. Replacing the MOCCCII based integrators and differentiator into SFG, the proposed Elliptic ladder low-pass filter can be written as Figure 10. Compare the proposed circuit to previous work [13], due to previous work use converted current gains the achieve low-pass function, but the proposed structure directly replace SFG functions by integrator or differentiator, which means the proposed structure has less complexity and deploys less components. The capacitors can be set to achieve condition of low-pass filter instead of previous work current gains.

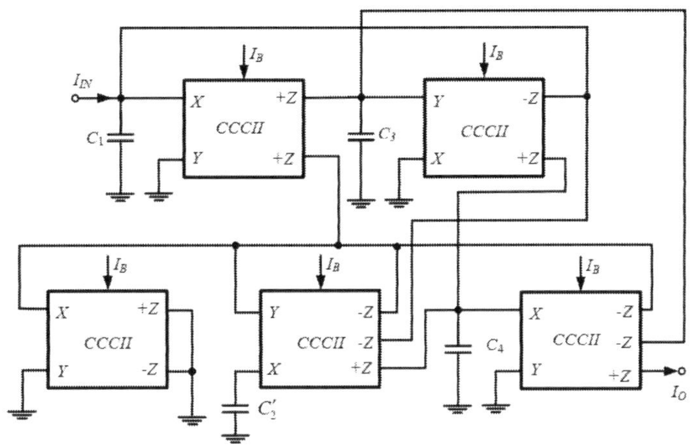


**Figure 10.** Proposed Elliptic ladder low-pass filter

## 4. Simulation Results

The simulation results are carried out by PSpice to confirm the theory. Supposed that, the RLC Elliptic ladder low-pass filter prototype operation frequency is set to 1 MHz ripple 0.1dB, the various passive elements are given by $R_S$=$R_L$=1Ω, $C_1$=$C_4$=158nF, $L_2$=172nH, $C_3$=9nF. Using scaling technique ($kf$=1k) to the proposed circuit by given the following parameters as $C_1$=$C_4$=158pF, $C'_2$=172pF, $C_3$=9pF and $I_B$=120μA.

The selected CMOS model in MOCCCII is MOSIS level 3 TSMC 0.25μm technology. The aspect ratio of NMOS and PMOS parameters are given in Table 1.

Table 1.NMOS and PMOS parameters

| Transistor | W(μm) | L(μm) |
|---|---|---|
| NMOS | 5 | 0.5 |
| PMOS | 15 | 0.5 |

The frequency responses of proposed Elliptic low-pass filter compared with prototype are shown in Figure 11. It can be seen that the passband gain of the proposed filter is mitigated from the passband of RLC prototype.

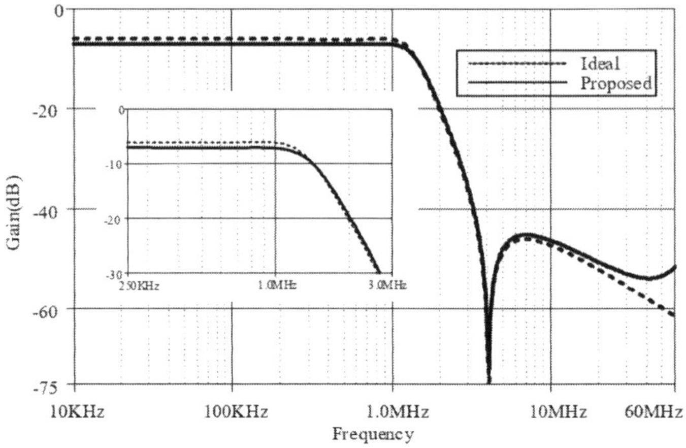


**Figure 11.** Frequency responses of proposed filter compared with RLC prototype

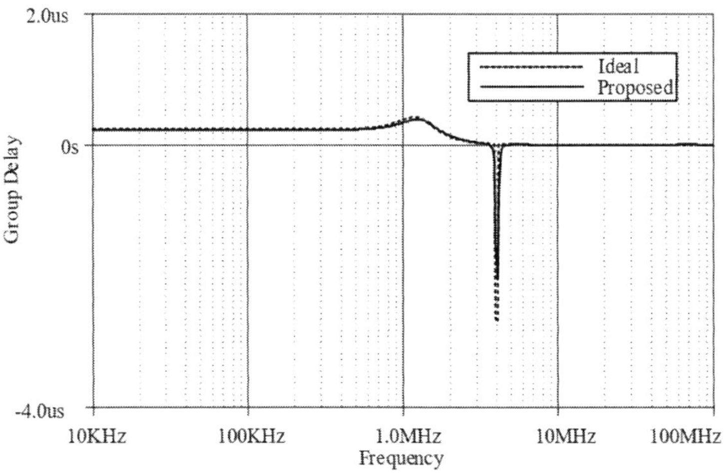


**Figure 12.** Group delay of proposed filter compared with RLC prototype

Group delay of proposed filter compared with RLC prototype is shown in Figure 12. It is obviously found that the group delay of proposed filter and RLC prototype are closely which are around 200ns along passband and suddenly changed at 4MHz. MOCCCII frequency responses can be tuned by using $I_B$ which relate to $R_x$ and can be explain in equations (5) and (3), the proposed circuit can be also tuned. Figure 13. shows the electronic tuning ability of the proposed circuit by varying $I_B$=1μA, 3μA, 10μA, 30μA and 100μA, the operation frequencies can be tuned between 80kHz and 1.2MHz.

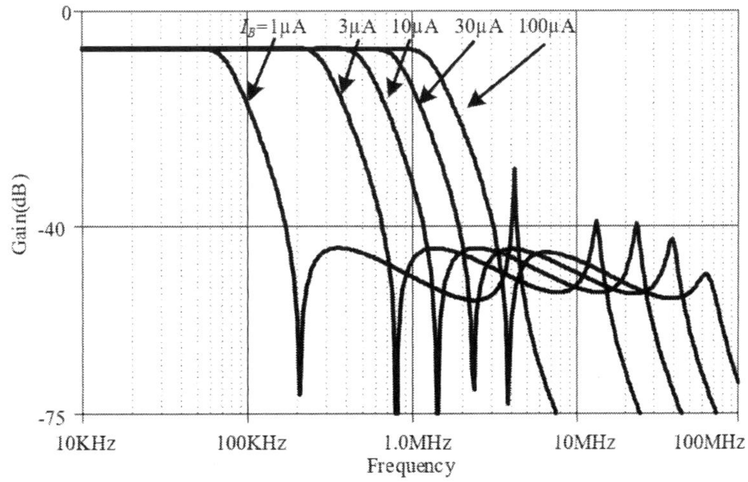


**Figure 13.** Frequency responses of proposed filter when varying $I_B$

Multi-tone test can be verified the performance of filter how good to remove unwanted frequency. The multi-tone result is shown in Figure 14. by adding various frequency sinusoidal signals 100kHz, 300kHz, 1MHz, 3MHz and 10MHz as inputs with bias current $I_B$=100μA. The outputs show only signals in passband that agreed well with the theory.

Filter performance can be verified by applying 3 input frequencies (1MHz, 3MHz and 5MHz) to the proposed LPF while bias current $I_B$=100μA. It is clear that, only 1MHz sinusoidal signal appear at the output, other higher frequencies are removed as shown in Figure 15.

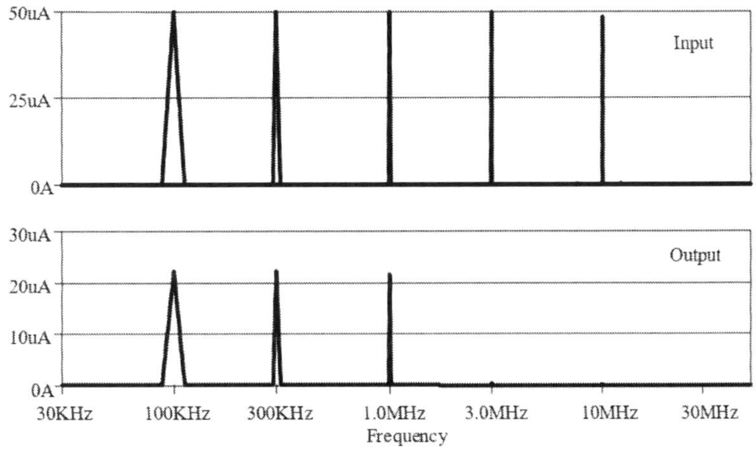


**Figure 14.** Multi-tone test of proposed Elliptic ladder low-pass filter

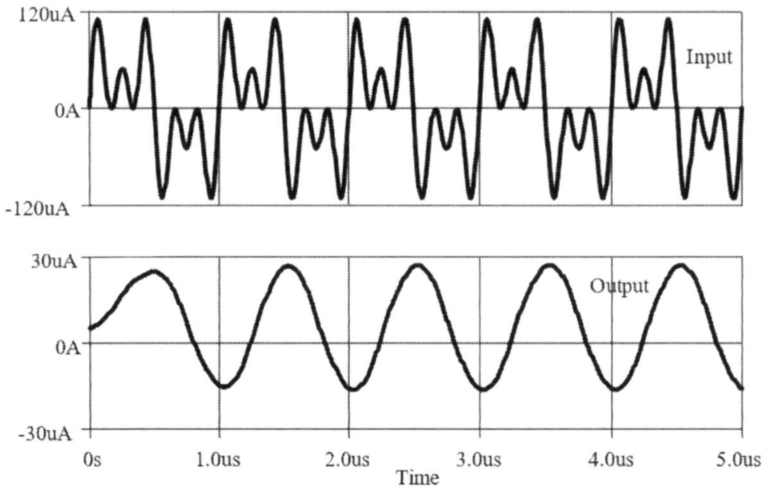


**Figure 15.** Output signal of 1MHz from low-pass filter

## 5. Conclusion

A new MOCCCII based Elliptic ladder low-pass filter is presented in this paper. Simple structure is designed based on RLC Elliptic ladder prototype. The SFG is used to synthesis the active signal processing building blocks. The proposed circuit consists of 2 lossy integrators, 1 lossless integrator, 1 lossless differentiator and 4 grounded capacitors. The proposed circuit has low complex structure suit to further integration. The operation frequencies can be electronically tuned between 50kHz and 1MHz by varying $I_B$ from 1μA to 100μA.

## References

[1] I. A. Khan, P. Beg, T.Muslim Ahmed, "First Order Current Mode Filters and Multiphase Sinusoidal Oscillators Using MOCCIIs," Microelectronics, pp. 146-149, 2006.

[2] S. Ozoguz, A. Toker, O. Cicekoglu, "First-order allpass sections-based current-mode universal filter using ICCIIs," IET Electronics letters, Vol.36, No.17, pp. 1443-1444, 2000.

[3] G.W. Roberts and A.S. Sedra, "All current-mode frequency selective circuits," Electronics Lett, vol. 25, pp.759-761, 1989.

[4] C.M. Chang, "Novel universal current-mode filter with single input and three outputs using only five current conveyors," Electronics Lett, vol.29, pp. 2005-2007, 1993.

[5] S. Minaei and S. Turkoz "New current-mode current-controlled universal filter with single Input and three outputs," Int'l J. Electronics, vol. 88, pp.333-337, 2001.

[6] J. Wu, E. El-Masry, Current-mode ladder filters using multiple output current conveyers. IEE Proceedings: Circuits, Devices and Systems, Vol.143, No.4, pp.218-222, 1996.

[7] J. Glinianowicz, J. Jakusz, S. Szczepanski and Y. Sun, "High-frequency two-input CMOS OTA for continuous-time filter applications IEE Proceedings: Circuits, Devices and Systems, Vol.147. No. 1. Feb. 2000.

[8] J. A. De Lima and C. Dualibe, "A Linearly Tunable Low-Voltage CMOS Transconductor With Improved Common-Mode Stability and Its Application to gm–C Filters," IEEE Trans. Circuits Syst. II, Vol. 48, No. 7, July 2001.

[9] G. Jacobs, D. Allstot, R. Brodersen, P. Gray, Design techniques for MOS switched capacitor ladder filters. IEEE Transactions on circuits and systems, Vol.25,No12, pp.1014-1021, 1978.

[10] F. Montecchi, Time-shared switched-capacitor ladder filters insensitive to parasitic effects. IEEE Transactions on circuits and systems, Vol.31, No.4, pp.349-353, 1984.

[11] A. Uygur, H. Kuntman, "Seventh-order elliptic video filter with 0.1 dB pass band ripple employing CMOS CDTAs," Int. J. Electron. Commun. (AEÜ) Vol.61, pp.320–328, 2007.

[12] P. Prommee, P. Thongdit, K. Angkeaw, "Log-domain high-order low-pass and band-pass filters," AEU - International Journal of Electronics and Communications, Vol. 79, pp.234-242, 2017.
[13] P. Prommee, T. Kunto, M. Somdunyakanok and T. Prommee, "CMOS-based current-mode tunable ladder Elliptic low-pass filter," Proc. of 38th International Conference on Telecommunications and Signal Processing, pp.142-145, 2015.
[14] A. Fabre, O. Saaid, F. Wiest and C. Boucheron, "High frequency applications base on a new current controlled conveyor," IEE Tran. Circuits and Syst. I, vol. CAS-43, pp. 82-91, Feb. 1996.
[15] P. Limpaibool, T. Kunto and P.Prommee, "Design of High-order Current-mode Ladder Low-pass Filters Using MOCCCIIs" International Symposium on Multimedia and Communication Technology, pp.34-37, 2015.
[16] P. Prommee, M. Somdunyakanok and K. Angkeaw, "CCCII-based Multiphase Sinusoidal Oscillator Employing High-pass Sections", ECTI-CON 2009, Vol.1, pp.530-533, 2009.
[17] T. Deliyannis, Y. Sun and J. K. Fidler, "Continuous-Time Active Filter Design," London: CRC Press, 1999.

# Modification fitness function of particle swarm optimization to improve the cluster centroid

**Chiabwoot Ratanavilisagul**

Department of Computer and Information Science of Applied Science, King Mongkut's University of Technology North Bangkok (KMUTNB), Bangkok, Thailand

chaibwoot@hotmail.com, chaibwoot@gmail.com, chaibwoot.r@sci.kmutnb.ac.th

**Abstract**. Clustering is a classification technique. The clustering is brought to create models. The model is applied with the new input data to identify the class of that data. The goal of clustering is finding the best cluster centroid. The process of searching the best cluster centroid can use Particle Swarm Optimization. Many researches use it and obtain good results. But, searching the best cluster centroid by PSO suffers the problem from calculating the fitness function, because the fitness function is calculated by distances from the cluster centroid to all data inputs. These fitness functions may lead to the bad cluster centroid. So, this paper proposed the fitness function is calculated by the results from classification. The results from classification can lead to the good cluster centroid and create a good model to apply with the new input data. The proposed technique is tested on seven datasets from the UCI Machine Learning Repository and gives more satisfied search results in comparison with PSOs for the data clustering problems.

Keywords: Data Clustering; Particle Swarm Optimization; Centroid Vectors; Fitness Function

## 1. Introduction

The clustering [1], [2] is optimization problem and classification technique. It is applied with many application areas, such as bioinformatics, machine learning, information retrieval, data analysis, pattern recognition and image processing [3], [4]. It is brought to create models. The model is applied to the new input data to identify the class of that data. So, the goal of clustering is creating a model to identify the class of new objects. The clustering technique composes as follows: The object (data input) comprises many attributes. The attribute can be used to identify the class of the object. The objects that have similarity attributes are in the same class. On the other hand, the objects that have differential attributes are in other classes. Particle Swarm Optimization (PSO) is inspired from the behavior of flying birds and their communication mechanism [5], [6]. PSO is a stochastic global optimization technique. PSO is used for solving optimization problems. The goal of PSO is finding the global optimum of a fitness function or an objective function. PSO has many pros such as rapid convergence, simplicity, and little parameters to be adjusted [7].

In comparison with several other population-based stochastic optimization methods, such as the genetic algorithm (GA) [8, 9] and the evolutionary programming (EP), PSO performs better in solving various optimization problems with fast and stable convergence rates [10 - 12].

IOP Conf. Series: Materials Science and Engineering **965** (2020) 012038 doi:10.1088/1757-899X/965/1/012038

The well-known technique for applying to clustering is the K-Means clustering (KM) [13]. The cons of KM are the problem of initializing the cluster centers and the problem of trapping in local optimum [3]. To overcome this cons, many researchers used PSO apply with KM. Applying PSO and KM is called that PSO-clustering [14]. PSO-clustering can increase distribution of population. Hence, it can decrease the problem of initializing the cluster centers and the problem of trapping in local optimum. The experiment results of these papers showed that these techniques can get better solutions than KM alone. Technique of PSO-clustering can be concluded as follows: PSO is used to find the cluster centroid and KM is used to calculate the fitness function.

Many researchers tried to improve the searching performance of PSO-clustering. Some researchers improved KM such as X. D. Liu [15] proposed a new clustering criterion function which combines intra-cluster distance and inter-cluster distance. C. C. Hung and H. Pumawan [16] proposed a hybrid rough K-means PSO algorithm. Fengqin Yang [17] proposed the K-harmonic means (KHM) added into the process of PSO. Some researchers improved PSO such as J. Sun and W. B. Xu combined Quantum-behaved Particle Swarm Optimization (QPSO) with KM for solving data cluster [18]. J. Y. Chen and H. Y. Zhang proposed a new K-means cluster algorithm based on improved velocity of PSO cluster algorithm (RVPSO-K) [19]. These techniques can improve the searching performance of PSO-clustering and get better answers.

The fitness function is guidance for searching of PSO. So, the main performance of applying PSO-clustering is the fitness function. It leads to classify classes. If the fitness function is defined to be a bad function, the results from searching get bad solutions. Hence, many researchers proposed the fitness function for PSO-clustering [14].

DW van der Merwe and AP Engelhrecht [14] proposed the fitness function uses the quantization error equation. It is the first paper which proposed PSO Clustering. So, it is the standard of PSO with clustering. This technique is called that particle swarm optimization with Clustering or PSO-clustering or PSOC. The results show that the solution of PSOC is better than that of KM. The source code of PSOC is proposed by Augusto Luis Ballardini [20].

Ching-Yi Chen and Fun Ye [21] proposed the fitness function uses the hard c-means which is created with the Euclidean norm. This technique is called PSO-clustering with the hard c-means or PSOCC. The result shows that the solution of PSOCC is better than that of KM and that of Fuzzy C-means.

Pei Zhenkui, Hua Xia, and Han JinFeng [22] proposed the fitness function uses the ratio of the maximum value from K-means to the minimum value from K-means. This technique is called the clustering algorithm based on PSO or CPSO. The results show that the solution of CPSO is better than that of KM.

R. J. Kuo, M. J. Wang, and T. W. Huang [23] proposed the fitness function uses K-means. In my paper, this technique is called that particle swarm K-means optimization algorithm or PSKO. The results show that the solution of PSKO is better than that of KM and that of PSOC.

PSOC, PSOCC, CPSO, and PSKO use the concept of KM which is calculated from distances between the input data and the cluster centroid. The main concept of KM, the cluster centroid vector has the least distances from all data inputs. That point obtains the best fitness value. However, that point may classify classes badly. The good fitness value may classify worse than the bad fitness value.

The goal of using PSO is finding the best point which can classify the most amounts of input data correctly. So, the fitness function which is calculated from distance is unsuitable. If the fitness function is calculated from the results from classification, it will get the point which can classify many amounts of input data correctly.

F-Measure is the results from classification [3]. If a point can classify many data correctly, it will get much value of F-measure. So, F-Measure can be used to guide searching of PSO. It can lead to the best classificatory point and it gets good model to apply with the new input data.

From previous mention, this paper proposed the fitness function is calculated from the result from classification or F-measure. The proposed technique is called that modification fitness function by F-measure for particle swarm optimization with clustering or FPSOC. A set of dataset in the UCI Machine Learning Repository [24] is used to compare PSOC [14] by source code from [20], PSOCC [21], CPSO [22], PSKO [23] and the proposed technique. The results show that the solutions quality of the proposed technique is better than other compared algorithms in all tested datasets.
The rest of this paper is organized as follows: Section 2 explains the basic theoretical for this paper such as the K-Means Clustering, the standard PSO and F-Measure. Section3 explains related works (PSOC, PSOCC, CPSO, and PSKO). Section 4 explains the proposed technique or modification fitness function by F-measure for particles swarm optimization with clustering (FPSOC). Section 5 explains the case studies of datasets, the experiment setup and presents the experiment results. Section 6 concludes the paper with a brief summary.

## 2. Background

### 2.1. *Particle Swarm Optimization*

In standard PSO, each member of the population is called a "particle". Each particle contains its position and its velocity. Each individual particle executes searching in the search space according to its velocity, the best position found in the whole swarm (GBEST) and the individual's best position (PBEST). PSO starts with randomizing particle positions and their velocities, and the evaluation of the position of each particle is calculated from the objective function of the optimization problem. In each generation, each individual particle updates its position and velocity according to the expression below:

$$V_{id}' = \varpi V_{id} + \eta_1 rand()(P_{id} - X_{id}) + \eta_2 rand()(P_{gd} - X_{id}) \tag{1}$$

$$X_{id}' = X_{id} + V_{id}' \tag{2}$$

Where $X_{id}$ is the previous positions of i particle and d dimension, $X'_{id}$ is the current positions of i particle and d dimension, $V_{id}$ is the previous velocity of i particle and d dimension, $V'_{id}$ is the current velocity of i particle and d dimension, $P_{id}$ is PBEST of i particle and d dimension, and $G_{id}$ is GBEST of d dimension. $\varpi$ is an inertia weight, $\eta_1$ and $\eta_2$ are acceleration constants, and rand() is the random number generated uniformly in the range [0, 1]. A limit velocity calls $V_{max}$. If calculate velocity of a particle exceeds $V_{max}$, it is replaced value of $V_{max}$.

### 2.2. *K-Means Clustering*

In standard KM, each data is called an object. The group of objects is called a class or the cluster centroid. Both class and object contain attributes. The attributes of an object and the attributes of classes are used to classify by Euclidean distance technique [14]. The main concept of KM is can be summarized as follows: the smallest Euclidean distances from the attributes of an object to the attributes of one class, this object is assigned into this class. The standard KM starts with randomizing attributes of classes and this step is applied only the first time. Then, the evaluation of each class is calculated from the expression below:

$$d(z_p, m_j) = \sqrt{\sum_{k=1}^{N_d} (z_{pk} - m_{jk})^2} \tag{3}$$

Where $N_d$ is the number of attributes, $Z_p$ is the object p, $Z_{pk}$ is the attribute k in the object p, $m_j$ is the class j, $m_{jk}$ is the attribute k in the class j, and $d(z_p, m_j)$ is the distance between the object p and the class j. Then, each class is recalculated from the expression below:

$$m_j = \frac{1}{n_j} \sum_{Z_p \in C_j} z_p \tag{4}$$

Where $n_j$ is the number of object in class j, and $C_j$ is all objects in the class j. This algorithm can be stopped when the maximum number of iterations has been exceeded.

### 2.3. *F-Measure*

F-Measure (FM) or F(i, j) is a famous evaluation measure. FM is the harmonic mean which combines the recall or R(i, j) and the precision or P(i, j). FM uses the ideas of precision and recall from information retrieval [3]. The precision is the most important measure of any classification system. It provides information on the probability that a prediction of a given category is correct. The recall is the relationship between the number of correctly detected intrusions and the total number of intrusions. True Positive ($T_P$) is the model correctly predicts the positive class. True Negative ($T_n$) is the model correctly predicts the negative class. False Positive ($F_P$) is the model incorrectly predicts the positive class. False Negative ($F_n$) is the model incorrectly predicts the negative class. FM, the precision and the recall for each class i of each cluster j is calculated as:

$$P(i,j) = \frac{T_p}{T_p + F_p} \tag{5}$$

$$R(i,j) = \frac{T_p}{T_p + F_n} \tag{6}$$

$$F(i,j) = \frac{2 \times P(i,j) \times R(i,j)}{P(i,j) + R(i,j)} \tag{7}$$

## 3. Related Works

### 3.1. *PSO with Clustering*

The standard PSO was proposed to solve the benchmark function problem [25]. To PSO can apply with the data clustering problem, the objective function and particle position are modified. A single particle represents the class. The fitness function of the panicle is measured as (8). In addition to the previous mention, the process of PSO with data clustering [14] is similar to the process of the standard PSO.

$$Fitness = \frac{\sum_{j=1}^{N_c} [\sum_{\forall z \in C_{ij}} d(z_p, m_j) / |C_{i,j}|]}{N_c} \tag{8}$$

Where $N_c$ is the number of classes, and $|C_{i,j}|$ is the number of object belonging to class j

### 3.2. *PSO-clustering with the hard c-means*

Process of PSO-clustering with the hard c-means (PSOCC) [21] is the same as PSO-clustering except the fitness function is used as follows:

$$J = \sum_{j=1}^{N_c} \sum_{p=1}^{N} \|Z_p - m_j\| \tag{9}$$

$$Fitness = \frac{k_c}{(J + J_0)} \tag{10}$$

Where N is the number of object, $k_c$ is a positive constant, and $J_0$ is a small-valued constant. The experimental result shows that PSOCC algorithm has better performance than KM algorithm. $\|Z_p - m_j\|$ is the Euclidean distance of $Z_p$ to $m_j$.

### 3.3. *The clustering algorithm based on PSO*

The clustering algorithm based on PSO (CPSO) [22] is the same as PSO-clustering except the fitness function is used as follows:

$$Fitness = \frac{\max(\overline{d_1}\ (Y_i))}{\min(d_2\ (Y_i))} \tag{11}$$

$$\max(\overline{d_1}\ (Y_i)) = \max_{a=1,\ldots N_c}\left(\sum_{\forall Z \in C_{ij}} \frac{d(z_p, m_j)}{|C_{ij}|}\right) \tag{12}$$

$$\min(d_2\ (Y_i)) = \min_{a=1,\ldots N_c, b=1,\ldots N_c, i \neq j}(d(m_i, m_j)) \tag{13}$$

Where $\max(\overline{d_1}\ (Y_i))$ is the maximum value of average values of distances within same classes in the classification plan which is showed by particle $Y_i$, $\min(d_2\ (Y_i))$ is the minimum value of distances between classes in the classification plan which is showed by particle $Y_i$, $m_j$ is the class j, and $m_i$ is the class i.

### 3.4. *particle swarm K-means optimization*

The particle swarm K-means optimization algorithm (PSKO) [23] is the same as PSO-clustering except the fitness function is calculated according to (3). Moreover, the process of recalculating the cluster centroid is added for each particle according to (4). From experimental results, PSKO which integrates both the PSO-clustering and K-means algorithm is able to give the particles both exploration and exploitation capabilities. The experimental results showed that performance of PSKO is better than those of other clustering algorithms.

## 4. Proposal Works

Duty of PSO is searching the cluster centroid. The process of PSO tries to get better fitness. So, the fitness function is guidance for searching of PSO and the main factor of searching the cluster centroid. For the data clustering problem, if the cluster centroid stays in a suitable area, it can classify data correctly. On the other hand, if the cluster centroid stays in an unsuitable area, it will rarely classify data correctly. These techniques (PSOC, PSOCC, CPSO, and PSKO) use the fitness function which is calculated by the distance from all input data to the cluster centroid. The effect from fitness function, PSO tries to decrease the lowest distance from all input data to the cluster centroid. So, the more fitness value is decreased, the better solution is obtained. It leads to the cluster centroid move to the space of two data groups and the cluster centroids stick together. The sticky cluster centroids are unsuitable to classify data because data of one group may be classified into another group as figure 1 (b). In fact, the cluster centroids should stay central in its group which it can classify data correctly as figure 1 (a).

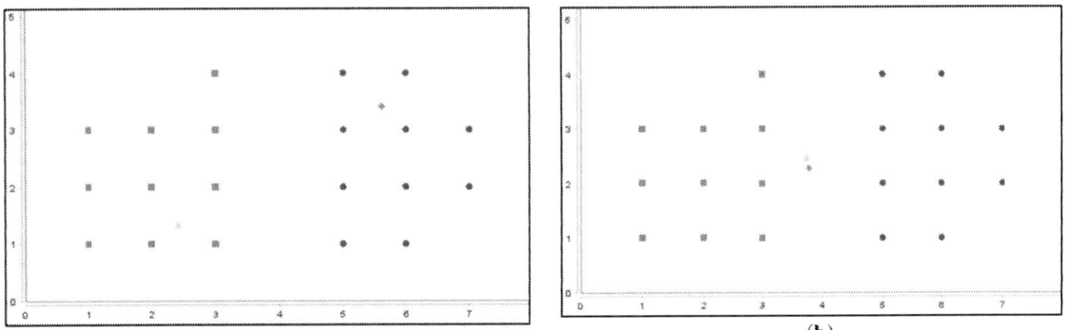

(a) (b)

**Figure 1.** This figure shows the results from searching of PSOC in two dimensions. The red point or the square point is the data of the first group which has ten points. The blue point or the circle point is the data of the second group which has ten points. The green point or the triangle point is the cluster centroid of the first group. The violet point or the diamond point is the cluster centroid of the second group.

From figure 1 (a), the result from calculation by fitness function as (8) gets value that is 1.303. The cluster centroid from this figure can classify data that give correct 100% or 20 points. The value of F-measure in this case is 1. So, this model classifies correctly and has performance and is suitable to be used. From figure 1 (b), the result from calculation by fitness function as (8) gets value that is 1.074. The cluster centroid from this figure can classify data that give correct 65% or 13 points. The value of F-measure in this case is 0.601. Some data in the first group is identified into the second group because they stay close to the cluster centroid of the second group more than that of the first group. So, this model classifies incorrectly and has no performance.
The result from fitness function as (8), PSO selects figure 1 (b) to set GBEST because fitness of figure 1 (b) is less than fitness of figure 1 (a). In fact, PSO should select figure 1 (a) to set GBEST because it is more classificatory than another. This case shows that distance may be unsuitable in some events to calculate fitness of PSO. On the other hand, the fitness function should use the results from classification. In this case, if the fitness function uses the results from classification or F-measure, it selects figure 1 (a) to set GBEST because the amount of correct data from figure 1 (a) is more than that from figure1 (b). The F-measure is calculated from the results from classification. So, F-measure can be used for guiding searching of PSO or the fitness function. Hence, this paper proposes the fitness function of PSO uses F-measure to solve the data clustering problem or modification fitness function by F-measure for particles swarm optimization with clustering (FPSOC). Pseudo code of FPSOC is shown below:

```
Initial position and velocity of each particle
Initial PBEST, GBEST
For gen = 0 to MAXGEN
        For i = 0 to N
                Evaluate particle fitness according to formula (7)
                If F(X_i) < F(PBEST_i)
                        PBEST = X_i
                End If
                If F(X_i) < F(GBEST)
                        GBEST = X_i
                End If
                Update particle position according to formula (1) and (2)
        End For
Next gen
```

Where N is the number of particles, $F(X_i)$ is fitness of i particle, $F(PBEST_i)$ is fitness from PBEST of i particle, $PBEST_i$ is PBEST of i particle, F(GBEST) is fitness of GBEST, MAXGEN is the maximum generation.

## 5. Experiments and Results

### 5.1. *Datasets*

The proposed algorithm is tested on seven datasets [24], listed in Table 1. The results of FPSOC on the datasets are compared with PSOC, PSOCC, CPSO, and PSKO.

**Table 1. Details of datasets**

| Order | Name | Datasets | Attributes | Classes |
|---|---|---|---|---|
| 1 | IRIS | 150 | 4 | 3 |
| 2 | WINE | 178 | 13 | 3 |
| 3 | GLASS | 214 | 9 | 6 |
| 4 | HEART | 270 | 12 | 2 |
| 5 | CANCER | 286 | 8 | 2 |
| 6 | ECOLI | 336 | 7 | 8 |
| 7 | YEAST | 1484 | 8 | 10 |

### 5.2. *The measures of algorithm performance*

The measures of algorithm performance in the experiments are as follows: F-Measure (FM), the average correct number (ACN) is the average of the correct number from the prediction of algorithm in the final generation from all running. FM and ACN indicate the clustering efficiency of an algorithm. The bigger FM and ACN are, the higher the quality of clustering is. Therefore, PSO is the stochastic algorithm. It has the process of randomness. So, FM and ACN may be enough to measure performance of PSO. To measure performance of PSO, this paper adds the standard deviation of ACN (SD). SD indicates the clustering reliability of an algorithm. The bigger SD is, the higher the reliability of clustering is.

### 5.3. *Parameters Setting*

Parameters for all experiments are as follows: acceleration constants of $\eta_1$ and $\eta_2$ are both set to be 1.45, $\omega = 0.75$, the number of population is 100. The number of experiments of each function is 25 runs. The maximum iteration is 1000. The non-PSO parameters for the compared algorithm parameters (PSOCC, CPSO, PSKO and FPSOC) are set according to suggested by the original papers. This research is conducted by a personal computer of AMD FX-8320 with 3.5 GHz CPU, 8 GB RAM and Visual C++ 2010 as the programming language. All datasets are used in experiment are from datasets [24].

### 5.4. *Experiment of the proposed algorithm*

**Table 2.** Comparative F-Measure (FM) of PSOC, PSOCC, CPSO, PSKO and FPSOC on datasets

| Datasets | Algorithms | | | | |
|---|---|---|---|---|---|
| | *PSOC* | *PSOCC* | *CPSO* | *PSKO* | *FPSOC* |
| IRIS | 0.75 | 0.70 | 0.72 | 0.70 | 0.99 |
| WINE | 0.57 | 0.60 | 0.60 | 0.61 | 0.77 |
| GLASS | 0.32 | 0.33 | 0.32 | 0.34 | 0.54 |
| HEART | 0.44 | 0.46 | 0.45 | 0.45 | 0.73 |
| CANCER | 0.53 | 0.51 | 0.50 | 0.51 | 0.70 |
| ECOLI | 0.32 | 0.28 | 0.30 | 0.32 | 0.60 |
| YEAST | 0.29 | 0.32 | 0.31 | 0.23 | 0.39 |

**Table 3.** Comparative the average correct number (ACN) of PSOC, PSOCC, CPSO, PSKO and FPSOC on datasets

| Datasets | Algorithms | | | | |
|---|---|---|---|---|---|
| | *PSOC* | *PSOCC* | *CPSO* | *PSKO* | *FPSOC* |
| IRIS | 117.00 | 110.47 | 114.84 | 111.88 | 148.48 |
| WINE | 115.36 | 117.44 | 115.24 | 117.96 | 137.68 |
| GLASS | 106.44 | 107.76 | 109.40 | 109.52 | 146.52 |
| HEART | 152.92 | 154.60 | 153.64 | 152.92 | 199.00 |
| CANCER | 187.56 | 188.12 | 188.80 | 183.32 | 215.68 |
| ECOLI | 178.04 | 166.04 | 169.76 | 174.00 | 238.28 |
| YEAST | 315.04 | 411.20 | 404.00 | 291.20 | 779.76 |

**Table 4.** Comparative the standard deviation (SD) of PSOC, PSOCC, CPSO, PSKO and FPSOC on datasets

| Datasets | Algorithms | | | | |
|---|---|---|---|---|---|
| | *PSOC* | *PSOCC* | *CPSO* | *PSKO* | *FPSOC* |
| IRIS | 15.44 | 17.55 | 15.30 | 7.20 | 0.51 |
| WINE | 8.65 | 10.13 | 10.00 | 9.83 | 3.73 |
| GLASS | 4.03 | 4.19 | 6.66 | 5.39 | 2.42 |
| HEART | 6.67 | 8.73 | 7.35 | 6.47 | 0.00 |
| CANCER | 15.20 | 10.10 | 16.54 | 20.31 | 9.10 |
| ECOLI | 0.09 | 0.07 | 0.08 | 19.39 | 0.06 |
| YEAST | 71.70 | 80.35 | 81.94 | 73.47 | 66.94 |

From the experimental results of Table 2 and Table 3 show that the quality solution of FPSOC is better than that of PSOC, PSOCC, CPSO, and PSKO because of its highest FM and ACN all datasets. From the experimental results of Table 4 show that the reliability of FPSOC is better than that of PSOC, PSOCC, CPSO, and PSKO because of its lowest SD all datasets. Hence, FPSOC outperforms PSOC, PSOCC, CPSO, and PSKO in case of reliability and quality solution. Hence, FPSOC outperforms PSOC, PSOCC, CPSO, and PSKO in case of quality solution. The experiments can prove that the fitness function affects searching of PSO and searching the centroid vector. PSOC, PSOCC, CPSO, and PSKO use the fitness function that is calculated from the distance. These techniques are calculated by the same concept so they get a few different solutions both the reliability and quality of solutions in all datasets. The solutions from searching of FPSOC are much better than solutions from searching of PSOC, PSOCC, CPSO, and PSKO. Hence, F-Measure can use the fitness function of PSO (FPSOC) to search the cluster centroid and it has performance better than the fitness function that is calculated by the distance (PSOC, PSOCC, CPSO, and PSKO). In the case of HEART dataset, SD of FPSOC equals 0 although the results of FPSOC are better than the results of other techniques. It shows that FPSOC encounters the problem of trapping in local optimum. In the future this problem should be solved to improve the searching performance of PSO.

## 6. Conclusions

The clustering is brought to create models. The model is applied to new input data to identify the class of that data. The clustering depends on the cluster centroid. PSO is brought to search the cluster centroid. The fitness function is the determinant of direction searching of PSO and the main factor of searching the cluster centroid. Many papers proposed that the fitness function is calculated by the distance from all input data to the cluster centroid. However, these methods may lead to bad solutions. To improve performances of PSO for searching the cluster centroid, this paper proposed the fitness function is calculated by the results from classification or F-Measure. The proposed technique is called that FPSOC. From the experimental results show that the proposed FPSOC outperforms PSOC, PSOCC, CPSO, and PSKO with regard to the reliability and quality of solutions in all datasets.

## References

[1] P.N. Tan, M.Steinbach, and V.Kumar. “Introduction to data mining,” Boston: AddisonWesley. pp.487–559, 2005.

[2] Jiawei han and Michelin Kamber. “Data mining concepts and techniques,” Elsevier. 2010.

[3] F. Yang, T. Sun, C. Zhang, “An efficient hybrid data clustering method based on K-harmonic means and Particle Swarm Optimization,” Expert Syst., pp.9847–9852, 2009.

[4] C.D. Manning, P. Raghavan, H. Schütze, “Introduction to Information Retrieval,” Cambridge University Press, Cambridge, Vol. 1, 2008.

[5] J. Kennedy and R. Eberhart, Particle swarm optimization, “Proceeding of IEEE International Conference on Neural Networks (ICNN’95),” Perth, Western Australia, IEEE, vol. 4, pp.1942–1947, 1995.

[6] J. Kennedy and R. Eberhart, “A new optimizer using particle swarm theory,” Sixth International Symposium on Micro Machine and Human Science, pp.39-43, 1995.

[7] Qinghai, B., “Analysis of Particle Swarm Optimization Algorithm,” Computer and Information Science, pp.180–184, 2002.

[8] J. H. Holland, *Adaptation in Natural and Artificial Systems*, University of Michigan press, Ann Arbor, pp.211, 1979.

[9] D. E. Goldberg, Genetic Algorithm in Search, Optimization and Machine Learning, Addison-Wesley, pp.372, 1989.

[10] K. E Parsopoulos, V. P. Plagianakos, G. D. Magoulas, M. N. Vrahatis, *Objective function stretching to alleviate convergence to local minima*, Nonlinear Analysis TMA 47, pp.3419-3424, 2003.

[11] R. Eberhart and Y. Shi, *Comparison between genetic algorithms and particle swarm optimization*, Proc. Seventh Annual Conference on Evolutionary Programming, pp.611-619, 1998.
[12] J. Kennedy and R. Eberhart, *Swarm Intelligence*, Morgan Kaufmann, San Mateo, 2001.
[13] J. Nayak, B. Naik, H.S. Behera, "Fuzzy C-Means (FCM) clustering algorithm," a decade review from 2000 to 2014. In: Jain, L.C. et al. (eds.) Comput. Intell. Data Min. vol. 2, Smart Innov. Syst. Technol. 32, vol. 2, pp.133–149, 2014.
[14] D.W.Van der Merwe. and A.P. Engelbrecht, "Data clustering using particle swarm optimization," Proceedings of IEEE Congress on Evolutionary Computation, pp.215-220, 2003.
[15] X.D. Liu,Q.F. Sha andYK Liu et.al, "Analysis of classifcation using particle swarm optimization," Computer Engineering, vol.32 (6), pp.201-202+217,2006.
[16] J. Komorowski,Z.Pawlak and L.Polowski,et al.,Rough sets: a tutorial, a new trend in decision making,Springer,Singapore Pte Ltd,1998.
[17] F. Q. Yang, T. E. L. Sun and C. H. Zhang, "An efficient hybrid data clustering method based on K-harmonic means and Particle Swarm Optimization," Expert Systems with Applications, vol.36, pp.9847- 9852, 2009.
[18] J. Sun, W. B. Xu and B. Ye, "Quantum-behaved particle swarm optimization clustering algorithm," Lecture Notes in Computer Science, Vol.4093, pp. 340-347, 2006.
[19] J. Y. Chen and H. Y. Zhang, "Research on application of clustering algorithm based on PSO for the web usage patter," International Conference on Wireless Communications, Networking and Mobile Computing, pp.3705-3708, 2007.
[20] Augusto Luis Ballardini, "A tutorial on Particle Swarm Optimization Clustering," arXiv:1809.01942, pp.1–13, 2016.
[21] Ching-Yi Chen, and FunYe, "Particle Swarm Optimization Algorithm and Its Application to Clustering Analysis", Proceedings of the 2004 IEEE International Conference on Networking, Sensing & Control Taipei. Taiwan, pp. 789 - 794 ,March 21-23, 2004.
[22] Pei Zhenkui, Hua Xia, and Han Jinfeng, "The Clustering Algorithm Based on Particle Swarm Optimization Algorithm," 2008 International Conference on Intelligent Computation Technology and Automation, pp. 148- 151, 2008
[23] R. J. Kuo, M. J. Wang, and T. W. Huang, "An application of particle swarm optimization algorithm to clustering analysis, " Soft Comput, vol. 15, pp. 533–542, 2009.
[24] Department of Information and Computer Science. University of California. "UCI Machine Learning Repository,"[Online].Available: http://www.ics.uci.edu/mlearn/MLRepository .html.1998.
[25] Molga, M., & Smutnicki, C. Test functions for optimization needs (2005). Retrieved June 2013.

IOP Conf. Series: Materials Science and Engineering **965** (2020) 012039 doi:10.1088/1757-899X/965/1/012039

# Long Range UAS Mission by LPWAN Communication

**Nuttawat Punpigul, Marut Muangkham, Parinya Anantachaisilp*, Sanga Srisuprapreeda, and Krongboon Singhanat**

Department of Electrical Engineering, Navaminda Kasatriyadhiraj Royal Air Force Academy, Bangkok, Thailand, 10220

*Coresponding Author: parinya_a@rtaf.mi.th

**Abstract**. One of the key challenges for long range Unmanned Aerial Systems (UAS) is the limitation of the communication capability. Many existing communication technics are conventionally adopted in UAS to improve the quality of the communication such as satellite communication, cellular networks, and relay aircrafts. However, the mentioned solutions often incur several drawbacks including costs, complexity, communication delay, or inefficient power consumption. LoRaWAN is one of the recently emerging communication technics used in the Low Power Wide Area Networks (LPWAN) technology. It has a promising potential to improve the coverage of UAS communication, since it can transmit/receive signals below the noise floor in a very long range with very low energy consumption by using the waveform modulation technique called Chirp Spread Spectrum (CSS). In this paper, we show the potential of LPWAN as the supplementary communication in UAS. The experiment results show that LoRaWAN can maintain the communication capability beyond 80 km where the popular RF device, RFD900x, has lost its ability to communicate at this distance.

## 1. Introduction

The communication subsystem of unmanned aerial system (UAS) is often the key limit factor of a long-range mission due to the required continuous update of the telemetry and mission data. Therefore, the extension of the UAS communication range is constantly attract attention from the research community. The key challenges are to maximize the coverage and data rate while minimizing the power consumption [1]. Extending the signal travelling distance results in a significant power attenuation along the path causing a very low power efficiency. Increasing the amount of transmitted data is achieved with the cost of heavy power consumption [2,3].

Besides the normal RF communication, there are multiple approaches to extend the range of the UAS mission including satellite communication, relay aircrafts, cellular network, and low power wide area network (LPWAN). In which, each approach has its own advantages and disadvantages. While relay aircrafts carrying the same type of RF transceivers as the mission aircraft to extend the distance of the mission seems to be simplest in implementation of the same communication protocol. But, the control of multiple aircrafts at the same time makes the mission more complicated [4,5]. Another existing solution to increase the range of UAS operation is to use the cellular networks. The drawback of cellular network is the lack of network infrastructure in some areas [6,7]. For the beyond line of sight (BLOS) scenario, the satellite communication is quite effective. However, there are subscription expenses to use the satellite services [8]. With all the mentioned limitations, the Low Power Wide Area Network (LPWAN) technology has a potential for improving the communication capacity in UAS because of its low energy expense and the capability to receive/transmit signals in a very long range below the noise floor [9]. The LPWAN technology has recently emerged following the trend of the Internet of Things (IoT) involving diverse applications [10]. Due to its advantages, we investigate the feasibility of the LPWAN technology as a communication subsystem for the long range UAS mission. This paper focuses on the LoRa communication technology, using the Chirp Spread Spectrum (CSS) as its waveform modulation because of its capability to support a very long-range data exchange below the noise floor. The implementation framework is mentioned and discussed, and the performances are evaluated in comparison with RFD900x.

The remainder of this paper is organized as follows: Section 2 summarizes the background of the LPWAN technology including the comparison of the available options of LPWAN for UAS. Section 3 overviews the system used for the experiment to prove the proposed concept as well as the concept of operation. Then, in Section 4, the experimental results are illustrated and discussed. Lastly, the conclusion of the work in this paper is summarized in Section 5.

## 2. LPWAN Background

LPWAN (Low Power Wide Area Network) is a technology that allows multiple endpoints to form a network over a long range with the characteristics of low-power consumption and low bit rate in a narrow bandwidth [11]. The LPWAN is originally designed for IoT applications and machine-to-machine (M2M) communication. The usual data capacity of LPWANs are 10 to 1000 bytes per transmission at the uplink speed up to 200 kbps, whose range varies from 1 to 1000 km, depending on the communication technics and the tradeoff between the data size and data rate.

### *2.1. Type of LPWAN*

There are three popular types of LPWANs as follows [11, 12]. The first type is Long Term Evolution for Machine Type Commutation (LTE-M) and Narrowband-IoT (NB-IoT). They are both 3rd Generation Partnership Project (3GPP) standards that operate on the licensed spectrum. NB-IoT, also known as CAT-NB1, is specifically on indoor coverage, low cost and low power. It has a data rate up to 200Kbps and limits the bandwidth at 200kHz. LTE-M, also known as CAT-M1, offers higher bandwidth than NB-IoT. The receive bandwidth of LTE-M around 1.4 up to 5 MHz. LTE-M consumes the highest bandwidth of all mentioned LPWAN technologies. However, The LTE-M's transmission power is equal to NB-IoT.

The next type is Weightless SIG, which has three different standards including the unidirectional Weightless-N, bidirectional Weightless-P, and bidirectional Weightless-W. Weightless-W is the most unpopular options because Weightless-W's battery life is shortest. Weightless types N and P use unlicensed spectrum in the sub-1 GHz but also support licensed spectrum operation using 12.5 kHz narrowband technology.

The last type is the LoRaWAN. LoRa runs in various unlicensed sub-gigahertz bands specified by LoRa Alliance. LoRa use chirp spread spectrum (CSS) modulation, allowing the communication in a very long range with less interference than other LPWANs. All users can program the LoRa chip with open source library from Semtech Corporation, which allows users to freely manage their desirable packet size. Nowadays, LoRa technology is implemented in many long-range communication projects because of its very low receiver sensitivity up to -148 dBm.

Table 1 shows the comparison of LPWAN technologies in many important aspects. LoRa possesses the highest link budget of all mention standards. Regarding the security aspect, LoRa offers 32-bits security technology, while the best contender only offers 16-bits technology. LoRa also implements CSS which allow the reception of the signal below the noise floor. Due to the operational characteristics of UAS, maintaining the communication system between Ground control Station (GCS) and Unmanned Aerial Vehicle (UAV) is critical for the survivability. Therefore, the communication subsystem of the UAV should offer a secured stable long-range connection. We chose LoRaWAN as the most potential candidate in this work because of several advantages on the communication range, network security and power efficiency.

**Table 1.** Comparing the capabilities of LPWANs [12]

| Feature | LoRa | Weightless | NB-IoT | LTE-M |
|---|---|---|---|---|
| **Modulation** | SS Chirp | GFSK/DBPSK | UNB/GFSK/BPSK | OFDMA |
| **Data Rate [bps]** | 290 –50 k | 100 | 100 | 200k–1 M |
| **Link Budget [dB]** | 154 | 146 | 151 | 146 |
| **Battery lifetime [years]** | 8 – 10 | 7 – 8 | 7 – 8 | 1 – 2 |
| **Power Efficiency** | Very High | Very High | Very High | Medium |
| **Security** | Yes (32 bits) | Yes (16 bits) | No | Yes (32 bits) |
| **Range [km]** | 25 (urban)<br>15 (suburban)<br>45 (rural) | 3 – 10 (urban)<br>30 – 50 (rural) | 1.5 (urban)<br>20 – 40 (rural) | 35 (2G)<br>200 (3G)<br>200 (4G) |

### *2.2. Class of LoRaWAN Devices*

LoRaWAN classifies its end devices three classes according to how the end device exchanges data with the server [13]:

Class A: The end device has a duty cycle of active/sleep interval, while the server remains active. The end device wakes up and transmit its data packet and wait for a potential data reception within the specified intervals. If the Server demands to send some message to the end device, the server must wait for the End device to wake up and transmit first.

Class B: The data transmission cycle is specified by the periodic beacon signal emitted by the server. The end device regulates its transmission windows using the beacon signal as the reference. Therefore, the server can control the speed and cycle of data transmission.

Class C: The end device has almost no sleep. It opens the slot for the data reception from the server all the time allowing a non-scheduling timely data transfer in exchange of energy expense.

The concept of operation of every class of LoRaWAN is illustrated in Figure 1.

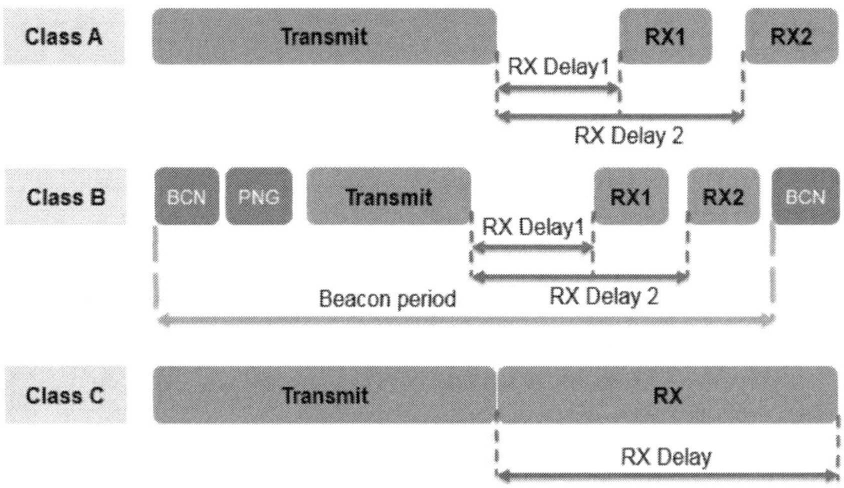


**Figure 1.** LoRaWAN classes

The class of end devices is determined based on the requirement of each application. For UAS application, class B seems to be the most suitable class where the power consumption and the success rate of transmitted are optimized. However, the LoRa class B driver is under development and will be available soon. So, the work presented in this paper will use only class A and class C devices for the experimentation.

### *2.3. LoRaWAN Topology*

According to the LoRaWAN topology as shown in Figure 2, there are four hardware components necessary for the operation.

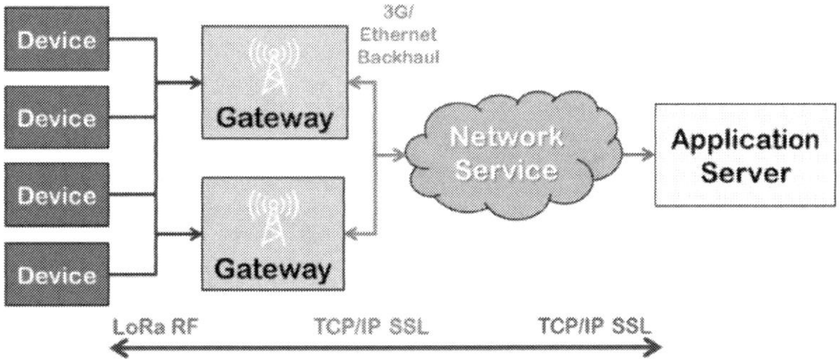


**Figure 2.** LoRaWAN topology [14,15]

*2.3.1 End Device*. It is the data transfer device that using the LoRa modulation technique to encapsulate the data and broadcast it to LoRa gateways.
*2.3.2* Gateway. It is the medium device between End-device and Network service. Generally, Gateway will send data received from End-device to Network service using internet. Gateway is also the point to schedule the Downlink signal coming from Network service to send to the End device.
*2.3.3 Network Service*. It is the LoRaWAN main network, is responsible for sorting and collecting data from the connected gateways. It also decodes the received data before forwarding to the Application Server via the internet. The network service usually in the form of a Cloud system so that the users can access to the network whenever the connectivity to the Internet is available.
*2.3.4 Application Server*. It is the user interface software using the Network service Application User Interface (API) to control Network service or Downlink data to End-device. Application Server usually is installed with Network service for simple and fast data transfer.

## **3. Concept of Operations and System Overview**

We have designed the concept of operations, hardware structure, and software structure of LoRaWAN technology for this research as follows.

### *3.1. Concept of operation*

In this system design, we use MAVLink [16] as the standard protocol for exchanging telemetry data between drones and GCS. The overall system will focus on checking the condition of the telemetry signal through MAVlink in order to be used as a condition to enable communication by LoRa. The system will start by checking the quality of the telemetry signal. If the quality of telemetry signal is lower than 10%, the warning message will first be downlink via LoRa system to alert and to check the stability of LoRa communication system to prepare LoRa system for the event of the telemetry signal loss. After the telemetry signal between UAV and GCS has been acknowledged as lost, UAV is still set to continue the assigned mission as the LoRa system begins the full back up operation. The process begins with the downlink of the telemetry warning message requesting the next command from the GCS whether to continue the mission, to put the UAV to return to land mode (RTL), or to stop the mission and return to the base. After GCS has sent the next command to the UAV, LoRa begins to downlink the UAV's location and check the condition of the main signal every 60 seconds so that the GCS can continue to monitor the operation and observe the movement of the UAV in absence of the main communication system. If the primary communication system gets back online, the system will switch to the main communication system automatically and the LoRa system will stop sending the UAV's information and get back to monitor the telemetry connection quality as usual. Figure 3 shows the flowchart of the UAV back-up communication CONOPS.

### *3.2. Hardware structure*

The hardware structure used in this research is divided into two parts, which are the hardware installed on the UAV and the hardware on the GCS as illustrated in Figures 4 and Figure 5. For the hardware installed on the UAV, the key component is the flight controller, which signal the control surfaces and actuators on the UAV. In this research, the Pixhawk 2 (The cube black) [17] is used as a flight controller. The Pixhawk 2 will be connected to the telemetry, device used to communicate between UAV and GCS, via UART.

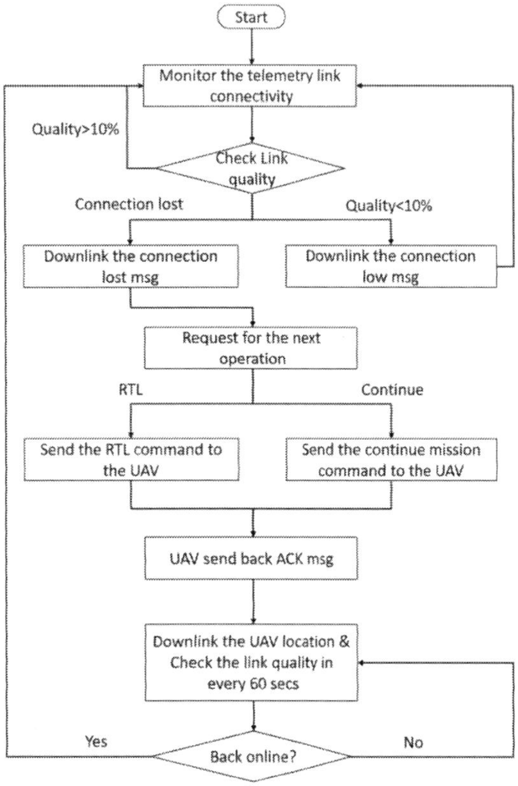


**Figure 3.** UAV back-up communication CONOPS

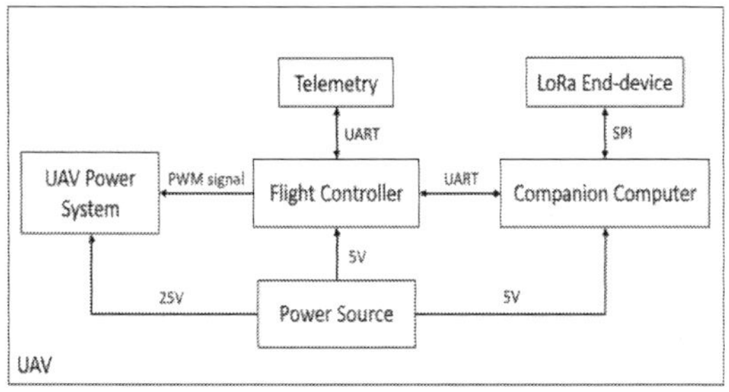


**Figure 4.** Hardware structure on the UAV

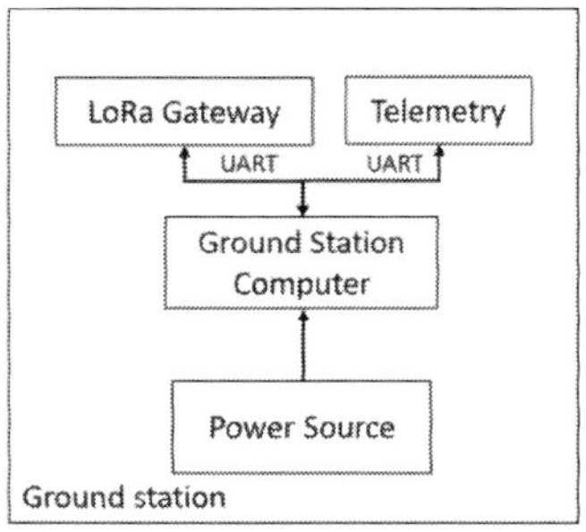


**Figure 5.** Hardware structure on the GCS

The RFD900x [18] is used as the main communication subsystem in this research. A companion computer is necessary as the interface between the flight controller and LoRa system. Raspberry Pi Zero [19] is chosen a companion computer that works alongside with the flight controller to receive and forward commands to the LoRa system. The RFM95W [20] is the UAV's LoRa End-device. The MAVlink protocol [16] is used as a communication protocol between Raspberry pi zero and the Pixhawk 2. The connection between each hardware installed on the UAV is shown in Figure 6.

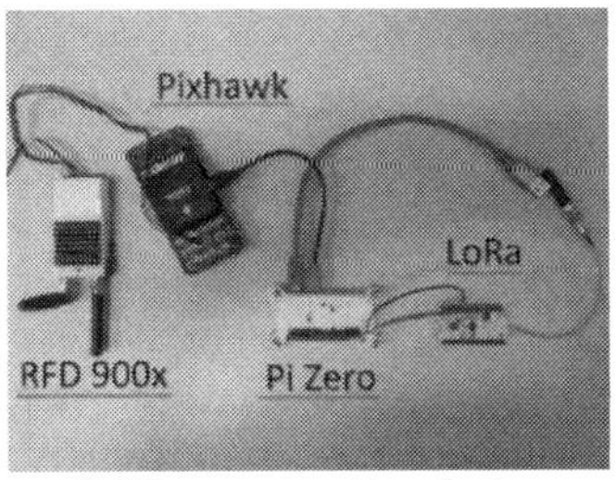


**Figure 6.** The connection between each hardware installed on the UAV

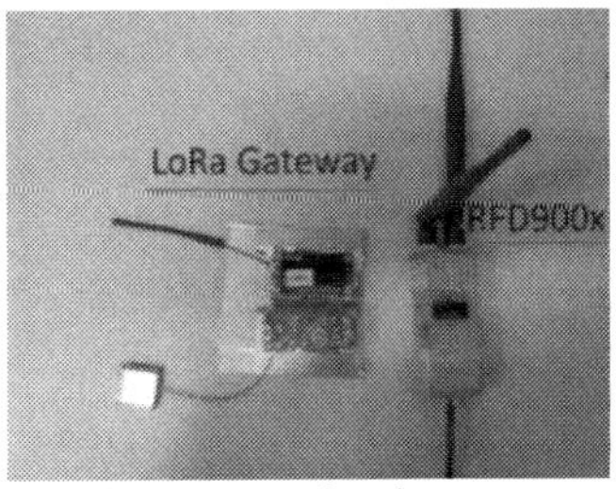


**Figure 7.** The connection between each hardware installed on the GCS

For the hardware installed on the GCS, the centre module for the GCS is the computer used for flight planning and commanding the UAV. The main communication device between the UAV and the GCS is implemented by RFD900x identical to the communication module on the UAV. For the LoRa device, the RAK2245 [21] is chosen, which is a LoRa gateway for the control of downlink commands for end device on the UAV. The frequencies used for communication with the RFD900x are in the range of 915 - 928 MHz and the LoRa frequencies are in the range of 923 - 925 MHz. The connection between each hardware installed on the GCS is shown in Figure 7.

### 3.3. *Software structure*

The software structure in this work is divided in to two parts, the LoRaWAN part and the system control part as shown in Figures 8 and 9. The software is installed inside the Raspberry pi zero which is used as a companion computer for the UAV. The details of the software components are as follows.

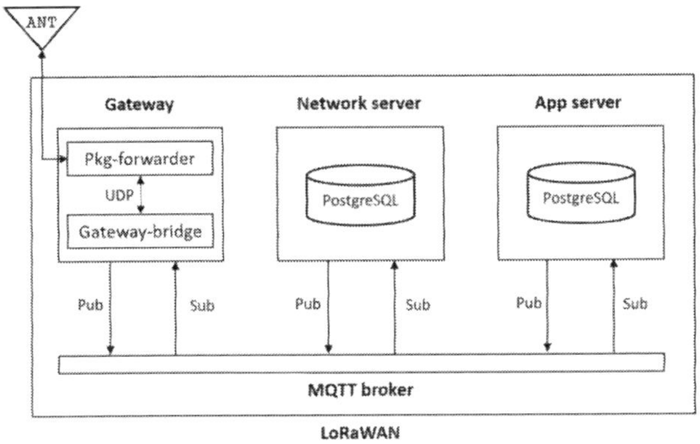


**Figure 8.** The connection between each hardware installed on the GCS

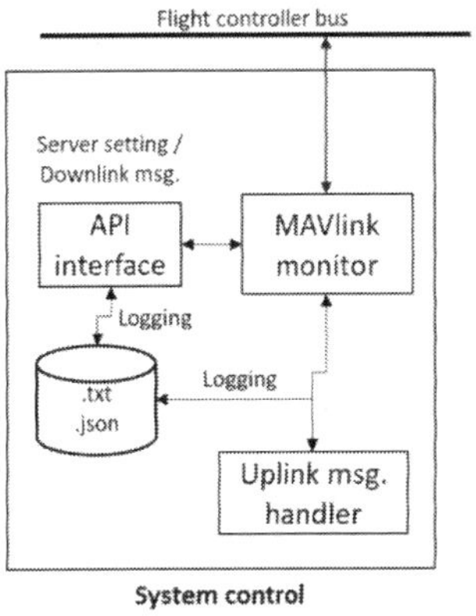


**Figure 9.** Software structure in the System control part

#### *3.3.1 The LoRaWAN*

Package forwarder: The Hardware Abstraction Layer (HAL) of the LoRa gateway. The program will get raw data of the LoRa message it received to LoRa gateway bridge. LoRa gateway bridge: The program which convert the raw data received from the package forwarder into a JSON (JavaScript Object Notation) object to match the LoRa server implementation model and publish it to the MQTT (Message Queuing Telemetry Transport) broker and also do the conversion of the JSON object received from the LoRa server back to its original form for sending to the End device as well.

MQTT broker: The communication medium software for connecting various programs within the Gateway together. It will operate in the form of publish /subscribe programs. The program that want to send data to other programs must publish the information and the information topic to the MQTT broker. After that, when any program wants to receive data from the publisher program, it must subscribe to that data by choosing from the topic of information. MQTT broker can support the program to connect to the broker at the same time to 1000 programs, including the ability to encrypt the data. Using MQTT as a medium for data transmission make it easier to send and receive data from various programs on the gateway and to support data receiving from more gateway in the future.

Network server: The program at the centre of the LoRaWAN network is responsible for handling redundant data that occurs during the transmission of data from the gateways to the server, managing data when there are many connections and prioritizing the delivery of downlink to end devices.

App server: The user interface program that works with the LoRa Network server. It will handle the join to server request message sent from end devices, managing the encryption and decoding of the data that the end devices send to the server and providing the API service to connect with third-party programs via the system API

PostgreSQL: Object-relational database management system using ORDBMS (Object-Relational Database Management System). It can use almost any type of SQL (Structured Query Language) code. It is also the most modern opensource database system.

*3.3.2 System control*

MAVlink monitor: the main program that will monitoring the MAVlink message to observe the UAV status and the connection quality. It will command the API interface to send the downlink message and observe the UAV's information from uplink message too.

API interface: Connect to the network server through the API service of the app server for change some setting of the network or sending the downlink message to the end device.

Uplink MSG handler: Subscribe to the uplink data topic in MQTT broker for logging the uplink message received from the end-device to a text file.

All software in the system control part are written in Python.

## 4. Experimental Setup and Results

There are two aspects that this work is focusing on the LoRaWAN technology for the long range UAS mission. The first one is the ability to extend the distance of the communication of the existing RF transceiver. Another aspect is on the variation of each class of LoRaWAN devices that suitable for UAS application.

*4.1. Communication range test*

The distance test is performed by simulating the distance of the communication using the signal attenuator as illustrated in Figures 10 and 11. There are 3 tested distance, which are 10, 30, and 80 km (as according to the specification of the RFD900x, the maximum communication distance is 80km, demonstrated by the Edge Research labs on a balloon [17]). To select the size of the attenuator, the principle of free space lost [22] are used in the calculation to find the optimal attenuator to simulate each tested distance. The format of the message to be used for RFD900x during the test is sending MAVlink messages as usual for communication between UAV and GCS. For the LoRa, only UAV coordinates will be sent during the test. However, the LoRa device has a limit on the size of messages that can be sent up to 51 bytes [23] with a low data rate, the size of messages to be sent with LoRa should be as small as possible. So, for the effective communication, the Military Grid Reference System (MGRS) positioning method instead of The World Geodetic System (WGS) is used stead in order to make the size of the UAV coordinates smaller. The Lora communication system on both the UAV side and the GCS side must have a program to convert these positions. The parameters used to measure the test results are Received Signal Strength Indicator (RSSI) of both sides. In the test, both types of devices are placed at 200 meters away from their pair before the test begun. The test will start with enabling the communication without any attenuator and then changing the attenuator according to the simulated distance.

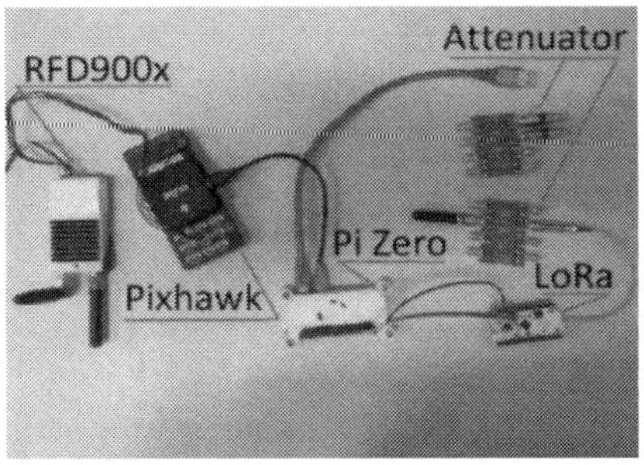


**Figure 10.** The connection between each hardware installed on the UAV with the attenuator

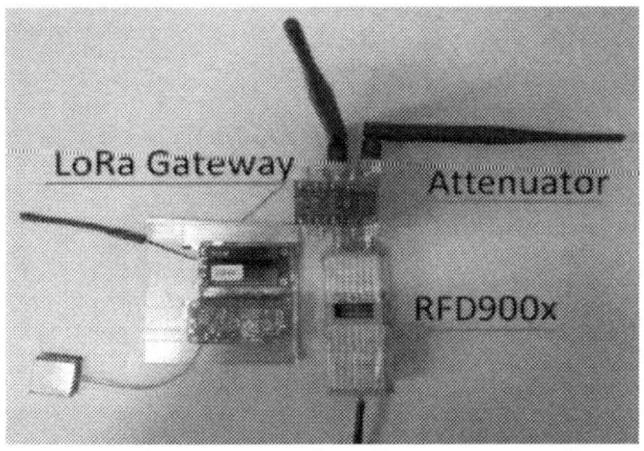


**Figure 11.** The connection between each hardware installed on the GCS with the attenuator

From the results of the Table 2, The RSSI of RFD900x reach 0 percent at the distance of 80 km, but LoRa keeps the RSSI in the operational range at 80 km and beyond. From the test, we can clearly see that the noise floor level in the test area (-90 to -100 dBm) has affect the communication directly, the RFD900x has lost its ability to communicate when the RSSI level is equal to or below the noise floor. The RFD900x cannot demodulate the transmitted data although its sensitivity is lower than the noise floor (-121 dBm [18]), while LoRa can demodulate the signal below the noise floor level [24] (Sensitivity @ -142 dBm) resulting in the ability to communicate over a longer distance. Hence, LoRa can be used as a supplementary module for UAS using RFD900x as a primary communication system.

**Table 2.** Communication range test results

| Distance Simulate (Km) | RFD900x Signal Strength (%) | LoRa Signal Strength (%) |
|---|---|---|
| 0 | 100 | 100 |
| 10 | 94.8 | 84.14 |
| 30 | 66.67 | 78.05 |
| 80 | 0 | < 78* |

*We cannot measure the true value due to the noise floor limit

### *4.2. LoRaWAN class verification test*

Using LoRa to exchange data requires the uplink from the end device downlink from the Gateway to exchange data between devices. When data is downlink, consideration of the LoRa End-device class selection is important. Due to the timing of opening of the RX slot of End-device in each class is different. It also results in different energy consumption. Therefore, we must test the amount of energy that LoRa consume to send a message every 60 seconds for 1 hour of class A and class C. Therefore, the number of messages sent are 60 in total. The testing End-device will transmit an Uplink message then the Gateway will answer the Uplink message with acknowledge Downlink message.

**Table 3.** Power consumption and success rate of transmission of class A and C LoRaWAN

| Class | Power consumption per hour (mA) | Downlink message received | Success rate of transmission (%) |
|---|---|---|---|
| A | 6600 | 47 | 78.33 |
| C | 87000 | 60 | 100.00 |

From the test results in Table 3, class A uses less power than class C, but there is a chance that the downlink message will not be successful because of the RX slot time mismatch problem. This is a major weakness that makes class A unsuitable for UAV missions that require high accuracy and consistency in data transmission. On the other hand, the weakness of class C is the power consumption, but in the other hand we got better communication quality. To implement Class C, we suggest optimizing the size of the battery by consider the amount of time to enable the LoRa system on the UAV.

## 5. Conclusions

This paper reports the adoption of another communication technics called LoRaWAN in a long range UAS mission where the distance can be extended with a low power consumption. The LoRa's capability of a long-range data exchange below noise floor using the Chirp Spread Spectrum (CSS) modulation appeals to the potential supplementary communication in UAS. The experimental results show that LoRaWAN can maintain the communication capability beyond 80 km where the popular RF device, RFD900x, has lost its ability to communicate at this distance. Moreover, the comparison between LoRaWAN Class C and Class A suggests that LoRaWAN Class C is more suitable for UAS application.

## 6. References

[1] Çuhadar, İsmet, and Mahir Dursun. Unmanned air vehicle system's data links. Journal of Automation and Control Engineering 4.3; 2016.
[2] Jain, Raj, Fred Templin, and Kwong-Sang Yin. Wireless Datalink for Unmanned Aircraft Systems: Requirements, Challenges, and Design Ideas. Infotech@ Aerospace 2011; 2011. p. 1426.
[3] Gupta, Lav, Raj Jain, and Gabor Vaszkun. Survey of important issues in UAV communication networks. IEEE Communications Surveys & Tutorials 18.2; 2015. p. 1123-1152.
[4] Dixon, Cory, and Eric W. Frew. Optimizing cascaded chains of unmanned aircraft acting as communication relays. IEEE Journal on Selected Areas in Communications 30.5; 2012.p. 883-898.
[5] Ono, Fumie, Hideki Ochiai, and Ryu Miura. A wireless relay network based on unmanned aircraft system with rate optimization. IEEE Transactions on Wireless Communications 15.11; 2016. p. 7699-7708.
[6] Hosseini, Nozhan, et al. UAV Command and Control, Navigation and Surveillance: A Review of Potential 5G and Satellite Systems. 2019 IEEE Aerospace Conference. IEEE; 2019.
[7] Zhang, Shuowen, Yong Zeng, and Rui Zhang. Cellular-enabled UAV communication: A connectivity-constrained trajectory optimization perspective. IEEE Transactions on Communications 67.3; 2018. p. 2580-2604.
[8] Autonomous Systems Enabling Global Line of Sight for Unmanned Vehicles in the Air, on Land, at Sea [Internet]. [cited 2020 Feb 10]. Available from: https://www.iridium.com/solutions/autonomous-systems/.
[9] Reynders, Brecht, and Sofie Pollin. Chirp spread spectrum as a modulation technique for long range communication. 2016 Symposium on Communications and Vehicular Technologies (SCVT). IEEE; 2016.
[10] Mekki, Kais, et al. Overview of cellular LPWAN technologies for IoT deployment: Sigfox, LoRaWAN, and NB-IoT. 2018 IEEE International Conference on Pervasive Computing and Communications Workshops (PerCom Workshops). IEEE; 2018.
[11] LPWAN (low-power wide area network) [Internet]. Margaret Rouse, [updated 2017; cited 2020 Feb 16]. Available from: www.internetofthingsagenda.techtarget.com.
[12] LPWAN COMPARISON [Internet]. Ubiik, [cited 2020 Feb 16]. Available from: https://www.ubiik.com/lpwan-comparisons.
[13] LoRaWAN® Specification (Classes) [Internet]. LoRaAlliance®, [updated 2016; cited 2020 Feb 16]. Available from: https://lora-alliance.org/about-lorawan.
[14] LoRaWAN® Specification v1.0.2 [Internet]. LoRaAlliance®, [updated 2016; cited 2020 Feb 16]. Available from: https://lora-alliance.org/sites/default/files/2018-05/lorawan1_0_2-20161012_1398_1.pdf.
[15] Basic communication protocol between Lora gateway and server [Internet]. Semtech, [updated 2017; cited 2020 Feb 16]. Available from: https://github.com/Lora-net/packet_forwarder/blob/master/PROTOCOL.TXT.
[16] D. Project. MAVLink introduction Dronecode Project; 2009.
[17] Pixhawk2.1 [Internet]. ProfiCNC, [updated 2017; cited 2020 Feb 17]. Available from: http://www.proficnc.com/content/13-pixhawk2.
[18] RFD900x Modem [Internet]. RFdesign.AU, [updated 2017; cited 2020 Feb 17]. Available from: http://www.proficnc.com/all-products/91-rfd900u.html.
[19] Raspberry Pi Zero W [Internet]. R. P. FOUNDATION, [updated 2018; cited 2020 Feb 17]. Available from: https://www.raspberrypi.org/products/raspberry-pi-zero-w/.

[20] WiFi LoRa 32 (V2) datasheet [Internet]. H. Automation, [cited 2020 Feb 19]. Available from: https://heltec.org/project/wifi-lora-32/.
[21] RAK2245 Stamp Edition - LoRaWAN™ Gateway Concentrator Module Shenzhen RAKwireless Technology Co. [Internet]. S. R. T. Co., [cited 2020 Feb 20]. Available from: https://doc.rakwireless.com/datasheet/rakproducts/rak2245-stamp-edition-datasheet.
[22] Free Space Path Loss: details & calculator [Internet]. A. Augustin, J. Yi, T. Clausen and W. Townsley, [updated 2016; cited 2020 Feb 22]. Available from: www.electronics-notes.com/articles/antennas-propagation/propagation-overview/freespace-path-loss.php.
[23] A. Augustin, J. Yi, T. Clausen and W. Townsley. A Study of LoRa: Long Range & Low Power Networks for the Internet of Things. Sensors; 2016. p. 1466.
[24] Semtech SX1276 [Internet]. Semtech, [updated 2019 Jan; cited 2020 Feb 22]. Available from: https://www.semtech.com/products/wireless-rf/lora-transceivers/sx1276.

IOP Conf. Series: Materials Science and Engineering **965** (2020) 012040 doi:10.1088/1757-899X/965/1/012040

# Case Study of Data Exchange using Cross-Boundary Direct Interconnection between highly constrained Wireless Sensor Networks

**Krongboon Singhanat[1] and Kantida Panchareon[2]***

[1] Department of Electrical Engineering, Navaminda Kasatriyadhiraj Royal Air Force Academy, Bangkok, 10220, Thailand
[2] Department of Electrical Engineering, Faculty of Engineering and Industrial Technology, Silpakorn University, Nakhon Pathom, 73000, Thailand

*Email: pancharoen_k@su.ac.th

**Abstract**. Internet of Things (IoT) connects multiple network entities together with different platforms and standards using the backbone network and Internet Protocol. However, some network interconnections may happen remotely at the edge of the network boundary between highly constrained platforms in absence of the connection with the backbone infrastructure. In this paper, we report the implementation of payload exchange between highly constrained multi-hop Wireless Sensor Networks (WSNs). The Constrained Application Protocol (CoAP) is accommodated by a native communication link using a payload fragmentation. The impacts of the hop distance and the fragmentation regarding the reliability and the latency are evaluated using a testbed showing less than 0.2 percent unsuccessful payload delivery.

## 1. Introduction

As the vision of IoT begins taking shape, many solutions have been emerged separately to support the diverse application. The interoperability between different hardware and distinctive network design is conventionally achieved by connecting each individual network with the Internet gateway using Internet Protocol [1]. However, in some cases, the communicating network entities may be deployed in a same spatial location which allows packet transfers between their physical platforms at the edge of the network under the authority consent [2], [3]. The concept of enabling interconnection has already been discussed by the previous works which report several benefits such as a data exchange, retrieving the lost segment and the extension of network lifetime via load distribution [2]–[5].

Enabling the connectivity between distinctive systems, the physical layer must be able to provide a mutual logical channel using the identical waveform for physical bit transfer. The link layer must adopt the mutual algorithm to access the common medium for timing the transmission and reception. Possible solutions on the link layer have been discussed by previous works [2], [3]. In the application layer, the payload format must be very compact and can be universally understood by all participants. The most well-known candidates for an application protocol in a constrained environment are MQTT (Message Queuing Telemetry Transport), AMQP (Advanced Message Queuing Protocol), XMPP (Extensible Messaging and Presence Protocol), and CoAP (Constrained Application Protocol) [6], In comparison with all of them, CoAP consumes the least resources [7]. CoAP is proposed by IETF using the constrained RESTful architecture (CoRE) for application data exchange in the form of resource access between client and server. Therefore, CoAP is mostly suitable for a constrained network which can be systematically integrated with HTTP REST for an elaborate system [7]. CoAP is designed and tested on top of IPv6 over Low-Power Wireless Personal Network (6LoWPAN) that can commonly be implemented on the up-to-date embedded platform. The 6LoWPAN standard is designed for IEEE 802.15.4 link layer with the goal to access every IoT device with IPv6 and requires at least 60-80 octets for payload accommodation [8]. However, a highly constrained WSN, deployed for sensing and report application, may possess a very compact frame size (less than 64 octets). Therefore, the network cannot directly support the 6LowPAN standard due to its insufficient maximum transmission unit (MTU). Nonetheless, forwarding application payload along the path with low-end devices may be used to provide an end-to-end communication for exchanging the data between neighbouring clusters or networks.

This work reports the implementation of the data exchange with a direct interconnection between highly constrained WSNs using CoAP on top of a native communication link on a testbed. The study assumes the scenario where neighbouring WSNs encounter each other and exchange the public resource. This paper is organized as follows: Section 2 elaborates the protocol design which allow the end-to-end exchange of CoAP messages on top of a customized link layer. In Section 3 the experiment setup is described. Section 4 discusses the results of the reliability measurement and end-to-end delay against the hop distance and degree of fragmentation.

## 2. Protocol Design

The main constraint in this design is the limitation on MTU which is imposed on hardware with low-computational performances. Therefore, the protocol design heavily focuses on the reduction of the frame space usage. The initial point of the protocol design considers the original convention and modifies the details to fit the design constraints.

### *2.1. Datagram Format*

CoAP is originally designed on top of UDP (User Datagram Protocol). The format of the datagram is illustrated as follows.

```
0      7 8     15 16    23 24     31
+--------+--------+--------+--------+
|     Source      |   Destination   |
|      Port       |      Port       |
+--------+--------+--------+--------+
|                 |                 |
|     Length      |    Checksum     |
+--------+--------+--------+--------+
|
|          data octets ...
+---------------- ...
```

**Figure 1.** UDP Format [9]

```
 0                   1                   2                   3
 0 1 2 3 4 5 6 7 8 9 0 1 2 3 4 5 6 7 8 9 0 1 2 3 4 5 6 7 8 9 0 1
+-+-+-+-+-+-+-+-+-+-+-+-+-+-+-+-+-+-+-+-+-+-+-+-+-+-+-+-+-+-+-+-+
|Ver| T |  TKL  |      Code     |          Message ID           |
+-+-+-+-+-+-+-+-+-+-+-+-+-+-+-+-+-+-+-+-+-+-+-+-+-+-+-+-+-+-+-+-+
|   Token (if any, TKL bytes) ...
+-+-+-+-+-+-+-+-+-+-+-+-+-+-+-+-+-+-+-+-+-+-+-+-+-+-+-+-+-+-+-+-+
|   Options (if any) ...
+-+-+-+-+-+-+-+-+-+-+-+-+-+-+-+-+-+-+-+-+-+-+-+-+-+-+-+-+-+-+-+-+
|1 1 1 1 1 1 1 1|    Payload (if any) ...
+-+-+-+-+-+-+-+-+-+-+-+-+-+-+-+-+-+-+-+-+-+-+-+-+-+-+-+-+-+-+-+-+
```

**Figure 2.** CoAP Message Format

```
0               1               2               3
0 1 2 3 4 5 6 7 0 1 2 3 4 5 6 7 0 1 2 3 4 5 6 7 0 1 2 3 4 5 6 7 0
|          SRC ADDRESS          |          DST ADDRESS          |
|     LENGTH    |   CHECKSUM    |          COAP MESSAGE          ...
```

**Figure 3.** Modified Local Datagram Format

According to Figure 1, UDP header occupies 8 octets containing Source (SRC) port, Destination (DST) port, UDP datagram length and checksum. Since the CoAP message is relayed by a local native communication link, separated from the Internet, SRC/DST port can be omitted and automatically filled by the default port number 5683 specified by CoAP at the DST endpoint [6]. Figure 3 shows the datagram for carrying CoAP messages after the modification. The header fields of the CoAP message are literally defined by RFC7252 [6]. The modification results in the minimum header size of the modified datagram at 10 octets.

### *2.2. Fragmentation*

Even though the framework for solving a limit frame size is recommended by Block-Wise Transfers RFC 7959 [8], it has a significant drawback. The block-wise transfer divides a CoAP message into blocks. However, each block must be separately requested causing excessive conversation. While these individual block requests can be handled by the systems recommended for 6LowPAN, the sheer number of block requests may pose a serious problem a bandwidth limited network. Therefore, the protocol design circumvents the drawback by using burst transmission of the message fragments which are reconstructed at the DST endpoint. Figure 4 illustrates the concept of the implemented fragmentation layer.

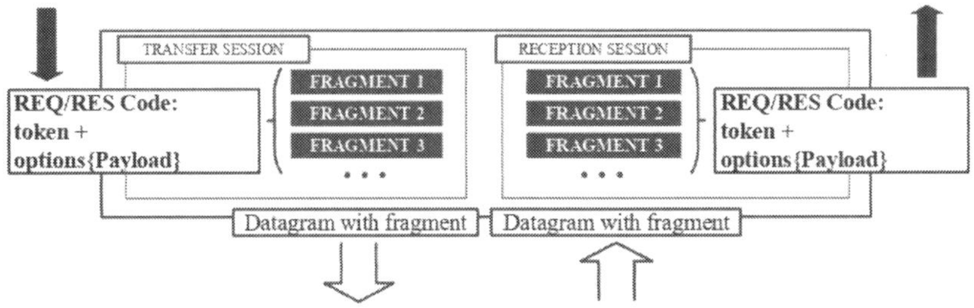


**Figure 4.** Implementation of fragmentation of payload message by using transfer/reception session to fragment and reconstruct messages

The fragmentation header occupies one octets described by following format: $FRAGMENTATION = (FRAGMENT\_NUM \ll 4) + FRAGMENT\_COUNT$. The reception process recognizes the total number of the message fragment from $FRAGMENT\_COUNT$. The $FRAGMENT\_NUM$ header specifies the order of the received fragment. Therefore, the maximum degree of fragmentation Since this design is directed to a highly constrained network, only one active session is permitted by default. The fragmentation knowingly introduces the exponential growth of the failed message reception $P_{failed} = 1 - (1 - PER)^n$ where PER is the probability of reception errors of the end-to-end packet transfer. To reduce $P_{failed}$, the reception session sets a timeout counter proportionally to the total number of fragments at the point of the first fragment reception to selectively request the lost fragments. The reception session is regarded as unsuccessful after a pre-set number of attempts. This protocol design can moderately support resource discovery via web linking. The payload content can be further compressed in the CBOR format (Concise Binary Object Representation). However, the payload content is application dependent. Hence, it is outside the scope of this paper.

## 3. Experiment Setup

For the experiment, the scenario where two co-located WSNs with interoperable hardware detect each other and exchange the information of the public available resources via web linking. The protocol is tested in a testbed of 12 EZ430-RF2500 motes [10]. Figure 5 illustrated the experiment setup.

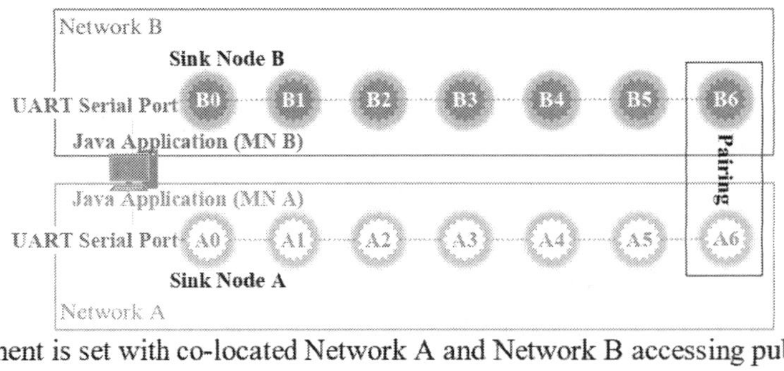


**Figure 5.** Experiment is set with co-located Network A and Network B accessing public resources via CoAP messages.

The experiment aims to study the effects of the fragmentation and hop distance which are the main key factors of message reception failures. Network A have been deployed in a sensing application in this local area. Later, Network B is deployed in the same Location. Each node periodically sends the sensing data packet to its own base station in which the application agent is programmed to exploit the sensing data. The EZ430-RF2500 mote is used for every sensor node in the experiment setup. However, Network A implements Low Power Polling Receiver-Initiated MAC protocol while Network B adopts Strobe Preamble Low Power Listening Sender-Initiated MAC Protocol. The end nodes (B6 and A6) at the edge of the network periodically exchange packets across a common channel. Exploiting this local connection, the application agent in the base station of each sensor network (emulated in the same PC for the sake of precise time measurements) accesses a useful resource provided by the neighbouring network using GET (0.01) request in the CoAP framework. The success rate is calculated every 10 messages and each measurement result is represented by 95-percent confidence interval of 10 samples. The end -to-end delay is represented by the 95-percentile of 1000 samples.

## 4. Experiment Results

### *4.1. Effects of Hop Distance*

The number of hop distances is proportional to the network scale. The experiment varies the number of hop distances between the base station and the pairing node while measuring the collective network performances. The results are illustrated in Figure 6.

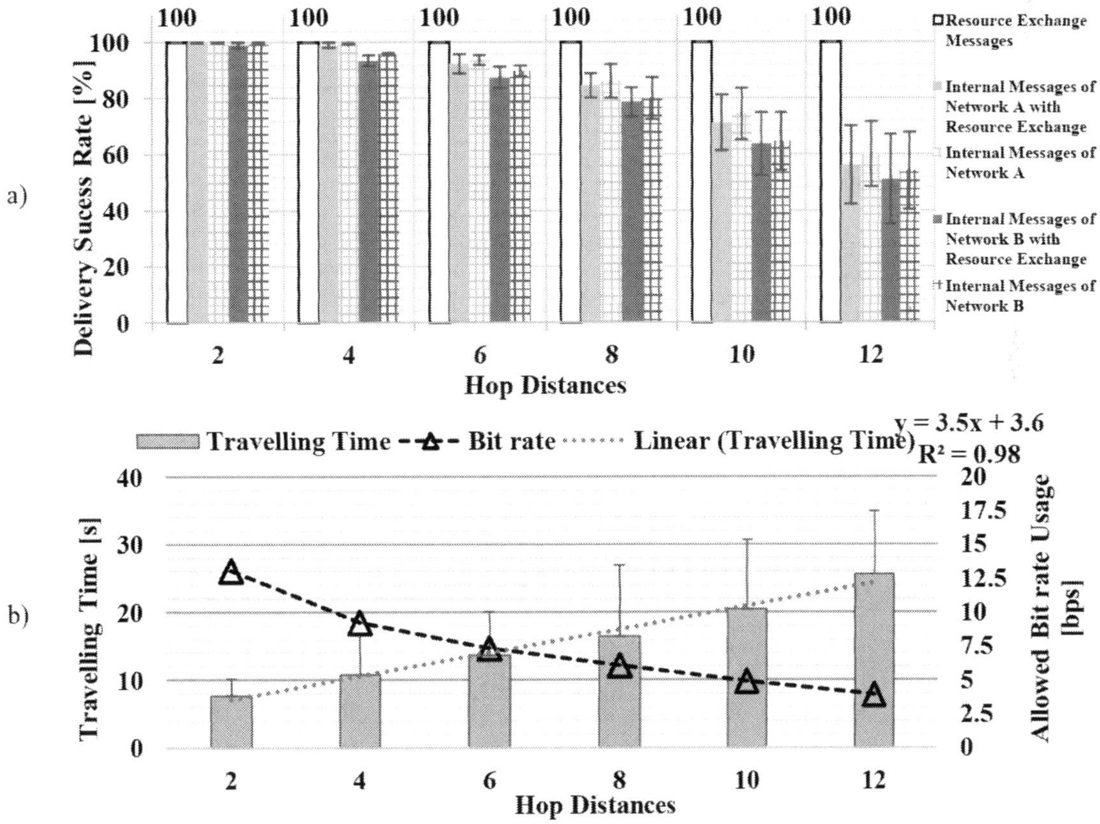


**Figure 6.** a) The measurement results show the success rate of the receptions of the resource exchange messages and periodical internal messages as the hop distance increases. b) The end-to-end delay and estimated data rate of the resource exchange messages as the hop distance increases.

The experiment shows the expected results. The link layer retransmits the packets until the packet buffer overflow forcing packet drops which is the only cause of packet losses. The resource exchange messages are always secured from the buffer overflow because the resource exchange message piggyback on the request and response in the form of CON messages in the CoAP framework which are prioritized. All packet drops are reflected in the reception failures of the periodic internal messages which are sent with a NON message in the CoAP framework. Additionally, the effect of carrying extra messages is relatively constant regardless of the hop distance as seen in the comparison between the cases of internal messages and the cases of internal messages with resource exchange. Most of the internal message failures are caused by the overflow due to the hop distance, not by the extra traffic from the resource exchange. That is because the extra traffic is limited to only one session. The effects of the hop distances on the end-to-end delay behaves as predicted. The travelling time proportionally increases with the hop distance approximately 3.5 seconds per hop.

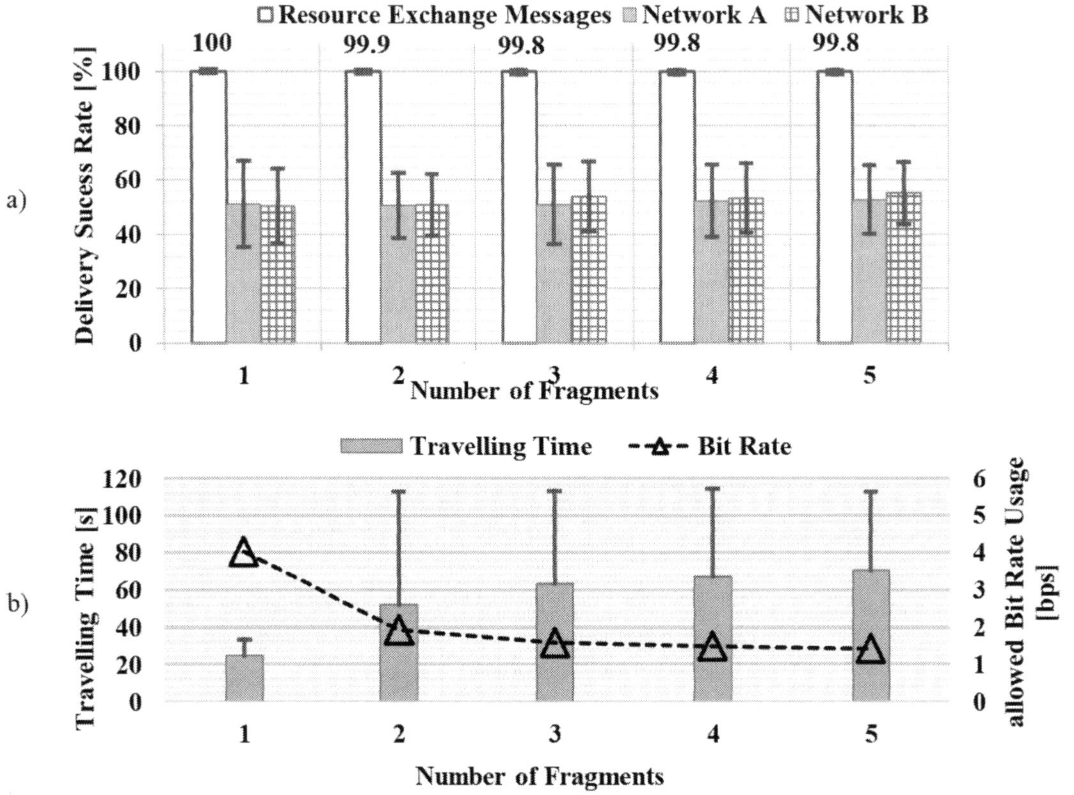


**Figure 7.** a) The measurement results show the success rate of the receptions of the resource exchange messages as the message fragment increases b) The end-to-end delay and estimated data rate of the resource exchange messages as the message fragments increases.

The fragmentation is particularly introduced to solve the constraint of the physical frame space. The affordable payload size depends on the implementation. In this case study, we use EZ430-RF2500 motes. The physical frame size is limited by the memory buffer (Tx/Rx FIFO) at 32 bytes, therefore a message fragment can accommodate only 11 octets payload. The packet buffer size is set at 10 packets, therefore, the maximum number of fragments per session must be set at 5, since the communication partner may use the rest available space for send his traffic simultaneously. The distance between two base stations is 12 hops. According to Figure 7, the fragmentation can cause some reception failures. This situation happens in the rare chance that only CON messages are stacked in the packet buffer of a node in the forwarding route until triggering the buffer overflow. However, the successful deliver rate in the case of 5 fragments is relatively reliable at 99.8 percent. In addition, the number of fragments does not introduce a sharp decline of the reliability and end-to-end delay because of the selective fragment request and message retransmission. Multiple fragment losses are corrected at the same time causing the slightly drop of performances instead of the exponential drops.

## 5. Conclusions

This paper reports the possibility of the data exchange with a local direct interconnection between neighbouring WSNs with highly constrained hardware. The CoAP framework is implement on top of a customized link layer. The protocol design heavily focuses on the compression of the datagram format to solve the frame space limitation. The fragmentation is introduced using burst transmissions and selective requests. The design is evaluated in a testbed of EZ430-RF2500 motes. The experiments show that the protocol offers a reliable data exchange up to 12 hop distances in the experimental conditions.

## 6. References

[1] P. Ray, "A survey on Internet of Things architectures," *J. King Saud Univ. - Comput. Inf. Sci.*, vol. 30, no. 3, pp. 291–319, 2018.

[2] K. Singhanat, N. R. Harris, and G. V. Merrett, "Experimental validation of opportunistic direct interconnection between different Wireless Sensor Networks," in *SAS 2016 - Sensors Applications Symposium, Proceedings*, 2016, pp. 36–41.

[3] T. Jiang, G. V. Merrett, and N. R. Harris, "Opportunistic Direct Interconnection between Co-Located Wireless Sensor Networks," in *2013 22nd International Conference on Computer Communication and Networks (ICCCN)*, 2013, pp. 1–5.

[4] T. Jiang, G. V Merrett, and N. R. Harris, "Enabling opportunistic energy trading between overlapping energy harvesting wireless sensor networks," in *Proceedings of the 2nd International Workshop on Energy Neutral Sensing Systems - ENSsys '14*, 2014, pp. 25–30.

[5] K. Singhanat, T. Jiang, G. V. Merrett, and N. R. Harris, "Empirical evaluation of OI-MAC: Direct interconnection between wireless sensor networks for collaborative monitoring," in *2015 IEEE Sensors Applications Symposium (SAS)*, 2015, pp. 1–5.

[6] Z. Shelby, K. Hartke, and C. Bormann, "The Constrained Application Protocol (CoAP)." Internet Engineering Task Force (IETF), pp. 1–112, 2014.

[7] V. Karagiannis, P. Chatzimisios, F. Vazquez-Gallego, and J. Alonso-Zarate2Nastase, "A Survey on Application Layer Protocols for the Internet of Things," *Trans. IoT Cloud Comput.*, 2015.

[8] C. Bormann, "Block-Wise Transfers in the Constrained Application Protocol (CoAP)." pp. 1–37, 2016.

[9] J. Postel, "User Datagram Protocol." IETF, 1980.

[10] M. Morales and Z. Shivers, "Wireless sensor monitor using the eZ430-RF2500," 2007.

# AUTHOR INDEX